Materials Issues In
Microcrystalline Semiconductors

MATERIALS RESEARCH SOCIETY SYMPOSIUM PROCEEDINGS   VOLUME 164

Materials Issues In Microcrystalline Semiconductors

Symposium held November 29-December 1, 1989, Boston, Massachusetts, U.S.A.

EDITORS:

Philippe M. Fauchet
Princeton University, Princeton, New Jersey, U.S.A.

Kazunobu Tanaka
Electrotechnical Laboratory, Ibaraki, Japan

Chuang Chuang Tsai
Xerox Palo Alto Research Center, Palo Alto, California, U.S.A.

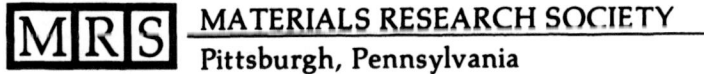 MATERIALS RESEARCH SOCIETY
Pittsburgh, Pennsylvania

CAMBRIDGE UNIVERSITY PRESS
Cambridge, New York, Melbourne, Madrid, Cape Town,
Singapore, São Paulo, Delhi, Mexico City

Cambridge University Press
32 Avenue of the Americas, New York NY 10013-2473, USA

Published in the United States of America by Cambridge University Press, New York

www.cambridge.org
Information on this title: www.cambridge.org/9781107410299

Materials Research Society
506 Keystone Drive, Warrendale, PA 15086
http://www.mrs.org

© Materials Research Society 1990

First published 1990
First paperback edition 2012

Single article reprints from this publication are available through
University Microfilms Inc., 300 North Zeeb Road, Ann Arbor, MI 48106

CODEN: MRSPDH

ISBN 978-1-107-41029-9 Paperback

Contents

PREFACE   xi

MATERIALS RESEARCH SOCIETY SYMPOSIUM PROCEEDINGS   xiii

PART I: MICROCRYSTALLINE SILICON: GROWTH

*ROLE OF SURFACE AND GROWTH-ZONE REACTIONS IN THE
FORMATION PROCESS OF μc-Si:H   3
    A. Matsuda and T. Goto

OPTO-ELECTRONIC PROPERTIES OF μc-Si GROWN FROM
SiF_4 AND H_2 BY PECVD   15
    Y. Okada, I.H. Campbell, P.M. Fauchet, and S. Wagner

FORMATION OF MICROCRYSTALLINE SILICON FILM BY RMS PROCESS   21
    Cheng Wang, G.N. Parsons, E.C. Buehler,
    R.J. Nemanich, and G. Lucovsky

PREPARATION AND CHARACTERIZATION OF HIGHLY CONDUCTIVE
(100 S/cm) PHOSPHORUS DOPED μc-Si:H FILMS DEPOSITED
USING THE VHF-GD TECHNIQUE   27
    Kshem Prasad, F. Finger, H. Curtins, A. Shah,
    and J. Bauman

COMPOSITION DETERMINATION OF MICROCRYSTALLINE TWO-PHASE
SILICON RICH OXIDES   33
    D.H. Bouldin, C.H. Lam, and K. Rose

*CHEMISTRY AND SOLID STATE PHYSICS OF MICROCRYSTALLINE
SILICON   39
    Stan Veprek

ELECTRONIC AND STRUCTURAL CHARACTERIZATION OF THE NEAR
SURFACE LAYER AND THE BULK IN μc-Si:H PREPARED WITH
HYDROGEN DILUTION   51
    Samer Aljishi, Shu Jin, Martin Stutzmann, and
    Lothar Ley

THE FORMATION OF SILICON CLUSTERS IN PLASMA-ENHANCED
CHEMICALLY VAPOUR DEPOSITED Si:O:H:F ALLOYS   57
    A.G. Dias and J. Figueiredo

PREPARATION OF CRYSTALLINE Si THIN FILMS BY SPONTANEOUS
CHEMICAL DEPOSITION   63
    Tohru Komiya, Akira Kamo, Hiroshi Kujirai,
    Isamu Shimizu, and Jun-ichi Hanna

MICROCRYSTALLINE SILICON FILMS PRODUCED BY RF MAGNETRON
SPUTTERING AND THE EFFECT OF DIFFERENT AMBIENTS ON
THEIR CONDUCTIVITY   69
    Ratnabali Banerjee, A.K. Bandyopadhyay, S.N. Sharma,
    A.K. Batabyal, and A.K. Barua

*Invited Paper

THE GRAPHITIZATION OF AMORPHOUS HYDROGENATED CARBON
FILMS DURING THERMAL ANNEALING   75
    Sam Shuhan Lin, Shuguang Chen, and Dang Mo

            PART II: SEMICONDUCTOR COMPOUNDS: NANOCRYSTALS

ELECTRODEPOSITION OF 3-D SIZE-QUANTIZED CdS AND CdSe FILMS   81
    G. Hodes, T. Engelhard, A. Albu-Yaron, and
    A. Pettford-Long

STRUCTURAL AND ELECTRICAL CHARACTERIZATION OF CdSe THIN
FILMS   87
    Miltiadis K. Hatalis, Fuyu Lin, and Michael R. Westcott

A STUDY OF THE PRESSURE-INDUCED PHASE TRANSITION IN BULK
AND NANOCRYSTALLINE CADMIUM SULFIDE   93
    Xue-Shu Zhao, John Schroeder, Peter D. Persans,
    and Enlian Lu

TIME-RESOLVED SPECTRA OF EDGE EMISSION OF CdSe MESOSCOPIC
CLUSTERS IN GeO$_2$ GLASS MATRIX   99
    Takao Inokuma, Mitsuru Ishikawa, Akinori Tanaka,
    and Toshihiro Arai

*OPTICAL PROPERTIES OF II-VI SEMICONDUCTOR DOPED GLASS   105
    P.D. Persans, An Tu, M. Lewis, T. Driscoll,
    and R. Redwing

FABRICATION AND CHARACTERIZATION STUDIES OF SEMICONDUCTOR-
IMPREGNATED POROUS VYCOR GLASS   117
    C.A. Huber, T.E. Huber, A.P. Salzberg, and J.A. Perez

SYNTHESIS, STRUCTURAL AND OPTICAL CHARACTERIZATION OF
II-VI SEMICONDUCTORS INCLUDED IN SODALITE-TYPE HOSTS   123
    K.L. Moran, W.T.A. Harrison, T.E. Gier,
    J.E. MacDougall, and G.D. Stucky

THE PREPARATION OF III-V SEMICONDUCTORS IN AQUEOUS
SOLUTION   129
    Toby J. Cumberbatch and Andrew Putnis

ROOM TEMPERATURE EXCITONIC ABSORPTION IN SMALL CdS
CRYSTALLITES   135
    D.K. Rai and Binod Kumar

PARTICLE-SIZE DISTRIBUTION OF CdSe QUANTUM DOTS
DETERMINED BY PHOTOLUMINESCENCE SPECTROSCOPY   141
    E.N. Prabhakar, C.A. Huber, and D. Heiman

THERMAL ANNEALING OF AMORPHOUS CoMnNiO FILM ON
OXIDIZED Si SUBSTRATE   147
    Tan Hui, Qin Dong, Tao Mingde, Lin Chenglu,
    and Zou Shichang

PREPARATION AND CHARACTERIZATION OF MOLYBDENUM DISULFIDE
MICROCRYSTALS IN COLLOIDAL DISPERSION   153
    E. Lu, P.D. Persans, A.F. Ruppert, and R.R. Chianelli

*Invited Paper

PART III: MICROCRYSTALLINE SILICON: PROPERTIES

*THE ROLE OF HYDROGEN IN SILICON MICROCRYSTALLIZATION   161
    S. Wagner, S.H. Wolff, and J.M. Gibson

MICROCRYSTAL Si FILMS PREPARED BY REMOTE PLASMA CVD   171
    Sung Chul Kim, Jung Tae Hwang, Seung Kyu Lee,
    Chang Young Jung, Sung Moo Soe, Sung Ok Koh,
    Kwan Soo Chung, and Jin Jang

FRACTAL-LIKE STRUCTURES PRESENT IN HYDROGENATED AMORPHOUS
AND MICROCRYSTALLINE SILICON   177
    M.J. Geerts, R.C. van Oort, and J.C. van den Heuvel

FAST-PULSE EXCIMER-LASER-INDUCED PROCESSES IN a-Si:H   183
    K. Winer, R.Z. Bachrach, R.I. Johnson, S.E. Ready,
    G.B. Anderson, and J.B. Boyce

HYDROGEN PASSIVATION OF DOPED AND UNDOPED MICROCRYSTALLINE
SILICON   189
    M. Stutzmann, C.P. Herrero, M. Ingels, and
    A. Breitschwerdt

*CONTROL OF CHEMICAL REACTIONS FOR GROWTH OF CRYSTALLINE
Si AT LOW SUBSTRATE TEMPERATURE   195
    Isamu Shimizu, Jun-ichi Hanna, and Hajime Shirai

EFFECTS OF HYDROGEN ATOMS ON PASSIVATION AND GROWTH OF
MICROCRYSTALLINE Si   205
    Toshimichi Ito, Tatsuro Yasumatsu, Hirokuni Watabe,
    Motohiro Iwami, and Akio Hiraki

A DISCUSSION OF ELECTRONIC OPTICAL ABSORPTION SPECTRA OF
NANOCRYSTALLINE SILICON THIN FILMS   211
    Etienne Bustarret and M.A. Hachicha

GROWTH OF MICROCRYSTALLINE SILICON IN ULTRATHIN LAYERS   217
    Y.-J. Wu, P.D. Persans, B. Abeles, and S.-L. Wang

PICOSECOND PHOTOMODULATION STUDY OF NANOCRYSTALLINE
HYDROGENATED SILICON   223
    M. Wraback, Lingrong Chen, J. Tauc, and Z. Vardeny

OPTICAL PROPERTIES OF MICROCRYSTALLINE SILICON   229
    Martin Ingels, Martin Stutzmann, and Stefan Zollner

TRANSPORT PROPERTIES OF B-, P-DOPED AND UNDOPED 50 kHz
PECVD MICROCRYSTALLINE SILICON   235
    M.A. Hachicha and Etienne Bustarret

SUPPRESSION OF ACCEPTOR DEACTIVATION IN SILICON BY
DISORDERED SURFACE REGIONS   239
    K. Srikanth and S. Ashok

*Invited Paper

PART IV: OPTICAL PROPERTIES

*PROPERTIES OF BINARY Si:H MATERIALS PREPARED BY HYDROGEN
PLASMA SPUTTERING   247
    Shoji Furukawa and Tatsuro Miyasato

CRITICAL REVIEW OF RAMAN SPECTROSCOPY AS A DIAGNOSTIC
TOOL FOR SEMICONDUCTOR MICROCRYSTALS   259
    P.M. Fauchet and I.H. Campbell

RAMAN SCATTERING FROM MICROCRYSTALLINE FILMS:
CONSIDERATIONS OF COMPOSITE STRUCTURES WITH DIFFERENT
OPTICAL ABSORPTION PROPERTIES   265
    R.J. Nemanich, E.C. Buehler, Y.M. LeGrice,
    R.E. Shroder, G.N. Parsons, C. Wang, G. Lucovsky,
    and J.B. Boyce

PHONON STATES IN SiC SMALL PARTICLES   271
    Y. Sasaki, C. Horie, and Y. Nishina

NONLINEAR OPTICAL PROPERTIES OF STRUCTURED NANOPARTICLE
COMPOSITES   277
    Meyer H. Birnboim and Wei Ping Ma

ENHANCED NONLINEAR OPTICAL RESPONSE OF COATED
NANOPARTICLES   283
    N. Kalyaniwalla, J.W. Haus, M.H. Birnboim,
    R. Inguva, and W.P. Ma

PART V: SILICON ALLOYS

*OPTOELECTRONICS AND PHOTOVOLTAIC APPLICATIONS OF
MICROCRYSTALLINE SiC   291
    Y. Hamakawa, Y. Matsumoto, G. Hirata, and H. Okamoto

THE EFFECT OF HYDROGEN ON THE STRUCTURE OF AMORPHOUS
AND MICROCRYSTALLINE SiC PREPARED BY THE POLYMER ROUTE   303
    C-J Chu, S-J. Ting, F. Bobonneau, and
    J.D. Mackenzie

POWER DENSITY EFFECTS IN THE PHYSICAL AND CHEMICAL
PROPERTIES OF SPUTTERED DIAMOND-LIKE CARBON THIN FILMS   309
    N.-H. Cho, K.M. Krishnan, D.K. Veirs, M.D. Rubin,
    C.B. Hopper, B. Bhushan, and D.B. Bogy

THE EFFECT OF HYDROGEN ON THE STRUCTURE AND ELECTRICAL
AND OPTICAL PROPERTIES OF SILICON-GERMANIUM ALLOYS   315
    C.M. Fortmann, D.E. Albright, I.H. Campbell,
    and P.M. Fauchet

RAMAN STUDIES OF MICROSTRUCTURAL CHANGES IN AMORPHOUS
SILICON-BORON ALLOYS DUE TO ANNEALING   321
    G. Yang, P. Bai, Y.J Wu, B.Y. Tong, S.K. Wong,
    J. Du, and I. Hill

*Invited Paper

PART VI: DEVICES AND APPLICATIONS

*PREPARATION OF HIGH-QUALITY poly-Si AND μc-Si FILMS BY
THE SPC METHOD   329
    T. Matsuyama, M. Nishikuni, M. Kameda, S. Okamoto,
    M. Tanaka, S. Tsuda, M. Ohnishi, S. Nakano, and
    Y. Kuwano

CHARACTERISTICS OF μc-Si:H FOR Si HETEROJUNCTION BIPOLAR
TRANSISTORS   341
    H. Fujioka, M. Ito, and K. Takasaki

DOPANT SEGREGATION AT POLYCRYSTALLINE SILICON GRAIN
BOUNDARIES IN DEVICE FABRICATION PROCESSES   347
    M. Itoh, I. Aikawa, N. Hirashita, and T. Ajioka

CORRELATIONS BETWEEN OPTICAL, ELECTRICAL, AND STRUCTURAL
PROPERTIES OF IN-SITU PHOSPHORUS-DOPED HYDROGENATED
MICROCRYSTALLINE SILICON - EFFECTS OF RAPID THERMAL
ANNEALING ON MATERIAL PROPERTIES   353
    David E. Kotecki, Shwu J. Jeng, Jerzy Kanicki,
    Christopher C. Parks, Werner Rausch, Krishna Seshan,
    and Jean Tien

SELECTIVE DEPOSITION OF N$^+$ DOPED MC-Si:H:F BY RF PLASMA
CVD ON Si AND SiO$_2$ SUBSTRATES   359
    K. Baert, P. Deschepper, H. Pattyn, J. Nijs,
    and R. Mertens

SELECTIVE GROWTH OF Si CRYSTALS OVER AMORPHOUS SUBSTRATES
SEEDED BY SOLID-STATE AGGLOMERATION OF PATTERNED Si   365
    K. Yamagata and T. Yonehara

CRYSTAL-AXIS-ROTATION OF LASER-RECRYSTALLIZED SILICON
ON INSULATOR   371
    K. Sugahara, T. Ippôshi, Y. Inoue, T. Nishimura,
    and Y. Akasaka

A NEW MODEL FOR THE POLY-SILICON THIN FILM TRANSISTOR
FOR USE WITH SPICE   377
    M.J. Izzard, P. Migliorato, W.I. Milne

ELECTRICAL PROPERTIES OF SIPOS FILMS DEPOSITED ON
CRYSTALLINE SILICON   383
    Tien-Min Chuang, Kenneth Rose, and Ronald J. Gutmann

DEVELOPMENT OF THE VERY THIN MICROCRYSTALLINE N-LAYER
AND ITS APPLICATION TO THE STACKED SOLAR CELL   389
    F. Nakabeppu, T. Ishimura, K. Kumagai, and
    K. Fukui

AUTHOR INDEX   395

SUBJECT INDEX   397

MATERIALS RESEARCH SOCIETY SYMPOSIUM PROCEEDINGS   401

*Invited Paper

Preface

This book contains the proceedings of the Symposium on Materials Issues in Microcrystalline Semiconductors held in Boston, from November 29 to December 1, 1989 during the 1989 MRS Fall Meeting. Microcrystals can be best defined by what they are not: they fall between single or large crystals and amorphous materials. When the dimension of the object becomes of the order of 10 nanometers, many electronic and structural properties are severely altered. It has been argued that microcrystals represent a new class of materials. The purpose of this symposium was to bring together physicists, chemists, materials scientists and engineers active in various areas of growth, characterization and device applications of microcrystalline semiconductors. Although a complete and definitive description of this class of materials is not available yet, we feel that, thanks to the symposium, a convergence of opinions in some areas has been achieved and areas of future investigation have been identified. It is also remarkable that more than 50% of the contributions came from Asia and Europe, making this symposium truly international.

For convenience, the papers have been divided into six sections:

1. Microcrystalline Silicon: Growth
2. Semiconductor Compounds: Nanocrystals
3. Microcrystalline Silicon: Properties
4. Optical Properties
5. Silicon Alloys
6. Devices and Applications

The organizers would like to thank all the authors for their active participation, the invited speakers for very stimulating presentations, the session chairs and referees for their good work, and Ms. K. Schnerr for her help in preparing this volume. The symposium was made possible by the financial support of the following organizations:

1. Sanyo Electric Company
2. Solar Energy Research Institute
3. Tonen Corporation
4. Xerox Corporation

February 1990

Philippe M. Fauchet
Kazunobu Tanaka
Chuang Chuang Tsai

Recent Materials Research Society Symposium Proceedings

Volume 145—III-V Heterostructures for Electronic/Photonic Devices, C.W. Tu, V.D. Mattera, A.C. Gossard, 1989, ISBN: 1-55899-018-6

Volume 146—Rapid Thermal Annealing/Chemical Vapor Deposition and Integrated Processing, D. Hodul, J. Gelpey, M.L. Green, T.E. Seidel, 1989, ISBN: 1-55899-019-4

Volume 147—Ion Beam Processing of Advanced Electronic Materials, N.W. Cheung, A.D. Marwick, J.B. Roberto, 1989, ISBN: 1-55899-020-8

Volume 148—Chemistry and Defects in Semiconductor Heterostructures, M. Kawabe, T.D. Sands, E.R. Weber, R.S. Williams, 1989, ISBN: 1-55899-021-6

Volume 149—Amorphous Silicon Technology-1989, A. Madan, M.J. Thompson, P.C. Taylor, Y. Hamakawa, P.G. LeComber, 1989, ISBN: 1-55899-022-4

Volume 150—Materials for Magneto-Optic Data Storage, C.J. Robinson, T. Suzuki, C.M. Falco, 1989, ISBN: 1-55899-023-2

Volume 151—Growth, Characterization and Properties of Ultrathin Magnetic Films and Multilayers, B.T. Jonker, J.P. Heremans, E.E. Marinero, 1989, ISBN: 1-55899-024-0

Volume 152—Optical Materials: Processing and Science, D.B. Poker, C. Ortiz, 1989, ISBN: 1-55899-025-9

Volume 153—Interfaces Between Polymers, Metals, and Ceramics, B.M. DeKoven, A.J. Gellman, R. Rosenberg, 1989, ISBN: 1-55899-026-7

Volume 154—Electronic Packaging Materials Science IV, R. Jaccodine, K.A. Jackson, E.D. Lillie, R.C. Sundahl, 1989, ISBN: 1-55899-027-5

Volume 155—Processing Science of Advanced Ceramics, I.A. Aksay, G.L. McVay, D.R. Ulrich, 1989, ISBN: 1-55899-028-3

Volume 156—High Temperature Superconductors: Relationships Between Properties, Structure, and Solid-State Chemistry, J.R. Jorgensen, K. Kitazawa, J.M. Tarascon, M.S. Thompson, J.B. Torrance, 1989, ISBN: 1-55899-029

Volume 157—Beam-Solid Interactions: Physical Phenomena, J.A. Knapp, P. Borgesen, R.A. Zuhr, 1989, ISBN 1-55899-045-3

Volume 158—In-Situ Patterning: Selective Area Deposition and Etching, R. Rosenberg, A.F. Bernhardt, J.G. Black, 1989, ISBN 1-55899-046-1

Volume 159—Atomic Scale Structure of Interfaces, R.D. Bringans, R.M. Feenstra, J.M. Gibson, 1989, ISBN 1-55899-047-X

Volume 160—Layered Structures: Heteroepitaxy, Superlattices, Strain, and Metastability, B.W. Dodson, L.J. Schowalter, J.E. Cunningham, F.H. Pollak, 1989, ISBN 1-55899-048-8

Volume 161—Properties of II-VI Semiconductors: Bulk Crystals, Epitaxial Films, Quantum Well Structures and Dilute Magnetic Systems, J.F. Schetzina, F.J. Bartoli, Jr., H.F. Schaake, 1989, ISBN 1-55899-049-6

Volume 162—Diamond, Boron Nitride, Silicon Carbide and Related Wide Bandgap Semiconductors, J.T. Glass, R.F. Messier, N. Fujimori, 1989, ISBN 1-55899-050-X

Volume 163—Impurities, Defects and Diffusion in Semiconductors: Bulk and Layered Structures, J. Bernholc, E.E. Haller, D.J. Wolford, 1989, ISBN 1-55899-051-8

Volume 164—Materials Issues in Microcrystalline Semiconductors, P.M. Fauchet, C.C. Tsai, K. Tanaka, 1989, ISBN 1-55899-052-6

Volume 165—Characterization of Plasma-Enhanced CVD Processes, G. Lucovsky, D.E. Ibbotson, D.W. Hess, 1989, ISBN 1-55899-053-4

Volume 166—Neutron Scattering for Materials Science, S.M. Shapiro, S.C. Moss, J.D. Jorgensen, 1989, ISBN 1-55899-054-2

MATERIALS RESEARCH SOCIETY SYMPOSIUM PROCEEDINGS

Volume 167—Advanced Electronic Packaging Materials, A. Barfknecht, J. Partridge, C-Y. Li, C.J. Chen, 1989, ISBN 1-55899-055-0

Volume 168—Chemical Vapor Deposition of Refractory Metals and Ceramics, T.M. Besmann, B.M. Gallois, 1989, ISBN 1-55899-056-9

Volume 169—High Temperature Superconductors: Fundamental Properties and Novel Materials Processing, J. Narayan, C.W. Chu, L.F. Schneemeyer, D.K. Christen, 1989, ISBN 1-55899-057-7

Volume 170—Tailored Interfaces in Composite Materials, C.G. Pantano, E.J.H. Chen, 1989, ISBN 1-55899-058-5

Volume 171—Polymer Based Molecular Composites, D.W. Schaefer, J.E. Mark, 1989, ISBN 1-55899-059-3

Volume 172—Optical Fiber Materials and Processing, J.W. Fleming, G.H. Sigel, S. Takahashi, P.W. France, 1989, ISBN 1-55899-060-7

Volume 173—Electrical, Optical and Magnetic Properties of Organic Solid-State Materials, L.Y. Chiang, D.O. Cowan, P. Chaikin, 1989, ISBN 1-55899-061-5

Volume 174—Materials Synthesis Utilizing Biological Processes, M. Alper, P.D. Calvert, P.C. Rieke, 1989, ISBN 1-55899-062-3

Volume 175—Multi-Functional Materials, D.R. Ulrich, F.E. Karasz, A.J. Buckley, G. Gallagher-Daggitt, 1989, ISBN 1-55899-063-1

Volume 176—Scientific Basis for Nuclear Waste Management XIII, V.M. Oversby, P.W. Brown, 1989, ISBN 1-55899-064-X

Volume 177—Macromolecular Liquids, C.R. Safinya, S.A. Safran, P.A. Pincus, 1989, ISBN 1-55899-065-8

Volume 178—Fly Ash and Coal Conversion By-Products: Characterization, Utilization and Disposal VI, F.P. Glasser, R.L. Day, 1989, ISBN 1-55899-066-6

Volume 179—Specialty Cements with Advanced Properties, H. Jennings, A.G. Landers, B.E. Scheetz, I. Odler, 1989, ISBN 1-55899-067-4

MATERIALS RESEARCH SOCIETY MONOGRAPH

Atom Probe Microanalysis: Principles and Applications to Materials Problems, M.K. Miller, G.D.W. Smith, 1989; ISBN 0-931837-99-5

Earlier Materials Research Society Symposium Proceedings listed in the back.

PART I

Microcrystalline Silicon: Growth

ROLE OF SURFACE AND GROWTH-ZONE REACTIONS IN THE FORMATION PROCESS OF μc-Si:H

A. MATSUDA* AND T. GOTO**
* Electrotechnical Laboratory, 1-1-4 Umezono, Tsukuba-shi, Ibaraki 305, Japan
** Nagoya University, Fuoi-cho, Chigusa-ku, Nagoya 464-01, Japan

ABSTRACT

The role of the surface reaction is discussed in the formation process of μc-Si:H in comparison to that of a-Si:H. It is suggested that the responsible radicals for the formation of μc-Si:H are SiH_3 as same in the case of a-Si:H depositions. On the top film-growing surface, a lot of H atoms reach the surface during the course of the μc-Si:H growth giving rise to the change in the surface condition, i. e. the loss probability of SiH_3 radicals is increased. At the same time, a full H-coverage of the surface is expected which enhances the surface diffusion of SiH_3 radicals, leading to the appearance of a μc nucleus. Moreover, it is speculated that the reaction in the growth zone is not necessary for the nucleation process in μc-Si:H.

INTRODUCTION

Plasma enhanced chemical vapor deposition (PECVD) produces a variety of Si-network structures at a constant temperature of 200-300C, i. e. amorphous (hydrogenated amorphous silicon; a-Si:H) or microcrystalline (hydrogenated amorphous-microcrystalline mixed phase silicon; μc-Si:H), when the starting-gas composition is changed from pure SiH_4 to SiH_4/H_2 mixture [1,2 and 3]. The formation process of μc-Si:H has attracted increasing attention since the formation of microcrystallites (μc) occurs at such low temperatures. However, the difference between the a-Si:H formation process and the μc-Si:H formation process is still ambiguous. Three possible roles of H atoms, abundant in the H_2-rich plasma, have been proposed to explain the growth of μc-Si:H from the SiH_4/H_2 glow-discharge plasma; (1) H atoms reaching the surface enhance the surface diffusion of ad-radicals such as SiH_3 through a sufficient coverage of the film-growing surface with H [1], (2) H atoms act as an "etchant" to form a volatile species and give a chemical equilibrium condition between the deposition and the etching on the film-growing surface [2,3], and (3) H atoms, which soak into several layers below the top growing surface (growth zone), promote the network-propagation reaction (chemical annealing) [4].

This paper describes, for the first time, the detection of key radicals in the plasma, both in the μc-Si:H and a-Si:H deposition conditions, using a diode-laser-absorption technique as well as optical-emission spectroscopy. We also discuss the difference in the surface conditions between the μc-Si:H growth process and the a-Si:H formation process on the basis of the values of the radical-reflection and the radical-sticking probabilities determined by a step-coverage experiment. Finally, the role of the reaction in the growth zone is speculated using the experimental result of a layer-by-layer deposition of the films.

EXPERIMENTAL

Measurement of SiH3 and SiH2 Radicals in the Plasma

Figure 1 shows the experimental setup for measuring the number density of SiH3 and SiH2 radicals in the radio-frequency (rf:13.56MHz) SiH4/H2 plasmas using a diode-laser-absorption technique [5]. Two parallel plate electrodes, 20cm in diameter, were placed at the center of the reactor chamber at a separation of 3cm. An rf power of 100W was applied to the upper electrode (rf electrode), which corresponds to a power density of $0.3W/cm^2$. Gas-flow-rate ratios of SiH4/H2=1/1 (a-Si:H), 1/19 and 1/49 (μc-Si:H) were used as the typical ratios for producing a-Si:H and μc-Si:H films, respectively, and the total pressure was kept constant at 80mTorr.

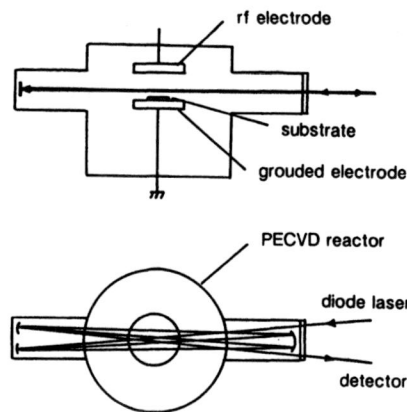

Fig.1. Experimental setup for measuring SiH3 as well as SiH2 radicals in the plasma.

The diode-laser beam traversed 40 times near the lower electrode (electrically grounded) by using a White-type multiple reflection arrangement as shown in the figure to detect SiH3 as well as SiH2 radicals in the plasma. The details for the measurement were reported in a previous paper [5]. A Corning #7059 glass substrate was placed on the grounded electrode, heated to 150C, to determine the deposition rate of the film in each discharge condition. At the same time, optical-emission spectroscopy was used for measuring Si* (288nm in wavelength) with a wide variety of SiH4/H2 ratios (1/1, 1/4, 1/9, 1/19, 1/29 and 1/49) in the rf plasma to find a relationship between the emission intensity and the deposition rate of the film on the substrate surface. The total pressure and the rf-power density were kept the same as the conditions used in the diode-laser-absorption experiment.

Determination of Sticking and Reflection Probabilities

The reflection probability of radicals reaching the film-growing surface was determined using a step-coverage experiment [6,7] under the a-Si:H and μc-Si:H deposition conditions. The triode-rf glow-discharge configuration was used as a mean of complete radical-separation [8] of SiH_3 as well as H from other radicals or atoms produced in the plasma. As shown in Fig.2, the discharge was confined between a 10cm-diameter electrode where the rf power was applied and a transparent mesh (#30) electrode, 4cm from the rf electrode. The substrate electrode was placed at 2cm from the mesh. The discharge parameters were set as follows: In the μc-Si:H formation condition, the rf power was 20W which corresponds to a power density of about $0.26W/cm^2$ on the rf electrode, the total pressure was kept constant at 70mTorr for a flow rate ratio of $SiH_4/H_2=1/49$. In the a-Si:H formation condition, the rf power density was set at $0.13W/cm^2$ and the pressure was kept at 30mTorr for a flow rate of 5SCCM of pure SiH_4. In these plasma conditions, films were deposited on the c-Si wafers where a trench structure (1μm or 3μm in width and 3μm in depth) was prepared on the surface, whose temperature was maintained at a constant value during the deposition of the film in the range from room temperature (rt) to 500C. The film-deposition rate was derived from the thickness measurement on the ridge surface in the trench pattern on the c-Si wafer. The reflection probability 1-β (β being the total loss probability of radicals) was estimated by the thickness profile in the trench structure on the c-Si wafer after the growth of the film by

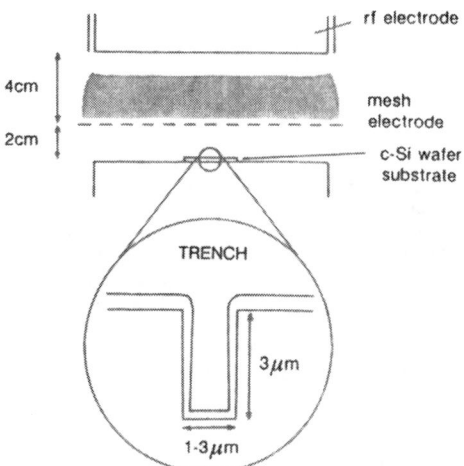

Fig.2. Triode configuration for separating SiH_3 and H from other radicals, and the trench structure prepared on the c-Si wafers.

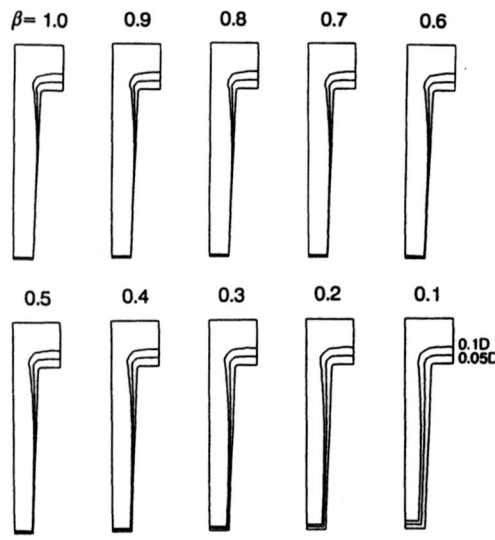

Fig.3. Step-coverage profile for each β calculated
using a Monte-Carlo simulation by assuming
a cosine reflection of radicals on the surface.

comparing to the profiles predicted by a Monte-Carlo simulation
[9]. Figure 3 shows the predicted thickness profiles for each β
in the trench structure calculated by a Monte-Carlo simulation.

Observation of the Growth-Zone Reaction

A layer-by-layer deposition was performed, in which a
sufficient amount of H-atoms were supplied after every several
monolayer growth of the film as shown in Fig. 4 in order to
simulate conditions for the reaction in the growth zone [4] which
is thought to be located just below the top growing surface within
several monolayers. Using the conventional diode-type PECVD
reactor, the surface of the film was exposed to the H_2 plasma for
two minutes keeping the substrate temperature constant at 400C
after every several monolayers (about 10A in thickness) deposition
carried out under the a-Si:H deposition condition, i. e., a flow
rate of 5SCCM of SiH_4, pressure of 30mTorr and rf power density of
$0.06W/cm^2$. In order to supply a lot of H atoms in the growth
zone, H_2-plasma condition was set as follows; a flow rate of 5SCCM
of H_2, the pressure of 70mTorr and the rf power density of
$0.26W/cm^2$. This layer-by-layer deposition was repeated several
hundred times, producing a film several hundred nm in thickness.
Since the surface-H-coverage factor starts decreasing from 350C
upwards in the a-Si:H deposition condition (as is mentioned later
on), each deposited 10A layer should be amorphous. If the
reaction in the growth zone plays an important role for the
formation of μc-Si:H in the presence of

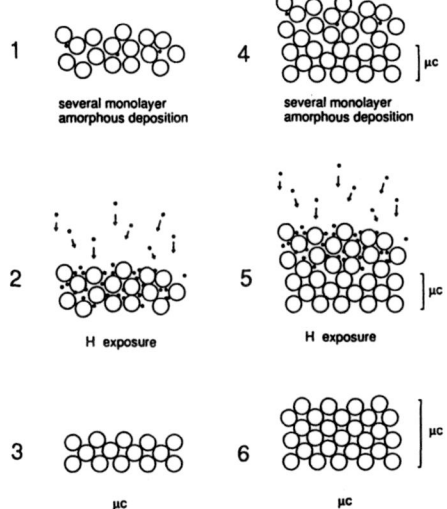

Fig.4. Layer-by-layer deposition and chemical annealing
processes to confirm the existence of the
growth-zone reaction.

sufficient amount of H, the film deposited by this procedure must
be μc-Si:H, because sufficient "chemical annealing" was performed
after each several monolayers growth (the thickness of the growth
zone).

Using the experimental data obtained as mentioned above, the
role of the surface reaction both on the top growing surface and
in the growth zone is discussed below.

RESULTS AND DISCUSSION

Responsible Species for the Growth of μc-Si:H

Figure 5 shows the deposition rate of the film on the
substrate placed on the grounded and heated electrode plotted
against the emission intensity of Si* and the number density of
SiH3 radicals determined by the diode-laser-absorption technique
during the growth of the film. It should be noted here that the
signal from the SiH2 radical was not observed in these
experimental conditions in spite that the detection limit of SiH2
in this system being as low as $6x10^9/cm^3$. The number density of
SiH3 is high enough to explain the actual deposition rate of the
films both in a-Si:H and μc-Si:H deposition conditions by assuming
a sticking probability of about 0.1 which is determined

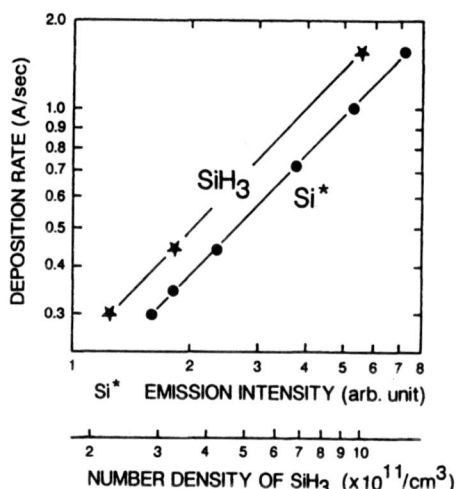

Fig.5. Deposition rate of the films plotted against the emission intensity of Si* and the number density of SiH3 (indicated by stars) and SiH2 radicals in a variety of plasma conditions.

in the next section, while the number density of SiH2 is too low to explain the actual deposition rate even if we take a maximum number density ($6x10^9/cm^3$) and a unity sticking probability. A linear correlation is clearly seen in the figure between the deposition rate and the emission intensity of Si* in the wide range of plasma conditions. The relationship between the emission intensity of Si* and the generation rate of SiH3 radicals in the plasma is derived from the following equations since the lifetime (10^{-8}sec) of Si* state is short enough to eliminate any collision with other species or molecules before emitting the light from Si* state. A stationary state density of Si* produced by one electron impact process in the plasma is expressed as

$$d[Si*]/dt = N_e \, \sigma_{Si*} \, v_e \, [SiH_4] - I_{Si*} = 0, \tag{1}$$

then,

$$I_{Si*} = N_e \, \sigma_{Si*} \, v_e \, [SiH_4], \tag{2}$$

where N_e is the electron density in the plasma, σ_{Si*} is the emission-cross section of Si*, v_e is the thermal velocity of electron, $[SiH_4]$ is the number density of SiH4 molecules and I_{Si*} is the emission intensity of Si*, respectively. On the other hand, the generation rate of SiH3 (Γ_{SiH_3}) in the plasma is described as

$$\Gamma_{SiH_3} = N_e \ \sigma_{SiH_3} \ v_e \ [SiH_4], \tag{3}$$

where σ_{SiH_3} is the dissociation-cross section for the SiH3 radicals. Therefore, from eqs. (2) and (3), the proportionality of Γ_{SiH_3} to I_{Si^*} is derived as

$$\Gamma_{SiH_3} = (\sigma_{SiH_3}/\sigma_{Si^*}) \ I_{Si^*} \ \propto \ I_{Si^*}. \tag{4}$$

Actually, a number-density ratio of SiH3 radicals determined by the diode-laser-absorption technique is just the same as the emission-intensity ratio of Si* between the two plasma conditions of SiH4/H2=1/1 and SiH4/H2=1/49 as seen in the figure. Therefore, the emission intensity can be thought to be a quantity proportional to the flux density of SiH3 radicals reaching the film-growing surface under the experimental conditions used in this experiment. Although we cannot make a quantitative discussion at the present stage since the density of SiH3 radicals was measured only at a fixed spatial position (7mm above the grounded electrode) between two electrodes, a good linearity between the deposition rate and the emission intensity of Si* in a wide range of SiH4/H2 ratios gives us three important suggestions: (1) The main film precursor is identical both for the growth of a-Si:H and μc-Si:H, and is not SiH2 radicals but SiH3 radicals. (2) The sticking probability of SiH3 radicals on the surface takes the same value both in the surface conditions of a-Si:H and μc-Si:H growth at least when the substrate temperature is kept constant at 150C. (3) The etching process can be ruled out in the formation process of μc-Si:H, although weak-bond breaking, which does not produce a volatile species, might occur during the growth of the film.

Surface Reaction Process on the Top Growing Surface

Figure 6 shows the total loss probability β (reflection probability being 1-β) of SiH3 radicals on the film-growing surface estimated by the step-coverage profile after the growth of the film in the trench structure by comparing to the computer simulated profiles both in a-Si:H and μc-Si:H formation conditions plotted against the substrate temperature (T). As shown in the figure, the total loss probability β in the a-Si:H formation condition does not show a strong temperature dependence up to 500C within the experimental precision. On the other hand, in the μc-Si:H formation condition, the total loss probability shows a higher value as compared to the a-Si:H formation condition in a lower temperature range and it tends to decrease toward higher temperatures reaching a similar value to the a-Si:H formation condition at 500C. Considering the fact reported previously [1] that μc formation was not observed at temperatures higher than 500C, the difference in the total loss probability between the μc-Si:H and a-Si:H conditions might give us an important clue to understand the difference of film-growing reactions on the surface between the μc-Si:H and a-Si:H growth.

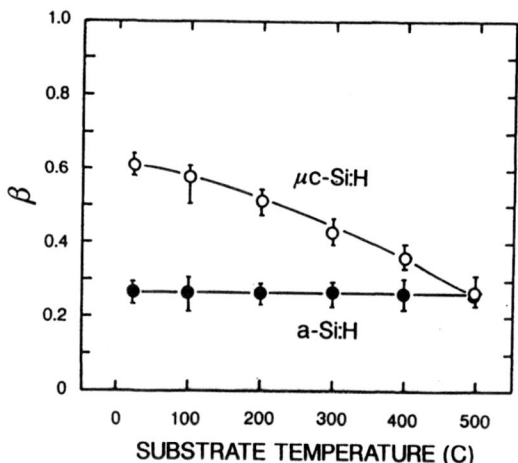

Fig.6. Loss probability (β) of radicals on the surface
plotted against the substrate temperature both
in the a-Si:H and μc-Si:H deposition conditions.

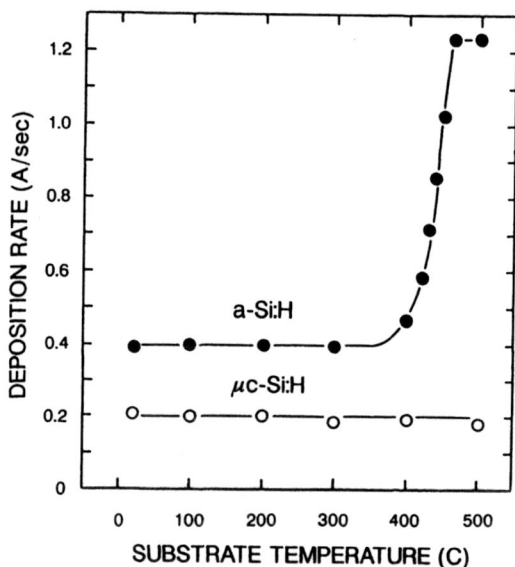

Fig.7. Deposition rate of the films on the ridge surface
of the trench structure plotted against the
substrate temperature both in the a-Si:H and
μc-Si:H deposition conditions.

Figure 7 shows the deposition rate on the ridge surface of the trench structure on the c-Si wafer both in a-Si:H and μc-Si:H deposition conditions plotted against the substrate temperature. In the a-Si:H deposition condition, the deposition rate shows no temperature dependence in the temperature range between rt and 300C. Then, above 350C and up to 460C, the deposition rate increases by a factor 3, although the value of β remains constant. Finally, the deposition rate reaches some saturation since there is no further increase from 460 to 500C. The constancy of β and the temperature dependence of the deposition rate can be understood in the light of a surface-reaction model reported elsewhere [7], in which the increase in the deposition rate above 350C with constant β is explained as an increase in the fractional appearance of the Si-dangling bond on the almost fully H-covered film-growing surface. Based on the model [7], β=s is expected in the temperature range between 460 and 500C so that the sticking probability s, in the case of the deposition of a-Si:H, can be represented in the β-T plane in Fig. 6 by fitting s to β. This is shown in Fig. 8.

On the other hand, in the μc-Si:H formation condition, the deposition rate has no temperature dependence in the whole temperature range from rt to 500C. The constancy of the deposition rate seems to be due to the reason that almost complete H-coverage is realized on the film-growing surface even at high temperatures by the termination of thermally exposed Si-dangling bonds with a lot of weakly chemisorbed H atoms in this condition. Since the sticking probability for the deposition of μc-Si:H is the same as that for the a-Si:H deposition which is suggested from the linear relationship between the deposition rate and the

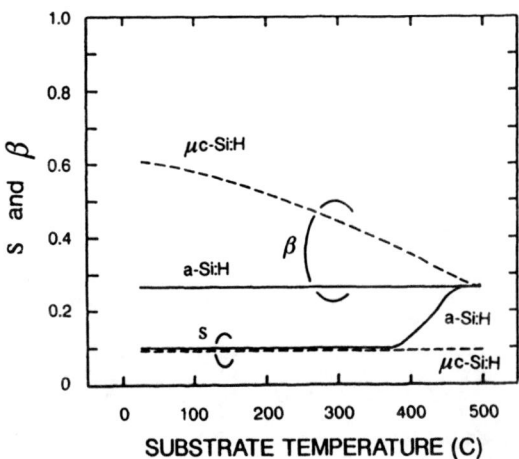

Fig.8. Loss (β) and sticking (s) probabilities of radicals on the surface as a function of the substrate temperature both in the a-Si:H and μc-Si:H formation conditions.

number density of SiH3 radicals in the plasma in all deposition conditions as shown in Fig. 5 as far as the substrate temperature is kept at 150C, the sticking probability in the case of the μc-Si:H deposition condition can also be superposed as shown in Fig. 8.

Figure 9 shows the schematic concept of the surface reaction process of SiH3 radicals reaching the film-growing surface. Part of incident SiH3 radicals on the film-growing surface are reflected into the gas phase keeping their initial form (reflection 1-β), and the other radicals are adsorbed on the surface (adsorption or surface loss β) where they start diffusing on the surface (surface diffusion) depending on the condition of the surface. Some part of diffusing radicals react with surface hydrogen atoms (weakly chemisorbed or bonded with Si) or with other adsorbed radicals and become stable molecules (recombination γ) which can desorb from the surface. Other diffusing radicals react with surface silicon (sticking s) forming Si-Si bonds, which contributes to the growth of the film. Therefore, we can write β=s+γ.

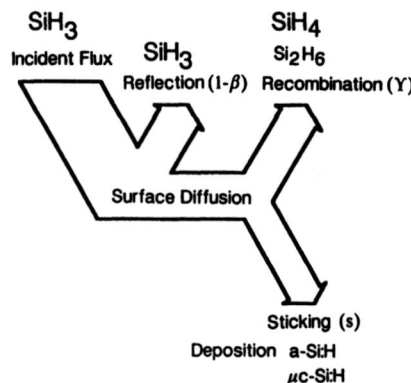

Fig.9. Schematic concept of the surface reaction process of SiH3 radicals reaching the film-growing surface.

Since the sticking probability of SiH3 radicals is always the same in the temperature range from rt to 300C both in a-Si:H and μc-Si:H growth conditions, the increase in γ in the μc-Si:H deposition condition is likely to be due to the existence of weakly chemisorbed H atoms on the surface which will also play an important role in realizing a H-coverage factor of unity on the surface. The occasional Si-dangling bond exposed during the growth process of the film is terminated by the weakly chemisorbed H atoms moving on the surface, leading to a complete coverage with H which helps the surface diffusion of SiH3 adsorbed radicals to find out the energetically favorable sites on the surface. The value of γ decreases gradually with an increase of the substrate temperature in the μc-Si:H formation condition as shown in Fig. 6.

This fact can be explained by the thermal desorption process of weakly chemisorbed H atoms from the surface by elevating the temperature. The surface condition at 500C even for the μc-Si:H deposition condition becomes the same as that for the a-Si:H deposition condition. Thus, the deposited film at 500C was amorphous [1].

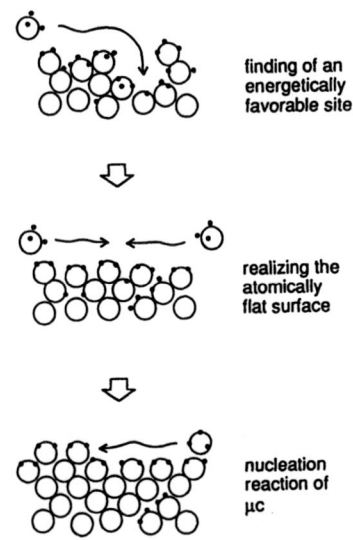

finding of an energetically favorable site

realizing the atomically flat surface

nucleation reaction of μc

Fig.10. Surface-reaction model for explaining the realization of flat surface and the nucleation reaction in the μc-Si:H deposition process.

The energetically favorable site for adsorbed SiH3 must be a deeper site which has lower potential energy in the surface atomic structure. Therefore, in the μc-Si:H formation condition, the surface tends to be flatter in the atomic scale during the growth of the film, which might result in the appearance of μc nuclei as shown in Fig. 10.

Reaction in the Growth Zone

Does the reaction in the growth zone also play an important role for the μc-Si:H formation process? In order to answer this question, the film growth was carried out using the layer-by-layer depositions with a shower of H-atoms after every several monolayer deposition as described before. Result is "NO !" The μc-Si:H was not formed even in the condition that sufficient amount of H atoms are supplied in the growth zone after the growth of every several amorphous monolayers. This fact suggests that the reaction in the growth zone does not play an important role for the nucleation of μc, although the growth of μc might occur in the growth zone.

Much clearer experiments are needed to know the role of the reaction in the growth zone.

CONCLUSION

The role of the surface reaction process in the formation of μc-Si:H was discussed in comparison to the formation process of a-Si:H, and the following items are suggested:
(1) The main film precursor is not SiH_2 but SiH_3 radicals in the growth of μc-Si:H as well as a-Si:H.
(2) In the deposition of μc-Si:H, weakly chemisorbed H atoms on the top growing surface tend to terminate Si-dangling bonds which are occasionally exposed on the surface during the growth of the film, which gives rise to the enhancement of the surface diffusion of SiH_3 radicals adsorbed on the surface. This might result in the ordered and stable network structure (μc-Si:H).
(3) It is not necessary to take into account the reaction in the growth zone for the formation process of μc nucleus.

ACKNOWLEDGEMENT

The authors wish to express their thanks to K.Nomoto, A.Suzuki, Y.Takeuchi and Y.Urano for their experimental assistance, to K.Sugawara for supplying us the c-Si wafers on which fine trench structures were prepared, to A.Yuuki for the Monte-Carlo simulation of the thickness profile in the trench structure, to N.Itabashi for measuring the SiH_3 radicals using a diode-laser-absorption technique and to G.Ganguly for critical reading of the manuscript.

References

1. A.Matsuda, J. Non-Cryst. Solids, 59&60 767 (1983).
2. S.Veprek, Chimia 34 489 (1980), and S.Veprek, Z.Iqbal, H.R.Oswald and A.P.Webb, J Phys. C 14 295 (1981).
3. C.C.Tsai, Amorphous Silicon and Related Materials, Vol.1, ed. H.Fritzsche (World Sci. Pub., Singapore, 1989) p. 123.
4. N.Shibata, K.Fukada, H.Ohtoshi, J.Hanna, S.Oda and I.Shimizu, Mat. Res. Soc. Symp. Proc. 95 225 (1987).
5. N.Itabashi, K.Kato, N.Nishiwaki, T.Goto, C.Yamada and E.Hirota, Jpn. J. Appl. Phys., 27 L1565 (1988).
6. C.C.Tsai, J.C.Knights, G.Chang and B.Wacker, J. Appl. Phys., 59 2998 (1986).
7. A.Matsuda, K.Nomoto, Y.Takeuchi, A.Suzuki, A.Yuuki and J.Perrin, to be published in Surf. Sci.
8. A.Matsuda and K.Tanaka, J. Appl. Phys., 60 2351 (1986).
9. A.Yuuki, Y.Matsui and K.Tachibana, Jpn. J. App. Phys., 28 212 (1989).

OPTO-ELECTRONIC PROPERTIES OF μc-Si GROWN FROM SiF$_4$ AND H$_2$ BY PECVD

Y. OKADA, I. H. CAMPBELL, P. M. FAUCHET AND S. WAGNER
Department of Electrical Engineering, Princeton University, Princeton, New Jersey 08544

ABSTRACT

Microcrystalline Si was grown from SiF$_4$ and H$_2$ by plasma-enhanced chemical vapor deposition. The films are almost completely crystalline with a crystallite size (determined from Raman spectra) of about 60 Å. The optical absorption and the electrical conductivity of these films were studied. With increasing hydrogen content in the films, the dark conductivity decreases strongly and the activation of the conductivity increases. We explain the conductivity qualitatively in terms of a grain boundary model.

INTRODUCTION

Microcrystalline silicon (μc-Si) grown by plasma-enhanced chemical vapor deposition (PECVD) exhibits opto-electronic properties that are clearly distinct from crystalline (c) or amorphous (a) silicon. μc-Si has a much higher doping efficiency than a-Si:H [1,2,3], and its optical gap does not narrow upon doping [2]. Furthermore, fine grained (20~30 Å) μc-Si shows a large band gap of 2.4 eV [4]. These are the reasons why μc-Si is used in the p-type layer of amorphous silicon solar cells [5,6]. Because the carrier mobilities in μc-Si are higher than in a-Si:H, μc-Si is under development for thin film devices. The observation that the band offset between a-Si:H and c-Si exists entirely in the valence band edge [7,8] makes μc-Si a candidate for the emitter in μc-Si/c-Si heterojunction bipolar transistors [9,10].

LeComber et al. [11] interpreted the electronic properties of doped μc-Si grown by glow discharge deposition in terms of a grain boundary model developed for polycrystalline silicon (poly-Si) [12,13]. As for the conductivity of undoped glow discharge deposited μc-Si, Mishima et al. [14] have reported that it is determined by the volume fraction of crystalline material, and Komura et al.[15] interpreted the conductivity by a percolation process. Konuma et al. [16] have shown that the conductivity increases with crystallite size.

Most μc-Si has been deposited from SiH$_4$ and H$_2$. Shimizu et al. [17,18] and we [19,20] have studied its growth from SiF$_4$ and H$_2$. In this paper, we report the optical gap and the electrical conductivity of μc-Si grown from SiF$_4$ and H$_2$ using PECVD. When μc-Si films are grown from SiF$_4$ with H$_2$ dilution and around a substrate temperature of 300 °C, they consist of only the microcrystalline phase [19]. We will describe and discuss the optical absorption and the charge transport in these films, and interpret the conductivity with a grain boundary model.

EXPERIMENTAL PROCEDURES

We prepared μc-Si films by glow-discharge decomposition of SiF$_4$ and H$_2$ molecules in a DC triode system [21]. The films were grown on p-type (100) Si substrates for IR measurements and on Corning 7059 glass substrates for conductivity measurements. The total flow rate of the gas mixtures was 30 to 60 sccm. The pressure during film growth ranged from 0.95 to 1.25 Torr, the substrate temperature from 230 to 355 °C, and the power density from 0.20 to 0.33 mW/cm^2. The film growth rates lay between 0.5 and 1 Å/sec. The important growth parameters, including SiF_4 and H_2 flow rates and substrate temperature, are shown in Table 1 with the opto-electronic properties of the resulting films.

Infrared spectra were measured in the range from 400 to 4000 cm^{-1} with a Digilab FTS-20C Fourier transform spectrometer or in the range 400 to 6500 cm^{-1} with a Nicolet 730 FT-IR spectrometer. The hydrogen content in the films was estimated from the intensity of the $Si-H_n$ stretching modes [22].

Raman spectra were acquired in the back scattering geometry using 514.5 nm (Ar-ion laser) excitation and 4 cm^{-1} spectral resolution. The full width at half maximum (FWHM) of the microcrystalline peak was taken as a measure of crystallite size [23,24].

Optical transmission spectra $\alpha(\lambda)$ at the absorption edge were recorded with a Hitachi U-3410 spectrophotometer. The sample thickness was determined from the interference fringes in the near-infrared region.

The electrical conductivity in the dark (σ_d) was measured with coplanar Cr contacts over the range from room temperature to 120 °C and the activation energy of the dark conductivity (E_a) was determined from an Arrhenius plot.

RESULTS AND DISCUSSION

Fig. 1 shows a Raman spectrum that is typical of the μc-Si film we studied. A crystalline peak is observed at 517 cm^{-1} with a broad tail at the lower energy side of the peak. The crystallinity of the film is close to unity with little volume fraction of an a-Si:H matrix. All samples in this experiment show this kind of Raman spectrum. The size of the microcrystals determined from the Raman FWHM [23,24] lies between 55 and 70 Å.

The optical absorption spectra $\alpha(h\nu)$ of four of our samples, together with typical spectra of a-Si:H,F and c-Si [25] are plotted in Fig. 2. The spectra were taken at room temperature. The absorption edge of μc-Si lies between those of c-Si and a-Si:H,F. The same observation was made by Matsuda et al. for p-type μc-Si [2] and by Mishima et al. for μc-Si embedded in an amorphous matrix [14]. Richter and Ley, on the other hand, found that μc-Si has a higher absorption coefficient at 2.0 eV than a-Si:H and c-Si [3].

Table 1 Deposition conditions and opto-electronic properties of μc-Si films

Sample No.	Gas flow SiF$_4$ (sccm)	H$_2$ (sccm)	Sub. temp. (°C)	Dark cond. (Scm^{-1})	Ea (eV)	Raman FWHM (cm^{-1})	Optical gap E$_{04}$ (eV)	E$_{Tauc}$ (eV)	H conc. (at.%)
X6 (*)	5	25	230	7.7E-8	0.55	12.2	1.99	1.76	11.4
X7 (*)	5	25	280	3.2E-7	0.53	12.3	1.98	1.79	8.9
X58	10	50	345	2.5E-6	0.31	10.4	2.02	1.73	4.4
X59	10	20	355	1.1E-2	0.11	11.8	2.06	1.84	2.4
X80	10	20	355	5.3E-3	0.115		1.98	1.65	
X153	10	50	300	1.6E-4	0.22	11.6	2.01	2.01	4.1
X154	10	20	300	2.0E-7	0.57	10.1	2.00	1.76	
X159	10	35	355	2.9E-7	0.61	11.1	2.07	1.83	
X160	10	35	310	1.6E-6	0.35		2.00	1.69	
X171	10	35	300	1.9E-4	0.20	12.5	2.10	1.92	3.6
X172	10	35	250	8.5E-5	0.21	12.9	2.10	1.82	3.2

(*) Samples X6 and X7 were grown on a different substrate holder.

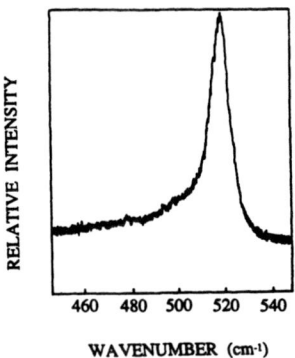

Fig. 1 Typical Raman spectrum of μc-Si.

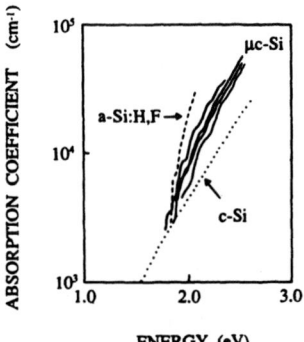

Fig. 2 Optical absorption coefficients of μc-Si, from left to right, samples X58, X159, X154, X153, (solid lines), a-Si:H,F (broken line) and c-Si (dotted line).

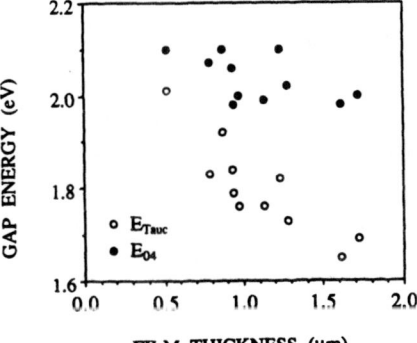

Fig. 3 The apparent effect of sample thickness on the optical gap energies, E_{Tauc} and E_{04}.

For μc-Si, a Tauc plot $((\alpha \cdot h\nu)^{1/2}$ vs. hν) shows a straight line around 2.0 to 2.5 eV. The optical gaps, E_{Tauc}, determined from a Tauc plot, and E_{04}, the energy where α is 10^4 cm^{-1}, are plotted vs. sample thickness in Fig. 3. The E_{Tauc} decreases drastically with increasing sample thickness, while the E_{04} lies in a small range and is nearly independent of the sample thickness. Therefore, we adopt E_{04} as a measure of the gap. The E_{04} of μc-Si is larger than that of a-Si:H,F but smaller than that of c-Si as shown in Fig. 2. Fig. 4 shows the the effect of the total hydrogen content on the E_{04}. Adding hydrogen decreases the E_{04}.

Fig. 5 is a plot of the dark conductivity against the Raman FWHM. Although the FWHM lies in a small range from 10 to 13 cm^{-1}, indicating almost identical grain sizes of ~ 60 Å, the dark conductivity at room temperature ranges from 10^{-2} to 10^{-8} S·cm^{-1}, the latter value being close to that of a-Si:H. The highest value that we obtain, 1.1 x 10^{-2} S·cm^{-1}, is comparable to the highest value of ~10^{-2} S·cm^{-1} [15] observed in films grown from SiH$_4$. This result suggests that the conductivity is not determined by bulk properties.

The dark conductivity decreases strongly with increasing total hydrogen content, as demonstrated in Fig. 6. The total hydrogen content, as determined from the IR spectra, varies from 2 to 11 at%, depending on the growth conditions. At high hydrogen content, Si-H and Si-H$_2$ bonds coexist but at low hydrogen content only Si-H$_2$ is found. The latter suggests that hydrogen is concentrated in intergrain regions. Also, the

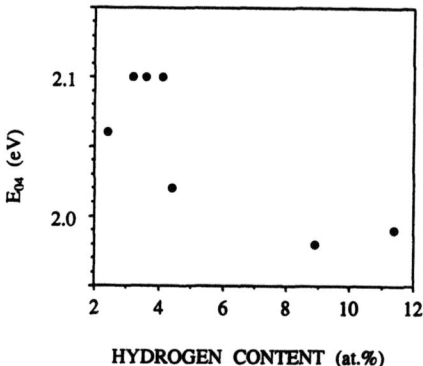

Fig. 4 The effect of total hydrogen content on E_{04}.

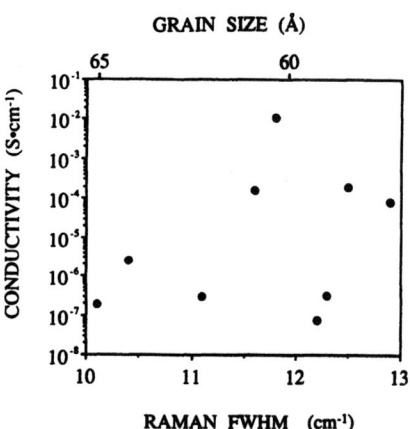

Fig. 5 Dark conductivity vs. Ramam FWHM.

activation energy of the conductivity increases with increasing hydrogen content (Fig. 6).

Our data show a large range in the dark conductivity of μc-Si films with identical grain size, and a strong correlation between σ_d, E_a and total hydrogen content. This observation suggests that the grain boundaries control the electrical conduction. Therefore, we apply Seto's grain boundary model [12]: when the dopant content is so low that the crystallites are completely depleted, the conductivity should be approximately proportional to $\exp(-E_a/kT)$. Fig. 7 is a plot of σ_d and E_a for eleven samples. The solid line in the figure is obtained from Seto's grain boundary model [12] for fully depleted grains. The model assumed traps located 0.37 eV above the valence band, a trap density of 3.34×10^{12} cm^{-2} and a grain size of 270 Å. The slope of this line is close to $-1/2.303kT$. Our experimental data exhibit a similar slope in the range from 0.1 to 0.4 eV of the activation energy. These results suggest that the grain boundaries control the electronic transport.

ACKNOWLEDGEMENTS

This work is supported by the Electric Power Research Institute under Contract No. RP 2824-2, by the Office of Naval Research (Y. O.) and by the National Science Foundation through the Presidential Young Investigator Program (I. H. C. and P. M. F.).

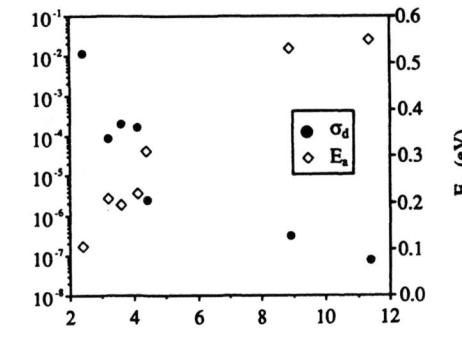

Fig. 6 Dark conductivity (left ordinate) and its thermal activation energy (right ordinate) vs. total hydrogen content.

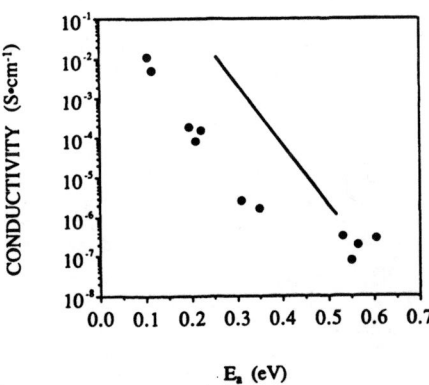

Fig. 7 Dark conductivity and its activation energy. Solid circles are the experimental data, solid line is Seto's theory.

REFERENCES

1. A. Matsuda, S. Yamasaki, K. Nakayama, H. Okushi, K. Tanaka, S. Iizima, M. Matsumura and H. Yamamoto, Jpn. J. Appl. Phys. 19, L305 (1980)
2. A. Matsuda, M. Matsumura S. Yamasaki,and H. Yamamoto, T. Imura, H. Okushi, S. Iizima and K. Tanaka, Jpn. J. Appl. Phys. 20, L183 (1981)
3. H. Richter and L. Ley, J. Appl. Phys. 52, 7281 (1981)
4. S. Furukawa and T. Miyasato, Phys. Rev. B38, 5726 (1988)
5. J. L. Guimaraes, R. Martins, E. Fortunato, I. Ferreira, M. Santos and N. Garvalho, Mat. Res. Soc. Symp. Proc. 118, 617 (1988)
6. R. Martins, M. Vieira, E. Fortunato, M. Santos, I. Ferreira, M. Lavado and L.Guimaraes, Conference Record of the International Topical Conference on Hydrogenated Amorphous Silicon Devices and Technology, p35, Yorktown Hights, New York, Nov. 21-23, 1988
7. M. Matsumura, Conference Record of the International Topical Conference on Hydrogenated Amorphous Silicon Devices and Technology, p115, Yorktown Hights, New York, Nov. 21-23, 1988
8. J. M. Essick and J. D. Cohen, Appl. Phys. Lett. 55, 1232 (1989)
9. H. Fujiioka, T. Deguchi, K. Takazaki and T. Takada, IEDM Tech. Dig., p574 (1988)
10. K. Sasaki, T. Fukuzawa and S. Furukawa, IEDM Tech. Dig., p186 (1987)
11. P. G. LeComber, G. Willeke and W. E. Spear, J. Non-cryst. Solids 59&60, 759 (1983)
12. J. Y. W. Seto, J. Appl. Phys. 46, 5247 (1975)
13. G. Baccarani, B. Ricco and G. Spadini, J. Appl. Phys. 49, 5565 (1978)
14. Y. Mishima, S. Miyazaki, M. Hirose and Y. Osaka, Phil. Mag. B46, 1 (1982)
15. S. Komura, Y. Aoyagi, Y. Segawa, S. Namba, A. Matsuyama, A. Matsuda and K. Tanaka, J. Appl. Phys. 56, 1658 (1984)
16. M. Konuma, H. Curtins, F. -A. Sarott and S. Veprek, Phil. Mag. B55, 377 (1987)
17. N. Shibata, K. Fukuda, H. Ohtoshi, J. Hanna, S. Oda and I. Shimizu, Jpn. J. Appl. Phys. 26, L10 (1987)
18. H. Tanabe, M. Azuma, T. Uematsu, H. Shirai, J. Hanna, I. Shimizu, Mar. Res. Soc. Symp. Proc. 149, 17 (1989)
19. Y. Okada, J. Chen, I. H. Campbell, P. M. Fauchet and S. Wagner, Mar. Res. Soc. Symp. Proc. 149, 93 (1989)
20. Y. Okada, J. Chen, I. H. Campbell, P. M. Fauchet and S. Wagner, J. Appl. Phys., February 1990
21. D. Slobodin, S. Alijishi, R. Schwarz and S. Wagner, Mat. Res. Soc. Symp. Proc. 49, 153 (1985)
22. C. J. Fang, K. J. Gruntz, L. Ley and M. Cardona, J. Non-Cryst. Solids 32, 405 (1979)
23. I. H. Campbell and P. M. Fauchet, Solid State Comm. 58, 739 (1986)
24. P. M. Fauchet and I. H. Campbell, CRC Crit. Rev. Solid State Mat. Sci. 14, S79 (1988)
25. "Handbook of Optical Constants of Solids", ed. by E. D. Palik, Academic Press, New York, 1985

FORMATION OF MICROCRYSTALLINE SILICON FILM BY RMS PROCESS

Cheng Wang, G.N. Parsons, E.C. Buehler, R.J. Nemanich and G. Lucovsky
Department of Physics, North Carolina State University, Raleigh, NC 27695-8202

ABSTRACT

We have deposited microcrystalline, μc-Si, silicon films by using RF reactive magnetron sputtering (RMS) at high substrate temperatures, $T_S > 500°C$, and at a relatively low partial pressure of hydrogen, $P_H = 0.40$ mTorr, and at low T_S ~200-300°C, but with a higher $P_H > 2$ mTorr. We have detected μc-crystallinity by Raman scattering and transmission electron microscopy. We discuss differences in the growth mechanisms for formation of μc-Si under these two deposition conditions.

INTRODUCTION

Hydrogenated amorphous silicon (a-Si:H) thin films have been used in several devices including p-i-n solar cells, and thin-film transistors. As these devices have evolved, it has become apparent that doped μc-Si thin films could be incorporated into the structures and yield improved performance. We have deposited electronic-grade a-Si:H by reactive magnetron sputtering (RMS) [1], and by remote plasma-enhanced chemical-vapor deposition [2], and have recently begun a study of the deposition of μc-Si thin films by these two techniques. This paper reports on our studies of the deposition of μc-Si by RMS.

We have been successful in depositing μc-Si by RMS under two different sets of deposition conditions: i) at high T_S, >500°C, independent of the of hydrogen P_H; and ii) at relatively low T_S, ~200°C-300°C, but only for high P_H, > 2.0 mTorr. For both depositions, $P_{Ar} = 2.0$ mTorr, and the RF power is 200 watts at 13.6 Mhz. We have detected the μ-crystallinity by Raman scattering, TEM bright- and dark-field imaging and electron diffraction. We discuss the growth mechanisms focusing on the factors that contribute to a transition region of a-Si, between the substrate (c-Si or fused silica) and μ-crystalline portion of the film structure.

EXPERIMENTAL PROCEDURES AND RESULTS

A UHV compatible reactive magnetron sputtering system (RMS) [1] has been used for the deposition of the μc-Si films. C-Si and fused silica substrates are loaded into the system via a load-lock substrate introduction chamber. Research purity Ar and H_2 comprise the sputtering ambient. P_{Ar} was fixed at 2.0 mTorr, P_H was varied between 0 and 4.0 mTorr, T_S was varied between "room-temperature, ~40°C, and 650°C, and the RF power to the magnetron sputtering target, polycrystal silicon, was 200 Watts at 13.6 MHz. μ-crystallinity was identified in the deposited films by Raman scattering, electron diffraction, and TEM imaging, in particular comparisons between bright and dark field images.

1. Raman Spectrum

Figure 1 displays the Raman spectra of the films deposited at $P_H = 0.40$ mTorr, and with T_S between 240°C and 650°C. The films deposited at 240°C: i) were "electronic-grade" a-Si:H; ii) contained ~13 at.% bonded hydrogen, predominantly in monohydride groups as determined by IR; and iii) had no crystalline features in their Raman spectrum. Films deposited at 450°C, were also amorphous, but contained no IR detectable bonded hydrogen. There was dramatic change in the character of the Raman spectrum for the films deposited at 540°C and 650°C. These films display a narrow feature at approximately 518 cm^{-1}, previously shown to be a spectral signature of μc-Si, and display no features indicative an a-Si "alloy" component [3]. The spectra of these films were clearly not a linear superposition of an a-Si spectrum, and a narrow μc-Si feature at 518 cm^{-1} [4]. The dominant feature in spectrum for the film deposited at T_S = 650°C is narrower than the corresponding feature in the film deposited at 540°C, and in addition the spectrum displays other features that are associated with 2-phonon scattering usually detected in polysilicon.

Similar changes in the Raman spectra occur for samples deposited at fixed T_S = 250°C, but with P_H varied from 0.7 mTorr to 4.0 mTorr. (see Fig. 2). Films deposited with P_H = 0.7 and 1.0 mTorr are amorphous, whereas films deposited with P_H = 2.0 and 4.0 mTorr are μ-crystalline. The spectra in Fig. 2 for P_H = 2.0 and 4.0 mTorr, display an distinct structure on the low wavenumber side of the spectral peak that is qualitatively different from any corresponding low wavenumber features displayed by the spectra in Fig. 1 for T_S = 540°C or 650°C.

2. Electron Diffraction and TEM Imaging

μ-crystalline features in the RMS films could also be detected by electron diffraction pattern and TEM imaging, especially via the complementary character of structure in the bright-field and dark-field images. Fig. 3 displays the electron diffraction, and the bright-field TEM image for the sample deposited at 540°C, with P_H = 0.40 mTorr. μ-crystallinity is clearly evident in both the diffraction pattern and in the TEM image. A film deposited with the same P_H, but at 450°C, also shows no evidence of μ-crystallinity by diffraction or TEM imaging. The films deposited at 540°C and 650°C have a transition region of a-Si between the c-Si substrate and the onset of the μc-Si: i) the boundary between the a-Si and μc-Si regions is planar and abrupt; and ii) similar transition regions occur with fused silica as well c-Si substrates. The total film thickness for the deposition at 540°C was ~1 μm, and the thickness of the transition layer was about 1000 Å.

Figure 4 shows the electron diffraction pattern and bright-field TEM images for a sample deposited onto c-Si

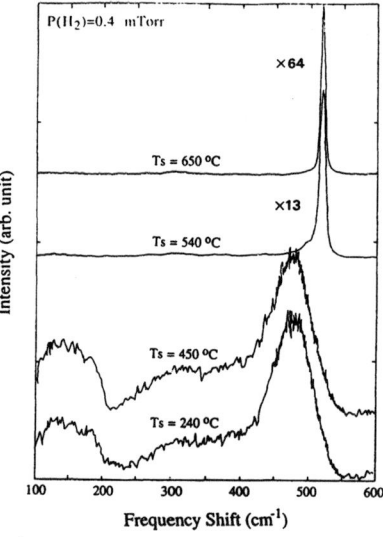

Figure. 1
Raman spectra for films deposited at constant P_H.

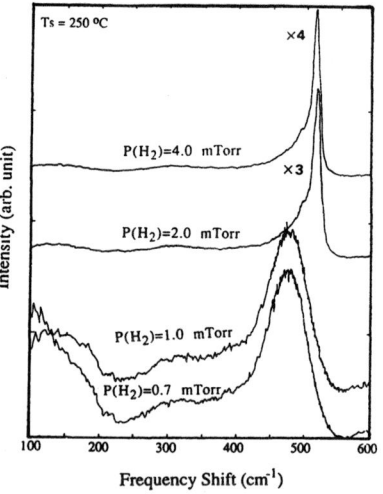

Figure. 2
Raman spectra for films deposited at constant T_S.

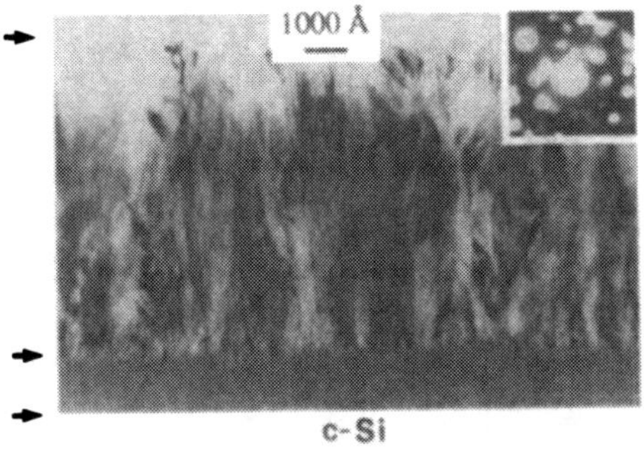

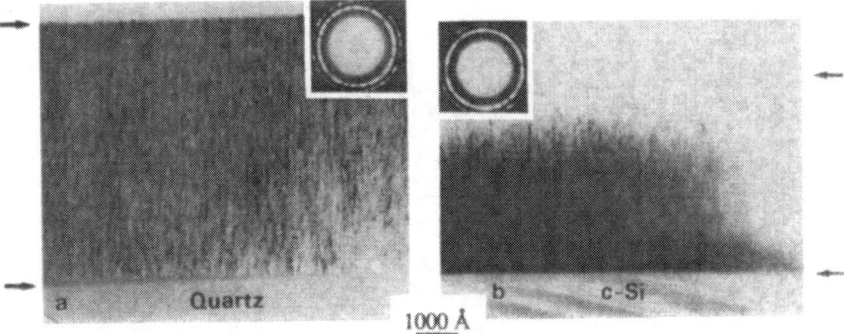

Fig.3: TEM bright-field image and electron diffraction: $T_S=540°C$ and $P_H=0.40$ mTorr.

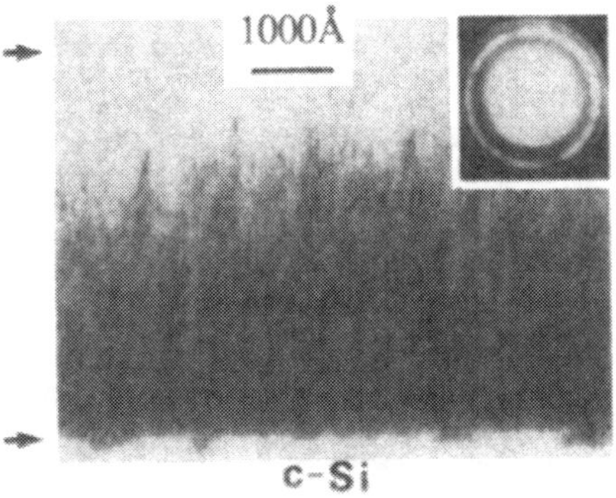

Fig. 4: TEM bright-field image and electron diffraction: $T_S=200°C$ and $P_H=4.0$ mTorr.

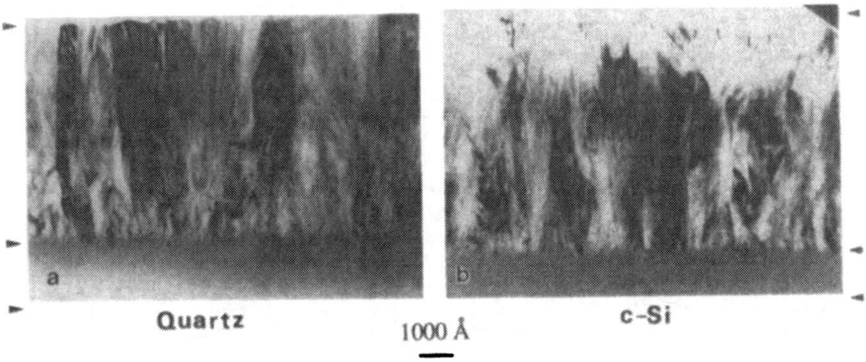

Fig. 5: TEM bright-field image: $T_S = 540°C$ and $P_H = 0.40$ mTorr (a) fused silica, and (b) c-Si substrates.

Fig. 6: TEM bright-field image and electron diffraction: $T_S = 300°C$ and $P_H = 4.0$ mTorr (a) fused silica, and (b) c-Si substrates.

substrate at 200°C with P_H = 4.0 mTorr. μ-crystalline features are evident in the diffraction pattern and bright-field image. In contrast to samples deposited at high T_S and low P_H, there is no transition layer between the substrate and the μc-Si region. In addition, the μ-crystallites are smaller, < 100 Å , as compared to >>100 Å in Fig. 3. The crystallites in Fig. 4 appear to be dispersed in a "connective" matrix, rather than being compact as in Fig. 3.

Figure 5 displays the TEM images of μc-Si films deposited on c-Si and fused silica at 540°C with P_H = 0.40 mTorr. These μc-Si films are similar in texture, thickness and both display transitions regions of a-Si between the substrates and the μc-Si. The transition region is about two times thicker for the film deposited on the fused silica. Figure 6 displays diffraction patterns and TEM of μc-Si films deposited on c-Si and fused silica at 300°C at a very high P_H = 4.0 mTorr. In contrast to Fig. 5, the thickness of these films is substrate dependent, and they display no detectable transition region between the substrates and the μc-Si regions.

3. Deposition Rates an Bonded-Hydrogen Concentrations

Figures 7(a) and (b) give the variations of the the deposition rate (Å/s) and the bonded hydrogen concentrations [H] (at.%) for: (a) films deposited at constant temperature, T_S = 250°C, and with a varying P_H; and (b) films with a fixed P_H = 0.38 mTorr, and with a varying T_S, 40°C to 650°C. For the films deposited at constant T_S = 250°C, the deposition rate decreases sharply between P_H = 0.4 and 2.0 mTorr and then levels off. [H] first increases with increasing as P_H from 0.4 mTorr to 1.0 mTorr, and then decreases abruptly by more than a factor of two from its maximum value of ~13 at.% to values of about 7-8 at.% for P_H = 2.0 and 4.0 mTorr. Referring to the Raman data in Fig. 2, we note that the films deposited with P_H = 1.0 mTorr or lower are amorphous, and those with P_H = 2.0 and 4.0 mTorr are μ-crystalline. In contrast films, for deposited at constant P_H =0.40 mTorr, [H] decreases approximately linearly from about 29 at.% to zero as T_S increases from 40°C to 400°C; films deposited with T_S > 400°C have no ir-detectable bonded hydrogen. The deposition rate shows vary little variation with T_S between 40°C and 400°C, and then falls off rapidly with increasing T_S. For these films, the transition between a-Si and μc-Si film occurs for T_S >500°C, so that films that are μ-crystalline have no ir-detectable bonded hydrogen.

DISCUSSION

The first successful depositions of μc-Si thin films using chemical transport in a hydrogen plasma were made by Veprek and Maracek [5]. These studies demonstrated that stable nuclei for μc-film growth were formed under plasma conditions in which chemical equilibrium between a deposition reaction and an etching reaction are approached at the plasma-solid interface. By adjusting the gas flow and plasma conditions it is possible to control the steady state kinetics and achieve deposition of μc-Si films. Several Japanese groups [6-8], as well as the group at Dundee [9], demonstrated that a similar near-equilibrium situation could be achieved by heavy dilution (>50:1) of the silane reactant by hydrogen in a glow discharge process. In addition, Shirafuji et al. [10] grew mixed phase (a-/μc-) Si films by RF sputtering in a hydrogen-rich plasma and suggested that μ-crystallinity was induced by bombardment of the substrate surface by hydrogen ions.

There are two types of deposition processes that can lead to crystallization: i) crystallization that occurs at the time of deposition and results from atomic rearrangements at the growth surface; and ii) deposition of a non-crystalline or amorphous phase, followed by a solid state recrystallization via a thermal annealing at the deposition temperature [11]. The first mechanisms is clearly operative in the chemical transport deposition of Veprek and Maracek [5], and can

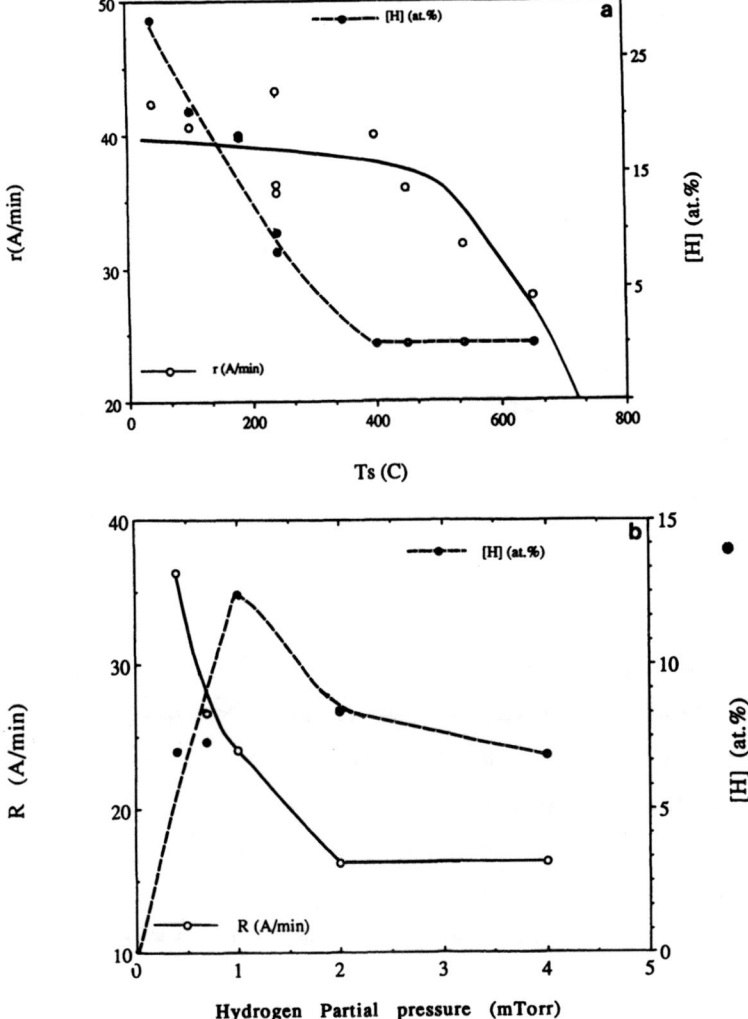

also occur under the high hydrogen dilution GD depositions [6-9]. The second mechanism will generally result in a non-homogeneous thin film wherein the material will display a layered structure comprised of both crystalline and amorphous material. Imura et al. [12] have found transition regions of about 300Å in materials deposited by several different techniques including sputtering and RF glow discharge. As discussed above, we have identified deposition conditions by RMS, that can either generate or eliminate transition regions, so that each of the mechanisms can apply.

The films deposited with a fixed, but relatively low partial pressure of hydrogen, 0.40 mTorr, are homogeneous a-Si:H alloys for low T_S, and for high T_S

Fig. 7. Deposition rates and bonded hydrogen vs: (a) P_H for films deposited at T_S = 250°C; (b) T_S for films deposited with P_H = 4.0 mTorr

they are μ-crystalline. The films deposited with T_S = 540°C and 650°C display a transition region, between an amorphous Si film in the vicinity of the substrate, and a crystalline character in the remainder of the film. This is indicative of a growth mechanism, in which annealing at the deposition temperature is required to promote nucleation. Once nucleation occurs at the growth surface, it continues for the remainder of the deposition. Films produced in this way for T_S > 500°C show no evidence of bonded hydrogen (via IR), and the crystalline material is homogeneous with crystallites that can be quite large; the material deposited at 650°C is better described as polycrystalline Si.

The materials deposited at T_S = 250°C, and at varying P_H are qualitatively different. The transition between a-Si:H and μc-Si occurs when when P_H is 2 mTorr or more. These films show no evidence via TEM images for a transition region between the substrate and the μc-Si film. The texture of these films is also different from the films deposited at high T_S, and in addition there is evidence by IR for bonded hydrogen incorporation. Analysis of the TEM bright- and dark-field images indicates relatively small crystallites, <100Å in size, and also suggests that the material is not as compact as the μc-Si deposited at high T_S. The Raman spectra (see Fig. 2) also display a larger tail, with a distinct shoulder at about 490 cm^{-1}, and differ in this respect from the spectra shown in Fig. 1 for the high T_S films. We have shown that the spectra in Fig. 2 cannot be decomposed into spectra characteristic of a μc-Si phase and an a-Si phase [4]. This means the "connective matrix" between the μ-crystallites is not simply a-Si:H, even though it contains the bonded hydrogen. We are continuing our studies of μc-Si using the Remote PECVD process, with emphasis on doping and defining deposition conditions that lead to the elimination of transition regions of a-Si between the substrates and μc-Si.

ACKNOWLEDGEMENT

This research has been supported by SERI under sub-contract XM-9-18141-2

REFERENCES

1. R.A. Rudder, J.W. Cook, Jr. and G. Lucovsky, Appl. Phys. Lett. 45, 887 (1984).

2. G.N. Parsons, D.V. Tsu and G. Lucovsky, J. Vac. Sci. Technol. A6, 1912 (1988).

3. Cheng Wang, G.N. Parsons, E.C. Buehler, G. Lucovsky and R.J. Nemanich, submitted to 1990 Spring MRS Meeting.

4. R.J. Nemanich et al, ICALS 1989, to be published in J. Non-Cryst. Solids (1989).

5. S. Veprek and V. Marecek, Solid State Electronics 11, 683 (1968)

6. S. Usi and M. Kikuchi, J. Non-Cryst. Solids 34, 1 (1979).

7. A. Matsuda et. al., Jpn. J. Appl. Phys. 19, L305 (1980).

8. T. Hamasaki et. al., Appl. Phys. Lett. 37, 1084 (1980).

9. W.E. Spear et. al., J. Physique 42, C.4, 257 (1981).

10. J. Shirafuji, H. Matsui, A. Narukawa and Y. Inuishi, Solid State Comm. 45, 577 (1983).

11. J. Narayan, O.W. Holland and B.R. Appleton, J. Vac. Sci. Technol. B1, 871 (1983).

12. T. Imura et al, Jpn. J. Appl. Phys. 23, 179 (1983).

PREPARATION AND CHARACTERIZATION OF HIGHLY CONDUCTIVE (100 S/cm) PHOSPHORUS DOPED μc-Si:H FILMS DEPOSITED USING THE VHF-GD TECHNIQUE.

Kshem Prasad, F. Finger, H. Curtins, A. Shah, J. Bauman*
Institute of Microtechnology, Rue Breguet 2, CH- 2000 Neuchatel, Switzerland.
*Physics Institute, University of Konstanz, D-7750 Konstanz, W. Germany

ABSTRACT

We report on the preparation and characterization of phosphorus doped μc-Si:H films produced by the very high frequency glow discharge (VHF-GD) at a plasma excitation frequency of 70 MHz. We present a systematic study of the deposition parameters i.e. hydrogen dilution of silane, VHF power density, gas phase doping ratio and deposition temperature and their influences on the electrical and structural properties of the material. In contrast to 13.56 MHz GD the VHF plasma conditions favour microcrystalline formation at low power densities; the resulting conductivities are significantly higher than those obtained at 13.56 MHz.

INTRODUCTION

Doped hydrogenated microcrystalline silicon (μc-Si:H) films have attracted much attention due to their potential applications in electronic and opto-electronic devices. High dilution of silane in hydrogen along with a high discharge power are considered to be the empirical conditions to obtain good quality μc-Si:H using the glow discharge (GD) technique [1]. The deposition conditions thus approach the partial chemical equilibrium (PCE) at the solid-plasma interface [2] under which the surface chemistry is modified due to the hydrogen rich plasma. This results in an equilibrium between the etching and the deposition process leading to μc-Si:H formation.

We have shown recently that the use of a higher excitation frequency for the GD process can change the growth conditions for a-Si:H significantly, such that device grade material can be produced at growth rates as high as 20 Å/s [3]. As this is believed to be due to a change in the plasma kinetics as a function of excitation frequency, it is of interest to study the influence of a higher frequency on the growth of μc-Si:H. First results of this investigation on phosphorus doped μc-Si:H will be presented in this paper.

EXPERIMENTAL

The films were deposited in a capacitively coupled VHF-GD system described elsewhere [3]. 70 MHz was selected as the discharge frequency as this was found to be the optimum for high rate deposition of intrinsic a-Si:H for our system configuration. Table I gives the details of the deposition parameters for the different series investigated in our system. All the samples were deposited on Dow Corning 7059 glass and have a thickness between 0.4 and 0.6 μm. The conductivity was measured using two coplanar aluminium contacts. The X-ray measurements were performed using a Cu K_α source. The average size of the crystallites δ was determined from the FWHM of the <111> peak and using Sherrer's equation. As the resolution was limited to 100 Å, it was not possible to determine δ in μc-Si:H films deposited using parameters in the neighbourhood of the amorphous to microcrystalline transition. Finally, the phosphorus content in the solid phase was determined using SIMS measurements.

Table I. VHF - GD deposition parameters for μc-Si:H formation.

PAR. SERIES	Dep.Temp (°C)	Pressure (mbar)	VHF power (mW/cm³)	SiH4 (sccm)	PH3* (sccm)	H2 (sccm)	SiH4 total	PH3 / SiH4	Dep. Rate R (Å/s)
"SiH4 dil.<i>"	300	0.4	15	4.2	0	45-130	3-10 %	0	0.5 - 2.2
"SiH4 dil. <n>"	"	"	"	4.3	52	0-90	3-8.5	6e-3	"
"VHF power"	"	"	15-125	4.5	"	90	3.2	"	0.6 -0.7
"Dop. Eff. "	300	"	"	"	0-130	120-0	3	var.	0.65
"Dep. Temp."	50-380	"	15	4	125	0	"	1.5e-2	"

* PH3: 500 ppm in H2

RESULTS AND DISCUSSION

Silane dilution with hydrogen

The influence of dilution of silane with hydrogen for the deposition of intrinsic films, at a constant VHF-power density of 15 mW/cm³, is illustrated in figure 1. The deposition rate decreases with increasing dilution. Below a silane concentration of about 7% there is a sudden increase in the dark conductivity by more than six orders of magnitude, reaching a maximum of 7×10^{-3} S/cm, which is comparable to that of doped a-Si:H. This is accompanied with the appearance of crystallite peaks in the X-ray measurements. On the other hand, for silane concentration higher than 8% the resulting amorphous films are of poor quality. The photosensitive gain is only 10^4, the defect density N_d and the Urbach edge E_0, as determined by PDS, are $N_d = 10^{17}$ cm^{-3} and $E_0 = 60$ meV, respectively.

In the case of phosphorus doping, part of the hydrogen has been replaced by a phosphine/hydrogen mixture (500 ppm of phosphine in hydrogen). Figure 2 gives the results. Once again a sharp change in conductivity is observed as the silane concentration is decreased below 6%. This is, again, accompanied by the appearance of the crystalline signature in the X-ray diffraction measurements. The average grain size tends to increase with increasing dilution. Although the gas phase doping ratio is constant, the solid phase incorporation of phosphorus decreases slightly in the microcrystalline phase. However, the conductivity is increased by four orders of magnitude. The highest conductivity obtained is 60 S/cm for a gas phase doping ratio of 6×10^{-3} and a silane concentration of 2.8%. The film was deposited at a rate of 0.65 Å/s which is quite high for the deposition of μc-Si:H. Thus, a typical doped layer of 200 Å for a solar cell can be deposited in about 5 minutes. This is compatible with the deposition time of an intrinsic layer deposited at 20 Å/s using the same VHF technology.

It is interesting to note that the transition from the amorphous to the microcrystalline phase occurs practically for the same dilution ratio, for both, intrinsic and phosphorus doped samples. However, for doped films the average grain size ∂ of the crystallites is smaller. This indicates that in our process, incorporation of phosphorus limits the microcrystalline growth to some extent.

We interpret these results, obtained at such low power levels using the model proposed earlier [4]: the change in the excitation frequency in the VHF -GD, as compared to the 13.56 MHz, is thought to influence the electron energy distribution function (EEDF) of the plasma. In fact, Optical Emission Spectroscopy studies of VHF-GD using SiH4 and H2 at 144MHz have revealed an increased generation rate of SiH and H$_\alpha$ radicals at relatively low VHF powers as compared to a 13.56 MHz plasma , favouring μc-Si:H formation at rather low power levels [5]. A possible explanation is that at higher excitation frequencies the tail of the EEDF contains a sufficient number of high energy electrons to enhance ionisation and dissociation. Thus, it is expected that the hydrogen diluted plasma is rich in atomic hydrogen and hydrogen ions which bombard the growing surface. In the case of high silane concentration (>8%) this ion bombardment also introduces morphological dislocations which finally result in a poor amorphous films. Under the conditions of a high dilution of silane in hydrogen, using the same low power, the surface - plasma interface approaches the PCE and influences the surface chemistry significantly due to higher concentration of ionised atomic hydrogen [2]. These produce a preferential etching of the weakly bonded hydrogen, thus, producing a new site for the oncoming silicon atom. The well-bonded hydrogen provides the necessary coverage to enable the surface migration of the radicals to the low energetic sites, resulting in the crystalline growth. This explanation is the same as that generally given for the formation of μc-Si:H using 13.56 MHz. However, the power levels required for this, in the VHF case, is a factor ten lower than those required in a 13.56MHz GD process [6]. Even in processes using a magnetic confinement of the plasma, the power levels are comparatively higher than in the VHF case [7].

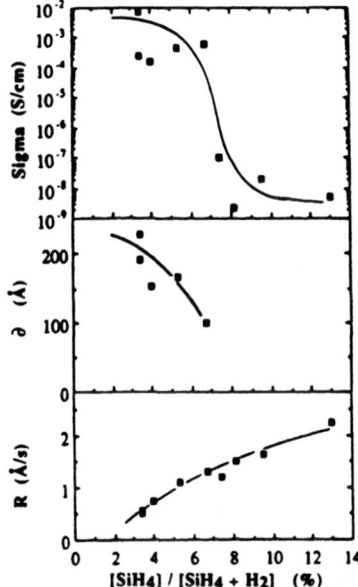

Fig. 1: Influence of silane concentration on the deposition rate, crystallite size and dark conductivity of intrinsic films.

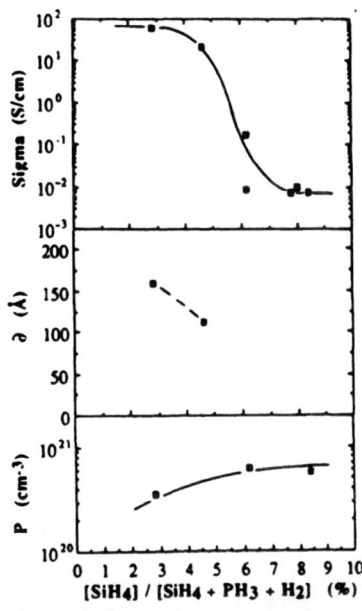

Fig. 2: Influence of silane concentration on the total phosphorus content, crystallite size and dark conductivity for doped films.

Influence of VHF power

The influence of an increase in VHF power is illustrated in figure 3. In general, in 13.56 MHz GD processes, besides a high dilution of silane, high RF power is necessary for the formation of μc-Si:H. Interestingly, in contrast to the general observation, best film quality is obtained at relatively low power; thereafter, the quality degrades with increasing power density: the conductivity, grain size and the P content are lower. Similar behaviour in the decrease of the conductivity and the grain size was also observed by us in intrinsic films deposited at the same silane dilution. Generally such a behaviour has been reported at much high RF power levels using the 13.56 MHz GD [8].

These results suggest that the behaviour of the VHF plasma at low power levels is similar to that of the 13.56MHz GD at high power. In this context, in magnetically confined plasmas μc-Si:H films have been deposited at intermediate power levels [7]. The magnetic confinement is understood to lower the electron temperature and increase the dissociation and ionisation of hydrogen due to higher collision rate, thereby favourising microcrystalline formation. We observe a self-confinement of the GD plasma in our system obtained due to the geometrical configuration and the optimised deposition parameters. This enables, even at such a low power densities , to obtain PCE conditions in the plasma needed for μc-Si:H growth. The decrease in the film quality for higher powers is believed to be due to an increase in the ion bombardment at the growing surface. This reduces its hydrogen coverage, reducing thereby the surface diffusion of the oncoming silicon radicals and hinders the crystallite growth. This is observed by the decrease in the grain size.

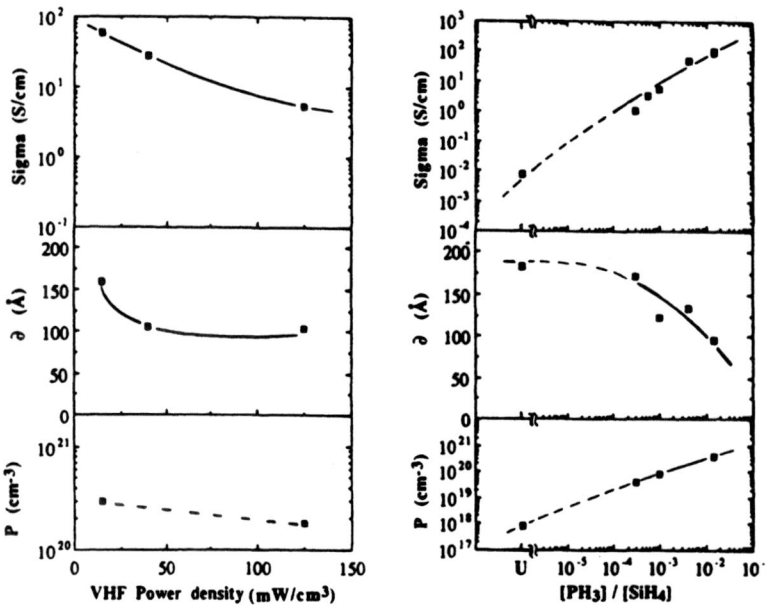

Fig. 3: Influence of VHF-power density on doped film properties.

Fig. 4: Influence of gas phase doping ratio.
U : Undoped

Doping Efficiency

Doping efficiency has been studied by varying the gas phase doping ratio : PH3/SiH4 from 3e-4 to 1.4e-2. The results are illustrated in figure 4. As expected, the phosphorus content and the conductivity increase monotonously with doping. The latter reaches a maximum of 100 S/cm. This is comparable to the best values obtained for polysilicon (doped in-situ) deposited at 600°C and annealed at 700°C [9]. A decrease in the grain size with increase in doping ratio, is observed, as previously reported by Hasegawa [10]. This suggests, as already conjectured, that the presence of phosphine during deposition limits the crystalline gowth at such low deposition temperatures.

Total incorporation of phosphorus, as measured by SIMS, is quite similar to the one reported by Kaya et al. [11] (Fig. 5). Therefore, the higher conductivities obtained in our samples cannot be explained on the basis of the phosphorus incorporation and need further investigation. A correlation with the crystallite size and the volume fraction of the crystallites Xc determined from the Raman diffraction spectra is presently under way.

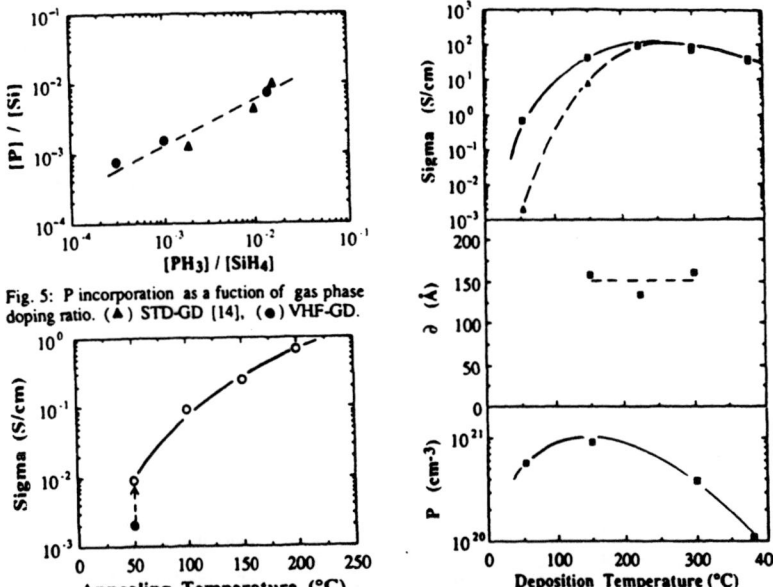

Fig. 5: P incorporation as a fuction of gas phase doping ratio. (▲) STD-GD [14], (●) VHF-GD.

Fig. 7: Influence of annealing on conductivity for a sample deposited at 50°C. (●) as deposited.

Fig. 6: Influence of deposition temperature. (▲) as deposited, (■) annealed at 220°C.

Influence of Deposition Temperature

Figure 6 illustrates the characteristics of the material as a function of deposition temperature. The conductivity of the samples, as deposited, increases rapidly with increasing temperature and reaches a maximum at 225 °C. This dependence on the deposition temperature is generally observed for the conductivity of a-Si:H and μc-Si:H prepared by GD process and is well understood in terms of the change of radical diffusivity at the growing surface. The optimum temperature is found to be about 250°C for the highest conductivity. Beyond this it decreases gradually. Phosphorus incorporation was a factor two lower in the optimum region (225-300°C).

It is surprising to note that the sample deposited at 50 °C has no crystalline signature in the X-ray diffraction and has a conductivity of 2×10^{-3} S/cm, but with successive annealing (Fig. 7) to higher temperatures the RT conductivity increases and reaches that of μc-Si:H. This suggests that even at such low temperatures (50 °C), the morphology of the as deposited film is not purely amorphous. However, preliminary measurements by Raman spectroscopy, of annealed samples, indicate very small crystalline signature which accounts for an estimated 5 to 10% of crystallites in the volume only [12]. This cannot explain the high conductivity. A possible explanation for the formation of films deposited at such low temperatures could be that the poor surface diffusivity which could result from low substrate temperature is partially compensated by the hydrogen rich plasma which improves surface coverage. In addition, the high energetic hydrogen ion bombardment could be responsible for an effective higher temperature at the film growing surface. However, these conditions do not satisfy, either, the conditions required to form a μc-Si:H film or that for an a-Si:H film. A further study of these depositions and film properties is under way.

CONCLUSION

The influence of the deposition parameters on the structural and electronic properties have been investigated. Highly conductive µc-Si:H films are prepared at power densities generally used to deposite a-Si:H and at low temperatures. These results are interpreted in terms of the change of the EEDF with the plasma excitation frequency. The hot electrons from the tail of the EEDF in the VHF plasma are responsible for obtaining the partial chemical equilibrium necessary for the formation of µc-Si:H. The highest conductivity obtained is 100 S/cm which is comparable to that of an in-situ deposited polysilicon. The possibility to deposit such high conductive layers, at deposition temperatures as low as 200°C, could open new possibilities for overlayers in microelectronic and sensor technology.

ACKNOWLEDGEMENT

The authors are thankful to Beate Leutz for technical assistance. The present paper forms a part of the Ph.D. thesis of Kshem Prasad. This work was supported by Swiss Federal Research Grant OFEN -REN (87) 9.

REFERENCES

[1]   Y. Osaka and T. Imura, in Amorphous Semiconductor - Technologies and   Devices.
      Ed. Y. Hamakawa, Publ. North-Holland, Amsterdam (1984) 80
[2]   S. Veprek, M. Heintze, F. A .Sarott, M. Jurcik-Rajman and P. Willmot, MRS Symp.
      Proc. 118 (1988) 3
      C. C. Tsai, Amorphous Silicon and Related Materials Ed. H. Fritzsche,
      World Scientific Co. Singapore (1989)pp 123
[3]   H. Curtins, N.Wyrsch, M.Favre, K.Prasad, M.Brechet and A.V.Shah, MRS Symp.
      Proc. 95 (1987) 249
[4]   H. Curtins, N.Wyrsch, M.Favre and A.V.Shah, Plasma Chem. and Plasma Process.
      7(1987) 267
[5]   S. Oda, J. Noda and M. Matsumura, MRS Symp. Proc. 118 (1988) 117
[6]   A. Matsuda, in Amorphous Semiconductor - Technologies and Devices.
      Ed. Y. Hamakawa, Publ. North- Holland, Amsterdam (1987) pp 111
[7]   M. Taniguchi, M.Hirose, T.Hamasaki and Y.Osaka, Appl. Phys. Lett. 37 (1980) 787
[8]   K. Nakatani, M.Yano, K.Suzuki and H.Okaniwa, J. Non-Cryst. Sol. 59/60 (1983) 827
[9]   S. Sze, in VLSI Technology, Mc Graw-Hill, New York (1983) pp.99
[10] S. Hasegawa, S.Narikawa and Y.Kurata, Phil. Mag. B 48 (1983) 431
[11] H. Kaya, Imura,T.Kusao,A.Hiraki,O.Nakamura,Y.Okayasu and M.Matsumura,
      Jpn. J. Appl. Phys. 23 (1984) L549
[12] K. Prasad and M. Schubert, to be published

COMPOSITION DETERMINATION OF MICROCRYSTALLINE TWO-PHASE SILICON RICH OXIDES

D.H. BOULDIN*, C.H. LAM* AND K. ROSE**
* IBM, 1000 River Road, Essex Junction, Vermont 05452
** Center for Integrated Electronics, Rensselaer Polytechnic Institute, Troy, New York 12181

ABSTRACT

HRTEM measurements of silicon rich oxides (SRO) show silicon microcrystals in an oxide matrix [1]. Simple, reliable characterization of this two phase material has been a problem. Ellipsometric measurement of the refractive index is a convenient method for characterizing SRO films. Film composition can be related to the refractive index by Bruggeman's effective medium approximation. In this paper we demonstrate correlation of film compositions obtained by this technique with those obtained by Auger electron spectroscopy (AES)and Rutherford back scattering (RBS). We further demonstrate regimes of LPCVD growth where simple correlation of film composition with $[N_2O]/[SiH_4]$ gas ratios is not reliable.

INTRODUCTION

The electrical behavior of silicon rich oxide (SRO) depends upon its composition. At one extreme it approaches polysilicon and, as semi-insulating polysilicon (SIPOS), is useful as a passivant for high-voltage devices [2] and as a bipolar emitter [3]. At the other extreme it behaves like silicon dioxide and, as an off-stoichiometry oxide (OSO), may be useful as an electrically alterable ROM (EAROM) insulator [4]. At intermediate compositions it is useful as an EAROM injector [5]. Often γ, the ratio of $[N_2O]$ to $[SiH_4]$ during deposition, is used to characterize film composition, with $\gamma < 2$ corresponding to SIPOS and $\gamma > 20$ corresponding to OSO. However, we will show that intermediate values of γ can be an unreliable guide to composition.

SRO films contain an excess of silicon, which is initially amorphous but becomes crystalline after high temperature annealing [6]. HRTEM measurements by ourselves [1] and others [6] provide convincing evidence for silicon crystallites. To a first approximation, SRO films can be regarded as a two-phase system with silicon crystallites embedded in an SiO_2 matrix. XPS data provides support for this view [7]. There may be a layer of disordered silicon at the surface of the silicon crystallites, as suggested by the Raman data of Olego and Baumgart [8].

A variety of techniques have been used to characterize the composition of SRO films. A review of the literature indicates considerable variation in the dependence of oxygen content, x in SiO_x, on γ as determined by different techniques [9]. There is reasonably good agreement between RBS measurements [10], and measurements by electron probe microanalysis and AES [11]. These measurements indicate that x rises rapidly from 0.1 to 1.0 as γ increases from 0.1 to 0.5. We will show that ellipsometry provides a convenient and reliable means of determining SRO composition.

FILM GROWTH AND CHARACTERIZATION

We report results on SRO films grown in LPCVD reactors. Most of our films were grown at deposition temperatures from 610 to 680°C and pressures from 0.4 to 0.5 torr. At RPI, two-inch p-type silicon wafers were RCA cleaned followed by a 30-second dip in 10% aqueous HF to reduce the thickness of the native oxide layer shortly before loading the wafers in the reactor. Wafers were loaded horizontally in the center zone of the reactor and displayed a 5% variation in film thickness. Typical SRO film thicknesses were 40 nm. After deposition, the wafers were annealed at 950°C in N_2 for 30 minutes to promote crystallization. A detailed analysis of SRO growth kinetics in our reactor has been reported elsewhere [12].

The refractive index and thickness of SRO films was measured by a microcomputer-controlled ellipsometer, the Rudolph Auto EL-II, at a wavelength of 632.8 nm and an incident angle of 70°. These measurements were made immediately after deposition and after isothermal annealing in nitrogen. The silicon substrate was assumed to have a refractive index of 3.858 and extinction coefficient of 0.0018. The effect of optical absorption due to the excess silicon in SRO films was investigated by Dong and coworkers using multiple angle ellipsometry [13]. They found less than ten percent error in refractive index and thickness resulted from neglecting the SRO extinction coefficient, provided film thicknesses were within the first ellipsometric period. Since our film thicknesses were well within the first ellipsometric period we have neglected extinction coefficient effects.

ELLIPSOMETRIC CHARACTERIZATION

Bruggeman's effective medium approximation (BEMA), is a simple effective medium theory that represents a heterogeneous dielectric mixture in terms of the dielectric constants of its components. It has been applied successfully to study the microscopic roughness of silicon films [14] and the microstructure of hydrogenated amorphous silicon [15]. The underlying assumptions of BEMA are a spherical inclusion geometry and only dipole interactions. Both are reasonable approximations for our crystallites in the absence of resonances and for layers which are relatively thick compared with atomic dimensions.

The general form of BEMA is

$$0 = f_1 \frac{\epsilon_1 - \bar{\epsilon}}{\epsilon_1 + 2\bar{\epsilon}} + f_2 \frac{\epsilon_2 - \bar{\epsilon}}{\epsilon_2 + 2\bar{\epsilon}} + \dots \tag{1}$$

$\bar{\epsilon}$ and ϵ_i are the complex dielectric constants of the effective medium and constituent i. f_i is the volume fraction of constituent i. Regarding SRO as a two-phase mixture of Si crystallites and SiO_2, (1) becomes

$$0 = f_{Si} \frac{n_{Si}^2 - n_{SRO}^2}{n_{Si}^2 + 2n_{SRO}^2} + (1 - f_{Si}) \frac{n_{SiO2}^2 - n_{SRO}^2}{n_{SiO2}^2 + 2n_{SRO}^2} \tag{2}$$

We have replaced ϵ by the square of the refractive index, n^2. (2) allows us to calculate the volume fraction of silicon from ellipsometric measurements of n_{SRO}. At 2 eV (λ = 630 nm) $n_{Si} \approx 3.9$ [16], corresponding to $\epsilon_{Si} = 15.2$; $n_{SiO2} = 1.46$ [17], corresponding to $\epsilon_{SiO2} = 2.13$. Because amorphous silicon has a dielectric constant of about 20 at 2 eV, equation (2) will tend to overestimate f_{Si} when amorphous silicon is present. Examination of (2) shows that f_{Si} is a linear function of ϵ_{SRO}.

$$f_{Si} = 0.0765\epsilon_{SRO} - 0.163 \tag{3}$$

Regarded as a two-phase mixture of Si and SiO_2, the SRO film has a stoichiometric form

$$SiO_x = (1 - \frac{x}{2})Si + \frac{x}{2}SiO_2 \tag{4}$$

where x ≤ 2. The volume of a unit of this composite material is

$$V_o = \frac{1}{N_{Si}}(1 - \frac{x}{2}) + \frac{1}{N_{SiO2}}\frac{x}{2}. \tag{5}$$

We take $N_{Si} = 5 \times 10^{22}$ atoms/cm^3 and $N_{SiO2} = 2.22 \times 10^{22}$ molecules/cm^3 which are the molecular concentrations of single crystal silicon and thermally grown SiO_2 respectively. This allows us to relate the volume fraction of silicon microcrystals to x in SiO_x.

$$f_{Si} = \frac{(2 - x)N_{SiO2}}{(2 - x)N_{SiO2} + xN_{Si}} \tag{6}$$

x is related to the atomic percentage of silicon by

$$At.\%Si = (1 + x)^{-1} \qquad (7)$$

x = 0 or 2 corresponds to 100 or 33 atomic percent silicon in SRO, respectively. Over this range the refractive index of SRO should vary from 3.9 for silicon to 1.46 for SiO_2.

Figure 1 compares the atomic percent silicon in SRO films obtained by Auger electron spectroscopy and Rutherford backscattering with $\bar{n} = n_{SRO}$ determined by ellipsometry. These SRO films were deposited in another LPCVD system under comparable conditions. The solid line is obtained by combining equations (3), (6), and (7). Our theoretical result provides a reasonable fit to this data, which displays considerable scatter. We conclude that ellipsometry provides a reliable guide to composition in SRO films. Because ellipsometry is a simple measurement with relatively high precision, it provides a convenient check of the effects of process variations on SRO composition.

We expect ellipsometric measurements to become more reliable as SRO film composition approaches SiO_2. An indirect check of this is provided by comparing film thicknesses derived by ellipsometry (d_e) and HRTEM (d_H). For a film with $n_{SRO} = 1.92$, $d_e = 50.5 \pm 1$ nm and $d_H = 50.9$ nm. For a film with $n_{SRO} = 2.25$, $d_e = 88.6 \pm 1.5$ nm and $d_H = 87.6$ nm. For a film with n > 2.5, the ellipsometer program could not obtain thickness and refractive index simultaneously. However, using $d_H = 133.5$ nm, we find $n_{SRO} = 2.90$. This data indicates the expected increase in precision with decreasing index of refraction.

Table I compares the compositions of these films on the basis of our theoretical formula and crystallite sizes obtained from HRTEM photographs. This shows that, as expected, crystallite size increases with increasing silicon content. Nevertheless, at these compositions, crystallite sizes remain small.

TABLE I: Composition of SRO Films [9]

n	x	At. % Si	Crystallite Size (nm)
1.92	1.5	39	1.5
2.25	1.2	45	4.3
2.90	0.6	64	4-10

CORRELATION WITH GAS RATIOS

The ratio of input reactants, $\gamma = [N_2O]/[SiH_4]$, is commonly used as a guide to SRO film composition. For small γ we approach polysilicon deposition while for large γ we approach silicon dioxide deposition. However, reliance on γ as a guide to composition can be treacherous. One reason is the tendency we have observed for both the silane and nitrous oxide mass flow controllers to clog with silicon dioxide, making γ measurements inaccurate. Another reason is the appearance of an anomalous deposition region at intermediate values of γ.

Figure 2 shows the variation of n_{SRO} with γ for SRO growth at substrate temperatures from 610 to 680°C. Care was taken to maintain the cleanliness and check the calibration of the mass flow controllers during these experiments. For $\gamma < 2$ the growth kinetics can be explained using Hitchman and Kane's model for SIPOS growth which assumes that adsorbed N_2O blocks surface sites for SiH_4 pyrolysis [18]. For $\gamma \gtrsim 4$ we were able to explain the growth kinetics by considering competition between the formation of Si-Si and Si-O bonds [12]. For $2 \lesssim \gamma \lesssim 4$ there is an anomolous growth region characterized by an **increase** in refractive index with increasing γ!

The SRO films described in Table I were grown at 650°C and 0.4 torr. The measured gas ratios were 0.3, 1, and 3 for films whose refractive indices were 2.90, 2.25, and 1.92. Comparison of these values of n and γ with Figure 2 suggests that they grew like SIPOS and that the observed gas ratios are overestimates.

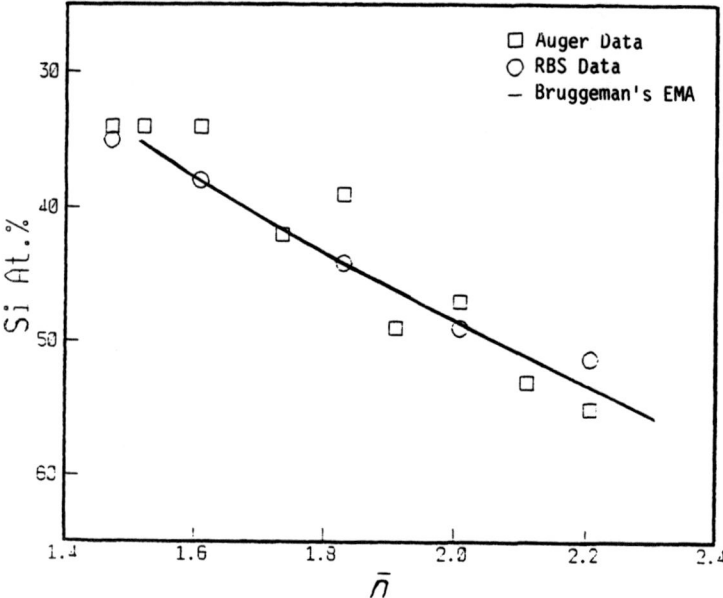

Figure 1. Bruggeman's Effective Medium Approximation compared to correlations of refractive index with composition.

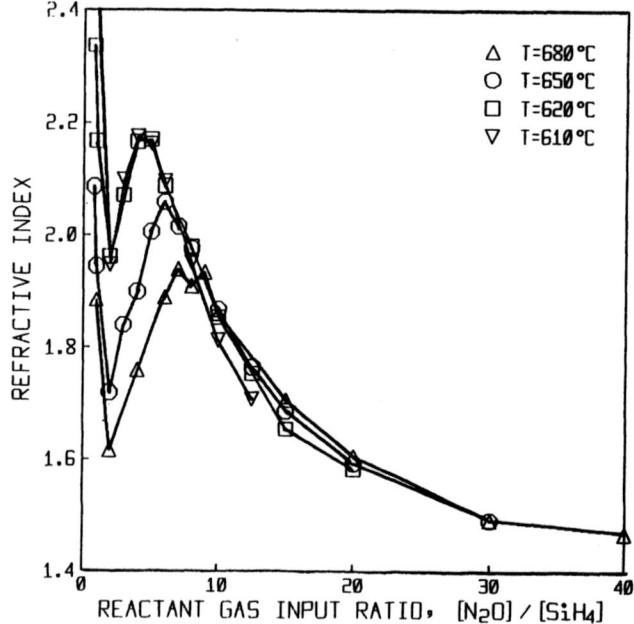

Figure 2. Dependence of refractive index on gas ratio for as-deposited SRO films.

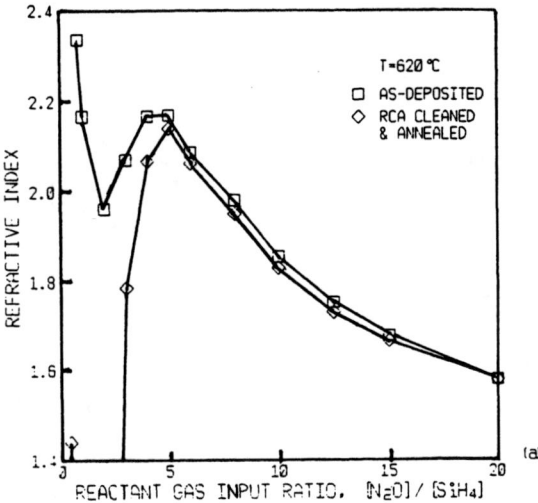

Figure 3. Changes in refractive index for SRO films after RCA cleaning and annealing.

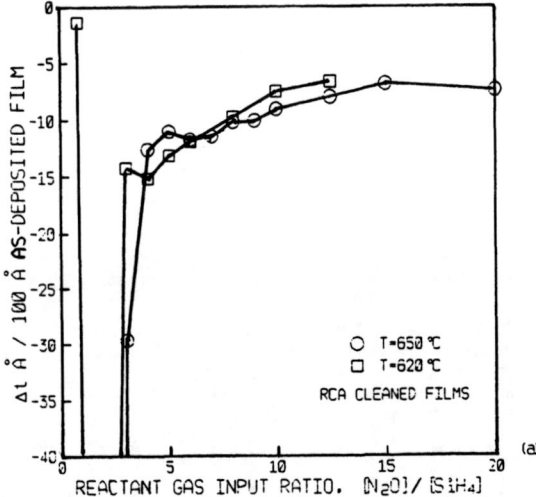

Figure 4. Changes in thickness for SRO films after RCA cleaning and annealing.

FILM BEHAVIOR IN THE ANOMALOUS REGION

SRO films deposited at 620 and 650°C were RCA cleaned before annealing to eliminate contaminants due to ellipsometric measurements and wafer handling. Ellipsometric measurements after cleaning and annealing generally exhibited a slight decrease (less than 3%) in refractive index and thickness. The decrease in thickness indicates densification while the decrease in refractive index may indicate conversion of amorphous to crystalline silicon.

On the other hand, films in the anomalous region were strongly etched after 10 minutes of immersion in the acid solution of the RCA cleaning process. Figures 3 and 4 show the effects on refractive index and thickness of RCA cleaning and annealing for SRO films deposited with different gas ratios. $\gamma = 2$ films vanished completely. This is particularly striking since the acid solution, dilute HCl in deionized water at 60°C, should not etch either silicon or silicon dioxide for any reasonable duration of contact.

Anomalously rapid etching suggests that films are deposited as a loosely packed, porous material in this region. Support for this view is provided by studies of SRO films deposited at 610 and 680°C which were cleaned instead in nitrogen areated deionized water (DIW). Outside the anomalous region similar small decreases in refractive index and thickness were observed after annealing. However, within the anomalous region, the thickness of DIW cleaned films **increased** by ten to fifteen percent. This anomalous region may correspond to deposition onto the substrate of products from a homogeneous gas phase reaction. It is interesting to note that the width of the anomalous region increases to higher values of γ at higher gas ratios.

ACKNOWLEDGEMENTS

The authors are grateful to K-T. Chang for his initial work on SRO composition, T. Marshall of Phillips Laboratories for the HRTEM measurements reported, and J. Haus of Rensselaer's Physics Department for helpful discussions about the effective medium interpretation of optical measurements. We are also grateful for the use of characterization equipment at IBM, Essex Junction.

REFERENCES

1. K.-T. Chang, C. Lam and K. Rose, MRS Proc., 105, 193 (1988).
2. T. Matsushita et al., Jap. J. Appl. Phys. Supplement, 15, 35 (1976).
3. J. Fujioka et al., IEEE IEDM Digest, 8.6 (1987).
4. D.J. DiMaria et al., J. Appl. Phys., 54, 5801 (1983).
5. D.J. DiMaria, K.M. DeMeyer, and D.W. Dong, IEEE Elec. Dev. Lett. EDL-1, 179 (1980).
6. J. Wong et al., Appl. Phys. Lett., 48, 65 (1986).
7. E.A. Irene et al., J. Electrochem. Soc., 127, 2518 (1980).
8. D.J. Olego and H. Baumgart, J. Appl. Phys., 63, 2669 (1988).
9. K-T. Chang, Ph.D. Thesis, RPI (1987).
10. B. Verstegen et al., J. Appl. Phys. 57, 2766 (1985).
11. J.H. Thomas and A.M. Goodman, J. Electrochem. Soc. 126, 1766 (1979).
12. C.H. Lam and K. Rose, MRS Proc., 131, 281 (1989).
13. D.D. Dong et al., J. Electrochem. Soc., 125, 819 (1978).
14. D.E. Aspnes et al., Phys. Rev. B, 20, 3292 (1979).
15. R.W. Collins et al. in Tetrahedrally-Bonded Amorphous Semiconductors, D. Adler and H. Fritzche Eds. (Plenum, New York, 1985) p. 63.
16. H.R. Philipp and E.A. Taft, Phys. Rev. 120, 37 (1960).
17. S.M. Sze, Physics of Semiconductor Devices, (Wiley, New York, 1981, 2nd Ed.) p. 852.
18. M.L. Hitchman and J. Kane, J. of Crystal Growth, 55, 485 (1981).

CHEMISTRY AND SOLID STATE PHYSICS OF MICROCRYSTALLINE SILICON

Stan Veprek
Institute for Chemistry of Information Recording,
Technical University Munich,
Lichtenberstr. 4, D-8046 Garching-Munich, F.R.G.

ABSTRACT

Various methods for the preparation of microcrystalline (nanocrystalline) silicon are summarized and compared with respect to the possibility of the control of the materials quality and scaling of the deposition process to large area applications. It is shown that the deposition of a pure microcrystalline material is achieved under conditions close to partial chemical equilibrium. The mechanism of the crystallization during the growth will be briefly discussed.

The second part of the paper deals with the physical properties of pure microcrystalline silicon which is free of any amorphous phase detectable by X-ray diffraction, i.e. less than about 1 vol%. Several aspects of electric conductivity, optical absorption and Raman scattering which have been frequently misinterpreted in the literature will be reviewed.

INTRODUCTION

Several review papers have summarized various aspects of microcrystalline solids [1] and semiconductors, in particular silicon (see [2] and references therein). The increasing importance of microcrystalline silicon, μc-Si, has been a subject of several papers presented at this symposium. μc-Si can be prepared by various techniques, such as thermal or plasma induced chemical vapour deposition, CVD, from silane diluted with hydrogen, from halosilanes and by sputtering. In this paper, we shall discuss the chemical parameters which control the formation of amorphous and crystalline phases during the plasma CVD, because this method allows one the best control of the deposition process and of the properties of the resulting material. In order to clarify at least some of the issues regarding the effects of the microcrystallinity and of the "amorphous tissue", the emphasis will be on the control of the structural and related physical properties of a pure microcrystalline silicon.

A number of papers published so far consider μc-Si as a two-phase system consisting of microcrystals imbedded in an amorphous tissue. It will be shown that such material is deposited under conditions where the chemistry of the system is unsufficiently controlled. The lack of such control in various papers reflects itself in a variety of the reported properties of the deposited material. Therefore, we shall start with the discussion of the chemistry of the $SiH_4/H_2/Si(s)$ -system under conditions of low pressure glow discharge plasma. It will be shown that a variety of the published experimental recipes for the preparation of μc-Si can be attributed to the same parameters which control the formation of either the amorphous or the crystalline phase. This crucial parameter is the departure of the system from the partial chemical equilibrium, PCE which will be discussed in the next section.

THE CHEMISTRY OF THE SiH$_4$/H$_2$/Si(s)-SYSTEM
AND THE CRYSTALLINE-TO-AMORPHOUS TRANSITION

The concept of PCE under conditions of non-equilibrium low pressure plasma has been introduced and analysed by the author in 1972 [3]. It corresponds to a state remote from thermodynamical equilibrium with vanishing chemical fluxes, i.e. where the actual PCE concentrations of the reactants and products are given by the balance of their formation and decomposition with a negligible flow term. Under conditions close to PCE, the rate of the deposition (or etching) of a solid is given by the difference of the rate of the decomposition and formation of the reactants (or vice versa). Quantitative data for the SiH$_4$/H$_2$/Si(s) -system together with a discussion of the approach of the system to PCE can be found in our earlier work [4,5]. Under typical conditions of the plasma CVD of silicon, the PCE-concentration of silane amounts to about 0.3 mol% which is five orders of magnitude larger than the equilibrium value at thermodynamic equilibrium.

The approach of the system to the PCE is achieved at a large value of the ratio of the dwell time of silane admitted to the reaction half time, τ_{dwell}: $\tau_{1/2} \geq 20$. The value of $\tau_{1/2}$ is proportional to the reciprocal value of the discharge current density (Fig. 1 from [4,5]).

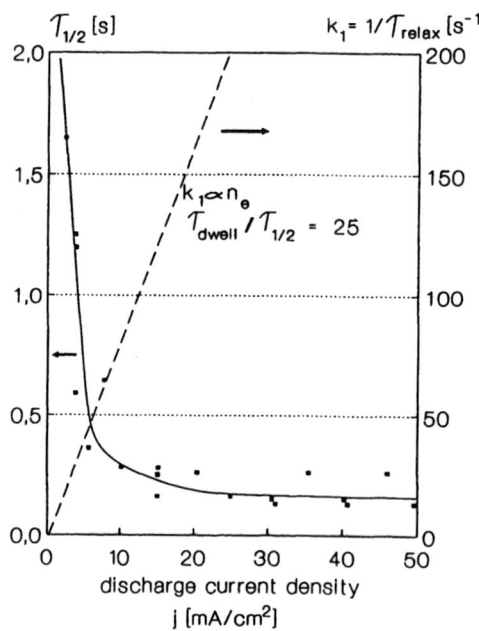

Fig.1: Dependence of the reaction half time and the rate constant for silane decomposition on the discharge current density [4,5].

Thus, the PCE is reached within a few seconds in an intense discharge which is typically used for the deposition of c-Si ($j_{disch} \geq 7$ mA/cm^2) whereas more than 20 seconds are needed in a weak discharge used for the deposition of a-Si. **Notice!** The dwell time is well defined only under conditions of the plug flow, i.e. in a long, narrow tube and fast flow rate. It is undefined in a typical parallel plate reactor with a complex flow and diffusion pattern, and a large "dead volume" in which also the gas composition may vary from place to place. This is neglected in most published papers.

The importance of an exact control of the gas composition near the substrate is documented by Fig.2 which shows the sudden transition from the amorphous to crystalline deposit when the silane concentration in the vicinity of the substrate reaches ≤ 0.5 mol %. The amount of the crystalline component in the film has been determined from the ratio of the integrated intensity of the (220) and (311) Bragg reflection to the total intensity in the region of the second maxima of the X-ray scattering from a-Si according to the procedure outlined in [6] and [8]. Let us emphasize that this is the only reliable way of determining the amount of the crystalline and amorphous component; the Raman scattering, which is frequently being used for this purpose, does not provide any meaningful data of this kind (see [9]).

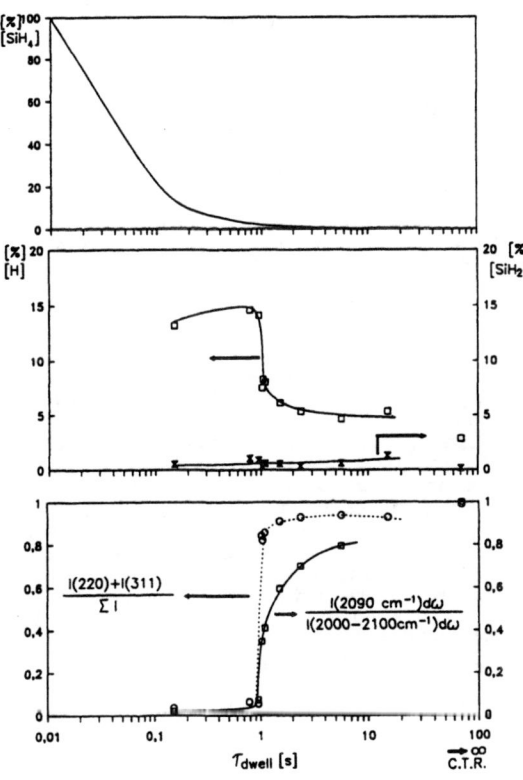

Fig.2: Dependence of silane concentration in the vicinity of the substrate, of the crystalline component, the total hydrogen content, the SiH$_2$-groups and the optical absorption at 2090 cm^{-1} in the deposited silicon on the dwell time of silane introduced into the discharge tube [6],[7].

The X-ray diffractograms which were used for the evaluation of Fig.2 can be found in ref. [6]. They clearly show that less than about 1 vol.% of crystalline component can be detected in a-Si (and vice versa), that a purely microcrystalline material without any detectable amorphous tissue (i.e. less than about 1 vol%) can be deposited by chemical transport and that a heterogeneous conglomerate of a-Si and μc-Si is obtained under deposition conditions just near the a-c-transition (see Fig.2).

Direct lattice images of the lattice planes in transmission electron microscope [10], detailed X-ray diffraction data and light scattering studies have shown that the material connecting the crystallites in the pure μc-Si consists of well ordered grain boundaries with somewhat expanded lattice plain distances [9].

This lattice expansion leads to the formation of spatially fluctuating electric dipoles which increase the optical absorption coefficient and the cross section for Raman and elastic light scattering. It also gives rise to grain boundary phonons seen as an amorphous-like feature in the Raman spectra around 480 cm^{-1} [11], which is in the current literature frequently incorrectly interpreted as the "amorphous tissue". We have shown that the cross section for this and for the crystalline Γ_{25}, central mode in Raman scattering, as well as the optical absorption coefficient decrease when the lattice expansion in the grain boundaries is reduced by applied compressive mechanical stress to the films. These results clearly show that the ratio of the intensity of the 480 cm^{-1} feature to the total intensity in the Raman data should not be used as a measure of the "amorphous tissue" in μc-Si (see ref. [9,11] and [18,19] for further details).

The lattice expansion within the grain boundaries originates from the anharmonicity of the interatomic potential [9]. It is the driving force for the discontinuous, first order amorphous-to-crystalline transition. Thus, there is a lower limit to the crystallite size of about 2,5 - 3 nm below which the a-Si is more stable than μc-Si [8]. Consequently, under conditions of the μc-Si deposition the thin film starts growing amorphous and, only when the thickness has reached a value larger than the minimum crystallite size, it transforms to μc-Si. This prediction of our earlier analysis has been confirmed by several researchers who used either Raman scattering [12] or ellipsometry [13] to study the initial stage of the growth.

Another parameter which influences the structural properties of the plasma deposited a-Si and μc-Si is the ion bombardment during the deposition (e.g. [14]). For both a-Si [7] and μc-Si [15] the ion bombardment induced desorption of the chemisorbed hydrogen from the surface of the growing film is the rate determining step. Thus, low energy ion bombardment is necessary for high deposition rate of good optoelectronic quality films [10,14,15,16,17]. However, formation of radiation damage and the concomitant degradation of the electronic properties occurs if the ion impact energy exceeds the threshold value for the damage. This is illustrated by Fig.3 which shows the dependence of the integrated one phonon density-of-states absorption in infrared versus the substrate bias voltage during the film deposition [9,14]. The onset of the absorption, which is forbidden in the Si-single crystal but allowed in a defect Si-lattice, at about 117 V agrees very well with the value of the threshold energy for radiation damage in silicon due to energetic H^+ ions [15]. Figure 3 also shows that there is almost no change of the relative intensity of the feature around 480 cm^{-1} in the Raman spectra up to about -800 V where a lattice damage induced amorphization occurs [9,14].

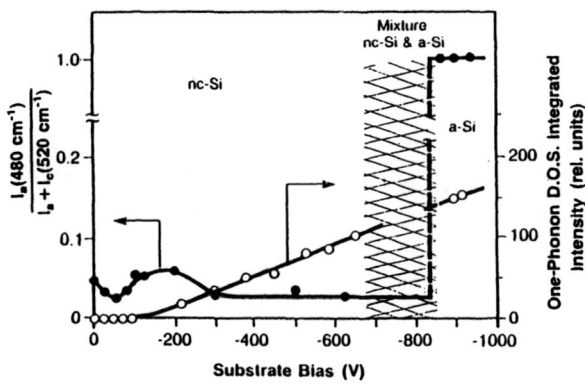

Fig.3: One-Phonon Density-Of-States integrated intensity and the relative intensity of the feature around 480 cm^{-1} in Raman spectra of μc-Si versus substrate bias during deposition. Chemical transport, T_d=256°C, p=0.3 mbar [9].

REACTION MECHANISM

Several research groups suggested that the dominant species responsible for the deposition of a-Si and μc-Si is the SiH$_3$ radical. However a critical consideration of the arguments given by these workers clearly shows that this suggestion is not substantiated by any unambiguous experimental data (see [20] and references therein). A detailed analysis of the recent kinetic data shows clearly, that the dominant channel of the high rate deposition of good quality a-Si proceeds via higher silanes with SiH$_2$ as a reactive intermediate species (see [5,7,21,22] for details):

$$SiH_4 \xrightarrow{e(v?)} SiH_2 + H_2 \qquad (1)$$

$$SiH_2 + SiH_4 \longrightarrow Si_2H_6 \qquad (2a)$$
$$SiH_2 + Si_2H_6 \longrightarrow Si_3H_8 \qquad (2b)$$

$$Si_2H_6 \longrightarrow 2\ a\text{-}Si + 3H_2 \qquad (3a)$$
$$Si_3H_8 \longrightarrow 3\ a\text{-}Si + 4H_2 \qquad (3b)$$

The deposition of μc-Si occurs at a low concentration of silane of ≤ 0.5 mol% where the probability of collision of SiH$_2$ with a silane molecule is relatively low. Using the values of the rate constant for the insertion reaction (2a) which were available at that time, Ensslen and Veprek calculated that the SiH$_2$ species should reach the surface of the growing c-Si film before undergoing the insertion reaction (2a) (see Fig.7 in [5]). However, more recent data of Jasinski and Chu [21] reveal that the rate constant for the insertion, eq. (2a), is about two orders of magnitude larger than the value of John and Purnell [22] used in our earlier paper [5]. Thus, the formation of disilane appears probable also under the conditions of c-Si deposition. Consequently, disilane (and not SiH$_2$) should be also the dominant species responsible for the plasma induced deposition of c-Si. Obviously, this question deserves further experimental studies.

The mechanism of the formation of μc-Si is still subjected to many speculations which are frequently accompanied by nice pictures but little substantial experimental data. The data discussed above suggest that the decisive microscopical parameter controlling the a-to-c transition is the ratio of the rate of deposition to that of etching. This has been utilized already in the first documented preparation of μc-Si and μc-Ge by Veprek and Marecek who used the chemical transport reaction [23]. Webb and Veprek have shown that the etch rate of silicon with atomic hydrogen in a plasma with parameters typical for the deposition of c-Si can reach values up to 1,6 nm/sec [24]. The etching removes preferentially the weakly bonded Si atoms from the surface thus promoting the formation of a regular crystalline lattice. This also explains why there is no amorphous tissue in μc-Si deposited by chemical transport close to PCE.

Another important factor associated with the a-to-c transition is the relatively large amount of energy dissipated within the topmost surface layer of the growing film. It has been pointed out earlier [8] that a large amount of the recombination energy of H-atoms, which is transformed into localized surface vibrations, may trigger the crystallization.

COMPARISON OF VARIOUS PREPARATION METHODS

Usui and Kikuchi [25], Matsuda et al [26] and Hamasaki et al [27] were the first researchers who have shown that μc-Si can be deposited from silane in a glow discharge at a relatively low temperature under conditions of a strong dilution with hydrogen and a relatively high power density. These and many other published papers somewhat differ regarding the exact value of the dilution which is necessary to achieve the formation of the microcrystalline phase. The above discussed approach of the system to the partial chemical equilibrium, which is determined by the ratio of the dwell and the half time, clearly shows that such differences are due to different values of the dwell time and of the "effective" half time in the different apparata used. (Remember that, in a typical parallel plate reactor used by these researchers neither the dwell time nor the half time are well defined due to the complex geometry.)

In any case, the decisive parameter determining the a-to-c transition is the silane concentration in the vicinity of the substrate and the degree of dissociation of hydrogen. Both have to be close to the values found in an intense discharge close to PCE, i.e. about 0.3 mol% of silane and 3 to 5 at% of atomic hydrogen . The dilution of the silane introduced into the reactor is not any unambigious parameter as it changes within the reactor approaching, more or less, the PCE value. A similar conclusion can be drawn for other preparation techniques reported in the literature, such as sputtering.

CHARACTERIZATION AND FUNDAMENTAL PHYSICAL PROPERTIES OF μc-Si

The optical and electronic properties of μc-Si are strongly dependent on the structural properties which are controled by the preparation conditions. In an extreme case, a pure μc-Si free of any detectable amorphous tissue can have properties which are considered to by typical of a-Si, such as a high optical absorption and low dark electric conductivity. It is therefore extremely important to measure and report always structural data on the material used in any study. Of primary interest and importance are the X-ray diffraction data which allow one to calculate the amount of amorphous tissue in the film (see above and ref. [6,8]), the mean crystallite size and its distribution, and the mechanical stress in the films [28].

The mean crystallite size and the mechanical stress in the films can be evaluated using either the Warren-Averbach analysis of the Bragg diffraction peak profiles [28] or a graphical method of Williamson and Hall [29]. The use of the simple Scherrer formula to calculate the crystallite size from the full width at half maximum (FWHM) of the Bragg reflection is allowed only if the films are stress free and the crystallite size is less than about 15 - 20 nm because of the following reasons:

Firstly, mechanical stress contributes to the broadening of the Bragg peaks and, if not corrected for, it simulates an apparently smaller crystallite size calculated from the Scherrer formula.

Secondly, the instrumental broadening becomes comparable with the FWHM of the Bragg reflection at a crystallite size $\geq 30 - 40$ nm and, if not corrected for by the Fourier deconvolution of the measured peaks, it simulates in a similar manner a smaller crystallite size.

This is illustrated by Fig.4 which shows the temperature dependence of the mean crystallite size determined by the Scherrer formula from the (111), (220) and (311) Bragg diffraction peaks (Fig.4a) and from the Warren-Averbach analysis [30].

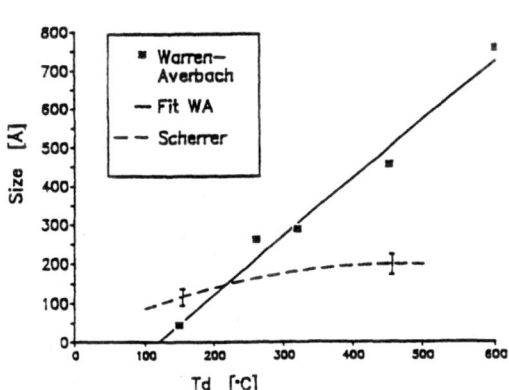

Fig.4: Dependence of the mean crystallite size, taken as an averaged value of the (100) cube edge as determined using the simple Scherrer formula from the (111), (220) and (311) Bragg reflections (dots and full line) [9] and using the Warren-Averbach analysis [30], on the deposition temperature. Chemical Transport, p=0.3 mbar, V_b=0 V.

Already a small negative bis applied to the substrate during the deposition has a pronounced effect on the structural and related physical properties of the films. Figure 5 shows the strong decrease of the mean crystallite size and increase of the compressive stress in the films with increasing value of the bias.

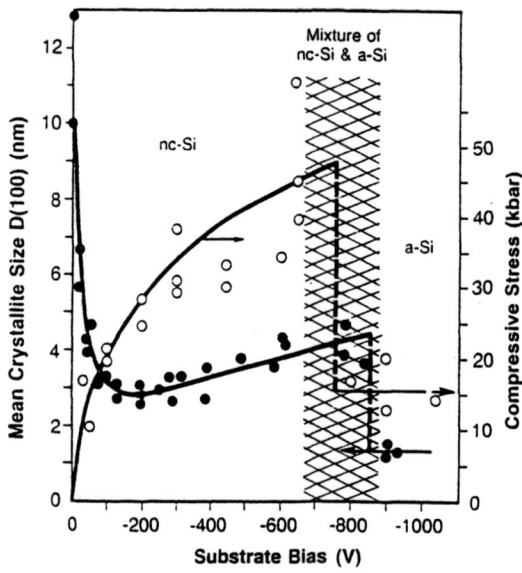

Fig. 5: Mean crystallite and compressive mechanical stress in the films versus bias. Conditions as in Fig.4, T_d=260°C [9].

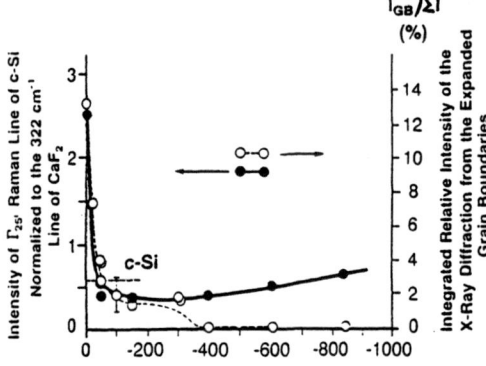

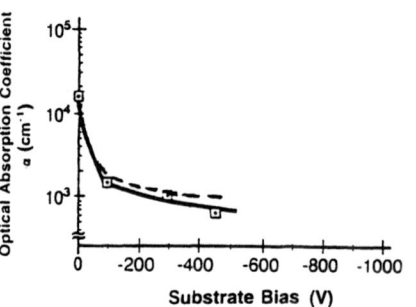

Fig.6: Lattice expansion of the grain boundaries (a, broken line), normalized intensity of the Raman Γ_{25},crystalline phonon line (Fig.a, full line), optical absorption (Fig.b, full line) and elastic light scattering (broken line) from μc–Si versus substrate bias.

The compressive stress decreases the above mentioned lattice expansion within the grain boundaries (see broken line in Fig.6a) which manifests itself in the decrease of the cross section for Raman scattering (Fig. 6a), for elastic light scattering and for optical absorption (Fig.6b) close to the values typical of single crystal silicon [9].

The dramatic change of the electrical dark conductivity with increasing bias is illustrated by the data shown in Fig.7. It is seen that the dark conductivity decreases five to six orders of magnitude when the bias increases from zero ("wall" potential, ca. 10 to 15 V negative with respect to the plasma) to -150 V.

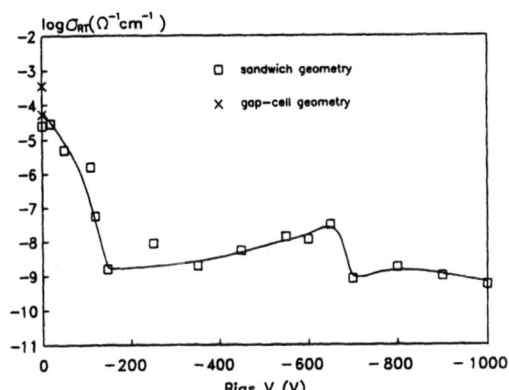

Fig.7: Dependence of electrical dark conductivity on substrate bias. Chemical transport, $T_d=260°C$, p=0.3 mbar [10].

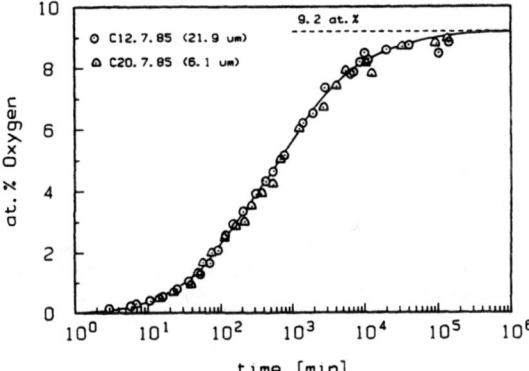

Fig.8: Uptake of oxygen by μc-Si films deposited at floating potential 10 to 15 V negative with respect to the plasma $T_d=260°C$ and subsequently exposed to air at about 260°C [32].

As a last example of the dramatic effect of substrate bias on the properties of the μc-Si films we show in Fig.8 the uptake of oxygen by the grain boundaries for films deposited at floating potential, i.e. having a tensile stress and expanded lattice within the grain boundaries [32]. The oxygen is incorporated within the grain boundaries and it dramatically changes the electronic properties of the films [16]. Such films can be used as selective gas sensors because their electrical conductivity changes by orders of magnitude due to adsorption of electronegative or -positive gases on their surface [33], but they surely do not represent the properties of intrinsic µc-Si. Application of a compressive bias of about -100 V is sufficient to densify the films, remove sufficiently the lattice expansion within the grain boundaries (see Fig.6a) and avoid the oxygen incorporation [16,32].

It should be remembered that in all these cases the deposited films are microcrystalline without any measurable amount of amorphous tissue (see ref. [6] and [10] for the experimental data). The physical interpretation of these results is beyond the scope of the present paper. The reader should consult the original papers for further information regarding these questions.

Here we should like to emphasize the necessity of a careful and relevant characterization of the preparation conditions and of the deposited films whenever reporting some data on the electrical and optical properties of µc-Si. Unfortunately, such characterization is missing in most papers. The examples given here show clearly that the optical and electrical properties of µc-Si can strongly differ if the preparation conditions are slightly changed. I leave it to the reader to make up his mind about the scientific quality of a paper in which the authors state that "the deposited films are amorphous because of their large value of optical absorption and low electrical conductivity".

CONCLUSIONS

The formation of the crystalline phase during plasma induced CVD of silicon from silane occurs when the heterogeneous system $SiH_4/H_2/Si$ approaches the partial chemical equilibrium. Under the discharge conditions typical for the deposition of c-Si the silane concentration in the plasma reaches about 0.3 mol% and the degree of dissociation of hydrogen amounts to about 2 - 5 at.%.

The amorphous-to-crystalline transition is a discontinuous first order phase transition which occurs suddenly. Pure µc-Si phase is formed under plasma conditions close to PCE where the deposition rate of 2 -6 Å per second is achieved as the difference between the high rates of silane decomposition by electron impact in the gas phase and etching of the deposited silicon by atomic hydrogen. A mixture of a-Si and µc-Si (or a µc-Si with an "amorphous tissue") is obtained if the deposition conditions are not carefully controlled.

The structural and related optical and electrical properties of pure µc-Si free of any amorphous tissue critically depend on the deposition conditions, in particular on the substrate bias during the deposition. Several selected examples are reported which illustrate this point.

ACKNOWLEDGEMENT

I should like to thank my wife and my coworkers for critical comments to the manuscript and to G. Ratz for help with the figures.

REFERENCES

[1] L. Genzel, in: "Festkörperprobleme" Vol. XIV, ed. H.J. Queisser, Vieweg, Braunschweig 1974, p. 183

[2] "Properties of Amorphous Silicon", 2nd ed., EMIS Datareviews Series No.1, INSPEC, The Institution of Electrical Engineers, London, 1989

[3] S.Veprek, J. Chem. Phys. 57,952(1972)

[4] J.J.Wagner and S.Veprek, Plasma Chem. Plasma Processing 2,219(1982); 3,219(1983)

[5] K.Ensslen and S.Veprek, Plasma Chem. Plasma Processing 7,139(1987)

[8] S.Veprek, Z.Iqbal and F.-A.Sarott, Phil. Mag. B 45,137(1982)

[9] S.Veprek, F.-A.Sarott and Z.Iqbal, Phys. Rev. 36,3344(1987)

[6] S.Veprek, M.Heintze, F.-A.Sarott, M.Jurcik-Rajman and P.Willmott, MRS Symp. Proc. Vol.118,3(1988)

[7] S.Veprek and M.Heintze, Plasma Chem. Plasma Processing 10,3(1990)

[10] M.Konuma, H.Curtins, F.-A.Sarott and S.Veprek, Phil. Mag. B 55,377(1987)

[11] Z.Iqbal and S.Veprek, J. Phys. C 15,377(1982)

[12] J.Richter and L.Ley, J. de Phys 43,C1-247(1982)

[13] R.W.Collins and B.Y.Yang, J. Vac. Sci. Technol. B 7,1155(1989)

[14] S.Veprek, Proc. Mater. Res. Soc. Europe, Strasbourg 1984, eds. P.Pinard and S. Kalbitzer, Les éditions de physique, Les Ullis, France (1984) p.425

[15] S.Veprek, F.-A.Sarott, S.Rambert and E.Taglauer, J. Vac. Sci. Technol. A 7,2614(1989)

[16] S.Veprek, Z.Iqbal, R.O.Kühne, P.Capezzuto, F.-A.Sarott and J.K.Gimzewski, J.Phys. C 16, 6241(1983)

[17] S.Veprek, M.Heintze, R.Bayer and M.Jurcik-Rajman, MRS Symp. Proc., San Diego, April 1989, Symp. E (in press)

[18] P.M.Fauchet and I.H.Campbell, Critical Reviews in Solid State and Material Sciences, Vol. 14, Supplement 1 (1988) p.S 79

[19] P.M.Fauchet and I.H.Campbell, MRS Symp. Proc., Boston November 1989 (this Volume)

[20] S.Veprek, Thin Solid Films 175,129(1989)

[21] J.M.Jasinski and J.O.Chu, J. Chem. Phys. 88,1678(1988)

[22] P.John and J.H.Purnell, J. Chem. Soc. Faraday Trans. 69,1455(1973)

[23] S.Veprek and V.Marecek, Solid State. Electron. 11,683(1968)

[24] A.P.Webb and S.Veprek, Chem. Phys. Lett. 62,173(1979)

[25] S.Usui and M.Kikuchi, J. Non-Cryst. Solids 34,1(1979)

[26] A.Matsuda, S.Yamasaki, K.Nakagawa, H.Okushi, K.Tanaka, S.Izima, M.Matsumara and Y.Yamamoto, Jap. J. Appl. Phys. 19,L305(1980)

[27] T.Hamasaki, H.Kurata, M.Hirose and Y.Osaka, Appl. Phys. Lett. 37,1084(1980)

[28] H.P.Klug and L.E.Alexander, "X-Ray Diffraction Procedures", 2nd. ed. John Wiley & Sons, New York 1974

[29] G.K.Williamson and W.H.Hall, Acta Met. 1,22(1953)

[30] F.-A.Sarott, Ph D Thesis, University of Zürich 1989

[32] H.Curtins and S.Veprek, Solid State Commun. 67,215(1986)

[33] F.Mattenberger and S.Veprek, CHEMTRONICS 1,107(1986)

ELECTRONIC AND STRUCTURAL CHARACTERIZATION OF THE NEAR SURFACE LAYER AND THE BULK IN μc-Si:H PREPARED WITH HYDROGEN DILUTION

Samer Aljishi[1], Shu Jin, Martin Stutzmann, and Lothar Ley[2]

Max-Planck-Institut für Festkörperforschung, Heisenbergstrasse 1, 7000 Stuttgart 80, F.R.G.

ABSTRACT

The near surface layer and the bulk of μc-Si:H prepared with hydrogen dilution are investigated by Raman, optical absorption, and total yield photoelectron spectroscopies. The results show that for low hydrogen dilution ratios, microcrystallites appear in the bulk while the growing surface layer remains amorphous, indicating that microcrystallite formation takes place primarily in the sub-surface layer. At high hydrogen dilution ratios, microcrystallites are detected at both the bulk and the near surface layer. The defect density and hydrogen bonding configurations at various hydrogen dilution levels are presented.

INTRODUCTION

Hydrogenated microcrystalline silicon μc-Si:H thin films have been a subject of much interest in the past few years because of their unique properties which fall between those of a-Si:H and monocrystalline silicon. Amongst those are a high electrical conductivity and low absorption coefficient, both of which make μc-Si:H and its alloys ideal materials for the doped window layers in photovoltaic tandem cells. Even though several of the important electronic and structural properties of these materials have been investigated [1,2,3], several issues remain unresolved especially those concerning the mechanisms of film growth, microcrystallite nucleation and formation, and the defect stucture in the various phases (amorphous, microcrystalline, and grain boundary) typically coexisting in the films.

In this work we combine electronic and structural characterization of the bulk through Raman and subgap optical absorption spectroscopies, with photoelectron yield spectroscopy characterization of the near surface layer to aid our understanding of the growth dynamics and defect structure in μc-Si:H materials prepared with hydrogen dilution.

TOTAL YIELD SPECTROSCOPY

In total yield spectroscopy [4], UV illumination (3.5 eV $<$ $\hbar\omega$ $<$6.4 eV) is used to excite electrons from occupied states to the vacuum level. As the excitation energy is scanned, the total number of photoemitted electrons is counted. Dividing the number of electrons over the photon flux results in a yield spectrum $Y(\hbar\omega)$ which in amorphous solids is proportional to a convolution of an initial occupied DOS g_v and a final unoccupied DOS g_c weighted by an average dipole matrix element $R(\hbar\omega)$

$$Y(\hbar\omega) \propto \hbar\omega \mid R(\hbar\omega) \mid^2 \int_{E_{vac}}^{\infty} g_v(E - \hbar\omega)g_c(E)dE, \qquad (1)$$

E_{vac} is the vacuum level which is the reference energy (E=0) in total yield spectroscopy. In the presence of k-selection rules, as in c-Si, Eq. 1 must include an additional term $D(E-E_{VAC})$ in the integral which is an escape function reflecting conservation of k parallel to the film surface for the photoemitted electrons.

---

[1] Alexander von Humboldt research fellow

[2] present address: Institut für Technische Physik, Universität Erlangen, Erwin-Rommel-str.1, 8520 Erlangen, F.R.G.

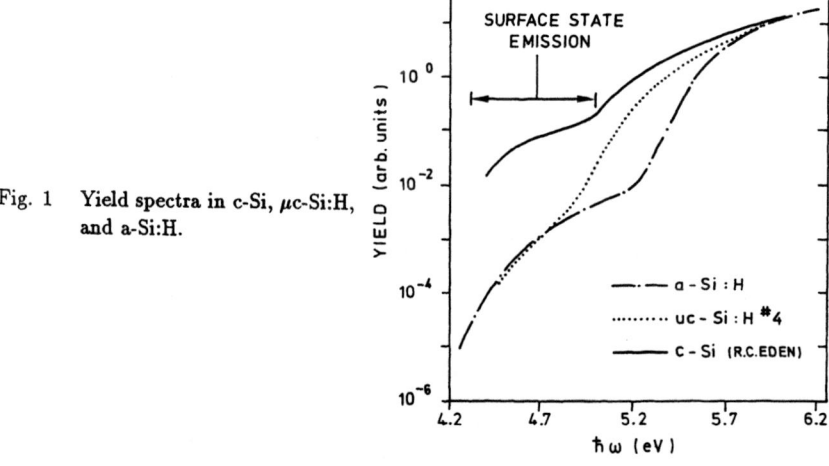

Fig. 1 Yield spectra in c-Si, μc-Si:H, and a-Si:H.

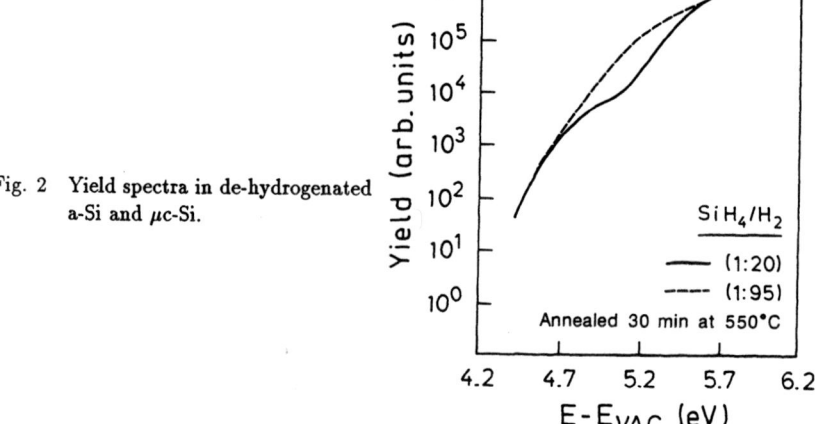

Fig. 2 Yield spectra in de-hydrogenated a-Si and μc-Si.

The probe depth of the total yield technique is limited simultaneously by the photon absorption and the photoelectron escape depths. These two factors set the probe depth to a region within 25Å to 100Å (depending on $\hbar\omega$) of the vacuum/film interface, resulting in a *spatial sensitivity confined exclusively to the top surface layer.*

EXPERIMENTAL DETAILS

Thin (<5000Å) a-Si:H and μc-Si:H films were prepared by conventional rf-diode glow discharge from a mixture of SiH_4 and H_2 gas. The hydrogen dilution ratio H_2/SiH_4 was varied between 0:1 and 200:1 in order to produce films with varying degrees of microcrystallinity. The deposition temperature was fixed at 250°C for all films. Other depostion parameters were: a total flow rate of 5 sccm, 0.4 to 0.6 mbar process pressure, and 13 to 14 Watt total rf power. Films were deposited on clean stainless steel substrates for yield measurements and concurrently on Herasil quartz substrates for optical characterization.

Raman spectra (with Ar^+ laser excitation at 514.5 nm) were measured in the range of 400 to 2200 cm^{-1}. The data was used to obtain a *rough estimate* [5] of the microcrystallite volume fraction ρ' in the films from the ratio of the areas under the 520 cm^{-1} and the 480 cm^{-1} features corresponding to the TO phonon vibrational modes in crystalline and amorphous silicon, respectively. Subgap optical absorption spectra $\alpha(E)$ were measured via photothermal deflection spectroscopy (PDS).

RESULTS

Since total yield measurements form the cornerstone for much of this work, it is useful to compare the yield spectra for μc-Si:H to those of a-Si:H and c-Si. Fig. 1 displays the yield spectra for an a-Si:H film grown with a H_2-dil. ratio of 5:1, for a μc-Si:H film grown with a H_2-dil. ratio of 95:1, and for monocrystalline Si [6]. The most significant differences between the three spectra lie in the values of the photoemission threshold energy E_T (alternatively called the top of the valence band, E_{TVB}) and in the magnitudes of the defect density. E_T in clean c-Si is measured at 5.1 eV. In a-Si:H, E_{TVB} lies typically between 5.4 and 5.5 eV. In μc-Si:H, we observe that the photoemission threshold energy shifts *abruptly* to approximately 5.2 eV, between that of c-Si and of a-Si:H. Part of this shift (from the value of a-Si:H) may stem from a lower hydrogen content in the microcrytalline material. However, Fig. 2 shows that even in a-Si and μc-Si films in which all of the hydrogen is evolved by a high temperature anneal (550° C), E_T is still lower in μc-Si (5.1 eV) than in a-Si ($\approx$ 5.3 eV). We are thus led to believe that the differences in E_T between a-Si:H and μc-Si:H are due to a combination of differences in hydrogen content and k-selection rules.

The defect density in μc-Si:H (given by the low energy shoulder of the yield spectra) is equivalent to that measured in a-Si:H ($\approx$ 4x10^{17} cm^{-3}). As can be seen from Fig. 1, this is more than a factor of 10^2 less than the surface-state defect density in c-Si. The difference between the defect densities in c-Si and μc-Si:H can probably be ascribed to hydrogen passivation of surface defects as occurs with a-Si:H.

The mechanism of microcrystallite formation during film growth has been one of the more controversial topics in the study of μc-Si:H. We can gain insight into the issue by comparing the structure of the films at the growing surface layer with that of the deeper lying bulk region. Figures 3,4, and 5 show the Raman, optical absorption, and yield spectra, respectively, for a series of films where the H_2 dilution ratio was varied between 1:20 and 1:70. The Raman spectra show that increasing the H_2-dilution ratio results in an increase in microcrystallite volume fraction. ρ' for the 4 films shown were estimated roughly at <5%, 20%, 40%, and 60% (in order of increasing H_2-dilution). The increase in

Fig. 3 Raman spectra for 4 films where
the H_2 dilution ratio
is varied from 20:1 to 70:1
producing an increase in the
microcrystallite volume fraction
ρ' from <5% to 60%.

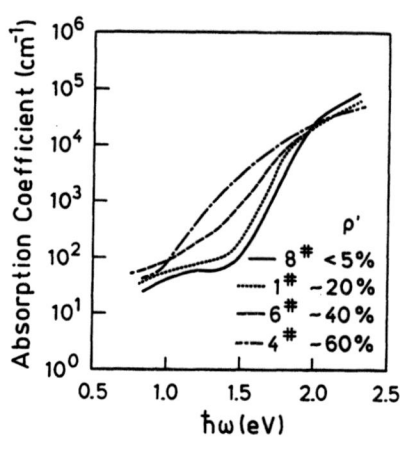

Fig. 4 Subgap optical absorption
spectra measured by PDS.

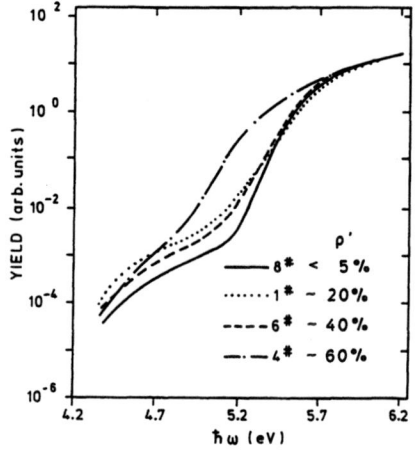

Fig. 5 Yield spectra for μc-Si:H
films where ρ' increases
from < 5% to 60% as the
H_2 dilution ratio is
increased from 20:1 to 70:1.

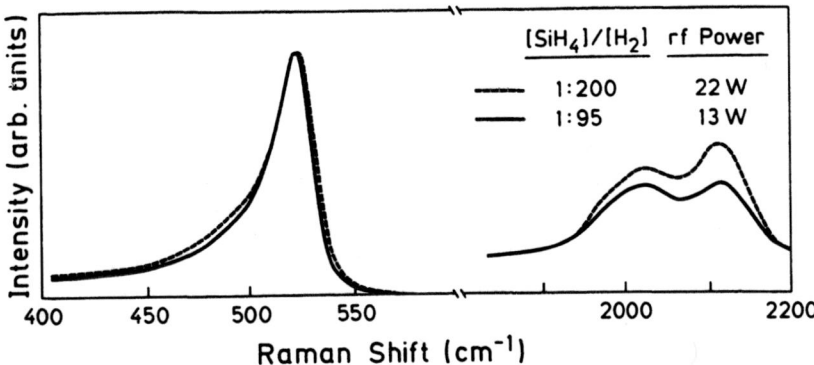

Fig. 6 Raman spectra for μc-Si:H films prepared with a hydrogen dilution
ratio of 200:1 and 95:1 at high and low power respectively.

Fig. 7 Yield spectra for several
different high hydrogen
dilution levels.

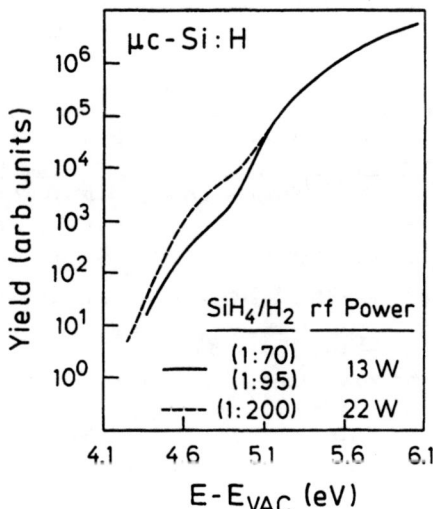

ρ' is accompanied by an increase in clustered hydrogen bonding. This is clearly revealed by the higher intensity of the 2100 cm^{-1} Raman mode (attributed to high order Si hydrides) relative to that at 2000 cm^{-1} (attributed to Si-H monohydrides) for high ρ'. The PDS-determined optical absorption spectra shown in Fig. 4 confirm the trends shown by the Raman data. We find that with increasing H$_2$ dilution, the $\alpha(E)$ spectra shift gradually from those characteristic of a-Si:H to those of a mixed phase a-Si:H/μc-Si:H material (low α at high energy and a broad absorption tail). The close correlation between the structure probed by Raman spectroscopy and the behaviour of the optical absorption measured by PDS is expected as both measurement techniques have a large spatial sensitivity extending throughout the majority of the bulk region in the films. Fig. 5 shows the yield spectra for the 4 films with ρ' between <5% and 60%. Unlike the Raman and optical absorption spectra where the amorphous to microcrystalline transition occurred gradually, the yield spectra undergo an abrupt shift when the H$_2$ dilution ratio is increased beyond 60:1. For dilution ratios less than or equal to 60:1, the yield spectra are characteristic of the amorphous phase (with E$_T$ at 5.5 eV) indicating that *even though there is a significant degree of microcrystallinity in the bulk, the near surface layer (probed by yield) remains amorphous.* Only at *high* H$_2$ *dilution ratios*($\geq$ 70:1) do the yield spectra shift abruptly to lower energy (with an E$_T$ of 5.2 eV) indicating the eventual nucleation and growth of microcrystallites at the near surface layer. The difference between the structure of the growing surface and the bulk observed for low H$_2$ dilution ratios indicates that *microcrystallite formation occurs primarily in the sub-surface or bulk layer rather than at the surface.* This argues against H$_2$ etching [1] as the process *solely* responsible for microcrystallite formation, since if that was the case, the surface would show the same structure as the bulk regardless of the H$_2$ dilution ratio.

Figures 6 and 7 show the Raman and yield spectra respectively for two μc-Si:H films prepared with high H$_2$ dilution ratios (95:1 and 200:1). We find that even though the calculated volume fractions are equivalent for both films ($\rho' \approx$60%), the (near surface) defect density is considerably different. The film prepared at a very high H$_2$ dilution ratio (200:1) and high power (22 W) displays a factor of ten higher defect density than the film prepared at lower power and lower dilution ratio. The increase in defect density is accompanied by an increase in the TO-phonon line width and an increase in clustered hydrogen bonding. These are indications that films prepared with very high dilution ratios contain smaller microcrystalline grains and highly defective interconnecting tissue, effects that are confirmed by previous studies as well [1].

We are grateful to Sigurd Wagner, Etienne Bustarret, and Martin Brandt for helpful discussions. S. Aljishi acknowledges support by the Alexander von Humboldt Foundation.

References

[1] A. Matsuda, J. Non-Cryst. Solids **59&60**, 767 (1983).

[2] C.C. Tsai, R. Thompson, C. Doland, F.A. Ponce, G.B. Anderson and B. Wacker, Mat. Res. Soc. Symp. Proc. **118**, 49 (1988).

[3] S. Veprek, M. Heintze, F.A. Sarott, M. Jurcik-Rajman and P. Willmott, Mat. Res. Soc. Symp. Proc. **118**, 3 (1988).

[4] K. Winer, I. Hirabayashi, and L. Ley, *Phys. Rev. B* **38**, 7680 (1988).

[5] R. Tsu, J. Gonzalez-Hernandez, S.S. Chao, S.C. Lee and K. Tanaka, Appl. Phys. Lett. **40**, 534 (1982).

[6] R.C. Eden, Ph.D. thesis, Stanford University (1967).

THE FORMATION OF SILICON CLUSTERS IN PLASMA-ENHANCED CHEMICALLY VAPOUR DEPOSITED Si:O:H:F ALLOYS

A. G. DIAS AND J. FIGUEIREDO
Centro de Física Molecular das Universidades de Lisboa, Instituto Nacional de Investigação Científica, Complexo I (I.S.T.), Av. Rovisco Pais, 1000 Lisboa, PORTUGAL.

ABSTRACT

The structural and optoelectronic properties of a new plasma-enhanced chemically vapour deposited Si:O:H:F alloy are reported taking into account respectively the results from X-ray diffraction, Raman and infrared absorption spectroscopy analysis and from dark and photo conductivity and optical absorption measurements. Films produced with a F_2/SiH_4 gas ratio (r) lower or equal than 0.4 are amorphous semiconducting Si:O:H:F alloys, whereas films deposited with $r \geq 0.5$ exhibit a multiphase structure consisting of silicon clusters embedded in an amorphous insulating fluorine doped silicon oxide tissue. The formation of silicon clusters is the result of a segregation process of the fluorine and oxygen atoms in the amorphous Si:O:H:F matrix.

INTRODUCTION

Fluorine doped amorphous silicon dioxide has aroused much interest in optical fiber preparation processes due to the lower refractive index of this material when compared with conventional silicate glasses[1,2,3]. A new plasma processing technique whereby fluorine is used to enhance the plasma oxidation and nitridation of silicon at substrate temperatures bellow 600° C have also received considerably attention[4], as well as fluorine enhanced silicon oxidation and photo-oxidation processes for low temperature growth ($\leq$600° C) of high quality dielectric materials[5,6].

As far as we are concerned, the structural studies of fluorine doped silicon oxide materials are of capital importance for a complete understanding of their properties, regarding the development of the above technological applications. In this work X-ray diffraction, Raman and infrared absorption spectroscopy analysis are used together with dark and photoconductivity and optical absorption measurements to study the influence of fluorine and oxygen incorporation on the structural and optoelectronic properties of plasma-enhanced chemically vapour deposited (P.E.C.V.D.) Si:O:H:F alloys.

EXPERIMENTAL DETAILS

The Si:O:H:F alloys were produced by radiofrequency glow discharge of $SiH_4/F_2/Ar$ gas mixtures at substrate temperatures of 300° C in a capacitively coupled apparatus described elsewhere[7]. The fraction of fluorine in the reactive mixture ($r=F_2/SiH_4$) was controlled by varying the fluorine and the argon-diluted silane flow rates at constant total flowrate and pressure. The deposition parameters listed in Table I were chosen to yield μc-Si:H films in the absence of fluorine in the gaseous mixture (r=0)[8]. The presence of oxygen in the reaction mixture, attributed to residual oxygen and oxygenated compounds from the fluorine cylinder (Matheson gas with 98.2% purity) and to the etching of the Pyrex wall of the deposition chamber by fluorine-active species, was previously discussed[9,10,11].

Table I - Deposition parameters of P.E.C.V.D. Si:O:H:F alloys.

Total flow rate	20 s.c.c.m.	Substrate temp.	300° C
Total pressure	3.8 torr	Base pressure	10^{-5} torr
r.f. frequency	12.9 MHz	Power density	0.35 W cm^{-3}
r = F_2/SiH_4	0 - 0.8	SiH_4/Ar	1:10

The infrared absorption spectra of samples deposited onto crystalline silicon substrates were recorded between 4000 cm^{-1} and 200 cm^{-1} using a Perkin-Elmer 683 double-beam spectrophotometer operated in the transmittance mode. The Raman spectra of these films were measured in the backscattering geometry through a triple monochromator at a maximum incident 5145 Å argon laser power of 50 mW. The X-ray diffraction analysis were performed at near grazing incidence ($\approx$0.3°) using the Cu-Kα radiation. Dark and photoconductivity measurements as a function of reciprocal temperature were performed in films deposited during the same run onto glass substrates using a coplanar configuration. A G. E. Quartzline ELH tungsten-halogen lamp was used to simulate the AM2 solar spectrum. The dependence of the absorption coefficient on wavelength in the visible-near infrared spectral range was deduced from the absorptance measurements carried out in a Cary 17E spectrophotometer.

RESULTS AND DISCUSSION

Structural properties

The films produced without F_2 in the gaseous mixture (r–0) are oxygen-free, as shown in the infrared (I.R.) spectrum of Fig. 1 [9,11]. The only vibrational modes present in the I.R. spectra of these films are the characteristic 2000-2100 cm^{-1} Si-H stretching modes and the corresponding wagging vibrations at 630 cm^{-1} [12]. The addition of fluorine to the reaction mixture results in the deposition of Si:O:H:F alloys, as seen in the I.R. spectrum of a film produced with r–0.8 also shown in Fig. 1. The absorption in the region between 1250 cm^{-1} and 1000 cm^{-1}, attributed to the Si-O-Si stretching modes of silicon oxygenated complexes [13], is indicative of a strong oxygen incorporation. The incorporation of fluorine results in the formation of $FSiSi_3$ and $FSiFSi_2$ complexes, whose Si-F stretching modes are centered respectively at 830 cm^{-1} and 930 cm^{-1} [14]. Nevertheless, the Si-H vibrational modes are still present in the I.R. spectra of films deposited with r$\geq$0.

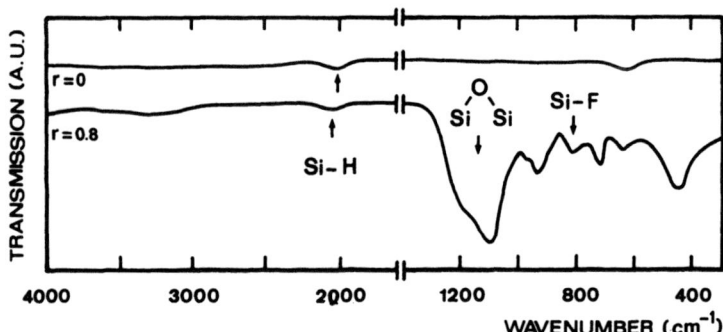

Fig. 1 - Infrared spectra of the r–0 (oxygen free) and r–0.8 samples.

In the I.R. spectra of Si:O:H:F films produced with r$\geq$0.5 we have observed a shift in the frequency of the Si-O-Si stretching mode to values higher than 1075 cm^{-1} (1090 cm^{-1} in the r–0.8 sample). This behaviour suggests the formation of oxyfluorinated complexes where one oxygen atom is substituted by a more electronegative specie like fluorine[15], being associated to the presence of a strong absorption band at 945 cm^{-1} attributed to the Si-F stretching modes of $FSiO_3$ complexes[2,10,16].

Figs. 2-a) and b) show respectively the Raman spectra and the X-ray diffraction patterns of films deposited at various gas ratios. The sample produced with r=0 yields the typical Raman spectrum of μc-Si:H films with two components in the main peak; a sharp feature at 505 cm^{-1} and a broader bump at 480 cm^{-1} [17]. The grain size estimated applying the Scherrer formula to the (111) X-ray diffraction peak is 140 Å. The addition of fluorine (and oxygen) to the reaction mixture results in the deposition of amorphous Si:O:H:F alloys in the range of fluorine/silane gas ratios between 0.1 and 0.4. The Raman spectra of these samples only display the characteristic asymmetrical hump of the amorphous silicon films around 480 cm^{-1}, whereas the corresponding X-ray diffraction peaks broaden. The increasing FWHM of the TO like mode and its slight shift to lower frequencies both indicate an increasing distortion of the silicon matrix[18].

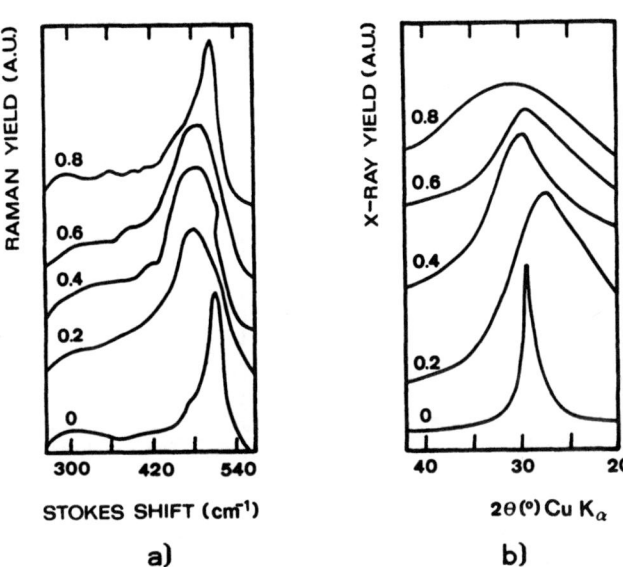

Fig. 2 - (a) Raman spectra and (b) X-ray diffraction patterns of films
deposited at various gas ratios (r).

Further increase in r from 0.5 to 0.8 strongly affects the local order of the Si:O:H:F films. The Raman spectra of these samples have two features around 500 cm^{-1} and 465 cm^{-1}, even though the corresponding X-ray diffraction patterns remain typical of amorphous films. This discrepancy between Raman spectroscopy and X-ray diffraction measurements results from the formation of small ($\leq$8 Å) silicon clusters in the amorphous Si:O:H:F tissue and is associated with the preferential sensitivities and different modes of formation of the Raman and X-ray signals [9,10,19].

The above I.R., X-ray and Raman data were correlated with previous Rutherford backscattering spectroscopy studies[11,20] which have given further support for the superstoichiometric nature of the great majority of our Si:O:H:F alloys and suggested an interpretation of their structure as inhomogeneous, consisting of a silicon rich Si:O:H:F amorphous phase and oxygen bubbles enveloped in an insulating fluorine doped silicon oxide tissue[10].

Optoelectronic properties

Figs. 3-a) and b) summarize the results from dark and photo conductivity and optical absorption measurements of Si:O:H:F alloys. As seen in Fig. 3-a), in spite of the strong oxygen incorporation detected in films produced with r>0 their optical gaps lay in the range of values characteristic of silicon hydrogenated alloys. Dark conductivities of films produced with $r \leq 0.4$, having activation energies (E_σ) of the order of 0.7-0.8 eV and pre-exponential factors (σ_0) around 10^3 Ω^{-1} cm^{-1} (see Fig. 3-b), are associated with an activated process of charge transport taking place in an amorphous semiconducting phase. The sligth increase of the optical gap (E_{opt}) from 1.6 to 1.7 eV observed in films deposited with $r \leq 0.4$ and the attainment of photo to dark conductivity ratios at room temperature (($\sigma_{ph}/\sigma_d)_{RT}$) of the order of 10^3 however indicate a small increase of oxygen incorporation in the amorphous semiconducting Si:O:H:F phase[21].

The formation of silicon clusters within this weakly oxygenated Si:O:H:F tissue deduced from X-ray diffraction and Raman spectroscopy studies, justifies the abrupt change in the conduction mechanism detected in films produced with $r \geq 0.5$, leading to lower activation energies and pre-exponential factors ($E_\sigma \simeq 0.5$ eV and $\sigma_0 \simeq 10^0$ Ω^{-1} cm^{-1}) and to the decrease in optical gap values towards 1.5 eV.

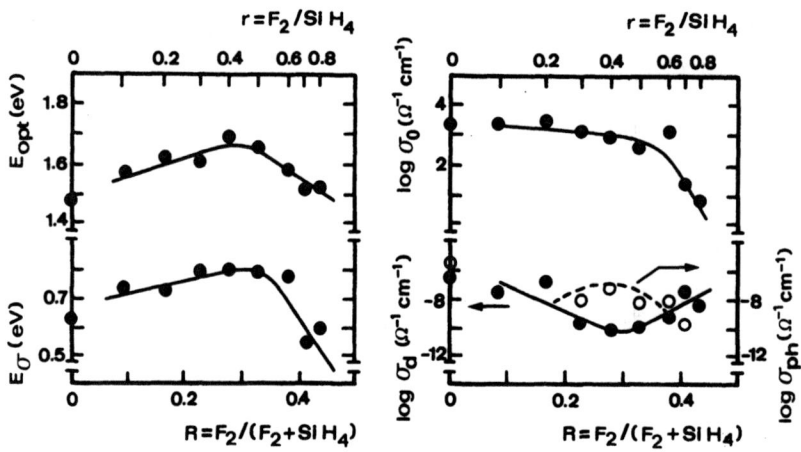

Fig. 3 - Electroptical parameters of films deposited at various gas ratios (r).

The optoelectronic properties of Si:O:H:F alloys are dominated by the contribution of the less-resistive and lower-gap phase, i. e. the amorphous Si:O:H:F tissue, being sensitive to the presence of silicon clusters in this phase.

CONCLUSIONS

Based on X-ray diffraction, Raman and infrared absorption spectroscopy analysis and on dark and photo conductivity and optical absorption measurements we have derived a multi-zone model for our P.E.C.V.D. Si:O:H:F alloys.

Films deposited with reduced fluorine/silane gas ratios ($r \leq 0.4$) are constituted by an amorphous semiconducting Si:O:H:F tissue which accomodates a relatively large volume fraction ($\simeq 50\%$) of oxygen bubbles (see Fig. 4-a)) [10]. The strong reduction on the fraction of molecular oxygen of films produced with higher concentrations of fluorine in the reactive mixture ($r \geq 0.5$) lead to a segregation process of the fluorine and oxygen atoms within the Si:O:H:F matrix, resulting in the formation of silicon clusters embedded in an amorphous insulating fluorine doped tissue (see Fig. 4-b)).

The optoelectronic properties of these Si:O:H:F alloys are controlled by the weakly oxygenated Si:O:H:F tissue being sensitive to the formation of silicon clusters within this semiconducting phase.

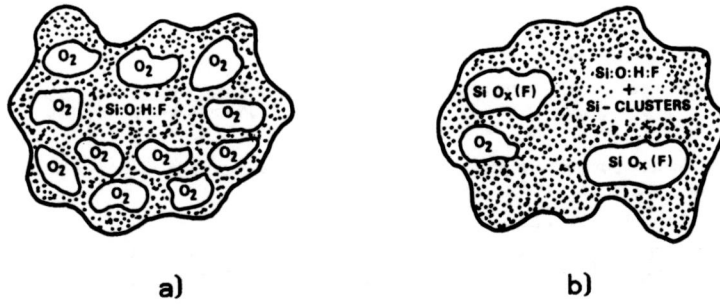

a)   b)

Fig. 4 - Structural model of the Si:O:H:F alloys produced with fluorine/ /silane gas ratios (r): (a) $r \leq 0.4$ and (b) $r \geq 0.5$.

Further work is being undertaken to study the gas phase dissociative processes and the surface reactions responsible for the deposition of amorphous Si:O:H:F alloys when small concentrations of fluorine ($r \leq 0.4$) are added to the reactive gas mixture, instead of the oxygen-free μc-Si:H films obtained in the absence of fluorine ($r=0$) [22], and the formation of silicon clusters in the amorphous Si:O:H:F tissue of films deposited with $r \geq 0.5$.

ACKNOWLEDGMENTS

The authors would like to thank Drs. E. Bustarret, D. Jousse, Y. Cros and M. Brunel from Centre Nacional de la Recherche Scientifique - Grenoble (France) for their interest, encouragement and many helpfull discussions throughout this work.

REFERENCES

1. E. M. Rabinovich, Phys. and Chem. of Glasses 24, 54 (1983).

2. P. Dumas, J. Corset, W. Carvalho, Y. Levy and Y. Neuman, J. Non-Cryst. Sol. 47, 239 (1982).

3. C. A. M. Mulder, R. K. Janssen, P. Bachmann and D. Lears, J. Non-Cryst. Sol. 72, 243 (1985).

4. R. P. H. Chang, C. C. Chang and S. Darack, Appl. Phys. Lett. 36, 999 (1980).

5. Morita, S. Aritome, M. Tsukude, T. Murakawa and M. Hirose, Appl. Phys. Lett. 47, 253 (1985).

6. M. Morita, S. Aritome, T. Tanaka and M. Hirose, Appl. Phys. Lett. 49, 699 (1986).

7. R. Martins, A. G. Dias and L. Guimarães, J. Non-Cryst. Sol. 57, 9 (1983).

8. A. G. Dias, M. A. Ochando and L. Guimarães, Proc. II Simpósio Ibérico de Física da Matéria Condensada, Sevilla, 1986, pp. 308-312.

9. A. G. Dias and E. Bustarret, Proc. II Simpósio Ibérico de Física da Matéria Condensada, Sevilla, 1986, pp. 317-320.

10. A. G. Dias, E. Bustarret and R. C. da Silva, in The Physics and Technology of Amorphous Silica, edited by R. Devine (Plenum Press, New York, 1988), pp. 353--358.

11. A. G. Dias, presented at the Colloque International sur la Science des Materiaux pour l' Energie, Trieste, 1986 (unpublished).

12. M. Cardona, Phys. Stat. Sol. (b) 118, 463 (1983).

13. S. S. Chao, G. Lucovsky, D. V. Tsu, S. Y. Lin, P. D. Richard, Y. Takagi, P. Pai, J. E. Keem and J. E. Tyler, J. Non-Cryst. Sol. 77-78, 929 (1985).

14. L. Ley, H. R. Shanks, C. J. Fang, K. Gruntz and M. Cardona, J. Phys. Soc. of Japan 49, Suppl. A, 1241 (1980).

15. G. Lucovsky, Sol. St. Comm. 29, 571 (1979).

16. A. G. Dias, L. Guimarães and M. Brunel, in The Physics and Technology of Amorphous Silica, edited by R. Devine (Plenum Press, New York, 1988), pp. 359--363.

17. Z. Iqbal and S. Veprek, J. Phys. C 15, 377 (1982).

18. D. Beeman, R. Tsu and M. F. Thorpe, Phys. Rev. B 32, 874 (1985).

19. E. Bustarret, A. Deneuville, H. Roux-Buisson and L. Brunel, J. Non-Cryst. Sol. 59-60, 205 (1983).

20. R. C. da Silva, A. G. Dias, L. Guimarães and M. F. da Silva, Proc. II Simpósio Ibérico de Física da Matéria Condensada, Sevilla, 1986, pp. 313-316.

21. M. A. Paesler, D. A. Anderson, E. C. Freeman, G. Moddel and W. Paul, Phys. Rev. Lett. 21, 1492 (1978).

22. A. G. Dias, Proc. 9th International Symposium on Plasma Chemistry, Pugnochiuso, 1989, pp. 1365-1370.

PREPARATION OF CRYSTALLINE Si THIN FILMS BY SPONTANEOUS CHEMICAL DEPOSITION

TOHRU KOMIYA, AKIRA KAMO, HIROSHI KUJIRAI,
ISAMU SHIMIZU*, AND JUN-ICHI HANNA

Tokyo Institute of Technology
Imaging Science and Engineering Laboratory
Graduate School at Nagatsuta*
Nagatsuta, Midori-ku, Yokohama, 227 Japan

ABSTRACT

The novel technique for the preparation of crystalline Si thin films termed "Spontaneous Chemical Deposition" has been proposed, in which silane decomposes spontaneously by gas phase reactions with fluorine at reduced pressure. The technique provided us the crystalline films in a wide range of the preparation conditions by a choice of the external parameters for the reactions. The films exhibited unique characteristics in the chemical structure and the optical and electrical properties, different from the conventional uc-Si:H thin films by the silane plasma processes.

The technique have been successfully applied for the homoepitaxial growth of Si thin films.

INTRODUCTION

Large-area thin films of the crystalline Si are one of the promising materials for the large area electronic devices, i.e., so-called "Giant Micro-electronics" such as TFT array in the LCD devices and line sensors, because of their high carrier mobilities and the stability. In fact, poly c-Si thin films have been utilized to fabricate these devices with advanced performances. However, the establishment of the technique for the film growth at low temperatures is essential to the practical application.

μc-Si:H thin films prepared by plasma and photo-CVD processes of silanes,[1],[2] which can be done at the temperature less than 400°C, may be one of the candidates for the materials, but their application to the devices have been limited so far to a p+-layer in the amorphous solar cells and a n+-layer to establish the ohmic contact with metal electrodes in TFT because of their heavy defect densities. Moreover the deposition rate is far from the practical application of the materials.

Another class of μc-Si:H has been prepared by plasma processes from fluorinated silanes.[3] Especially the thin films by HR-CVD (Hydrogen Radical enhanced CVD)[4] exhibited high photoconductivity and improved carrier mobilities due to the reduction of defects densities, by which the high deposition rates over 10A/sec could be achieved.

We have proposed a new technique for the preparation of Si thin films featured by a non-plasma and purely chemical process, in which silane was decomposed oxidatively by the gas phase reactions with fluorine at the reduced pressure.[5] This technique allows us to prepare not only a-Si:H but also uc-Si:H thin films in a wide range of the preparation conditions.

In addition to this advantage, the films show unique characteristics in the chemical structures and properties. In this paper, we will describe the general feature of the new technique termed "Spontaneous Chemical Deposition" on the film growth of crystalline Si and a preliminary results on its application for the homoepitaxial growth of Si.

EXPERIMENTAL

The experimental setup is illustrated elsewhere.[5] The reactor was very simple and consisted of a Pyrex glass tube for the reaction chamber, a substrate holder and a gas nozzle for mixing SiH_4 with F_2, backed by mechanical booster and rotary pumps. Pure SiH_4 and F_2 diluted with He down to 10 vol.% were used as the raw materials. The substrates were Corning 7059 glass plates for optical and electrical measurements and c-Si(111) wafers for the ir study. In a series of the experiments for the high temperature film growth, c-Si(111) wafers were used without any treatments as a substrate to establish the high substrate temperature. c-Si(100),(110) wafers were used for the experiments of the epitaxial growth, in which the wafers were treated by the RCA cleaning method and rinsed with diluted HF solution to remove the oxide layer on the surface just prior to the film growth. The substrate temperature, Ts was monitored by a thermocouple attached to the substrate directly.

A typical preparation condition is similar to that in the preparation of Si thin films by rf-glow discharge of silane: the reaction pressure was 550mtorr, silane flow rate 15 to 60 sccm while F_2 flow rate was fixed at 30sccm in all the experiments. The substrate position from the nozzle was one of the important parameters for the film deposition in the present technique and fixed at 40mm.

Optical absorption, Raman scattering, and X-ray diffraction studies and SEM and TEM observations were carried out to elucidate the structures in the films. Hydrogen contents in the films, C_H were estimated from the absorption at 630cm^{-1}. Electrical properties were evaluated by measuring the current-voltage characteristics in the surface type cells with aluminum electrodes. The crystallinity of the epitaxial films was investigated by RHEED and Rutherford backscattering and channeling experiments.

RESULTS AND DISCUSSIONS

In the present technique, crystal growth was induced in the selection of various preparation parameters such as the gas flow ratio(SiH_4/F_2), reaction pressure, the substrate temperature and so on. These parameters were complementary each other and we could adapt the other parameters in a wide rage to carry out the crystal growth when one parameter was fixed.

A typical preparation condition for the crystalline thin films is as follows; the gas flow ratio, SiH_4/F_2 is 30/30sccm; the pressure 550mtorr; the nozzle-substrate distance 40mm; the substrate temperature 340-390°C. In this condition the uc-Si thin films were deposited at a fast rate of 12A/sec on the glass substrate.

Fig. 1 shows a Raman spectrum in the typical film prepared in this condition. The peak at 520cm^{-1} was very sharp, characterized by the FWHM of $8cm^{-1}$ and slightly asymmetric indicating little contribution of the amorphous phase and a fluctuation of the crystalline lattice.

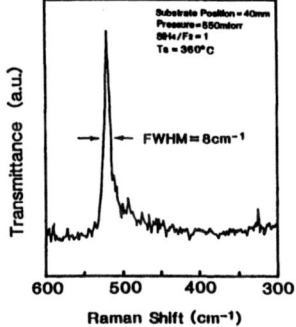

Fig.1 Raman spectrum of a typical μc-Si:H thin film by gas phase reaction of SiH₄ with F₂

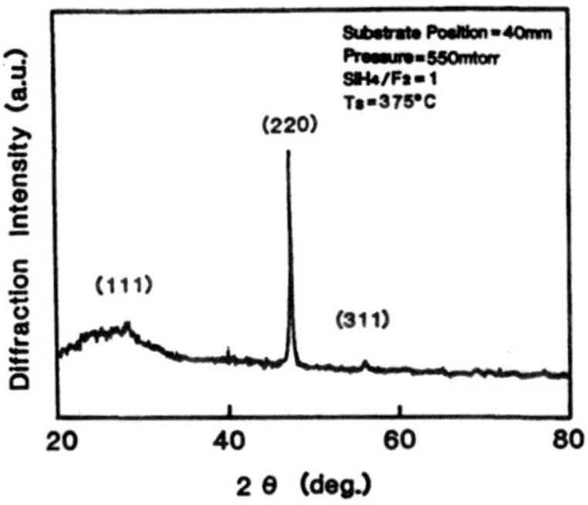

Fig.2 X-ray diffraction pattern of typical μc-Si:H film on glass substrate

The crystals exhibited a strong orientation to (110) direction as is clearly seen in Fig.2. The FWHM of the diffraction peak at 47.3° by Si(220) face were 0.3 to 0.5° and the crystalline size was estimated to be 400 to 450A by the application of Scherrer's equation.[6]

Fig.3 shows a SEM photograph of the cross section of the uc-Si thin films prepared by the typical condition. The columnar structure normal to the substrate could be observed and its diameter estimated to be 1000 to 2000A, which showed a goog agreement with the TEM observation in spite of the disagreement of the estemated crystalline size.

The optical absorption was weak in the visible region and has larger coefficient more than 10^3 in the infra-red region. The estimated band gap was 1.4 to 1.6eV.

Fig.3 SEM photograph of cross section of a typical μc-Si:H thin film

The uc-Si:H films have rather small dark conductivity of 10^{-8} to 10^{-6} S/cm characterized by the activation energy of 0.4 to 0.6eV compared with the conventional μc-Si:H films by plasma processes[7],[8] and high photoconductivity of 10^{-6} to 10^{-5} S/cm under AM-1 illumination. This results indicate lower defect density in the midgap.

Fig.4 shows an ir spectrum of the typical μc-Si film. A marked difference in the Si-H stretching modes from the conventional uc-Si:H films was seen in which the absorption peak for Si-H stretching modes is dominated by $2100cm^{-1}$ attributed to SiH₂ bonds.[7],[9]

Hydrogen content, C_H was estimated to be 2 to 4at%. In this technique the Si-network structure changed dramatically from "amorphous" to "crystalline" with a little change in the substrate temperature as shown in the Raman Spectra of Fig.5. The ir spectra of these films, however did not change at all and were dominated by 2000cm⁻¹ attributed to SiH bonds as shown in Fig.6. On other hand the hydrogen content was reduced from 13at.% to 1at.% with an increase in the substrate temperature from 250° C to 420°C and the reduction could be attributed to that inthe SiH bonds.

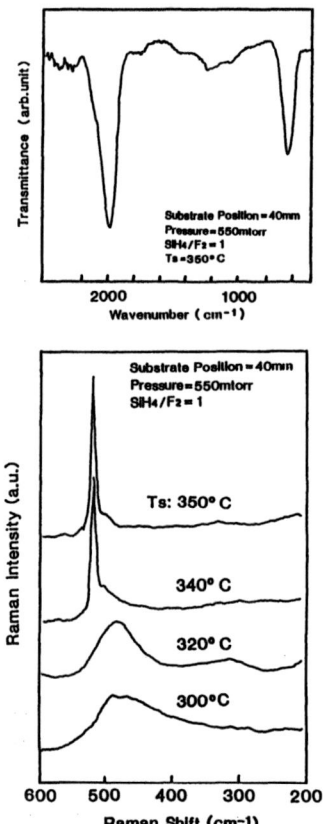

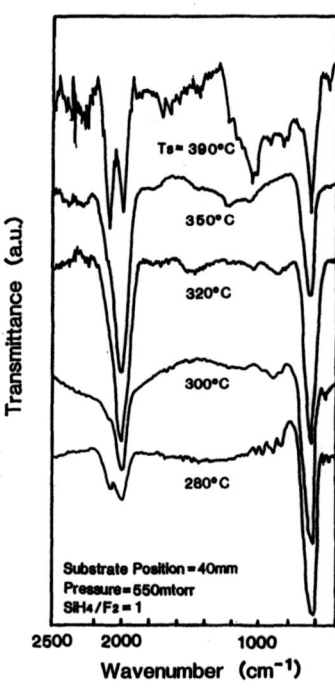

Fig.4 ir spectrum of typical μc-Si:H film by gas phase reaction of SiH₄ wuth F₂

Fig.5 Raman spectra of films prepared at different substrate temperatures

Fig.6 ir spectra of the films prepared at different substrate temperatures

As described above the film growth of the crystalline Si could be carried out in a wide range of the preparation condition. As far as the μc-Si:H thin films from silane plasma many groups have pointed out that the crystallinity was degraded with an increase in the substrate temperature in a range of >450°C and resulted in amorphous films at 500°C.[8],[10]) Then we tested a feasibility of the crystal growth at higher substrate temperature than 450°C.

Fig.7 shows Raman spectra in the films prepared at the substrate temperatures of 425°C to 505°C at excess flow ratio of silane. With an increase in the substrate temperature the crystallinity of the films was not degraded at all but much improved. It should be noted that the film prepared even at 465°C exhibited a very sharp absorption peak at 2000cm⁻¹ as shown in Fig.8, whose hydrogen content was estimated to be 1.5at.%.

As far as the lower substrate temperature we could prepare μc-Si thin films at 230°C so far at the excess flow ratio of F₂.

We applied the technique for the homoepitaxial growth of Si thin films, which is a sensitive probe for qualifying the technique itself as a reliable process for the crystalline materials.

Fig.9 are photographs of RHEED patterns in the films grown on (100) and (110) c-Si wafers under the typical condition described before. Both of the films were single crystals judging from spots and streaks in the patterns, whose deposition rates were 12A/sec and did not change at all compared with those on the glass substrate. Especially the film on (100) c-Si wafer exhibited clear Kikuchi lines and bands as seen in the photograph, indicating the good crystallinity and surface flatness.

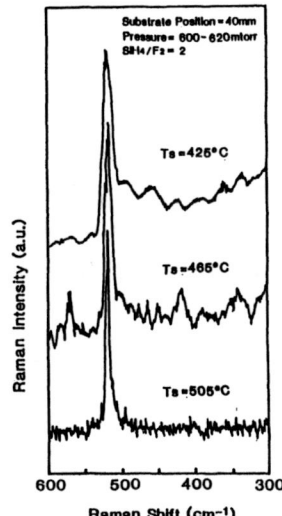

Fig.7 Raman spectra of the films prepared at high substrate temperatures

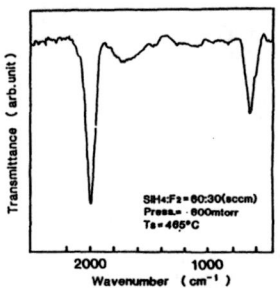

Fig.8 ir spectrum of the film prepared at high substrate temperature of 465°C

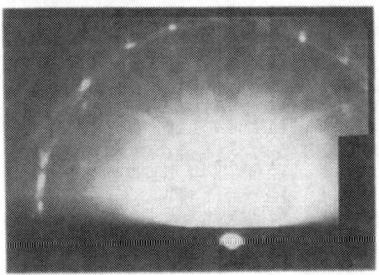
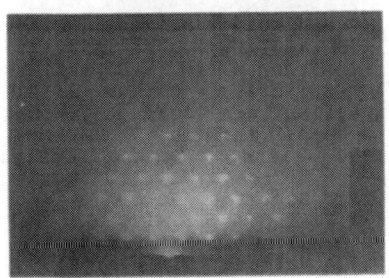

Fig.9 RHEED patterns of the epitaxial films grown on (100) and (110) c-Si wafers at 370°C. The deposition rate was 12A/sec.

The hydrogen and fluorine contents in the epitaxial film were analyzed by means of SIMS and proved to be $1 \times 10^{18} cm^{-3}$ and $6 \times 10^{17} cm^{-3}$, respectively. Further evaluation of the films was carried out by Rutherford backscattering and channeling experiments and revealed that the channeling yields, X_{min} were less than 10% and detectable defects and contaminations were not accumulated at the interface between the substrate and the epitaxial layer.

Since we have not collected the experimental data to clarify the growth mechanism in this present technique yet, we have no conclusive explanation on the mechanism. However as we reported before,[5] fluorine does contribute only to the gas phase reactions resulting in the generation of the precursors but also to the surface reactions for the film growth. We think fluorine may play an important role to establish a chemical equilibrium between the formation and dissociation of the Si-Si bonds locally in the growth zone which promotes the crystallization at such a low temperature that the thermal equilibrium can not be established in the whole Si-network structure.

CONCLUSION

We have demonstrated high potentialities and unique features of the new technique for the film growth of crystalline Si termed "Spontaneous Chemical Deposition". The technique enabled us to prepare crystalline Si thin films at the low temperature and high deposition rates over 10A/sec. The films exhibited high photoconductivity and the unique structures in terms of the Si-H bonding mode.

The technique was successfully applied to the epitaxial growth of device grade Si thin films on c-Si(100) face, keeping high deposition rate over 12 A/sec.

ACKNOWLEDGMENT

The authors thank gratefully to Dr. T. Namikawa of Tokyo Institute of Technology, Graduate at Nagatsuta for the TEM and RHEED measurements and to Central Glass Co.Ltd. for the supply of F_2 gas.

REFERENCES

1) T.Hamasaki, H.Kurata, M.Hirose and Y.Osaka:
   Appl.Phys.Lett., 37 (1980) 1084
2) S.Nishida, H.Tasaki, M.Konagai and K.Takahashi:
   J.Appl.Phys., 58 1427 (1985)
3) R.Tsu, M.Izu, S.R.Ovshinsky, F.H.Pollak:
   Solid State Commun. 36 817 (1980)
4) N.Shibata, K.Fukuda, H.Ohtoshi, J.Hanna, S.Oda
   and I.Shimizu: Jpn.J.Appl.Phys., 26 (1987) L10
5) J.Hanna, A.Kamo, T.Komiya, H.D.Nguyen, I.Shimizu and
   H.Kokado, Mat.Res.Soc.Symp.Proc., (1989) in press.
6) P.Scherrer: Gottingen Nachs., 98 (1918)
7) Y.Mishima, S.Miyazaki, M.Hirose and Y.Osaka:
   Phil.Mag., B46 (1982) 1
8) A.Matsuda: J.Non-cryst.Solids 56/60 (1983) 767
9) J.C.Knight: Jap.J.Appl.Phys., 18, Suppl. 18-1,101
10) C.C.Tsai, G.B.Anderson, R.Thompson, B.Wacker and C.Doland
    Mat.Res.Soc.Symp.Proc., (1989) in press

MICROCRYSTALLINE SILICON FILMS PRODUCED BY RF MAGNETRON
SPUTTERING AND THE EFFECT OF DIFFERENT AMBIENTS ON
THEIR CONDUCTIVITY

RATNABALI BANERJEE, A.K.BANDYOPADHYAY,S.N.SHARMA,A.K.BATABYAL & A.K.BARUA
Energy Research Unit,
Indian Association for the Cultivation of Science,
Calcutta : 700 032, India.

ABSTRACT

Results on characterisation of undoped μc-Si:H films prepared
by rf magnetron sputtering technique are presented. Highly conducting films
(10^{-3} $\Omega^{-1} cm^{-1}$) were obtained at fairly low rf power density (1.2W/cm²).
Critical parameters for obtaining microcrystalline phase were identified. The
effect of humid ambient on film properties was looked into.

INTRODUCTION

Microcrystalline silicon films have generated considerable interest
as an electronic material with high conductivity and low visible absorption
as compared to amorphous silicon films. While doped μc-Si:H films find
use as contact layer in solar cells, a possible application of the undoped
material has been indicated [1] in using μc-Si/a-Si multi-layer as the active
layer in photovoltaic devices.

There have been several reports on microcrystalline silicon films
prepared by different methods, including the sputtering technique [2-5]. This
paper reports on the characteristics of undoped μc-Si:H films produced by
rf magnetron sputtering. The critical parameters are identified. This work
also delves into the effect of ambients on this mixed phase material, necessi-
tated by the changeable relative humidity of the atmosphere.

EXPERIMENTAL

μc-Si:H films were prepared in two rf magnetron sputtering systems
(to be referred henceforth in the text as S1 & S2). In the former, a single
crystal silicon wafer (4" dia) was taken as the target and in the latter,
a hard pressed poly-Si target (8" dia) was used. The sputtering gas was
a mixture of argon and hydrogen. Parametric variations studied include the
hydrogen flow rate (R_H), the power density (P), the target voltage (V_T),the
system pressure (p), the substrate temperature (T_s) and the substrate to
target distance (1).

The films were characterised by the conductivity, optical absorption
spectra and infra-red vibrational spectra in addition to Raman spectra,
transmission electron microscopy and X-ray diffraction studies.

RESULTS AND DISCUSSION

Fig.1 shows the variation in the dark conductivity, $\tilde{\sigma_D}$,of hydrogenated silicon films with R_H. For both S1 and S2, the character of the variation is the same although there is difference in the absolute values. For S2, the parameters are yet to be fully optimised. Curve 1(S1) shows a peak value of 3.9×10^{-3} Ω^{-1} cm^{-1} obtained at a fairly low power density of 1.2W/cm². For magnetron sputtering configuration, a power density of 2.6W/cm² was reported by Saito et al [3] for the preparation of microcrystalline silicon films. The target voltage increased with increasing R_H for the same P. Increasing V_T at low R_H and decreasing it at high R_H causes a drop in σ_D as shown by dotted line in fig.1. In S2, higher system pressure (30mTorr) was utilised as compared to that in S1 (18mTorr). Curves 2 & 3 (S2) show the σ_D variation for P=1.5W/cm² and 0.9 W/cm², respectively. At the peak values of curves 1 and 2, the deposition rates were nearly the same (0.4Å/S).

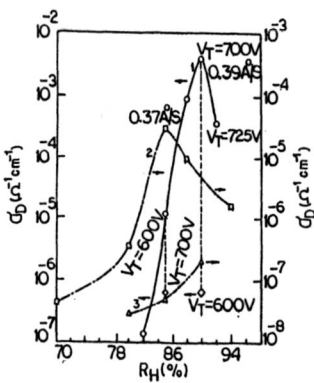

Fig.1. Variation of σ_D with R_H for hydrogenated silicon films.

Curve 1:S1,T_s=250°C,P=1.2W/cm², p=18mTorr
Curve 2:S2,T_s=220°C,P=1.5W/cm², p=30mTorr
Curve 3:S2, T_s=220°C,P=0.9W/cm², p= 30mTorr

Fig.2 shows the change in the α versus photon energy as R_H is varied from 20 to 85% (S2). Curve 1 represents an amorphous film, with $\sigma_D = 10^{-9}$ Ω^{-1} cm^{-1} and photosensitivity $\sim 10^4$. There is a marked difference in the slope between curve 2 (R_H =70%) and curve 3 (R_H =80%) which is concomitant with the increase in σ_D (fig.1). Changes in film properties for some more parametric variations were studied. A significant point is the fall of σ_D from 3.1×10^{-5} Ω^{-1} cm^{-1} to 1.7×10^{-9} Ω^{-1} cm^{-1} when 1 is decreased from 7.3 cm to 4.8cm(S2).

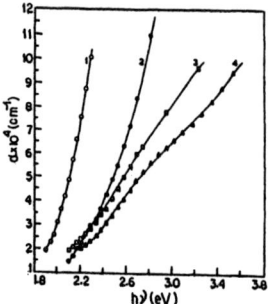

Fig.2. Variation of with photon energy,(S2).

1: R_H=20%, p=5mTorr.
2: R_H=70%, p=30mTorr.
3: R_H=80%, p=30mTorr.
4: R_H=85%, p=30mTorr.

There is also a drop in σ_D from the first mentioned value to $5.4 \times 10^{-8} \Omega^{-1}$ cm^{-1} when the power density is decreased to 0.9W/cm².

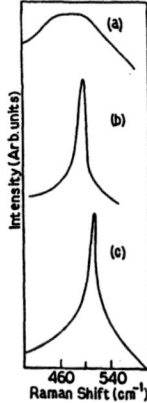

Intensity (Arb.units)

(a)

(b)

(c)

460   540
Raman Shift (cm⁻¹)

Fig.3. Raman backscattering spectra (of 5145Å excitation) of hydrogenated silicon films, (S2).

(a) R_H = 85%, 1 = 4.8cm.

(b) R_H = 80%, 1 = 7.3cm.

(c) R_H = 85%, 1 = 7.3cm.

Fig.3 shows the Raman backscattering spectra (of 5145Å excitation) of hydrogenated silicon films for some parametric changes in S2. While the shift in the Raman peak when R_H is decreased from 85% to 80% is curiously large, the total absence of a sharp peak near 520cm⁻¹ for R_H = 85% with the lowering of 1 (4.8), renders the substrate to target distance a critical parameter for obtaining microcrystalline phase. It is interesting to note that for 1=7.3cm, R_H =85%, a film produced with a power density as low as 0.9W/cm² shows presence of microcrystalline phase. A rough estimate of the volume fraction of microcrystallites using the effective medium

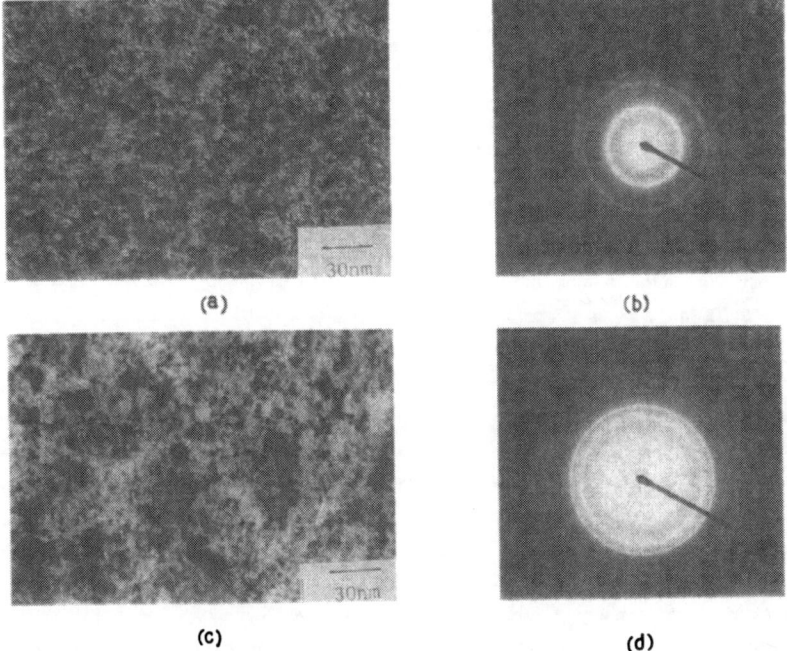

(a)   (b)

(c)   (d)

Fig.4. Bright field TEM pattern and corresponding TED pattern μc-Si:H films for variation in R_H (S1).

(a) & (b) - R_H =85%   (c) & (d) - R_H = 90%

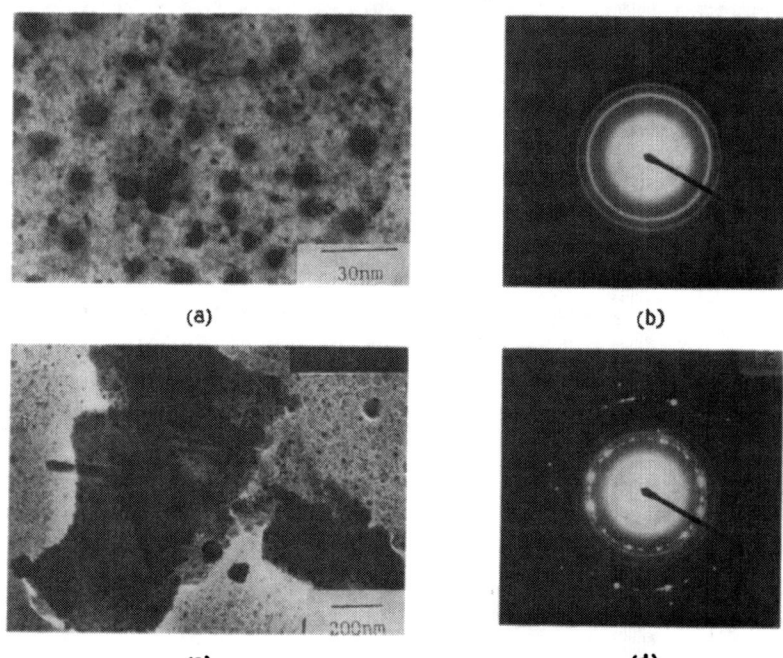

Fig.5. Bright field TEM pattern and corresponding TED pattern of μc-Si:H film (S2)

(a) & (b) - R_H = 85%

(c) & (d) - R_H = 85%, isolated large crystals.

theory [6] gives values of about 62% and 47% respectively for the peak regions in σ_D^- for curves 1 and 2 in Fig.1.

The bright field TEM image and the TED pattern for films deposited with R_H = 85% (S1) are shown in fig.4(a) and (b), respectively. The microcrystalline nature is obvious with a uniform distribution of grains of size 40-60Å. With increase in R_H, a matrix similar to the previous one is studded with larger crystals of size 100-200Å (fig.4(c)). A further increase in R_H results in a decrease in the number density of the larger crystals with concurrent decrease in σ_D^-. In S2, for R_H =85%, large grains of 100-200Å co-exist with a distribution of smaller grains, 30-60Å (fig.5(a)) which are strewed less densely however as compared to the matrix of Fig.4(c).This supports the Raman data. Some isolated, large crystalline features are found in these films (fig.5(c) & (d)). X-ray diffraction spectra show peaks corresponding to (111), (220) and (311) orientations of crystalline silicon powder.

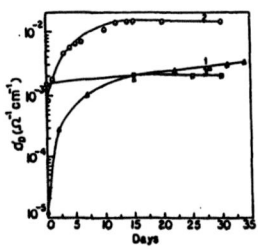

Fig.6: Change in σ_D for μc-Si:H films for variation in R_H (S1), when exposed to humid atmosphere(rel. humidity 95%).

Curve 1 : R_H = 85%

Curve 2 : R_H = 88%

Curve 3 : R_H = 90%

An interesting feature observed for films produced in both S1 and S2 is that a temporal increase shows up in the conductivity of the sample when left in the atmosphere (relative humidity 70%). The effect is stalled in vacuum and accelerated in a simulated humid atmosphere (relative humidity ~95%). Fig.6 shows the increase in σ_D of samples left in humid atmosphere. There is a concomitant increase in the refractive index. With increasing R_H, films seem to get less and less sensitive to the ambient, conductivity-wise.

In order to understand this ambient effect better, infra-red vibrational spectra were taken of representative films for the wagging and stretching modes of Si-H as well as the spectra in the wavenumber range 700-1250cm^{-1}. Whereas the changes in the former are too marginal to be commented upon, in the latter it is considerable (fig.7). Like σ_D, the effect is most drastic during the first couple of days, after which it gradually levels off. Further, the effect is not reversed on annealing in vacuum at 150°C for six hours although there is a marginal decrease in the absorption in the wavenumber range 700-925cm^{-1}. The dark conductivity value does not show resilience but rather an increase.

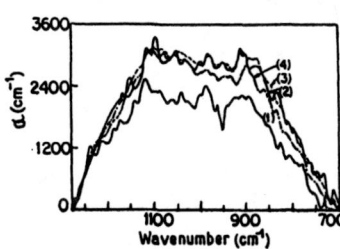

Fig.7. Infrared absorption spectra from 700 to 1250^{-1} of μc-Si:H film (S1, R_H =89%).

Curve 1 – as deposited

Curve 2 – Humid atm., 2 days.

Curve 3 – Humid atm.,12 days.

Curve 4 – Annealed at 150°C, 6hrs, vacuum, after exposure to humid ambient for 12 days.

It is difficult to pinpoint whether the increase in absorption in the wavenumber range 700-1250cm^{-1} is due to the increase in either surface-like Si-O or bulk Si-O or Si-OH or-and this is most likely-a bit of each. Dissociative chemisorption of H_2O on crystalline silicon has been reported earlier [7]. Microcrystalline silicon films often consist of defective grain boundary regions through which water molecule can traverse into the bulk of the film and on dissociation passivate some of the dangling bonds giving a less defective grain boundary region and facilitating conduction. Thus, the effect is one of absorption and unlike in the case of a-Si:H films [8], it is not reversed on annealing at 150°C. However this ambient effect does not appear to bring about any notabale changes in the volume fraction of microcrystallites. Finally, encapsulation by sputtering $\sim 1000\overset{\circ}{A}$ layer from a quartz target on the μc-Si:H layer protects it from the humid atmosphere and associated conductivity changes.

CONCLUSION

Highly conducting $(3.9 \times 10^3 \ \Omega^{-1} \ cm^{-1})$ μc-Si:H films can be produced at fairly low power density $(1.2 W/cm^2)$ with suitable adjustment of the hydrogen flow rate. The substrate to target distance is a critical parameter. μc-Si:H films with lower conductivity values show temporal increase in conductivity which gets accelerated with increasing relative humidity in the atmosphere. There is a concomitant increase in absorption in the wave number range $700-1250 cm^{-1}$. The effect is not reversed on annealing. Susceptibility to humid ambient may be an indication of the quality of network surrounding the microcrystals.

REFERENCES

1. D.Kruangam, K.Hamaki, S.Nonomara, H. Okamoto and Y. Hamakawa, Technical Digest of the International Photovoltaic Science and Engineering Conf.,Kobe,1984,Japan Convention Services,Tokyo,P437 (1984).

2. T.D.Moustakas, Tetrahedrally-Bonded Amorphous Semiconductors, Edited by David Adler and Hellmut Fritzsche (Plenum Publishing Corp.), P93 (1985).

3. N.Saito, H.Sannomiya, T.Yamaguchi and N.Tanaka, Appl. Phys. A35, 241 (1984).

4. H.Ishida, M.Noda and H.Shimizu, Jpn.J.Appl.Phys., 22, L73 (1983).

5. J.Shirafuji, H.Matsui, A.Narikawa, Y.Inuishi and N.Tanaka, Solid State Commun. 45, 577 (1983).

6. R.Landauer, J. Appl. Phys. 23, 779 (1952).

7. H.Ibach, H.Wagner and D.Bruchmann, Solid State Commun. 42, 457 (1982).

8. M.Tanielian, M.Chatani, H.Fritzsche, V.Smid and P.D.Persans, J. Non-Cryst. Solids, 35 & 36, 575 (1980).

ACKNOWLEDGEMENTS

This work was carried out under a project jointly funded by United Nations Development Program, India and Dept. of Non-Conventional Energy Sources, Govt. of India.

THE GRAPHITIZATION OF AMORPHOUS HYDROGENATED CARBON FILMS DURING THERMAL ANNEALING

SAM SHUHAN LIN,* SHUGUANG CHEN,** and DANG MO**
*University of Missouri-St. Louis, Department of Physics, St. Louis, MO 63121
**Zhongshan University, Department of Physics, Guangzhou, People's Repulic of China

ABSTRACT

Effects of thermal annealing on the properties of amorphous hydrogenated carbon films have been studied with spectroscopic ellipsometry, electrical conductivity, and infrared and Raman spectroscopy. The results show that not only the optical and electrical properties, but also the infrared and Raman spectra, change significantly after annealing above 400°C. We suggest that thermal annealing makes amorphous hydrogenated carbon films more graphite-like, and the observed changes in the microstructure cause the changes in the optical and electrical properties.

INTRODUCTION

Amorphous hydrogenated carbon (a-C:H) films have been of great interest to scientists in recent years because of their outstanding wear-resistance, hardness, chemical inertness, and electrical resistivity. However, whether or not these properties are stable and what is the mechanism for their stability are still unsolved problems[1,2]. In our last paper [3], we found that the complex index of refraction of a-C:H films changed after annealing at temperatures higher than 400°C. The present study is a continuation of our last paper.

In the present work, we use infrared and Raman spectoscopy to study the microstructure of the films, as well as ellipsometry to study the optical properties and electrical resistivity to study the electrical properties. The changes in the infrared and Raman spectra are in good agreement with the changes in the optical and electrical measurements, and in good agreement with our previous work. Based on these results, we suggest that the changes in the microstructure of these films are the cause of the changes in the optical and electrical properties of the films.

EXPERIMENTAL PROCEDURE AND RESULTS

The a-C:H films for IR and Raman measurements were deposited on high resistivity single-crystal silicon wafers, and the films for electrical resistivity measurements were deposited on quartz, using rf glow-discharge decomposition of benzene. Both films on silicon and on quartz were used in ellipsometry measurements. The deposition conditions were: rf frequency, 13.56 MHz; power density, about 0.5 W/cm²; substrate temperature, 100°C; presure, 46 Pa; feedstock gas flow rate, 0.4 SCCM. Annealing was carried out in a temperature range from 300°C to 500°C. The same sample was annealed at each temperature for an hour.

The optical properties, the complex index of refraction, the absorption coefficient, and the complex dielectric constant, can be deduced for photon energy between 2.0 and 4.0 eV from the ellipsometry data[3]. Fig. 1 shows the absorption coefficient α as a function of photon energy $h\nu$ of the same film after various annealing temperatures. We can see that α does not change much after annealing at 300°C, but it increases rapidly after annealing at 400°C and 500°C. We also observe that α follows an exponential function of $h\nu$ at low values of $h\nu$, $\alpha = \alpha_0 \exp(\Gamma h\nu)$, where Γ is the slope of the exponential absorption edge. After annealing at 500°C, α increases about one order of magnitude, but Γ decreases about 0.2 eV⁻¹. The decrease of Γ indicates the broadening of bandtail states after annealing.

The infrared absorption spectra were obtained in a 5-DX FTIR spectrophotometer with a resolution of 2 cm⁻¹. Fig. 2 shows the infrared absorption spectra of an

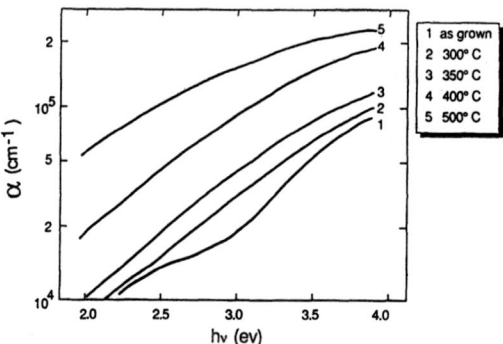

Fig. 1 Absorption coefficient α as a function of photon energy hν. Curves (1) to (5) are for as-grown film, and after annealing at 300°C, 350°C, 400°C, and 500°C, respectively.

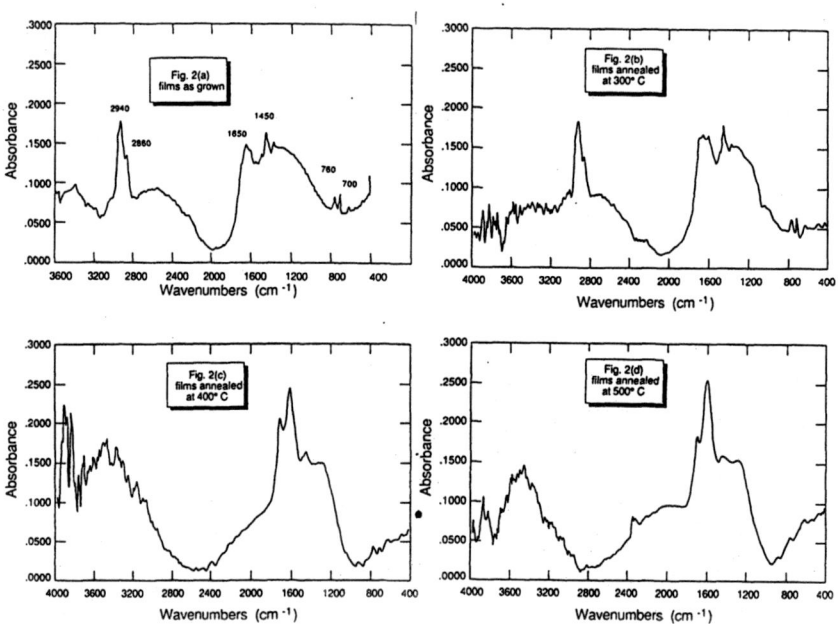

Fig. 2 Infrared absorption spectra of a-C:H films. Fig. 2(a) to 2(d) are for as grown film, and after annealing at 300°C, 400°C, and 500°C, respectively.

α-C:H film unannealed and annealed at 300ºC, 400ºC, and 500ºC. In the spectrum of the unannealed film, the observed absorption peaks are the same as reported by other workers[4,5]. Fig. 2(a) and (b) are very similar, showing that almost no change occurs after annealing at 300ºC. Fig. 2(c) is most interesting, because the absorption due to C-H stretching modes at 2850-3000 cm^{-1} dramatically decreases to an undetectable level, even though the two peaks located at 700 and 760 cm^{-1} due to C-H bending modes are still observable. On the other hand, the absorption due to C=C at 1620 cm^{-1} increases tremendously. Fig. 2(d) shows that no C-H absorption modes are detectable after annealing at 500ºC, suggesting that at this annealing temperature most of the hydrogen has effused out of the films, leaving primarily a carbon structure.

Further evidence for the graphitization of the a-C:H films due to high temperature annealing comes from the change in the Raman spectra. Fig. 3 shows the Raman spectra of the unannealed film and the same film annealed at 500ºC. There are two lines in both spectra, one located at 1360 cm^{-1} indentified as the D line, and the other located at 1590 cm^{-1} indentified as the G line, according to previous workers[6]. The peak positions do not change, but both lines become narrower due to annealing at 500ºC; the FWHM of the G line decreases from 95 cm^{-1} to 80 cm^{-1}, indicating an increase in crystallinity of the film. The integrated intensity ratio of the D line to the G line increases from 0.238 to 0.774, consistent with graphite crystalization in the annealed film.

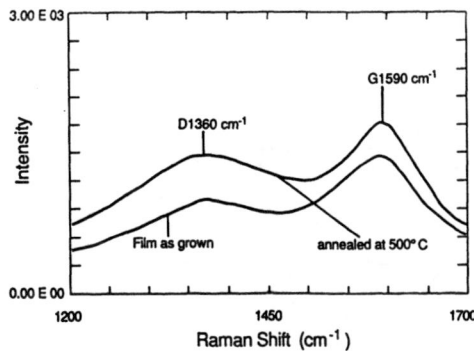

Fig. 3 Raman spectra of the a-C:H as grown film and after annealing at 500ºC.

Fig. 4 shows the DC resistivity as a function of annealing temperature. The resistivity is little changed after annealing at 300ºC, but it decreases about 5 orders in magnitude after annealing at 500ºC. This is consistent with the increased graphitization of our annealed films, because graphite has a very low resistance and more and bigger crystallites of graphite in our films will lower the resistivity of our films.

DISCUSSION

What is the atomic scale mechanism for the hydrogen effusion out of our annealed films and how do the C-H bonds change into C-C bonds forming crystallites of graphite? Two possible mechanisms are shown in Fig. 5. Both mechanisms in Fig. 5 generate evolved H_2 gas and form a carbon double-bond structure, but only mechanism (2) generates C=C symmetrical conjugate double bonds which is consistent with the strong infrared absorption peak at 1620 cm^{-1} in the annealed films. For this reason, we suggest that mechanism (2) is the way that hydrogen effuses and graphite crystallites form. It is these graphite crystallites which cause the observed changes in the optical and electrical properties of a-C:H films.

78

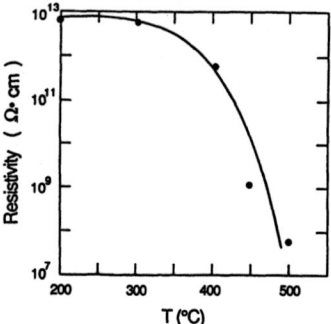

Fig. 4 The DC resistivity of a-C:H films as a function of annealing temperature.

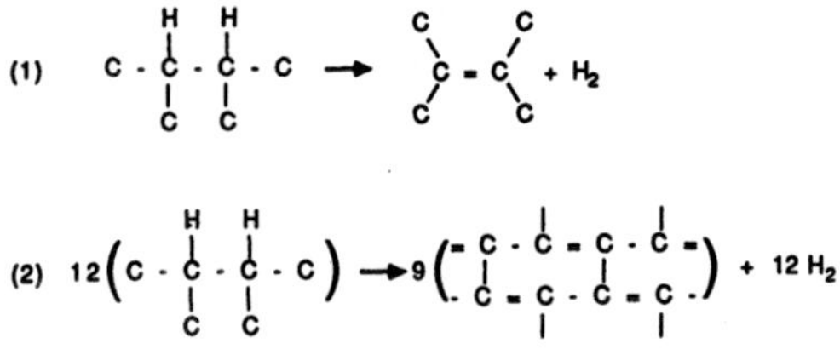

$$(1) \quad C - \overset{\overset{\displaystyle H}{|}}{\underset{\underset{\displaystyle C}{|}}{C}} - \overset{\overset{\displaystyle H}{|}}{\underset{\underset{\displaystyle C}{|}}{C}} - C \quad \longrightarrow \quad \overset{C}{\underset{C}{\diagdown}}C = C\overset{C}{\underset{C}{\diagup}} + H_2$$

$$(2) \quad 12\left(C - \overset{\overset{\displaystyle H}{|}}{\underset{\underset{\displaystyle C}{|}}{C}} - \overset{\overset{\displaystyle H}{|}}{\underset{\underset{\displaystyle C}{|}}{C}} - C \right) \longrightarrow 9\left(\begin{array}{c} - C - \overset{|}{C} = C - \overset{|}{C} - \\ - C = C - \overset{|}{C} = C - \end{array} \right) + 12\, H_2$$

Fig. 5 Two possible mechanisms for the hydrogen effusion and the formation of graphite crystallites.

ACKNOWLEDGEMENTS

We would like to thank Bernard Feldman for his valuable discussions and critical reading of the manuscript and Jin Zengsun for his Raman measurements. This work is financially supported by the CNSF.

REFERENCES

1. F.W. Smith, J. Appl. Phys. 55, 764 (1984).

2. A. Azim Khan, D. Mathine, and J.A. Woollam, Phys. Rev. B 28, 7229 (1983).

3. Shuhan Lin and Shuguang Chen, J. Mater. Res. 2 (5), 646 (1987).

4. B. Dischler, A. Bubenzer, and P. Koidl, Solid State Commun. 48, 105 (1983).

5. M.A. Tamor, C.H. Wu, R.O. Carter,III, and N.E. Lindsay, Appl. Phys. Lett. 55, 1388 (1989).

6. R.O. Dillon and John A. Woollam, Phys. Rev. B 29, 3482 (1984).

PART II

Semiconductor Compounds: Nanocrystals

ELECTRODEPOSITION OF 3-D SIZE-QUANTIZED CdS and CdSe FILMS

G. HODES*, T. ENGELHARD*, A. ALBU-YARON** and A. PETTFORD-LONG***
* Dept . of Materials Research, Weizmann Institute of Science, Rehovot, 76100 Israel
** ARO , The Volcani Center, P.O.B. 4, Bet-Dagan 50-250, Israel.
***Dept. of Metallurgy and Science of Materials, University of Oxford, OX1 3PH, UK.

ABSTRACT

CdS and CdSe films electrodeposited from chalcogen-containing non-aqueous electrolytes are shown to possess a nanocrystalline structure and exhibit 3-D quantum size effects. A large variation in optical transmission spectra, due for the most part to this size-quantization, is shown. Photoelectrochemical activity of the as-deposited films is discussed in terms of their nano-structure. Extension of the crystallite size effects to a metal-insulator transition in MoO_2 deposited from a similar electrolyte is mentioned.

INTRODUCTION

Since its introduction by Baranski and Fawcett [1], electrodeposition of CdS (and, to a lesser extent, other metal chalcogenides) from non-aqueous electrolytes containing dissolved elemental S (or Se) has attracted interest as an alternative to the more established method using aqueous electrolytes containing the chalcogen in a high oxidation state. In particular, the films tend to have a smoother morphology which renders them potentially better for photovoltaic purposes [2,3,4]. For the most part, however, these films have shown poorer photovoltaic properties than those deposited from aqueous solutions.

Baranski et al. have carried out an extensive characterization of these films [2,3]. They note that the one important structural property which they were unable to measure was grain size [3]. They also note the discrepancy between the estimated band gap (Eg) for their CdSe films deposited from DMSO and those deposited from aqueous solutions (1.94 and 1.74 eV respectively, the latter corresponding to the normal bulk CdSe band gap), as well as the increase in Eg for their CdS_xSe_{1-x} films compared to other reported values [4]. They suggest that the increased values of Eg result from structural differences between their films and other reported films.

This increase in Eg for CdS, and some other metal chalcogenide semiconductors, electrodeposited from non-aqueous electrolytes containing either elemental or zero-valent chalcogen, has been reported, either explicilty or implicitly, in a number of other papers [3,5-8]. It has either been ascribed to experimental error, or not discussed further.

Recently, we showed that chemically deposited CdSe films consisted of nanocrystalline grains (ca. 5-7 nm dimensions) which resulted in 3-dimensional quantization in the individual nanocrystals making up the film and an increase in the effective Eg of the CdSe [9]. This 3-D quantization was subsequently found by us to occur in chemically deposited PbSe and also in CdS and CdSe which was electrodeposited from non-aqueous electrolytes containing elemental S or Se, in all cases resulting in an increased value of effective Eg compared with bulk material [10].

In this paper, we present more detailed results on CdS and CdSe electrodeposited from various non-aqueous electrolytes with emphasis on optical and microstructural properties and further evidence for size quantization in some of these films. The photoelectrochemical (PEC) activity of the different layers is briefly compared. We also present some preliminary results on MoO_2 electrodeposited from a similar electrolyte, which suggests that this method may be rather general for preparing nanocrystalline films.

EXPERIMENTAL

CdS and CdSe films were cathodically electrodeposited on SnO_2 conducting glass from 4 different electrolytes: $DMSO/CdCl_2$, $DMSO/Cd(ClO_4)_2$, $DEG/CdCl_2$ and $DEG/Cd(ClO_4)_2$ (DMSO - dimethyl sulfoxide; DEG - diethylene glycol). Typical concentrations were: CdX - 50 mM; S - 100 mM; Se - saturated ($\approx$10 mM). The anode was graphite (we found appreciable traces of Pt in the films in many cases when a Pt anode was used). The quoted thickness is approximate (±25%) and measured by weight and/or quantity of electricity passed during deposition. For TEM measurements, the films were removed from the substrate with dilute HCl. Photoelectrochemical (PEC) measurements were carried out in polysulfide electrolyte (typically 0.4M Na_2S, 0.2M S) with illumination equivalent to approximately AM1. For measurement of Eg, see text.

RESULTS AND DISCUSSION

We will treat the CdS and CdSe separately.

CdS

Many of the important details of the 4 different set-ups are given in tabular form in table 1. For the most part, these numbers are averaged over many samples. They should be considered, however, as typical rather than absolute - particularly for the PEC response, where a considerable scatter of results was obtained.

Table I - CdS

	DMSO		DEG	
	$CdCl_2$	$Cd(ClO_4)_2$	$CdCl_2$	$Cd(ClO_4)_2$
Eg shift[1] low T[2]	0.2 eV	< 0.1 eV	0.2 eV	<0.1eV
high T[3]	0.15 eV	0	<0.1 eV	-0.1eV
Crystal size (nm)[4]	4 - 7	-	4 - 7	7 - 15
Effect of light on plating	small	none	none	large
PEC response[5]	V_{oc}[6] I_{sc}[7]	V_{oc} I_{sc}	V_{oc} I_{sc}	V_{oc} I_{sc}
low T[2]	30 30	300 100	200 20	300 15
high T[3]	300 50	very variable	350 20	very variable

(1) See text.
(2) (T = temperature), ca. 100-120°C.
(3) ca. 140-160°C.
(4) Typical dimensions; some crystals outside these ranges can always be seen.
(5) In polysulfide electrolyte - averaged over typically 5-6 samples.
(6) mV.
(7) mA/cm^2.

The two most important parameters are effective band gap (Eg) and crystallite size measured from TEM. While the transmission spectra could often be fit to a direct Eg $\{(\alpha h\upsilon)^2 \alpha (h\upsilon)\}$, in many cases, such a fit could not be found. In theory, the spectrum of a monodisperse collection of size-quantized semiconductor crystallites should consist of sharp absorption peaks corresponding to discrete transitions. For a mixture of different crystallite sizes, the spectrum will be broadened with a shape dependent on the concentration of different crystal sizes. For this reason, we measure the effective Eg as the value of wavelength (in eV) corresponding to the point 2/3 up the steep part of the spectra; for samples which do give a good fit to a direct Eg, the value of Eg is approximately at this point.

For the two $CdCl_2$-containing elecrolytes at low deposition temperatures, a shift of ca. 0.2 eV to higher values of Eg is found (as we reported previously [10]). According to simple theory, this increase in Eg should occur for a a crystallite size of ca. 5 nm [10], which is approximately the size measured by TEM (Fig.1a). At high deposition temperatures, this shift becomes less, as expected, since crystal size increases with increasing temperature.

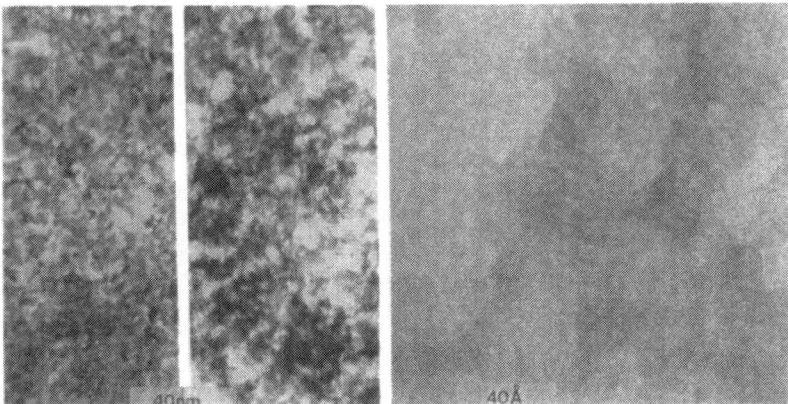

Figure 1 *(a)* *(b)* *(c)*
TEM micrographs of CdS deposited films: (a) and (c) DMSO/CdCl₂ (low temp.)
(b) DEG/Cd(ClO₄)₂ (low temp).

The two $Cd(ClO_4)_2$-containing electrolytes behave differently. Other differences between the Cl^- and ClO_4^--containing electrolytes - mainly structural ones measured by SEM and XRD as well as electronic data were described previously [3]. The shifts in E_g are considerably less or negligible, and in the $Cd(ClO_4)_2/DEG$ electrolyte at high temperature, orange films with an apparent E_g lower than that of bulk CdS by ca. 0.1 eV were obtained. It is probable that this is due not to a lowering of E_g, but rather to an impurity level or band. The smaller shifts in E_g for the ClO_4-containing electrolytes are consistent with a somewhat larger crystal size for films deposited from these solutions (Fig. 1b). At 7-8 nm, the shift in E_g already becomes negligible from simple particle-in-a-box theory including coulomb effect [10]. Also, appreciable amounts of CdO were found in the $DEG/Cd(ClO_4)_2$ films by electron diffraction. Fig. 1c shows a high resolution TEM micrograph of CdS deposited from $DMSO/CdCl_2$ which exhibits clearly the lattice planes of the individual crystallites. This shows the nanocrystalline structure of the films, the random orientation of the crystallites (ref. [3] finds, from XRD, oriented films from this electrolyte, although at somewhat higher deposition temperature), and also the crystalline perfection (lack of dislocations or other physical defects) of the individual crystallites. The crystal structure of the CdS from this (and all the other) electrolytes was found to be hexagonal from electron diffraction measurements.

The changes in transmission spectra of the various CdS layers are, to a large extent, summed up in Fig. 2. In Fig. 2a, the dotted line shows the spectrum of an annealed (400°C 1 hr in Ar) film deposited from $DMSO/CdCl_2$. It is shown both as a reference for 'normal' CdS (it gives a good fit to a direct transition of 2.44 eV) and also to show that annealing these films, thereby increasing effective crystal size, destroys the size quantization effect. It is noteworthy that the shifts in spectra to the red occur gradually with increasing annealing temperature consistent with a gradual increase in effective crystal size, as was shown previously for chemically deposited CdSe [9]. The solid line in Fig. 2, for a low-temperature $DEG/Cd(ClO_4)_2$ set-up, shows a relatively small blue shift. The broken line shows the considerably larger shift for a Cl^--containing electrolyte. Again, this spectrum shows two facts: the blue-shift for Cl^--containing electrolytes (this spectrum is typical of $CdCl_2$-containing electrolyte) and also the large blue shift is caused by Cl^- ions, since there is no apparent difference between this spectrum $(Cd(ClO_4) + LiCl)$ and one from a $CdCl_2$ electrolyte. The effect of the Cl^- compared with the ClO_4^- ion has been discussed previously [11] in terms of increased adsorption of Cl^- on the cathode or complexation of Cd^{2+} by Cl^-

ions. We find that using CdI$_2$ in place of CdCl$_2$ gives similar results to CdCl$_2$ as might be expected.

    Fig. 2b shows spectra from the DEG/Cd(ClO$_4$)$_2$ set-up. This electrolyte provides the greatest variety of layers. The red shift in the spectra with increasing temperature can be clearly seen. Also notable, and apparent with films deposited from this electrolyte, (and, to a lesser extent, from DMSO/Cd(ClO$_4$)$_2$ electrolyte) is the strong red shift with increasing thickness. This is a real shift, and not just due to increased absorbance by thicker layers. It suggests an inhomogeneity of the film through its thickness - either structural or chemical. Unfortunately, the thick high-temperature films, which are clearly orange in color, cannot be analysed in plan view by TEM due to their thickness (in principle, cross-sectional TEM is possible). Impurities (or excess Cd) incorporated in the CdS could in principle cause the red shift due to an impurity band in the band gap. This red shift is most apparent with a fresh electrolyte and gradually decreases as electrolysis proceeds. This suggests removal of impurities as electrolysis proceeds (as in pre-electrolysis) if the red shift is assumed to be caused by impurities.

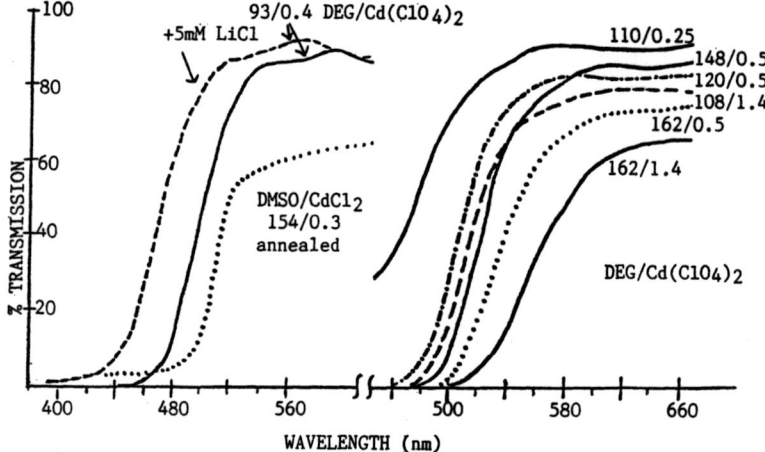

Figure 2   (a)   (b)
Transmission spectra of CdS deposited from various electrolytes. The numbers before and after the oblique lines show temperature of deposition ($^\circ$C) and film thickness (μm) respectively.

    We note briefly the effect of illumination of the CdS during plating as shown by a decrease in polarization of the cathode upon illumination. This gives an indication of photoconductivity of the films, since their resistance increases with thickness (we find all of the films to be highly resistive, although we did not measure this in a systematic way). Photoconductivity of these films has been discussed previously [3].

    Finally, the PEC reponse, in particular short circuit current (Isc), is of interest. The very high resistance of the films precludes appreciable solid state photovoltaic activity of the as-deposited films (annealing lowers the resistance). Because of the intimate contact between individual crystallites and liquid, however, appreciable PEC activity can be obtained as we showed for chemically-deposited (CD) CdSe films [9]. The activity of the electrodeposited CdS (and CdSe) films is much inferior to that of CD CdSe, but, in most cases, still appreciable, and may be useful to study mechanisms of charge transport in the films or (since the films can be considered as an aggregate of immobilized colloidal particles) to study photo(electro)chemical catalysis and photoconversion reactions.

CdSe

    We will discuss CdSe much more briefly than was the case for CdS, pointing out important similarities and differences with the CdS films.

As for CdS, the CdSe films, particularly those deposited from $CdCl_2$ electrolytes, show blue shifts of the spectra (table II, Fig. 3) and nanocrystallite structure (Fig. 4). The dark field TEM mode (Fig. 4b) shows more clearly the crystallite size, which is rather dispersed, but with a large population in the 5-7 nm region. The larger sizes, which probably dominate the spectra, are equivalent to an effective Eg of 1.9 eV [10], in general agreement with the measured spectra. The shapes of the spectra differ from those of CdS. They are, in general, less steep and rather variable. They do not fit to a direct gap, but closer to $(\alpha h\nu) \propto (h\nu)$. The less steep and variable shape is consistent with the relatively large scatter of crystallite sizes seen in these films.

Table II - CdSe

		DMSO		DEG	
		$CdCl_2$	$Cd(ClO_4)_2$	$CdCl_2$	$Cd(ClO_4)_2$
Eg shift	low T	0.2 eV	<0.1 eV	does not plate well	
	high T	0.1 eV	0.05 eV	0.2 eV	-0.1 eV
PEC response		V_{oc} $\quad$ I_{sc}	V_{oc} $\quad$ I_{sc}	V_{oc} $\quad$ I_{sc}	V_{oc} $\quad$ I_{sc}
	low T	150 $\quad$ 70	100 $\quad$ ~50	-	-
	high T	150 $\quad$ 100	170 $\quad$ 20	220 $\quad$ 150	350 $\quad$ 150

Explanation of terms as in Table I.

The films deposited from $Cd(ClO_4)_2$ solutions show, like CdS, smaller shifts or even, as for the high temperature DEG/$Cd(ClO_4)_2$ CdS system, red shifts of the spectra compared with normal CdSe (Fig. 3). As for CdS, the crystal structure of the films is predominantly hexagonal CdSe, although some cubic CdSe is sometimes seen. The large variety of spectra obtained is clearly seen from Fig. 3.

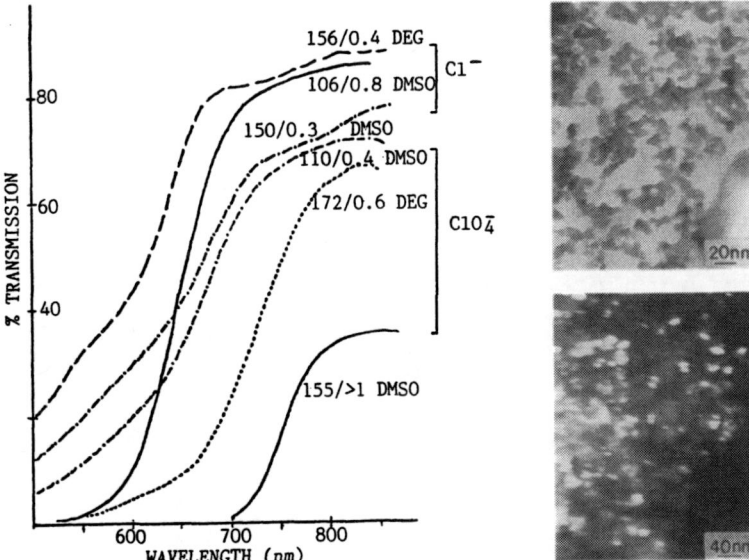

Figure 3
Transmission spectra of CdSe deposited from various electrolytes. Deposition temperature and film thickness shown as in Fig. 2.

Figure 4
Bright field TEM (top, 4a) and dark field (bottom, 4b) of CdSe deposited from low temperature DMSO/$CdCl_2$ electrolyte.

The PEC response is also given in table II. The overall response is similar to that of CdS, taking into account the lower Eg and greater light absorption of CdSe.

We mention, in conclusion, that unsuccessful attempts to electrodeposit MoS_2 from similar electrolytes containing ammonium molybdate and S gave brown, transparent layers indentified by electron diffraction, as MoO_2. TEM showed the films to be made up of nanocrystalline grains. The very high resistance of these films (>20MΩ) contrasts with the normal metallic conductivity of MoO_2, suggesting a size-induced metal-insulator transition. This shows the potential versatility of this method for obtaining nanocrystalline materials.

REFERENCES

1) A.S. Baranski and W.R. Fawcett, *J. Electrochem. Soc.*, **127**, 766 (1980).

2) A.S. Baranski, W.R. Fawcett, A.C. McDonald, R.M. de Nobriga and J.R. MacDonald, *J. Electrochem. Soc.*, **12A**, 963 (1981).

3) A.S. Baranski, M.S. Bennett and W.R. Fawcett, *J. Appl. Phys.*, **54**, 6390 (1983).

4) A.S. Baranski, W.R. Fawcett, K. Gatner, A.C. McDonald, J.R. MacDonald and M. Selen, *J. Electrochem. Soc.*, **130**, 579 (1983).

5) E. Fatas, P. Herrasti, T. Garcia, F. Arjona and E. Garcia Camarero, *Maters. Chem. and Phys.* **13**, 497 (1985).

6) S. Preusser and M. Cocivera, *Solar Energy Mats*, **15**, 175 (1987).

7) B.W. Sanders and M. Cocivera, *J. Electrochem. Soc.*, **134**, 1075 (1987).

8) K. Mishra, K. Rajeshwar, A. Weiss, M. Murley, R.D. Engelken, M. Slayton and H.E. McCloud, *J. Electrochem. Soc.*, **136**, 1915 (1989).

9) G. Hodes, A. Albu-Yaron, F. Decker and P. Motisuke, *Phys. Rev. B.*, **36**, 421 (1987).

1 0) G. Hodes and A. Albu-Yaron, *Proc. Electrochem. Soc.*, **88-14**, 298 (1988).

1 1) A.S. Baranski and W.R. Fawcett, *J. Electrochem. Soc.*, **131**, 2509 (1984).

STRUCTURAL AND ELECTRICAL CHARACTERIZATION OF CdSe THIN FILMS

Miltiadis K. Hatalis*, Fuyu Lin* and Michael R. Westcott**
*Lehigh University, Dept. of Computer Science and Electrical Engineering, Bethlehem, PA 18015
**Litton Systems Canada Ltd., Display Systems Engineering, Rexdale, Ontario, M9W-5A7, Canada

ABSTRACT

The structural and electrical properties of thin films of undoped and indium-doped cadmium selenide deposited on glass substrates have been investigated. The as-deposited films were found to be microcrystalline with grain size less than 10 nm. Grain growth occurred upon annealing. Enhanced grain growth was observed in the indium doped films. Transmission electron microscopy of in-situ annealed films revealed the formation of large single crystal areas having cubic structure with <111> as the dominant orientation. The resistivity and the effective electron mobility of polycrystalline cadmium selenide films were investigated as function of annealing conditions and device channel length. Reduction of the electrical resistivity and increase of the electron mobility was observed in devices with channel lengths less than 25 µm.

INTRODUCTION

Polycrystalline films of CdSe are receiving attention for a wide variety of applications such as flat panel displays, photovoltaics and photoelectrochemistry. Two major advantages of this material are the relatively high electron mobility and the low process temperatures which are required to fabricate electronic devices in CdSe. As a result good performance devices can be fabricated on low cost substrates. An important emerging field is that of flat panel displays. Thin film transistors fabricated in CdSe can be used in active matrix liquid crystal displays and prototype systems have already been reported [1].

The structure of CdSe thin films has been found to be hexagonal (wurtzite), cubic (sphalerite), or a mixture of hexagonal and cubic. Structural studies of this material have shown that as-deposited material has a fine grain size and that grain growth occurs during annealing [2],[3]. It has been reported that the performance of TFTs fabricated in indium-doped CdSe is superior to that of devices in undoped films [4]. The structural and electrical properties of indium doped CdSe films have not yet been well studied. The purpose of this study was to investigate the structural and electrical properties of undoped and indium-doped CdSe thin films.

EXPERIMENTAL PROCEDURE

Two different sets of samples were prepared. The first set consisted of the thin film sequence SiO_2-CdSe-SiO_2 which was deposited onto Corning 7059 glass substrates. The thickness of the base and cover oxides were 100 nm and 40 nm respectively and both oxides were deposited by e-beam evaporation. The CdSe films were deposited by thermal evaporation from a 0.5 at. % indium - doped cadmium selenide single crystal. All three films (base-SiO_2, CdSe, cover-SiO_2) were deposited in a single pump down operation. Undoped CdSe films were also deposited for comparison purposes. The undoped films were deposited either by thermal evaporation or by sputtering. The base oxide for the undoped films was deposited by sputtering. The thickness of all CdSe films was 25 nm. All samples from this set were annealed in a nitrogen ambient, unless otherwise noted.

The second set of samples was prepared in a similar way except that a chromium layer was deposited, and patterned into long parallel strips, prior to the deposition of CdSe films. This set was used for determining the resistivity of CdSe films. The electrodes were each 250 μm wide and 11.5 cm long with separations in the range of 5 μm to 400 μm. This set of samples was annealed either in nitrogen or in a simulated air ambient (20% O_2 and 80% N_2). A modified version of this set had an additional top or bottom metal gate structure for measuring the effective electron mobility.

Structural characterization of the CdSe thin films was performed by Transmission Electron Microscopy (TEM). TEM samples were prepared by immersing the glass substrates in a 10% HF in H_2O solution. The effect of in-situ annealing was also investigated. Electrical measurements were performed on the samples with the metal strips, in order to determine the film resistivity and the effective electron mobility. Thin film transistor structures having a bottom gate structure were also used for mobility measurements.

STRUCTURAL CHARACTERISTICS OF CdSe THIN FILMS

The TEM studies on as-deposited CdSe revealed that the 25 nm thick films are microcrystalline with average grain size less than 10 nm. Figure 1 shows a TEM micrograph of as-deposited indium doped film. Distinctively different grain growth has been observed in the In-doped and undoped CdSe films.

Grain growth in the undoped CdSe was found to slow down when the grain size became equal to the film thickness. This was true for both thermally evaporated and sputter deposited undoped films. An average grain size ~25 nm was observed after annealing the thermally evaporated film at 350 °C for 2 hours as can be seen in figure 2. Similar grain size was observed in the sputter deposited film after annealing at 400 °C for 2 hours. It was only after a prolonged annealing, 10 h. at 400°C in air, that a grain size greater than the film thickness was observed. Our results are in agreement with those reported in ref. [2]. It has been found that when undoped CdSe films are annealed in the temperature range 350 °C to 400°C, grain growth occurs until the grain size becomes approximately equal to the film thickness. A strong driving force for grain growth is the lowering of the free energy of the polycrystalline film. It has been proposed that the as-deposited undoped CdSe films contain high angle boundaries which provide the driving force for grain growth. Grain growth, however, reduces substantially when the grain size becomes approximately equal to the film thickness. It is assumed that most of the boundaries are then low angle ones [2].

In contrast to the results of the undoped films, we observed that the grain growth of the indium doped CdSe did not stop when the grain size reached the film thickness. After annealing for 2 hours at 350 °C the average grain size

Fig. 1 TEM micrograph of an as-deposited 25 nm thick indium doped CdSe film.

Fig. 2 TEM micrograph of a 25 nm thick undoped CdSe film after annealing at 350°C for 2 hours.

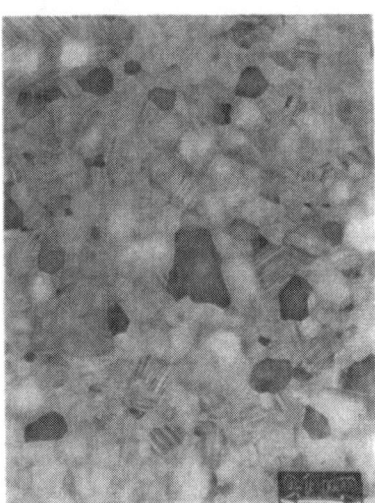

Fig. 3 TEM micrograph of a 25 nm thick In-doped CdSe film after annealing at 350°C for 2 hours.

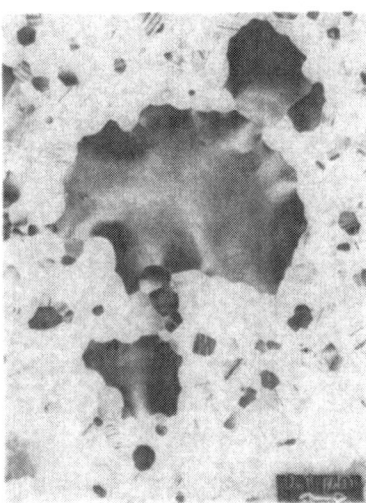

Fig. 4 TEM micrograph of a 25 nm thick In-doped CdSe film after annealing at 400°C for 4 hours.

became ~50 nm which is twice the thickness of the film. Figure 3 shows a TEM micrograph of an In-doped sample after annealing. We have found that grain growth can occur even at 200 °C. After annealing an as-deposited doped film for 4 hours at 200°C, we observed a ~50% increase in the grain size. It was also observed in the doped films, with the Cr lines present during annealing, that grains as large as one micron were present after annealing at 400°C for 4 hours. These grains were surrounded with grains having grain size less than 0.1 μm, as can be seen in figure 4. It is clear from our results that indium enhances grain growth. It is known that chemical inhomogeneities at the grain boundaries can provide a strong boundary force for grain growth [5]. Segregation of indium at the grain boundaries can, thus, be a possible cause for the observed enhanced grain growth. The role of Cr contacts on the grain growth is at present under investigation.

A series of in-situ annealing experiments were also performed on both doped and undoped samples. Two types of grain growth processes were observed. A normal grain growth that resulted in uniform grain size and an abnormal grain growth that resulted in isolated large grains, similar to the one shown in figure 4. Upon further in-situ annealing at 400°C the large grains continued to grow until they consumed all the small grains in between them. The result of this growth was the formation of large area almost single crystal material. The main defects in these areas were low angle boundaries and dislocations, as can be seen in figure 5. Electron diffraction has shown that these areas have cubic structure with <111> orientation. This abnormal growth is attributed to surface energy driven recrystallization (SEDR), where the driving force is the reduction of surface free energy. When SEDR occurs the lowest energy planes are exposed [6]. In a cubic structure these are the {111} planes [6], in agreement with our observations.

ELECTRICAL RESULTS

The two point probe method was used to determine the resistivity of the CdSe films. Low contact resistance was achieved between the Cr metal strips and the indium-doped CdSe films, on both as-deposited and annealed films. In contrast high contact resistance was present in the undoped samples.

Figure 6 shows the trend in resistivity for the In-doped CdSe films as a function of electrode spacing. Only the temperature of the anneal is indicated since no systematic dependence on anneal time was found in the time range 0.5 hours to 4 hours nor on the use of nitrogen or simulated air (20% O_2, 80% N_2) as an annealing atmosphere. For channels longer than 25 μm, the resistivity and the ratio of light to dark resistivities increased following anneals at 300°C and 400°C; increasing mostly for the higher temperature anneal. Resistivities were observed to decrease with shorter channel lengths.

The increase in resistivity of the long channel devices is likely due to electron capture by oxygen diffused into the grain boundaries during anneal, generating a depletion region in the CdSe grains and establishing a barrier to carrier movement. This effect presumably overrides any increase in mobility resulting from grain growth. The reduction in resistivity in short channel devices could be due to increase in mobility due to a network of very large grains developed in the short channel. Alternatively this could result from diffusion of contact material

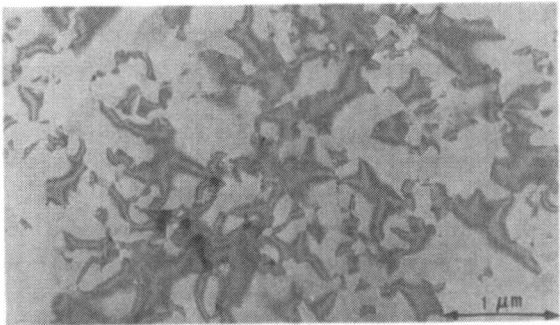

Fig. 5 TEM micrograph of a 25 nm thick indium doped CdSe film after an in-situ anneal for 75 min at 325°C followed by 30 min at 400°C.

which can either increase the doping level or could react with oxygen present at grain boundaries and thus decrease the intergrain barrier height.

The effective electron mobility was found to depend upon the channel length. For channels longer than 25 μm, the measured mobilities were much lower than 1 cm^2/V-sec. In the case of shorter channels, mobilities increased considerably following anneal to typically 20 cm^2/V-sec. Figure 7 shows the output characteristic of a thin film transistor fabricated in a 25 nm thick indium-doped CdSe. The channel of this device was 20 μm long and 15 μm wide.

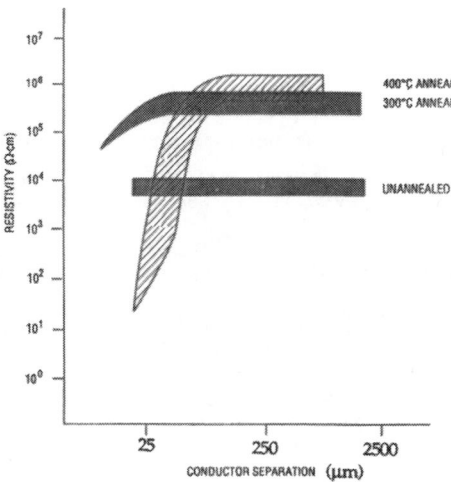

Fig. 6 Resistivity of 25 nm thick indium doped CdSe film as a function of electrode spacing.

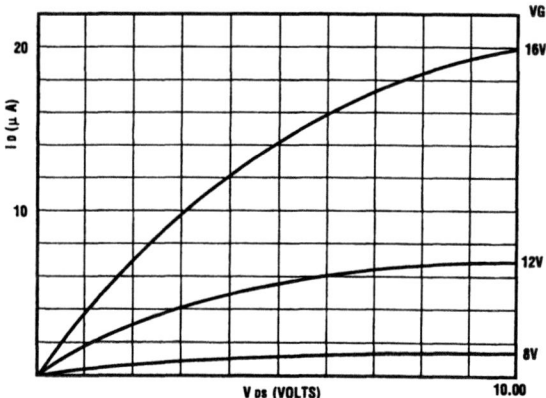

Fig. 7 Characteristics of TFT fabricated in 25 nm thick indium doped CdSe film having L=20 μm and W=15 μm.

SUMMARY

We have investigated with TEM the grain size of undoped and indium doped CdSe films. The grain growth in undoped films during annealing slows down when the average grain size becomes approximately equal to the film thickness. The doping of CdSe films with indium enhances the grain growth during annealing. The average grain size of the doped films can be larger than the film thickness. Abnormal grain growth was found to result in grains with sizes much larger (>20) than the film thickness. TEM observations of in-situ annealed films have shown that large almost single crystal areas can be formed having a cubic structure with a <111> predominant orientation.

The resistivity and the effective electron mobility of indium doped CdSe thin films were characterized. The resistivity was found to decrease and the electron mobility to increase in devices with channel lengths less than 25 μm. An electron mobility of 20 cm²/V-sec was measured in TFTs having 20 μm long channels.

REFERENCES

1. F.-C. Luo, in <u>Comparison of Thin Film Transistor and SOI Technologies</u>, edited by H.W. Lam and M.J. Thomson (Mater. Res. Soc. Proc. <u>33</u>, Pittsburgh, PA 1984) pp. 239-245.
2. K. H. Norian, Thin Solid Films, <u>47</u>, 195 (1977).
3. M. K. Hatalis and M. R. Westcott, presented at the 1989 Electrochem. Soc. Spring Meeting, Los Angeles, CA, 1989, in Extended Abstracts, <u>89-1</u>, pp. 377-378.
4. A. Van Calster, A. Vervaet, I. De Rycke, J. De Baets and J. Vanfleteren, J. Crystal Growth, <u>86</u>, 924 (1988).
5. C.R.M. Grovenor and D.A. Smith, in <u>Thin Films and Interfaces II</u>, edited by J.E.E Baglin, D.R. Campbell and W.K. Chu (Mat. Res. Soc. Symp. Proc., <u>25</u>, Pittsburgh, PA 1984) pp. 305-310.
6. C. G. Dunn and J. L. Walter, <u>Recrystallization, Grain Growth and Textures</u>, (American Society for Metals, Ohio, 1965) p. 461.

A STUDY OF THE PRESSURE–INDUCED PHASE TRANSITION IN BULK AND NANOCRYSTALLINE CADMIUM SULFIDE

XUE–SHU ZHAO, JOHN SCHROEDER, PETER D. PERSANS, AND ENLIAN LU
Department of Physics, Rensselaer Polytechnic Institute, Troy, NY 12180–3590

ABSTRACT

We have used Resonant Raman scattering induced by pressure tuning to study the phase transition and electronic states of bulk and 60±20 Å colloidal microcrystallite CdS. The experimental results show that bulk CdS undergoes a well–defined first order phase transition at 27 kbar and that the intensity of the Raman scattering increases sharply when the level of the intermediate state (bound exciton I_2) is close to the photon energy. After the phase transition no Raman scattering and photoluminescence can be observed. However, the phase transition in the colloidal CdS is quite different from the bulk CdS and the complete phase transition occurs above 60 kbar. Both bulk and colloidal CdS reverse to the original wurtzite phase after releasing the pressure.

INTRODUCTION

According to a group theoretical analysis first order Raman scattering is forbidden in the rock salt phase of CdS which has an inversion symmetry center. Brafman [1] and Briggs et al. [2] studied the pressure dependence of LO phonon spectra in CdS and found that up to the phase transition LO phonon Raman scattering can be observed, whereas beyond the phase transition all Raman spectra disappear. However, recently Venkateswaren et. al. [3] have observed that the intensity of LO phonon first order Raman scattering changes very little soon after the phase transition and steadily decreases by factor of three between 30 kbar and 42 kbar both in heavily doped and pure CdS, this provides evidence for a mixed wurtzite–rock salt structure at least up to 42 kbar.

In recent years semiconductor microcrystallites with diameters of 20–200 Å have been actively investigated to understand the size dependence of their electronic and phonon structure.[4,5] Potential applications include electronic charge storage and non–linear optical devices. When the microcrystallite particle diameter is comparable to or smaller than the bulk free exciton, the quantum size effect should transform the continuous bands of the bulk crystal into a series of discrete states. There should be two major effects. Firstly, quantum confinement will increase the kinetic energies of electrons and holes this blue shifts the energy gap. Secondly, confinement increases in the attractive potential between electrons and holes, which red shifts the energy gap. In small particles the surface to volume ratio increases, so surface reconstruction and surface tension will have important effects on the electronic, phonon and phase transition properties. Theoretical work [6,7] indicates that the effect of finite size is to broaden the transition region for first and second order phase transitions. Recent high pressure Raman scattering studies on colloidally–prepared CdS microcrystallites suggest that a complete phase transition from wurtzite to rock salt is similar to that observed by Venkateswaren, et. al. in bulk CdS based on the observation of the disappearance of the Raman peak.[4]

EXPERIMENTAL RESULTS AND DISCUSSION

In order to examine the difference between the phase transitions of bulk crystal and colloidally prepared microcrystallite CdS, we present high pressure resonance Raman scattering in bulk crystal and 60±20 Å diameter colloidally – prepared microcrystallite aggregates. The bulk CdS used in our experiment was prepared by a vapor phase transport method.[8] The sample has strong I_2 luminescence coming from the emission of an exciton bound to a neutral donor. At room temperature and atmospheric pressure the photoluminescence peak due to the exciton bound to a neutral donor is at 2.452 eV.[8] The pressure coefficient is 5meV/kbar for heavily—doped CdS. We chose the 4765 Å line of the Ar⁺ laser as incident light which has a photon energy of 2.602 eV. The band gap of CdS at 30 kbar and room temperature should be 2.602 eV if there is no phase transition in the sample. Thus we can use resonance Raman Scattering induced by pressure to study the phase transition of CdS.

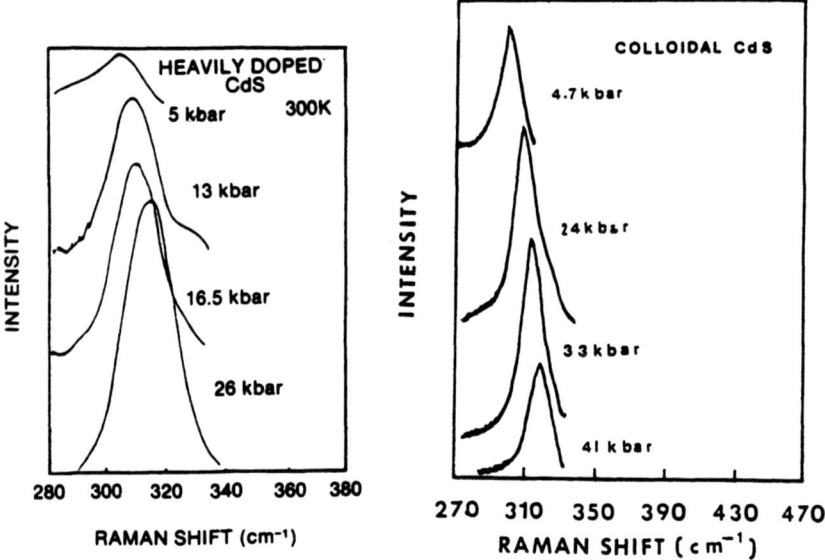

Fig. 1a) The Resonant Raman Scattering at several pressures for heavily doped bulk CdS in wurtzite phase after the phase transition.

Fig. 1b) Resonant Raman Scattering at several pressures for 60±20Å colloidal CdS.

Typical Raman spectra are shown in Fig. 1a and 1b. At atmospheric pressure the Raman shift of the bulk—CdS LO phonon is 304 cm⁻¹. The phonon frequency fitted as a

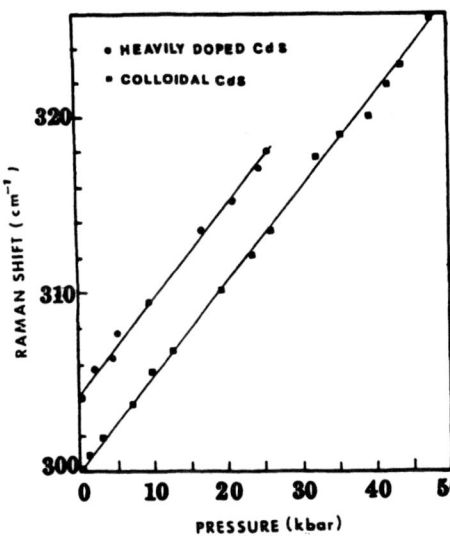

Fig. 2) Energy of the LO phonon as a function of pressure (•) line and (■) line represent Raman shift of LO phonon with pressure in heavily doped CdS and colloidal CdS respectively.

linear function of pressure, shifts at the rate of 0.50± 0.03 cm^{-1}/kbar, [Fig. 2]. Before the phase transition there is a strong background for the Raman scattering spectra of the LO phonon due to luminescence of the polaritons and high energy electrons, especially at high pressure. When the pressure is close to 27 kbar the band gap increases from 2.452 eV to 2.587 eV thus nearing the exciting photon energy of 2.602 eV. At this pressure the Raman intensity has increased dramatically due to a resonance effect (Fig. 3). When the pressure is slightly above 27 kbar both the photoluminescence and the LO Raman peak suddenly disappear and the color of the sample changes from yellow to dark orange. A 20% decrease in crystal volume is simulutaneously observed. If the phase transition is not complete, that is if there is a mixed wurtzite—rocksalt phase in the sample, then we expect to observe a reduced Raman peak for the LO phonon proportional to the wurtzite volume fraction or some new mode of the zincblende structure due completely to this resonance effect. But we have not observed any Raman spectra up to 50 kbar. This means that for bulk CdS the sample undergoes a complete first order phase transition from the direct band gap Wurtzite phase to the indirect band gap rock salt phase.

Upon returning to atmospheric pressure after having reached 50 kbar, there is no luminescence, probably due to the large number of nonradiative centers induced by the phase transition. The free exciton becomes very weak due to scattering by defects induced by the phase transition but the Raman scattering intensity of the LO phonon is almost the same as that measured before the phase transition. This proves that the I_2 bound exciton states are important intermediate states for resonance Raman scattering in heavily doped CdS. It also shows that CdS returns to its original wurtzite band structure.

Impurity states relax Raman selection rules because they break the translational and point group symmetry. According to Fröhlich's electron—phonon coupling theory, the wave vector of the LO phonon involved in Raman scattering are on the order of the inverse of the bound exciton radius. This broadens the width of the LO phonon Raman scattering peak (≈ 22 cm^{-1}).

The colloidal microcrystallite CdS used in our study was prepared by injection of a 20ml 9.3x10^{-3} M CdSO$_4$ aqueous solution into 100 ml of rapidly stirred 5x10^{-3}M Na$_2$S aqueous solution at room temperature. During the preparation for the pressure dependent work, the suspension precipitated as a result of aggregation. The particle diameter measured by TEM is 60±20 Å. The fresh colloidal suspensions were loaded into a diamond anvil cell and the pressures were measured by the ruby R$_1$ fluorescense. Up to 70 kbar the relative shift between the R$_1$ and R$_2$ lines is no larger than one wave number, indicating that hydrostatic pressure was well maintained. The 4765Å line from an Ar$^+$ laser was chosen as the exciting line. Typical resonance Raman spectra at various pressures are shown in

Fig. 1b. Fig. 2 shows the frequency shift of the LO phonon Raman scattering with pressure in bulk and colloidal CdS. At atmospheric pressure the Raman shift of the LO phonon lines of colloidal CdS is 300 cm^{-1}. The LO phonon frequency increases linearly with the pressure at the rate of 0.48 cm^{-1}/kbar which is basically in agreement with the bulk results. The colloids in aggregate form used in this experiment have an absorption edge at 2.49 eV at atmospheric pressure and room temperature. We observe that intensity of the LO phonon Raman scattering in colloidal CdS increases slowly with increasing pressure reaching a maximum at about 24 kbar. The intensity decreases slowly above 30 kbar as pressure is increased and the Raman intensity reduces to 50% of the 1 bar Raman intensity when the pressure is higher than 50 kbar (Fig. 3).

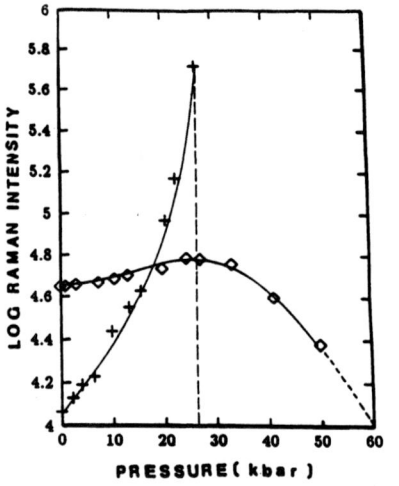

Fig. 3) Raman intensity of the LO phonon as a function of pressure. + curve and (□) curve represent the change of Raman intensity with pressure in heavily doped CdS and colloidal CdS respectively.

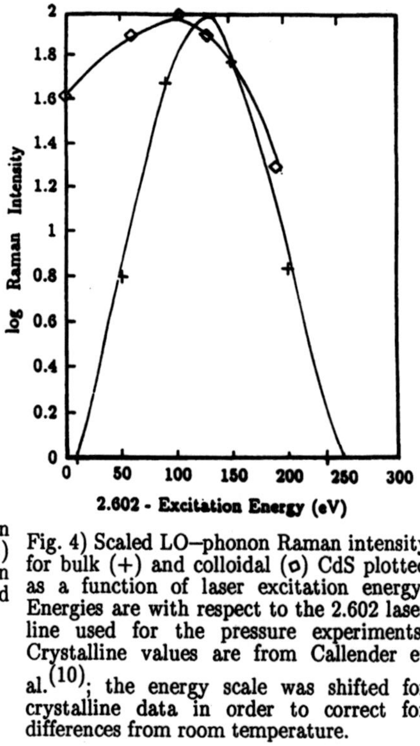

Fig. 4) Scaled LO–phonon Raman intensity for bulk (+) and colloidal (○) CdS plotted as a function of laser excitation energy. Energies are with respect to the 2.602 laser line used for the pressure experiments. Crystalline values are from Callender et al.[10]; the energy scale was shifted for crystalline data in order to correct for differences from room temperature.

In order to intepret pressure–tuning effects, we also measured the excitation spectrum of the Raman intensity in colloidal CdS. In Fig. 4 we show the scaled [9] Raman intensity for several excitation energies measured on Cds colloids dried onto a silicon substrate.[13] We observe that the Raman cross–section has a broad peak centered at about 2.50 eV with a FWHM of about 20 meV. We have also plotted the LO phonon excitation spectrum of bulk CdS taken from the literature[11]. The colloidal CdS spectrum is slightly wider and higher in energy than bulk CdS. If the pressure dependence of the band–gap for colloidal CdS is the same as that of bulk CdS (5 meV/kbar) then we can compare the energy–tuned to the pressure–tuned intensity dependence by scaling the energy axis at 5 meV/kbar in Fig. 4. We should thus predict that the intensity measured with the 4765Å laser line would increase with increasing pressure up to about 21 kbar and then would decrease smoothly to about one–fifth of its peak value as pressure is increased to 40 kbar. This is roughly the behavior that is observed in this work. We conclude that evidence for a pressure induced phase transition at pressures below 40 kbar is weak. The complete loss of LO phonon scattering above 60 kbar suggests that the transition occurs in this range.

The width of the intensity maximum suggests that the Wannier 1S exciton, which is the intermediate state for resonant Raman scattering[9], is spread over a wide energy range in colloidal CdS. This could be due to quantum size shifts and a spread in particle size or to enhanced electron–photon interactions in nanoparticles[10].

The phase transition induced by pressure in colloidal CdS is quite different from that occurring in bulk CdS. We suggest that the phase transition of the colloidal CdS occurring over a large pressure region up to 60 kbar is caused by surface effects of small CdS particles. Almost half of the atoms contained in these small particles are in the surface layer. We expect that surface reconstruction and large surface distortion energy will result in a change in the relative energies of different phases and thus must affect the phase transition of colloidal CdS. This experimental result is also consistent with theoretical analysis which predicts that finite size broadens the phase transition for both first and second order phase transitions.[6,7]

CONCLUSION

In conclusion, we have used resonant Raman scattering induced by pressure to study the phase transition of bulk CdS and 60±20Å fresh colloidal CdS. By measuring the shifts and intensity of the LO phonon Raman scattering as a function of pressure we conclude that bulk CdS undergoes a well–defined first order phase transition at about 27 kbar. In contrast with bulk CdS, the phase transition in the colloidal microcrystallite CdS occurs over a large pressure region up to 60 kbar. Both Bulk and colloidal CdS reverse to their original phase after pressure release. We attribute the phase transition occurring over a large pressure region (40–60kbar) in colloidal CdS to the large surface energy of the small particles. These results support other theoretical analyses and experimental results for noble metal particles.[12]

The authors gratefully acknowledge the support of this work by the National Science Foundation under NSF Grants DMR–88–01004, DMR–87–14634 and EET–87–14842.

REFERENCES

1.   O. Brafman and S.S. Mitra, in proc. 2nd international conference on light scattering in solids p. 287 (1971).

2.   R.J. Briggs and A.K. Ramdas, Phys. Rev. B 13, 5518 (1976).

3.   V. Venkateswaran, M. Chandrasekhar, and H.R. Chandrasekhar, Phys. Rev. B. 30, 3316 (1984).

4.   B.F. Variano, N.E. Schlotter, D.M. Hwang, C.J. Sandroff, J. Chem. Phys. 88, 2848 (1988).

5.   A.P. Alivisatos, T.D. Harris, L.E. Brus and A. Jayaraman, J. Chem. Phys. 89, 5979 (1988).

6.   M.E. Fisher and V. Privman, Phys. Rev. B 32, 447 (1985).

7.   S. Singh and R.K. Pathria, Phys. Rev. Lett. 55, 347 (1985).

8.   X.S. Zhao, J. Schroeder, T. Bilodeau, L. G. Hwa,Phys. Rev. B 40, 1257 (1989).

9.   R. Rossetti, S. Nakahara, and L.E. Brus, J. Chem. Phys. 79, 1086 (1983).

10.   R. H. Callender, S. S. Sussman, M. Selders and R. K. Chang, Phys. Rev. B, 7, 3788, (4.3).

11.   J. B. Renucci, R. N. Tyte, and M. Cardona, Phys. Rev. B, 11, 38885, (1975).

12.   C. Solliard and M. Flueli, Surf. Sci. 156, 487 (1985).

13.   The intensity was scaled by simultaneously measuring the first order intensity from a crystalline Si substrate, normalizing the Si intensity to its known Raman cross–section[11], and dividing this into the measured CdS intensity. This removes ω^4, power differences, and volume effects from the excitation spectrum.

TIME-RESOLVED SPECTRA OF EDGE EMISSION OF CdSe MESOSCOPIC CLUSTERS IN GeO$_2$ GLASS MATRIX

Takao Inokuma, Mitsuru Ishikawa·, Akinori Tanaka, and Toshihiro Arai

Institute of applied physics, University of Tsukuba, Tennodai, Tsukuba, Ibaraki 305, Japan.
·Tsukuba Reserch Laboratory, Hamamatsu Photonics K. K., Tokodai, Tsukuba, Ibaraki 300-26, Japan.

ABSTRACT

Time-resolved edge emission spectra of CdSe clusters were measured at 80K. Cluster size and excitation intensity dependences of the decay behavior were discussed in terms of the quantum size effect for electronic states.

Introduction

The effects of crystal size on optical properties of mesoscopic semiconductor clusters are of considerable interest and have been investigated extensively. Efros and Efros [1] have studied theoretically such a so-called quantum size effect on optical absorption spectra as a function of crystal size relative to the Bohr radii of electron and hole. The theoretical approaches for strong confinement limit of carriers have been made by Brus [2] in consideration of coulombic and depolarization effects, and by Schmitt-Rink, Miller, and Chemla [3] for the nonlinear optical properties. Experimentally, many authors [4-8] have studied the optical properties of small semiconductor crystals in glasses, polymers etc., and observed modifications of optical spectra due to the quantum size effects for electrons and holes. In particular, after Jain and Lind [9] found that the CdSSe clusters in a glassy matrix showed large optical nonlinearities, those materials are attracting much interest in application for optical processing devices that have a fast response time. At present, understanding of their fundamental physical properties are required.

In this paper we report time-resolved edge emission spectra of CdSe mesoscopic clusters embedded in a germanate glassy matrix. The obtained photoluminescence decay curves will be decomposed into two or three components and discussed in terms of modification of electronic states due to carrier confinement.

Experimental

Glassy samples which contain CdSe clusters are made by melt-quenching of the mixture of relevant materials and thermal treatments of the as-quenched glass. The composition of host glass is (GeO$_2$)$_{92}$(Na$_2$O)$_8$. Such a simple composition was used in order to reduce and simplify the effect of impurities that diffuse into CdSe clusters from the host glass. Cd and Se pure elements of 0.3 molar percent are mixed with the materials of the host glass and sealed in a quartz ampoule after evacuation. The ampoule was heated up to 1180° C

Mat. Res. Soc. Symp. Proc. Vol. 164. ©1990 Materials Research Society

for melting and rocked for 6 hours at that temperature. The melt was quenched in air and a colorless glass was obtained. The as-quenched glass underwent the various thermal treatments and CdSe clusters grew depending on the treatment. The average sizes, d, of clusters contained in each sample were 3-16nm in diameter. These values were estimated, for d>7nm, by observation with a transmission electron microscope and, for d<7nm, by applying Brus's theory [2] to the energy of the peak position in absorption spectrum.

For the excitation of photoluminescence, we used the 514.5nm line of a mode-locked Ar-ion laser which has about 50ps pulse duration and 82MHz repetition rate. The average power, I_{Ex}, for excitation was 0.6-60mW/mm². The luminescence signal was dispersed by a polychrometer and its temporal response was detected with a synchronous-scanning streak camera. The measurement of time-resolved spectra was performed at liquid-nitrogen temperature, ca. 80K.

Experimental results

Fig. 1 shows the steady-state photoluminescence and absorption spectra of 4.8nm CdSe clusters. The luminescence spectrum of CdSe clusters has two peak structures. The high-energy-side peak around the absorption edge is thought to be the emission due to radiative recombination of electron-hole pairs. The broad peak found in lower energy region is thought to be the emission related to defect states that may exist mainly on the cluster surface. Our present interest is in the high-energy-side emission peak, namely, the edge emission. The temporal response of the peak wavelength of edge emission spectra for samples with various cluster sizes are shown in Figs. 2 and 3. Fig. 2 is for I_{Ex}=6.0 mW/mm², and Fig. 3 is for

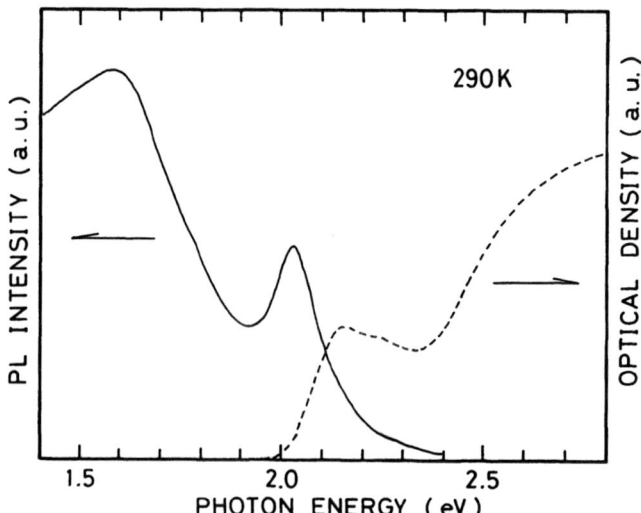

Fig. 1. Steady-State Photoluminescence and absorption spectrum of CdSe clusters of d=4.8nm.

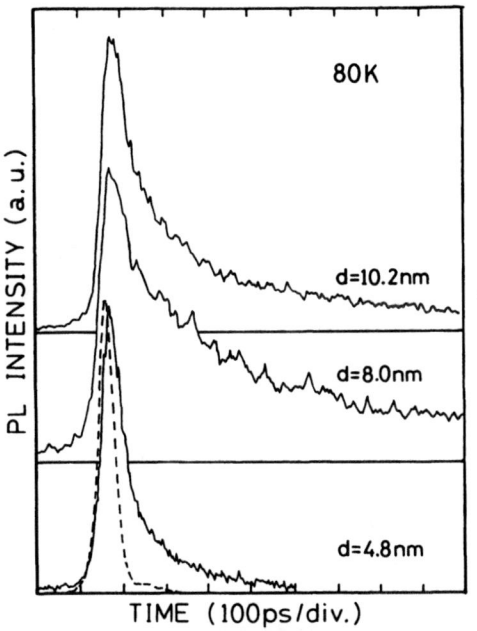

Fig. 2. Temporal profiles of edge emission of CdSe clusters under 6.0mW/mm^2 excitation. Broken curve represents the temporal profile of excitation laser pulse.

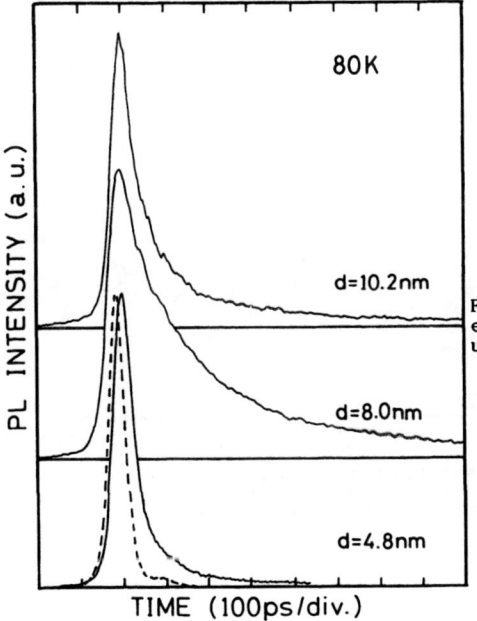

Fig. 3. Temporal profiles of edge emission of CdSe clusters under 60mW/mm^2 excitation.

I_{EX} =60 mW/mm^2. It is found that these decay curves can not be expressed by single exponential. Furthermore, when we compare Fig. 2 and Fig. 3, the relative intensity of slower decay components seems to decrease with increasing excitation power. Although we checked the time evolution of spectra, no substantial change in spectral shape was found except for the decrease of the emission intensity. This fact means that the high-energy-side tail of defect-related emission does not contribute to the slow decay component seen in Fig. 2, 3.

The experimental decay curves were fitted with the convolution of the temporal profile of excitation pulse and the sum of two or three exponential decay curves by employing a least-square-fit routine for a microcomputer. As a result of the fitting, the decay time and relative intensity of respective components can be obtained. All decay curves can be fitted very well under the assumption of two decay times except for that of d≃8nm. Therefore three decay times were assumed for that decay curve. In Fig. 4 obtained decay times are plotted as a function of the cluster size. The solid circles and the open circles correspond to I_{EX} =6mW/mm^2 and 60mW/mm^2 respectively. These decay times can be classified into three groups, and we labeled the respective group as component F, S_1 and S_2 (see Fig. 4). It is found that the decay times tend to become shorter on the whole with increasing excitation intensity. Also, the relative intensity of slower components, S_1 or S_2, decreases with increasing excitation intensity, as anticipated from Fig. 2 and Fig. 3.

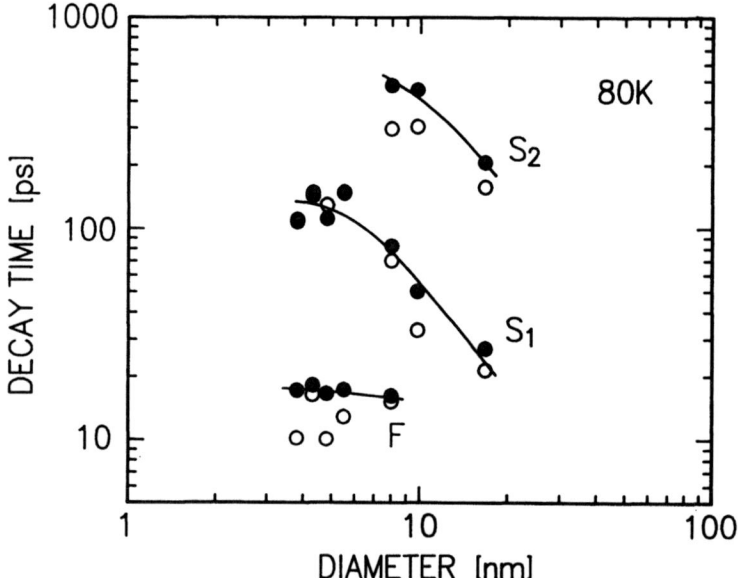

Fig. 4. Size dependence of decay times. Solid circles and open circles show 6.0mW/mm^2 and 60mW/mm^2 excitation respectively.

Discussion

The effective Bohr radius, $a_B{}^*$, of excitons in bulk CdSe is known to be 4.5nm , and therefore the cluster size in our experiment was varied across the bulk exciton size. In the case of $d \gg 2a_B{}^*$, it is a valid approximation that center-of-mass motion of the exciton is quantized. Then the blue shift of excitation energy is small and does not exceed the binding energy of the bulk exciton. The energy separation between the lowest excited state and higher excited states of center-of-mass motion is also small, but the transition oscillator strength is distributed to higher excited states with decreasing cluster size [10]. On the other hand, in the case of $d \ll 2a_B{}^*$, we can consider that electrons and holes are confined individually and form well-separated sublevel structure with the strong blue shift of the lowest exciting energy. In the limiting case, the optical transition is probable only between electron and hole sublevel that have same quantum number. As pointed out by Kayanuma [11], the transition from the former case to the latter case would occur in the narrow size range around $2a_B{}^*$. In fact, our results shown in Fig. 4 exhibit the clear difference in the decay behaviors and the size dependence between sizes larger and smaller than $2a_B{}^* \sim 9.0$nm.

For $d > 10$nm, the components S_1 and S_2 can be seen. The decay time of these components increase with decreasing cluster size. This may be due to redistribution of the oscillator strength of the bulk exciton as stated above. Under the condition of exciton confinement, even at 80K, generated excitons would be scattered into many closely-lying excited states by lattice vibrations and could not occupy the lowest exciton state dominantly. This effect is thought to cause the component S_2. When the excitation intensity is increased, the exciton-exciton interaction would become larger and enhance the recombination rate and nonlinear effects. This is the reason for the relative decrease of the slower component.

For $d < 6$nm, the component F and S_1 are clearly observed but the component S_2 is not. That may be because the energy separations between sublevels are sufficiently large and excited carriers dominantly occupy the lowest energy sublevel of electron or hole. Although the component F was not found for larger clusters, it may be observable by measurement at lower temperature. Weak size dependence for the decay times in this size range may be because the overlap integral of electron and hole wavefunctions for the lowest excited state would not vary substantially.

We expressed the decay profiles assuming a few exponential decay curves for convenience' sake. However, the decay mechanism should include many phonon excitations because carriers are excited to higher excited states and then relax to the lowest excited state which is related to the edge emission. So decay curves may be composed of many decay components. The transition probability of exciton confinement regime and individual-particle confinement regime may differ each other. So the decay behavior is different between both confinement regimes, that is, the component S_2 appears for $d > 10$nm and the component F does for $d < 6$nm.

In conclusion, the energy separation between the lowest excited state and other higher excited states is thought to be a important factor when we discuss the edge emission decay of semiconducter cluster at finite temperature. Furthermore, the exciton-exciton interaction may be considerably large in mesoscopic clusters.

Acknowledgement

We thank the Minstry of Education, Science and Culture of Japan for financial support under Grant-in-Aid 59460030 for Scientific Research.

References

1. Al. L. Efros and A. L. Efros, Fiz. Tekh. Poluprovodn. 16, 1209 (1982) [Sov. Phys. Semicond. 16, 772 (1982)].
2. L. E. Brus, J. Chem. Phys. 80, 4403 (1984).
3. S. Schmitt-Rink, D. A. B. Miller, and D. S. Chemla, Phys. Rev. B35, 8113 (1987).
4. A. I. Ekimov, Al. L. Efros, and A. A. Onushchenko, Solid State Commun. 56, 921 (1985).
5. N. F. Borrelli, D. W. Hall, H. J. Holland, and D. W. Smith, J. Appl. Phys. 61, 5399 (1987).
6. T. Ito, Y. Iwabuchi, and M. Kataoka, phys. stat. sol. (b) 145, 567 (1988).
7. B. G. Potter, Jr. and J. H. Simmons, Phys. Rev. B37, 10838 (1988).
8. T. Arai et al., Jpn. J. Appl. Phys. 28, 484 (1989).
9. R. K. Jain and R. C. Lind, J. Opt. Soc. Am. 73, 647 (1983).
10. Y. Kayanuma, Phys. Rev. B38, 9797 (1988).
11. Y. Kayanuma, Solid State Commun. 59, 405 (1986).

OPTICAL PROPERTIES OF II-VI
SEMICONDUCTOR DOPED GLASS

P. D. Persans, An Tu, M. Lewis, T. Driscoll, R. Redwing

Physics Department and Center for Integrated Electronics
Rensselaer Polytechnic Institute, Troy NY 12180-3590

ABSTRACT

We review structural and optical properties of $CdS_x Se_{1-x}$ semiconductor nanoparticles embedded in an insulating glass matrix. Vibrational Raman scattering and x-ray diffraction can be used to determine the composition of the crystallites for all X and sizes. Debye-Scherrer broadening of x-ray diffraction peaks from the crystallites yields an average grain size in the semiconductor crystallites of 60Å for the series studied here. Small angle x-ray scattering reveals that the average particle diameter is close to 120Å. Optical absorption, photoluminescence, and photomodulated absorption spectra are interpreted within a spherical quantum well model. Electron-phonon coupling and size distribution effects on the spectra are also discussed.

INTRODUCTION

Semiconductor-glass composites made up of small semiconductor particles embedded in an insulating glass matrix have been the recent focus of intensive research in non-linear optics. Work on commercially available materials (first prepared by Corning as cut-off filters) suggests that these materials may have a large third-order susceptibility [1-4] with a response time of less than twenty picoseconds. In order to better understand the linear and non-linear optical properties of this class of materials it is important to more fully understand the relationship between structural and electronic properties. The ultimate goal of our recent work has been to understand fundamental limitations to the width of the optical band edge of these materials and hence understand some of the limitations to the application of these materials for nonlinear optics. We shall review the preparation, optical properties, structural characterization and size effects in $CdS_x Se_{1-x}$. We attempt to separate size-distribution effects from electron-phonon effects by analysis of the temperature dependence of the optical absorption, the excitation energy dependence of photoluminescence (PL) and photomodulation of absorption (PA).

EXPERIMENTAL DETAILS

Materials for the present studies were prepared from long-pass filters manufactured by Schott Glass Technologies, Inc. The series of $Schott^R$ glasses from GG495 to RG715 consist of crown glass in which $CdS_x Se_{1-x}$ microparticles with x varying from 0 to 1 are embedded. Unless otherwise noted, code numbers such as RG645 denote the 50% internal transmission wavelength for 1 mm thick as-received $Schott^R$ filters.

Semiconductor doped glass is prepared [5,6] by mixing silicate (crown) glass and semiconductor additives (CdS and CdSe) and melting the mixture at about 1100°C until the components are dissolved in the glass. The solution is quenched and then annealed at 600°C to 800°C to nucleate and grow the semiconductor crystallites. Some of our experimental glasses were prepared by melting as received Schott glass filters and re-annealing to grow crystals of varying sizes. Some experimental glass composites with pure CdS and CdSe additives were supplied by scientists at Corning Glass Works.

X-ray diffraction measurements were made on an automated step-scanned Philips diffractometer using unresolved Cu Kα (λ=0.154 nm) radiation in order to check the crystallite composition and average size. The instrument resolution at $2\theta=45°$ was $2\theta=0.15$ degrees. A crystalline Si powder reference was used for angular and resolution calibration. Typically, the step size was $2\theta=0.05$ degrees and the counting time was 30 to 300 seconds on each point.

Small angle x-ray scattering (SAXS) was employed to measure the average particle size. The x-ray source consisted of a fixed-tube Cu point source with pinholes for collimation. The detector was a multichannel diode array x-ray detector. Samples for SAXS were prepared from thick glass filters by polishing to about 100-200 μm thickness to optimize x-ray transmission and scattering from the sample.

Optical absorption measurements were carried out using a Cary 17 dual-beam spectrophotometer in the absorptance mode. A small correction for reflectance was made assuming that the sample index was 1.48, independent of wavelength. Variable temperature measurements were carried out by clamping the sample to a Cu cold-finger in an evacuated cryostat with quartz windows.

Photoluminescence and Raman scattering measurements were carried out using a one meter focal length scanning double monochromator (J-Y HG2000) coupled to an RCA 31034A photomultiplier. All measurements were made in the photon counting mode with the tube cooled to -20C (dark count rate < 2 cps). The excitation source for Raman scattering and most photoluminescence measurements was a Spectra-Physics 165 Ar ion laser operated at λ= 457.9, 488.0, or 514.5 nm. A Spectra-Physics 375B dye laser with R6G dye and optics was pumped with the Ar+ laser to produce excitation energies from 570 to 650 nm. Some photoluminescence measurements were carried out with 1 mW HeNe laser excitation. All measurements were carried out in a near-backscattering configuration. Plasma lines from the laser were filtered by the use of a double Pellin-Broca prism dispersing stage.

STRUCTURAL CHARACTERIZATION

A detailed structural picture of the samples must be generated before we can interpret optical data. The most important structural parameters are : i.) crystallite composition; ii.) average microparticle size; iii.) average crystallite grain size within microparticles; and iv.) particle size distribution.

The most straightforward way to measure crystalline composition is by x-ray diffraction [5,12]. We used the position of the (110) line for composition determination on most as-delivered samples. We fit the peak and background by least squares to a Gaussian plus a straight line background. A typical x-ray diffraction scan through the (110) peak of sample RG695 is shown in Figure 1. The peak position could typically be determined to ±0.05° and thus the composition could be determined to ±2%. The width was less certain than the position due to the low signal to noise level in the x-ray data. Applying the Debye-Scherrer formula $(d=0.9\lambda/\Delta(2\theta)cos\theta)$ we find an average crystallite diameter of 65Å for RG645. The range for all glasses was 50-80Å average diameter. The peak width was about the same for (200) and (103) diffraction peaks, indicating that the crystallites were isotropic.

The average particle size was determined by SAXS to be ≈120Å for as-received RG645. This is considerably larger than the size determined by XRD widths. We believe that this difference is due to defects or grain boundaries within the particles. This difference between particle size and crystallite size can lead to problems in the interpretation of optical quantum size effects and can lead to an overestimate of the electron-hole effective mass [12].

Although XRD is reliable, there are many situations in which it is not a useful probe of composition due to the low signal to noise level in doped glasses. In the present experiments

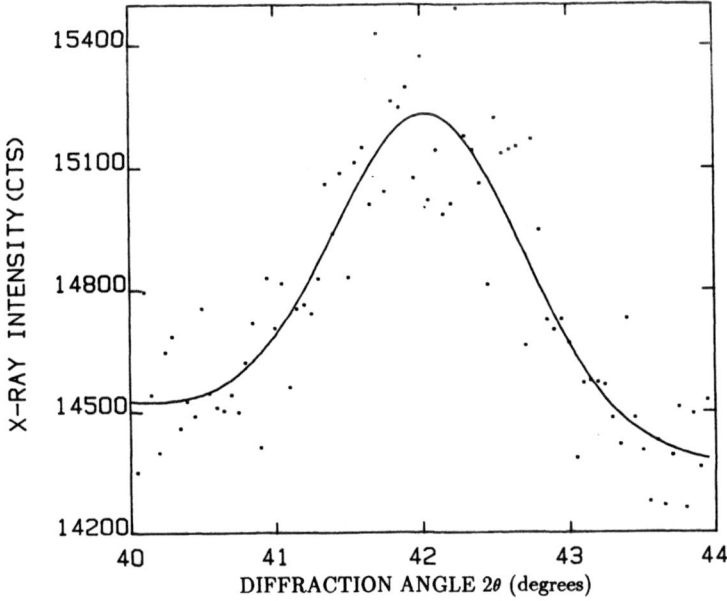

Figure 1- X-ray diffraction data for Schott filter glass RG695. Counting time on each point was 300 seconds.

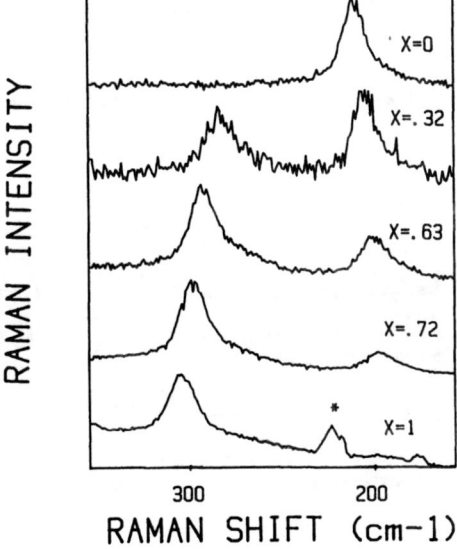

Figure 2- Raman scattered intensity plotted against Raman shift for several as-received Schott glasses with code numbers as denoted in the figure.

the crystallites make up only 0.1% of the sample volume. At the (110) x-ray peak the diffracted crystal intensity is only 10% of the background scattered by the glass. The crystal peak intensity drops inversely with the peak width (linearly with particle size) and hence the signal to noise also decreases. Data collection time for a 2θ scan of $4°$ of sufficient quality to determine position and width to the limits stated above is several hours. The necessary collection time goes up as the square of the peak intensity. We were unable to detect the x-ray diffraction peak from sample GG495 even after 12 hrs of data collection. For this reason we have developed a second, more easily implemented technique for the measurement of crystallite composition - vibrational Raman scattering.

In Figure 2 we show the vibrational Raman scattering Stokes spectrum for several as-received filters. We observe two clear modes at about 300 cm^{-1} and 200 cm^{-1} corresponding to CdS optic modes and CdSe optic modes respectively [7,8]. The CdS mode energy shifts downward and the CdSe mode shifts upward with decreasing S content. This behavior is also observed in bulk crystals [7,8]. Mode positions were measured by fitting both Stokes and anti-Stokes lines with a Lorentzian and averaging the shifts with respect to the laser line. We find that there is a small shift of all lines about 1 cm^{-1} closer to the laser than expected for the same composition bulk crystals and thus a straightforward use of crystal peak positions for composition calibration yields a systematic difference between composition determined from CdS mode position and CdSe mode position. We found however that the difference in position between the two modes compared quite well to the crystal values. We suggest that the slight systematic downshift in position is due to short phonon coherence length effects [9,10] or to strain induced by thermal mismatch to the glass [11]. In either case the CdS and CdSe modes are expected to shift nearly parallel to one another and the mode-difference approach should yield a more robust parametrization for composition determination. Our calibration curve of peak position difference ($\Delta\omega$) is plotted against x-ray composition in Figure 3. The correlation can be fit with a straight line given by $X = ((\Delta\omega - 60)\times 1.65 + 5)\%$.

OPTICAL CHARACTERIZATION

Optical Absorption and Size Effects

In Fig. 4 we show the optical absorption coefficient α plotted against photon energy for Schott RG645 for several temperatures . At room temperature we observe an algebraic absorption region from 2.0 eV to 2.3 eV which can be fit to a parabolic density of states $\alpha^2 = A^2(h\nu - E_G)$ with $E_G = 1.94$ eV. For α below 10 cm^{-1} the absorption is nearly exponential with $\alpha(E) = \alpha_0 exp((E - E_O)/E_S)$. E_S (= 24 meV) is substantially larger than that predicted by the Urbach relation (≈ 12 meV) for CdSe [13]. We focus first on the room temperature data.

In Figure 5 we plot the band gap E_G (found by plotting α^2 against $h\nu$ and extrapolating to $\alpha^2 = 0$) against composition for several as-received filters. We have also plotted (solid line) the band gap for CdS-CdSe solid solutions from the literature [14]. The band gaps for microparticles are systematically higher than the bulk crystal band gaps by about 60 meV. This is indicative of either a small size effect or compressive strain. Since compressive strain is not consistent with the Raman mode positions we believe that the shift must be due to a quantum size effect. If we take an average particle size of 120Å and employ the simple quantum-well expression for shift of the lowest excited state in an infinite spherical box then we can estimate the effective mass necessary to observe a shift of 60 meV. From ref. 15 we get :

$$E_{shift} = E_i - E_G = (\frac{h^2}{8\pi^2\mu a^2})X^2 \tag{1.}$$

where X=3.142 is the first zero of the Bessel function $J_{1/2}$, μ is the reduced effective mass, and a is the particle radius. We find an effective reduced mass μ of 0.1 m$_e$; which is quite close

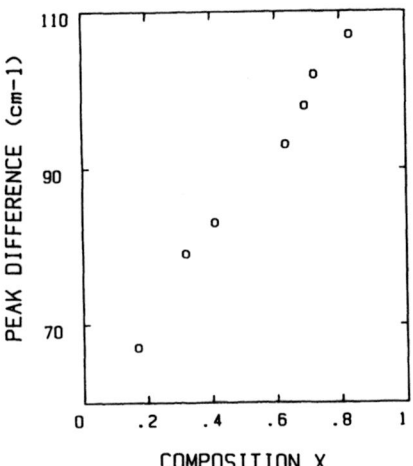

Figure 3- Difference Δω between Cds-like and CdSe-like optic peaks determined by Raman scattering plotted against composition determined by x-ray diffraction.

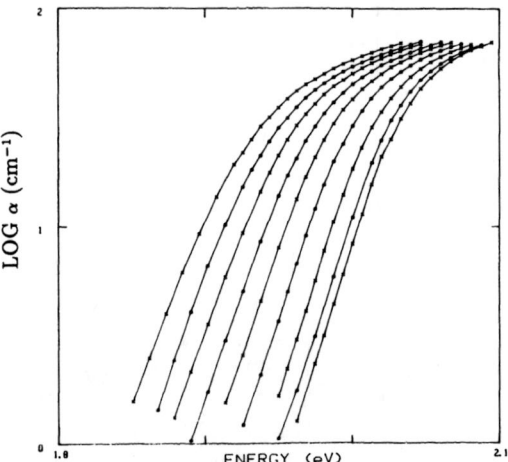

Figure 4- Optical absorption coefficient α plotted against photon energy for Schott glass RG645 for several measurement temperatures. Note the logarithmic scale.

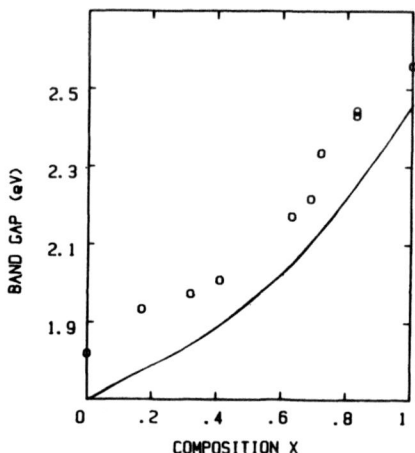

Figure 5- Energy gap E_G determined from extrapolation of α^2 plotted against crystallite composition. The solid line is the gap for single crystal CdS-CdSe solid solutions from the literature [14].

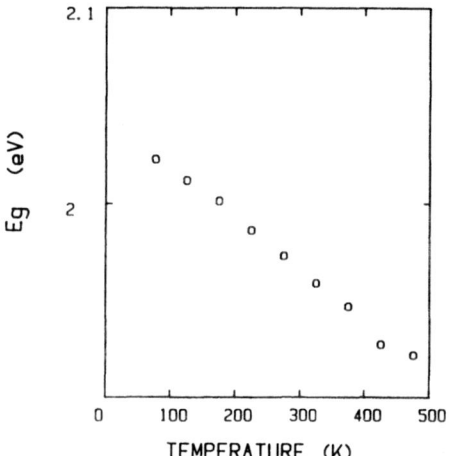

Figure 6- Energy gap E_G from optical absorption extrapolation plotted against temperature for sample RG645.

to the reduced mass expected for weakly bound excitons [6]. This effective mass is smaller than our previous estimate [12] because we are using the particle diameter in this calculation as opposed to the crystallite diameter. In our earlier work [12] we noted discrepancies in apparent effective mass for Corning and Schott as-received samples and suggested that the use of XRD widths (crystal size) instead of particle size might be one possible origin of the problem; we now believe this to be the case.

Electron-Phonon Interactions

It has been suggested that electron-phonon interactions will modify the width of the absorption edge in nanoparticles [16-18]. We explore the magnitude of the electron-phonon interaction here. In Fig. 4 we plotted the optical absorption spectra for RG645 for several temperatures. In Fig. 6 we plot the optical gap E_G found by extrapolation of α^2. We observe that E_G decreases with increasing temperature with a good fit to the form $E_G = 2.045 - 2.65 \times 10^{-4}$ T for 80K < T <450K. At lower temperatures we expect the band-gap to change more slowly with temperature. The temperature dependence of the band-gap is due to two effects; thermal expansion and electron-optical phonon coupling [13], hence it is a measure of the coupling strength. The temperature dependence of E_G for bulk CdS and CdSe can also be fit to straight lines in T about room temperature with slopes -5.0 and -4.6×10^{-4} eV/K respectively. We expect that alloys will fall between the two end members and hence we conclude that the temperature dependence of E_G for RG645 is smaller than that expected for the bulk material of similar composition, but it is of the same order of magnitude. Extending measurements to lower temperature would allow us to deduce the phonon energy which most affects the gap [16].

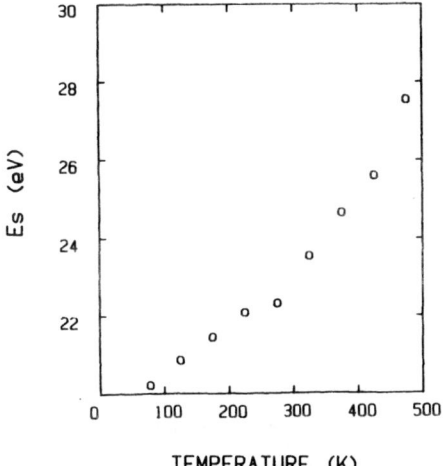

TEMPERATURE (K)

Figure 7- Exponential tail slope E_S plotted against temperature for sample RG645.

Another manifestation of the electron-phonon interaction in crystals is the Urbach tail, characterized by an exponential absorption edge $\alpha = \alpha_0 exp((E - E_O)/E_S)$ where $E_S = kT/\sigma$

$\sigma = \sigma_O(2kT/h\omega)$ $tanh(h\omega/2kT)$, where ω is a phonon energy and σ_O is a constant. Measurements on bulk CdS and CdSe [13] indicate that the tail-width parameter E_S is of the order of 10 meV at room temperature with a zero-point width of a few meV for both materials. We observe in Fig. 7 that E_S for sample RG645 is much larger than this prediction but that the temperature dependence is consistent with the equations above. Between 80K and 450K the exponential slope changes by only 5 meV. We conclude that the major part of the absorption tail in commercial glasses is not due to electron-phonon coupling. Electron -phonon coupling strength in these materials appears to be of the same order of magnitude as in bulk materials.

Size Distribution

We concluded in the previous section that there was significant excess broadening of the absorption spectrum above that due to thermal effects. This behavior is also observed in photoluminescence (PL) and photomodulated absorption (PA) spectra. We suggest that it can be explained by the fact that the observed spectrum is the sum of the spectra of many particles with slightly different optical properties - due primarily to size effects. Since PL, PA and absorption spectra depend in different ways on factors such as excited state occupation we expect to be able to combine analyses of different measurements to deduce the particle size distribution [12].

We show the PL emission spectrum for RG 645 at room temperature, excited by $h\nu=2.5$ eV and $h\nu=1.96$ eV light in Figs. 8a. and 8b. respectively. E_G is 1.94 eV for this sample. The PL peaks near E_G and drops rapidly on either side of the peak. Note that the PL intensity drops more rapidly at high energies for 1.96 eV excitation. Information from the PA spectrum will help us to analyze the PL spectra. In PA we clearly observe absorption bleaching near the band-edge[19]. We attribute this to band filling. From the magnitude of $\Delta\alpha/\alpha$ at the maximum bleaching we estimate the occupation of states at the band edge to be about 10^{-3}. Assuming that the band-edge is in equilibrium with deeper tail states we conclude that the trap- quasi Fermi level is at least a few tenths of an eV below the PA peak. This should also be the case for PL since it was measured under similar pump conditions. PL near the band edge is thus representative of thermalized carriers.

The PL spectrum in Fig. 8a. can be modelled by assuming a distribution of particles with different lowest excited states. Above the lowest excited state we expect the spectrum of an individual particle to become more featureless with increasing energy due to phonon replicas, short higher excited state lifetimes, and electron-phonon coupling [16]. The emission from a particle with band-gap E' is given by [12,20]:

$$R(E, E')dE = CF\alpha_{exc}\alpha(E)V(E')\tau E^2 e^{-\beta(E-E')}dE \qquad (2.)$$

where C is a constant, $F\alpha_{exc}V$ is the generation rate when the particle diameter is smaller than α^{-1}, and τ is the free carrier lifetime, and the exponential is an occupation factor scaled to the lowest excited state energy. The time-averaged excitation energy is scaled to each particle separately.

If we were simply to take $\alpha(E)$ from the data in Fig. 3 then the predicted PL spectrum would peak well below E_G and the PL intensity would fall off exponentially with $1/e$ energy of 26 meV above the edge. This is not the case, which indicates that the spectrum must be the composite of the spectra of many particles:

$$PL(E)dE = \int f(E')R(E, E')/V dEdE' \qquad (3.)$$

where $f(E')$ is the fraction of the particle volume occupied by particles with band gap E'. For energies well above the PL peak we expect α to be featureless and hence the particle size distribution can be estimated [12].

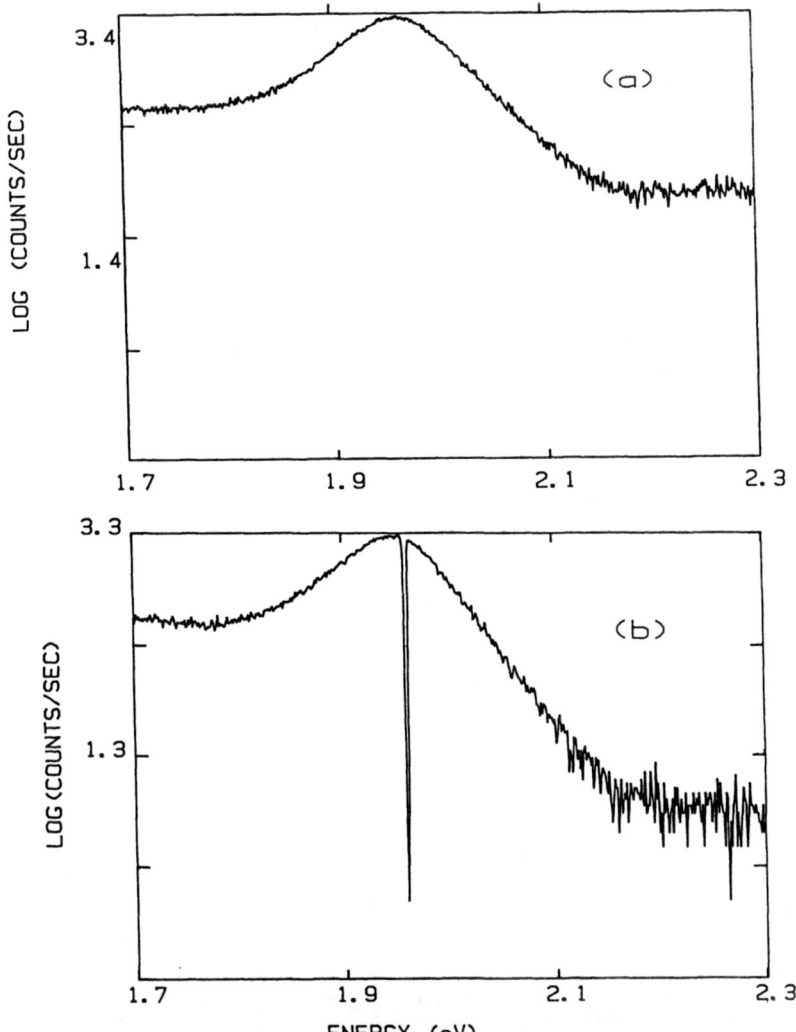

Figure 8- (a)- Photoluminescence intensity plotted against photon energy for Schott glass RG645. Excitation was with the 488nm line of the Ar laser. Note the logarithmic scale. (b)- Photoluminescence intensity plotted against photon energy for Schott glass RG645. Excitation was with the 632.8 nm line of a HeNe laser. Note the logarithmic scale.

The generation rate in eq. 2 depends on the absorption coefficient of the particle at the laser energy. For 1.96 eV excitation the absorption of small particles (large band gaps) is negligible and these particles do not contribute to the spectrum. The more rapid fall-off in PL intensity with increasing energy thus establishes that the excess PL width in the 2.5 eV excitation spectrum is due to either a particle size distribution or hot carrier PL.

From analysis of the data in Figs. 4 and 8a. using eqs. 1-3 and assuming $\mu=0.1m_e$ we deduce that the particle size distribution is peaked at 120$\mathring{A}$ and has a faily wide distribution with FWHM of 40 $\mathring{A}$. A more detailed description of the application of eqs. 1-3 can be found in ref. 12.

SUMMARY

We have reviewed the structural and optical properties of CdS-CdSe composite micropar- ticles embedded in glass. Structurally, we find that the average crystallite size can be much smaller than the particle size. In many cases, for smaller particles especially, x-ray diffraction does not yield sufficiently high quality diffraction data for accurate compositional analysis. We have calibrated the vibrational Raman peak positions and found an empirical rule for compositional determination for these cases. The temperature dependence of optical ab- sorption spectra indicate that electron-phonon coupling in 120$\mathring{A}$ nanoparticles is similar to coupling in bulk materials. Broadening of the optical absorption edge is attributed to particle size distribution and quantum size effects. Techniques for the analysis of photoluminescence, photomodulated absorption and optical absorption spectra which allow us to learn about the particle size distribution have been discussed.

ACKNOWLEDGEMENTS

We thank D. Hall and N. Borrelli of Corning Glass Works for helpful discussions and for supplying CdSe composite glasses.

This research was supported in part by NSF under grant numbers DMR-8714634 and EET-8714842 and by Rensselaer Polytechnic Institute. M. Lewis and T. Driscoll were also supported in part by the Undergraduate Research Participation program at Rensselaer.

REFERENCES

1. R.K.Jain and R.C.Lind, J. Opt. Soc. Am., 73, 647, (1983).
2. S.S.Yao, C.Karaguleff, A.Gabel, R.Fortenberry, C.T.Seaton, and G.I.Stegemann, Appl. Phys. Lett., 46, 801,(1985).
3. P.Rousignol, D.Ricard, J.Lucasik, and C.Flytzanis, J.Opt. Soc. Am., B 4, 5, (1987).
4. H.M.Gibbs, et al., Appl. Phys. Lett., 41, 221, (1982).
5. N.F.Borrelli, D.W.Hall, H.J.Holland, and D.W.Smith, J. Appl. Phys., 5399, (1987).
6. B.G.Potter, Jr., and J.H.Simmons, Phys. Rev. B, 10838, (1988).
7. R. K. Chang J. M. Ralston and D. E. Keating, "Light Scattering in Solids", edited by Wright (Springer-Verlag Inc., New York, 1969), p369.
8. J. F. Parrish, C. H. Perry, O. Brafman, I. F. Chang and S. S. Mitra, "II-VI Semicon- ducting Compounds 1967 International Conference", p11164, D. G. Thomas (ed.), W. A. Benjamin, New York, 1967.
9. E. Lu, P. D. Persans, K. Rajan, Phys. Rev. B, submitted.
10. H. Richter, Z. Wang, and L. Ley, Sol. St. Commun., 39, 625, (1981).
11. X.-S. Zhao, J. Schroeder, P. D. Persans, and E. Lu, Phys. Rev. B, submitted ; and these proceedings.
12. P. D. Persans, A. Tu, T.-J. Wu, M. Lewis, J. Am. Opt. Soc., B 6, 818, (1989).

13. M. V. Kurik, Phys. Stat. Sol., (a), $\underline{8}$, 9 (1971).
14. Khansevorov, Ryvkin and Ageeva, J. Tech. Phys. USSR, $\underline{28}$, 480(1958).
15. L. E. Brus, J. Chem. Phys. $\underline{80}$, 4403, (1984).
16. A. P. Alivisatos, T. D. Harris, P. J. Carroll, M. L. Steigerwald, L. E. Brus, J. Chem. Phys., in press.
17. P. Roussignol, D. Ricard, and C. Flytzanis, Phys. rev. Lett., $\underline{62}$, 312, (1989).
18. S. Schmitt-Rink, D. A. B. Miller, and D. S. Chemla, Phys. Rev. B, $\underline{35}$, 8113, (1987).
19. R. Redwing, Master's Thesis, Rensselaer Polytechnic Institute, 1988.
20. R.B.Stephens, Phys. Rev. B, $\underline{29}$, 3283, (1984).

FABRICATION AND CHARACTERIZATION STUDIES OF SEMICONDUCTOR-IMPREGNATED
POROUS VYCOR GLASS

C. A. HUBER, T. E. HUBER, A. P. SALZBERG, and J. A. PEREZ
Department of Physics, University of Puerto Rico, Rio Piedras, PR 00931

ABSTRACT

Porous Vycor glass has been impregnated with semiconductors by
pressure forcing the nonwetting semiconductor melt into the interconnected
pores. Dense semiconductor mesh-like microstructures with a characteristic
size of 50 Å can be fabricated by this technique. Measurements are
reported which show the composites are suitable for both optical and
transport studies, particularly those addressing quantum confinement of
carriers and unusual electrical transport phenomena in this new class of
materials.

Introduction

As a result of their reduced dimensions, semiconductor crystallites
display unusual physical properties which make them of current interest.
This is more so since optical studies have shown them to have potential
device applications in nonlinear optics. Microcrystallite systems
investigated include colloidal semiconductors [1], glasses and polymers
doped with semiconductors [2,3], and semiconductor clusters stabilized in
the cavities of zeolites and in porous glass [4,5]. We report here on the
fabrication and characterization of novel semiconductor-silica composites
which are also amenable to studies addressing microcrystallite effects.
The composites are prepared by impregnating a porous glass with
semiconductors. This is achieved in practice by pressure-forcing the
semiconductor melt into the interconnected pores. A new class of
semiconductor composites suitable for both electrical and transport
studies can be fabricated in this way.

Fabrication

The porous Vycor glass employed (Corning code 7930) is made by acid
leaching the boron-rich phase in a phase-separated borosilicate glass. The
result is a high content silica glass containing an interconnected network
of pores typically less than 100 Å diameter. Its 0.28 porosity exceeds the
critical value of 0.16 for continuous percolation in three dimensions. The
large pore surface area and fairly narrow pore-size distribution of porous
Vycor glass have made it a material of choice for studies of the physics
of adsorbates, the properties of confined fluids and, more recently, of
fractal structures. A standard porosimetry analysis was performed on the
porous glass material used for our studies. Mercury intrusion and nitrogen
adsorption measurements indicate average pore diameters of 37 and 56 Å,

respectively. The main reason for the different average pore diameters is that the pores are not cylinders, as assumed by these techniques, but shaped as "ink bottles" [6]. Pore volume distributions from nitrogen adsorption/desorption data show that essentially all the pores are in the 30–80 Å diameter range. Scanning electron micrographs show a random-packed assembly of silica clusters about 300 Å in size, close to the resolution limit. The optical transmission spectra of porous Vycor glass in the visible–near infrared region of interest is shown in Figure 1. The glass is essentially transparent from about 0.3–4.0 μm except for a broad absorption band in the 2.5–3.5 μm range due to surface vibrations of OH groups and of physically adsorbed water, its width and peak position depending on the degree of hydration [7]. Weak absorption bands are present at 1.4, 1.9 and 2.3 μm and correspond to overtones and combinations of OH, H_2O, and Si–OH modes [8].

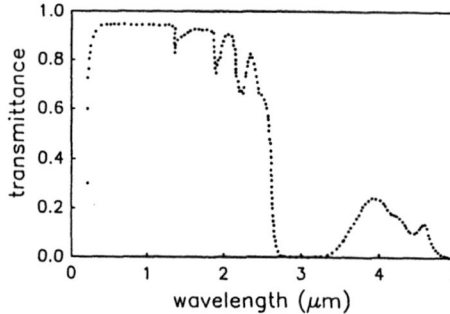

Figure 1. Transmission of a 0.07 cm-thick sample of porous Vycor glass.

Using the porous glass as a matrix, we have fabricated composites of selenium, tellurium and gallium antimonide, the initial choice of these semiconductors being based on their relatively low melting point. For these cases (and also for various metals investigated), it is found that the semiconductor melt does not wet the glass. An externally applied pressure is needed to overcome surface tension effects which prevent the semiconductor from entering the narrow pores. Assuming pores of circular cross section, an estimate for the diameter d of the smallest pore filled at pressure P can be obtained from [9]:

$$d = -4\gamma \cos\alpha \, /P \qquad\qquad (1)$$

where γ is the surface tension of the liquid and α is the contact angle between the liquid and the glass ($\alpha > 90°$ for nonwetting liquids). Experimental values for the surface tension of some liquid semiconductors are available in the literature, they span the 100–600 dyne/cm range (88 and 180 dyne/cm for selenium and tellurium, respectively) [10]. From Equation 1 and using a mid-range surface tension of 300 dyne/cm, an upper limit for d of 28 Å is obtained for a pressure of 60,000 lb/in² (4.1 kbar).

The apparatus employed for impregnation is based on a reactor which allows for externally heated operation up to 65,000 lb/in^2 at 850 °C, and on a pressure pump using water as pressure-transmission medium. Details about its construction and operation and about the sample preparation process have been reported elsewhere [11]. Briefly, the reactor is first heated to a temperature above the semiconductor melting point (217 °C for Se, 450 °C for Te, and 712 °C for GaSb). Water is then introduced and its pressure is gradually raised, the molten semiconductor being pushed inside the glass pores by the hydrostatic pressure exerted by the water/water vapor. When impregnation is complete the reactor temperature is lowered and the semiconductor solidifies inside the pores, the high pressure is then removed. In this way we have fabricated composite samples in the form of cylinders 3.6 mm in diameter and up to 30 mm long. From gravimetric measurements performed on various impregnated samples a filling factor of 92%-97% of the pore volume is obtained.

Characterization

X-ray diffraction analysis shows that the selenium inside the pores is amorphous. This is not surprising since moderately-sized selenium single-crystals are quite difficult to obtain by standard crystal growth techniques. Tellurium and gallium antimonide, on the other hand, retain their crystalline structure (trigonal and cubic zinc-blende, respectively) and their lattice parameters remain unchanged from the corresponding bulk values. An estimate for the average crystallite size D can be obtained in these cases from the broadening of the x-ray diffraction lines [12] :

$$D = K\lambda/\beta\cos\theta \qquad (2)$$

where λ is the x-ray wavelength (1.542 Å), β is the contribution to the linewidth due to the finite crystallite size, 2θ is the diffraction angle, and K is a factor depending on the crystallites shape and is taken here to be one. The contribution to the full-width-at-half-maximum (FWHM) of the x-ray diffraction lines resulting from instrumental resolution was determined from the diffraction spectra of powder standards made from crystalline GaSb and Te, and then appropiately subtracted from the diffraction linewidths of the composites. From the measured value for β of 0.22° at 2θ = 22.7° for tellurium-impregnated Vycor, a value for D of 400 Å is obtained. The same analysis applied to the diffraction line at 2θ = 82° with β = 1.2° of the gallium antimonide composite gives D = 97 Å. The accuracy with which Equation 2 can be applied in these cases is limited by the absence of information on the crystallite shape (K could assume numerical values ranging from as little as 0.7 to as much as 1.7), and on the particular definition of D, the crystallite dimension, that is adopted. Still, the overall rather small widths of the x-ray diffraction lines suggest that the average crystallite size in these composites is not necessarily limited by the 50 Å average pore diameter.

Elementary transport measurements performed on the composites indicate that the semiconductor inside the pores is interconnected, at least through some paths, thus affording the possibility of studying electrical transport phenomena in this new type of mesoscopic system. The composites mechanical properties such as hardness are qualitatively like those of the glass. Selenium is observed to extrude from the glass, the extruded selenium taking the form of spheroids typically 1µm in size after a few days. That is not the case for the tellurium or the gallium antimonide composites which do not show any evidence for extrusion over a period of months. Extrusion is related to the surface tension of the solidified semiconductors and to the low flow stress of selenium at room temperature. It is prevented by storing the selenium composites at liquid nitrogen temperatures.

The composites appear to be of good optical quality under visual inspection. This was confirmed by optical measurements. For this purpose, samples were mechanically polished using standard techniques. The absorption spectra of the composites do not show significant contribution from elastic scattering. For the tellurium composite optical measurements are not straightforward since the fundamental absorption edge of tellurium lies at 3.75 µm in a wavelength region where the porous Vycor itself can be quite opaque from absorption by physisorbed water. The absorption and reflectivity spectra of the selenium-impregnated Vycor have been discussed by us previously [11]. Those spectra do not show clear evidence for quantum-size effects such as a blue shift of the fundamental absorption edge because the carrier effective masses in selenium are too large to produce sizable confinement energies.

In contrast to the selenium composite, the optical absorption of the gallium antimonide-impregnated Vycor is significantly different from that of bulk GaSb. Figure 2 shows the room temperature absorption spectra of the GaSb composite and of bulk GaSb. The absorption coefficient of the composite has been calculated by considering an effective GaSb thickness 0.28 times the sample thickness. The minimum thickness attainable by polishing determined the highest absorption that could be measured. Bulk GaSb shows a sharp absorption edge at photon energies corresponding to the direct energy gap at 0.71 eV. The absorption spectrum of the GaSb-impregnated Vycor is clearly blue-shifted from that of bulk GaSb. The long absorption tail exhibited by the composite, probably related to a density of defect states and to a crystallite size distribution, makes a precise determination of an energy gap difficult in this case. Possible ways of preparing thinner samples of the composite which will allow us to determine the optical absorption at higher photon energies are presently being investigated.

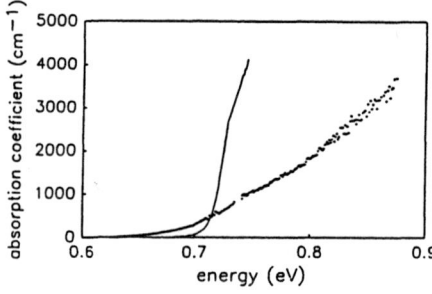

Figure 2. Absorption of gallium antimonide-impregnated porous Vycor glass (dots) and of bulk gallium antimonide (solid line) at room temperature.

An estimate for the magnitude of carrier confinement energies in this case can be obtained by considering that the Vycor average pore radius, R = 25 Å, is smaller than or comparable to the effective Bohr radii of electrons (150 Å), holes (20 Å and 120 Å for heavy and light holes, respectively) and of excitons (170 Å and 270 Å for the heavy—hole— and light—hole—exciton) in GaSb. In this limit, the lowest energy of an electron—hole pair is larger than that in the bulk due to independent size—quantization of the electron and of the hole motion [13]. If a simplified model of cylindrical pores is assumed and possible polarization effects arising from the dielectric discontinuity between the semiconductor and the silica are disregarded, the absorption threshold can be calculated using the energy levels of a particle in an infinite cylindrical potential well. It is blue—shifted by an amount E_c = $5.78\hbar^2/2\mu R^2$. With $\mu=0.036m_0$ the reduced electron—heavy hole effective mass in GaSb, $E_c = 1.5$ eV. The results of preliminary measurements on gallium antimonide—impregnated Vycor shown in Figure 2 seem to indicate that confinement energies are substantially smaller than the above simple estimate. This likely results from a crystallite—size distribution, and from the Vycor pore structure which is quite intricate as opposed to the simple assumption of cylindrical pores.

Summary

We have shown that novel microcrystalline structures can be fabricated by pressure—impregnating porous Vycor glass with semiconductors. Composites prepared in this manner are of good optical quality and their mechanical properties are qualitatively like those of the glass. The semiconductor inside the glass pores is interconnected, allowing also for transport measurements on this new mesoscopic system. The impregnation technique can be applied to other semiconductors (and metals) and other porous hosts. For porous Vycor glass in particular, an upper limit on the melting temperature of the impregnating material of about 900 °C is established by the sintering which takes place at these temperatures.
We thank C. Thieme for assistance with the x—ray diffraction measurements.This work was supported by the National Science Foundation through its Experimental Program to Stimulate Competitive Research.

References

1. L.E. Brus, J. Phys. Chem. 90, 2555 (1986); C.J. Sandroff, D.M. Hwang, and W.M. Chung, Phys. Rev. B33, 5953 (1986).
2. N.F. Borrelli, D.W. Hall, H.J. Holland, and D.W. Smith, J. Appl. Phys. 62, 5299 (1987).
3. Y. Wang, A. Suna, W. Mahler, and R. Kasowski, J. Chem. Phys. 87, 7315 (1987).
4. V. N. Bogomolov, V.V. Poborchii, S.V. Kholodkievich, and S.I. Shagin, Pis'ma Zh. Eksp. Teor. Fiz. 38, 439 (1983) [Sov. Phys. JETP Lett. 38, 533 (1983)]; Y.Wang and N. Herron, J. Phys. Chem. 91, 257 (1987).
5. J.C. Luong, Superlatt. and Microstruc. 4, 385 (1988).
6. J.H.P. Watson, Phys. Rev. 148, 223 (1966).

7. T.H. Elmer, I.D. Chapman, and M.E. Nordberg, J. Phys. Chem. 66, 1517 (1962).
8. T.E. Huber and C.A. Huber, J. Phys. Chem., in press (1990).
9. See, for example, A.W. Adamson, Physical Chemistry of Surfaces (Wiley, New York, 1982), p. 338.
10. V.M. Glazov, S.N. Chizhevskaya, and N.N. Glageleva, Liquid Semiconductors (Plenum, New York, 1969).
11. C.A. Huber and T.E. Huber, J. Appl. Phys. 64, 6588 (1988).
12. H.P. Klug and L.A. Alexander, X-Ray Diffraction Procedures for Polycrystalline and Amorphous Materials (Wiley, New York, 1954).
13. Al.L. Efros and A.L. Efros, Fiz. Tekh. Poluprovodn. 16, 1209 (1982) [Sov. Phys. Semicond 16, 772 (1982)]; E. Hanamura, Phys. Rev. B37, 1273 (1988).

SYNTHESIS, STRUCTURAL AND OPTICAL CHARACTERIZATION OF
II-VI SEMICONDUCTORS INCLUDED IN SODALITE-TYPE HOSTS

K. L. Moran, W. T. A. Harrison, T. E. Gier, J. E. Mac Dougall, G. D. Stucky

Department of Chemistry, University of California, Santa Barbara, CA 93106

ABSTRACT:

Solid-state chemistry has been used to control both the size and the interconnection distance of small II-VI semiconductor moeities incorporated in zeolitic hosts with the sodalite-type structure. Structural characterization was carried out using X-ray Rietveld powder methods, and optical properties of these materials were also measured. These novel materials show quantum superlattice effects as evidenced by blue shifts in their optical absorption spectra.

INTRODUCTION:

Small particle semiconductors, either prepared by colloidal methods,[1] included in porous hosts,[2] or formed in multiple quantum wells,[3] have been shown to have a host of novel physical properties including blue shifts in optical absorption spectra, increased photooxidation potentials, and unusual nonlinear optical properties.[4] Synthesis of these small clusters in porous crystalline hosts, namely zeolites, has allowed for the study of both their optical and structural properties.[5,6]

Zeolites are crystalline aluminosilicates with pore sizes on the order of atomic dimensions, i.e. 4-14 Å, and these structures have been used as hosts for the self assembly of small clusters of II-VI semiconductors with well defined size, shape, and interconnectivity. The optical properties of the cube-like clusters formed in zeolites A and Y are governed by the way they are able to interact through the double 6 and 4 ring windows of the zeolite host.[5] This study has focused on the idea of using both careful control of the semiconductor 'particle size' and the periodicity of the 3-dimensional quantum-well superlattice, which defines the interaction between the particles, to control optical properties. In particular, in sodalite-type phases, individual tetrahedral clusters of II-VI semiconductors of precisely specified dimensions and connectivity may be prepared *in situ*. The framework provides a well-defined 3-dimensional superlattice for inter-cluster interations in terms of cluster-cluster distance.

The approach has been to build into a structure, during the synthesis of the host, discrete "clusters" of M_4X^{6+}, where M = Zn or Cd and X = S or Se. By changing M or X, or by doping different amounts of M or X atoms into a particular composition, the local connectivity and thus the physical properties are changed. This chemistry has been carried out in structural analogues of sodalite (zeolite HS: $Na_8(OH)_2(AlSiO_8)_6$). These analogues are boralites or helvites, in which the aluminum or silicon in the tetrahedral framework sites are replaced with boron or beryllium, the sodium by zinc or

124

cadmium, and finally the hydroxide with sulfur or selenium. The composition per unit cell is $M_8X_2(BO_2)_{12}$ for boralite or $M_8X_2(Be_3Si_3O_{12})_2$ for helvite, where M=Zn or Cd and X=S or Se. Having only boron atoms or both beryllium and silicon atoms in the tetrahedral zeolitic framework raises its overall negative charge, allowing the incorporation of both di- and trivalent cations as charge compensation. The structure, illustrated in Figure 1, is made up of cages which are packed sharing faces in an all-space filling, cubic framework to give the closest possible approach between the cage contents. Structural and optical aspects of these materials are reported below.

Figure 1. Several unit cells of the $Zn_4X(BO_2)_6$ structure, showing inter-cage connectivity of the tetrahedral semiconductor clusters.

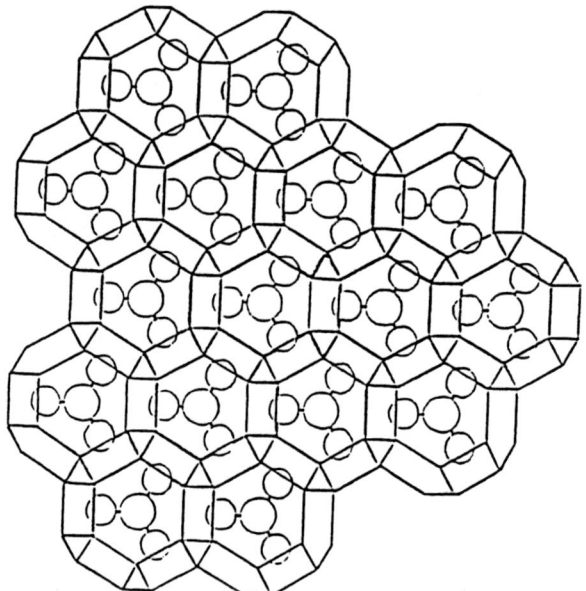

EXPERIMENTAL:

$Zn_8S_2(BO_2)_{12}$[6] was synthesized by solid-state reaction in a sealed quartz ampule. Stoichiometric mixtures of ZnS, ZnO and ZnB_4O_7[7] were flame sealed inside an evacuated quartz ampule, heated to 550°C, and held overnight. The resulting white powder was recovered, and X-ray powder diffraction identified the phase as boralite. The zinc seleno, mixed Zn/Cd, and mixed S/Se boralites were made similarly. Helvite analogues were prepared hydrothermally according to literature methods.[8]

High resolution powder diffraction data were collected for each sample using a Scintag Pad X automated X-ray diffractometer using Cu Kα radiation. In all cases, a highly crystalline sodalite-type phase dominated the pattern. For boralite-type phases, the patterns could be indexed on a body centered cubic lattice of cell dimensions ~7.6Å, as was found for zinc oxo boralite[9]. Helvites gave a primitive-cubic unit cell, as reported earlier. Full X-ray Rietveld structural analysis was carried out for some of these phases using the Rietveld analysis program GSAS[10], using the zinc oxo boralite structure as a starting model. Refining the usual structural and profile parameters resulted in satisfactory refinements for each phase: the use of bond-distance restraints was found to be unnecessary, as rapid and stable convergence of the structural parameters was achieved. Some important structural parameters for the boralite phases are listed in Table 1. Optical absorption spectra were measured for these materials and the results are summarized in Fig. 2A and B.

Table 1. Bond distances (Å) and angles (°) for members of the boralite series.

$Zn_4X(BO_2)_6$ Series

X =	d(Zn-O)	d(ZnX)	θ(O-Zn-O)	θ(O-Zn-X)
O	1.957(3)	1.982	95.74(8)	121.09(6)
S	2.006(8)	2.260(3)	101.0(3)	117.0(3)
$S_{0.5}Se_{0.5}$	1.970(5)	2.3184(25)	103.2(3)	115.2(2)
Se	2.008(7)	2.3699(25)	103.0(3)	115.4(2)

$Zn_3CdS(BO_2)_6$

	d(Zn-O)	d(ZnX)	θ(O-Zn-O)	θ(O-Zn-X)
	2.002(8)	2.221(4)	101.0(4)	116.97(29)

RESULTS:

The results of Rietveld analysis demonstrate that powder samples and a laboratory X-ray source are sufficient to gain useful structural information in well-ordered systems for which there is a good starting model. In the cases of zinc oxo, sulfo, and seleno boralites, all of the zinc atoms are crystallographically equivalent; the relevant bond distances and angles are given in Table 1. In the mixed systems, $Zn(S_{0.5}Se_{0.5})$ both the S and Se atoms randomly occupy the same crystallographic site and are therefore indistinguishable, thus the Zn X bond distance is intermediate to those in the ZnS and ZnSe boralites. Optical absorption spectra for some small particle, $(MX)_4$ in zeolite Y, 1 μm, and bulk semiconductors are summarized in Fig. 2.

126

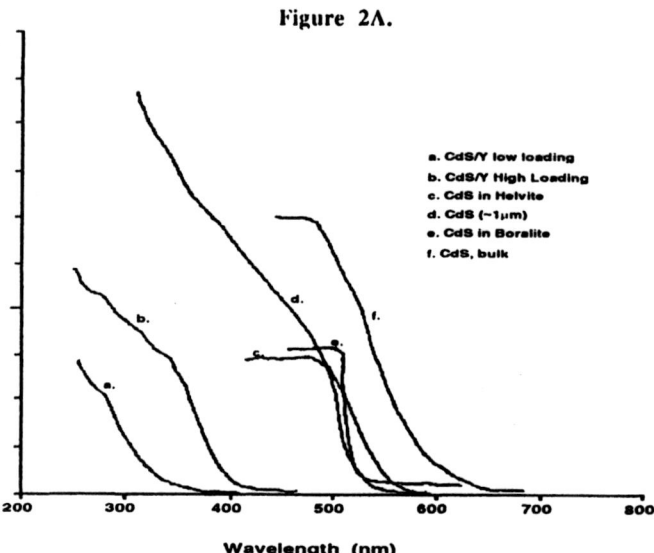

Figure 2A.

Comparison of the absorption spectra of CdS in several hosts and the bulk.

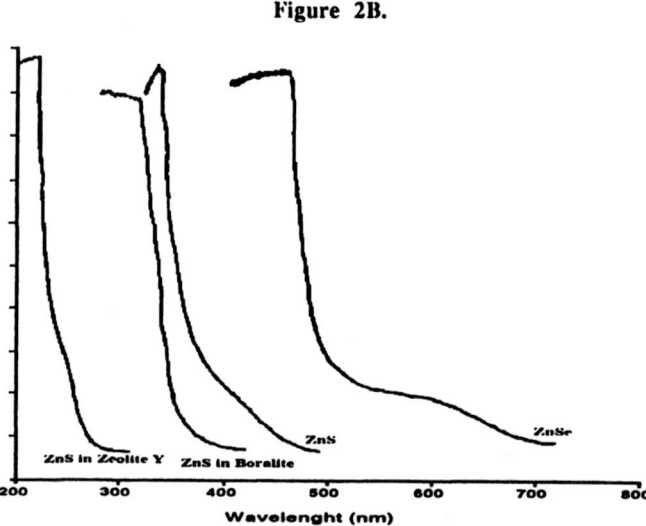

Figure 2B.

Comparison of the absorption spectra of ZnS in several hosts and the bulk.

Several features are noteworthy in the comparison of these systems. By varying the inter-cluster Cd-Cd atom contact from 6.2 Å in CdS-zeolite Y to the closest next-nearest neighbor Cd-S distance of 4.4 Å in CdS-helvite, we observe a red shift from 380 nm to 505 nm in the absorption edge. All of the small particle semiconductors have absorption edges which are blue shifted compared with the bulk semiconductor. Importantly, the CdS helvite, CdS boralite, and the ZnS boralite have the sharpest absorption bands, even sharper than those of the corresponding bulk semiconductors. The sharpness of this absorption corresponds to the fact that the semiconductor "units" are monodisperse throughout the entire structure with complete control over their cluster-cluster contact distances.

CONCLUSION:

We have demonstrated that II-VI semiconductor particles can be formed with essentially identical size in well ordered periodic arrays. Results of Rietveld analysis indicate that tetrahedral M_4X^{6+} clusters are formed in the cages of the sodalite-type host and are allowed to interconnect through channels in the host framework. Extremely sharp absorption features are observed in these materials, consistent with a narrow distribution in the "particle" size, and regular 3-dimensional quantization with small potential barriers. The controlled size and 3-dimensional periodicity of these materials make them unique candidates for applications where a sharp, specific absorption is required, shifted into the blue relative to the bulk material. In all other preparations, a size distribution is produced, and this distribution complicates the analysis of optical results. Since all of these materials crystallize in a noncentrosymetric space group, second order nonlinear optic properties are currently being studied.

ACKNOWLEDGEMENT:

GDS acknowledges the support provided by the Office of Naval Research and the DuPont company. We thank N. Herron and Y. Wang (DuPont) for assistance with the optical measurements.

REFERENCES:

1. For a recent review of this area see: A. Henglein, Top. Curr. Chem. 143, 113 (1988).

2. J. E. Mac Dougall, H. Eckert, G. D. Stucky, N. Herron, Y. Wang, K. Moller, T. Bein, D. Cox, J. Am. Chem. Soc. 111, 8006 (1989), and references therein.

3. For a recent review see: K. Ploog, Angew. Chem. Int. Ed. Engl. 27, 593 (1988).

4. Y. Wang, N. Herron, W. Mahler, A. Suna, J. Opt. Soc. Am. B. 4, 808 (1989).

5. Zeolite Y is structurally derived from sodalite units·which are joined through double 6 rings, and this arrangement yields the formation of "Supercages" with diameters of 13Å, and pore opening of 8Å. Like wise zeolite A is formed by sodalite units, that are joined through double 4 rings, while this arrangement gives rise to the formation of "α" with 4Å openings and 10Å. See Figure 1 for a description of the sodalite structure and for a complete introduction to zeolites see: D. W. Breck, Zeolite Molecular Sieves, (John Wiley & Sons, New York, 1974).

6. This composition has previously been reported, JCPDS card #24,1439. The ZnSe boralite and all the mixed systems are reported for the first time.

7. Note that ZnB_2O_4, JCPDS card #9,107, could not be reproduced by solid state reaction. The stoichiometry to produce ZnB_2O_4 gave a mixture of ZnO and ZnB_4O_7 at 600°C as judged by X-ray powder diffraction.

8. O. K. Mel'nikov, B. N. Litvin, and S. P. Fedosva, in Hydrothermal Synthesis of Crystals, edited by A. N. Lobachev (Consultants Bureau, New York, 1971) p. 119.

9. P. Smith-Verdier, S. Garcia-Blanco Z. Kristallogr. 151, 175 (1980).

10. A. C. Larson, R. B. Von Dreele, GSAS Users Guide, Los Alamos Report, LAUR 86-748 (1988)

THE PREPARATION OF III-V SEMICONDUCTORS IN AQUEOUS SOLUTION

TOBY J. CUMBERBATCH* AND ANDREW PUTNIS**
*Department of Engineering, University of Cambridge, Trumpington Street,
Cambridge CB2 1PZ, England.
**Department of Earth Sciences, University of Cambridge, Cambridge CB2 3EQ, England.

ABSTRACT

Indium arsenide colloids have been prepared by passing arsine through an aqueous solution of the metal nitrate. Similar experiments with gallium solutions produced an extremely fine precipitate which has been tentatively identified as amorphous gallium arsenide. Attempts to precipitate aluminium arsenide from aqueous solutions or mixed organic solvent systems were unsuccessful. The materials were characterised by transmission electron microscopy, X-ray powder diffraction and and optical absorption spectroscopy.

INTRODUCTION

The electronic properties of the Group III metal arsenides have made them very attractive candidates for optoelectronic applications. In particular, gallium arsenide possesses an energy gap for which the photovoltaic conversion efficiency is a maximum with AM2 solar radiation. This investigation was stimulated by the possibility of being able to deposit thin films of Ga(Al)As from colloidal suspension for the economic fabrication of thin film solar cells.

There are very few reported investigations into the synthesis of the arsenides from aqueous solution although Cotton and Wilkinson [1] say that "these hydrides are strong reducing agents and react with solutions of many metal ions .. to give the phosphides, arsenides .. or a mixture of these with the metals". In 1923, Brukl [2] prepared copper arsenide by passing arsine through a cupric salt solution. Later work found that the stoichiometry of the precipitate was critically influenced by the pH of the solution [3]. Kulifay [4] has reported an aqueous route for the preparation of some arsenides in which reducing solutions were used to produce elemental As and Cu or Ni from their oxides. We were unable to repeat these results.

The only reported solution synthesis of the arsenide semiconductors is an US Patent filed by Stahl [5] in 1975. This states that the entire III-V materials family and their intermediate compounds may be prepared by passing the appropriate Group V hydride through an aqueous solution of the Group III metal. It is difficult to understand some of the claims in view of the spontaneous decomposition of bismuth hydride at temperatures exceeding -45°C.

This investigation into the preparation of aluminium, gallium and indium arsenides using solution chemistry has found that only InAs can be synthesised with any degree of certainty. The results reported here suggest that the behaviour of the constituent ionic species in aqueous solution leads to a low material yield and colloidal suspensions with short lifetimes.

CHEMISTRY

The low aqua-ligand exchange rates for the three Group III metals of interest and the low aqueous solubility of AsH_3 result in very slow reaction rates. Furthermore, the virtually unknown properties of the aqueous arsenide ion and the sparsely documented aqueous chemistry of arsine make it difficult to optimise the colloid preparation conditions. Data are available for the three dissociation reactions [6] which, in conjunction with a reported solubility of 200ml of AsH_3 per litre of water, yield initial molar concentrations of 1.53×10^{-27} for the As^{3-} anion, 3.20×10^{-21} for HAs^{2-} and 3.23×10^{-13} for H_2As^-. Hence, the probability of finding an As^{3-} anion is extremely small.

$$H_3As \rightarrow H^+ + H_2As^- \qquad [K_1 = 5.62 \times 10^{-3}]$$

$$H_2As^- \rightarrow H^+ + HAs^{2-} \qquad [K_2 = 1.70 \times 10^{-7}]$$

$$HAs^{2-} \rightarrow H^+ + As^{3-} \qquad [K_3 = 3.95 \times 10^{-12}]$$

In aqueous solution, the Group III metals are present as hydrated cations, $[M(H_2O)_6]^{3+}$, and related species after proton loss. Their aqua exchange rates are reproduced in Table I together with those for methanol and a selection of other solvents (DMFA - Dimethylformamide)[7].

Table I: Ligand Exchange Rates

Cation:	Solvent:	Exchange Rate $\log_{10}k_{25}$: s^{-1}
Al^{3+}	H_2O	-0.8
	MeOH	3.6
	DMSO	-1.2
	DMFA	-0.8
In^{3+}	H_2O	4.3
	MeOH	?
Ga^{3+}	H_2O	3.3
	MeOH	4.0
	DMFA	0.2
Cd^{2+}	H_2O	8.2

We initially used 0.01M Cd^{2+} solutions, for which the aqua-ligand exchange rate is high (see above), to explore the use of arsine for the precipitation of metal arsenides from aqueous solution. A single phase material was precipitated rapidly, whose X-ray powder pattern indexed to that for Cd_3As_2. Our preliminary investigation into the preparation of III-V colloids from 0.01M aqueous solutions of the metal nitrates indicated that the reaction rate was limited by the cation ligand exchange rate and not the supply of As^{3-} anions as might be expected. Further experiments revealed that the reaction rate between AsH_3 and aqueous solutions of In^{3+} was moderate, for Ga^{3+} it was very slow whilst no reaction at all was observed in Al^{3+} solutions. Mixed solvent systems were therefore investigated as a means of increasing the reaction rate. Pure organic solvents have the disadvantage of a reduced dielectric constant and consequent influence on the colloid stability,

Methanol appeared to be an ideal solvent since the ligand exchange rate for Al^{3+} increases, by a factor of more than 10^4, and that for Ga^{3+} also shows a significant increase. However, AsH_3 reacts with MeOH in 0.01M solutions, precipitating an impurity compound which was also found in concentrated aqueous solutions. It is interesting to note that the reaction rate with the solvent was extremely slow; an 'incubation' period of approximately 30 minutes was required before a precipitate was observed in Ga^{3+} and Al^{3+} solutions. For In^{3+} solutions, the reaction rate was similar to that observed in aqueous solution. For H_2O/MeOH solutions, the aqua-ligand exchange rate was observed to dominate from which it was inferred that the cations were solvated preferentially by H_2O.

Data for the behaviour of solvated Group III cations in other solvents are very sparse. We investigated the solubility of $Al(NO_3)_3$ in ethanol, diethyl ether, DMSO and acetonitrile; of these, only ethanol proved a suitable host. A 0.02M mixed solvent solution was prepared in which the EtOH was diluted 9:1 with MeCN to reduce the possibility of reaction with the alcohol; it having already been established that MeCN was unconditionally stable in the presence of AsH_3. However, no reaction was observed, presumably because the exchange rate was too slow.

Attempts to produce ternaries of the type $Ga_xAl_{1-x}As$ from aqueous solutions were unsuccessful. The Al^{3+} appears to exert no influence on the reaction rate in either aqueous or H_2O/MeOH solutions.

COLLOID PREPARATION AND BEHAVIOUR IN AQUEOUS SOLUTION

Arsine was generated by dripping concentrated HCl onto Zn_3As_2 (2 - 3g) in a conical flask and passed into a nitrogen line for transport to the precursor solution. A 500ml triple necked flask, placed on a magnetic stirrer, contained 250ml of the metal precursor solution at room temperature into which the gas was introduced through a fine mesh gas distribution tube. Prior to generation of the hydride, the system was purged with nitrogen for about 30 minutes to remove any dissolved oxygen from the solution. The nitrogen flow rate was then reduced to a minimum when the hydride generation was started. Different reaction rates in the indium and gallium solutions called for a specific procedure for each cation.

Indium arsenide was the easiest material to synthesise with the highest yield and fastest reaction rate. Precursor solutions of concentration $\leq 0.01M$ were found to produce a uniform single phase precipitate. Attempts to increase the yield by increasing the solution concentration to 0.1M resulted in the deposition of an arsenic mirror, on the inner surface of the reaction flask, and the formation of impurities. An indication of the reaction rate was obtained by recording the intensity of the 90° scattering of a He-Ne laser beam after introduction of the hydride. The appearance of colloidal particles in a 0.008M solution was observed approximately 3 minutes after the introduction of the hydride. The solution continued to darken for a further forty five minutes, long after the hydride supply was exhausted. Flocculation occurred within another hour.

The behaviour of gallium solutions was markedly different to that which would be expected from a simple consideration of the ligand exchange rates. An 'incubation' period of several hours was required before the presence of any colloidal particles could be detected. Indeed, the particles were so fine that they could only be observed by the scattering of a He-Ne laser beam and not by the naked eye. In dilute solutions ($\leq 0.01M$), these particles subsequently disappeared, presumably due to an increase in HNO_3 activity. Further additions of the hydride after extended periods of time did not produce a lasting solid phase. In more concentrated solutions ($>0.05M$), a very small quantity of material was recoverable after the solution had been allowed to stand for two to three days.

Thin films of these materials were deposited onto stainless steel and tin oxide coated glass substrates by electrophoresis; the colloidal particles deposit on the cathode, which reveals that they are positively charged. The films were thin and not very adherent due to the simultaneous evolution of gas by electrolysis. Attempts to reduce the ionic concentration through the addition of MeOH resulted in colloid flocculation and, at high potentials, the tin oxide was attacked electrochemically.

MATERIAL CHARACTERISATION AND DISCUSSION

The particles from each experiment in aqueous and non-aqueous solution were carefully washed in deionised water and dispersed in ethanol prior to examination in the transmission electron microscope (TEM). The principal product, in all but the most dilute aqueous In^{3+} and Ga^{3+} solutions, was characterised by an extremely fine dispersion of clustered particles overlaid with material which was very beam absorbent. The spacings from the diffuse rings in the electron diffraction pattern matched those obtained from the X-ray powder pattern from the same material (Figure 1). This is an impurity phase which remains to be identified. The spacings, at 5.74Å, 2.88Å and 1.77Å do not index to an arsenate, arsenic oxide or elemental arsenic as might be expected. Moser and Brukl [8] reported the slow formation of a brown precipitate in aerated water containing

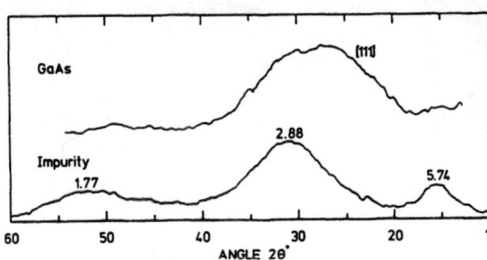

Figure 1. X-Ray Diffractometer Powder Patterns from GaAs and the unknown Impurity

dissolved AsH_3 which they attributed to solid As_2H_2. It has not been possible to find the powder pattern for this in the standard powder diffraction files.

Gallium arsenide presented more of a problem in view of the very small quantity of material available. However, sufficient precipitate was available from a 0.1M aqueous solution to generate the X-ray diffractometer powder pattern reproduced in Figure 1. This compares very favourably with the results obtained from amorphous films deposited by DC sputtering [9] although it bears little relation to crystalline GaAs. The width of the {111} peak, centred at $2\Theta \approx 27.5°$, is identical to that from the amorphous material whilst the amplitude of the second, just visible at $2\Theta \approx 49°$, lying in between the {220} and {311} reflections, is smaller than that observed by Mahavadi [9]. Since the

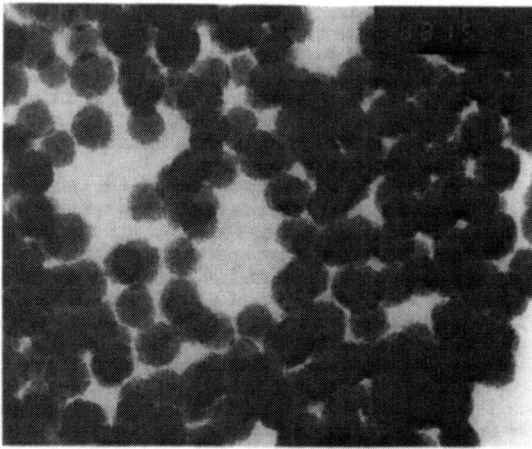

Figure 2. Colloidal InAs Particles

principal peak does not index to any of the possible products from the two precursors or the liquid medium, this spectrum is ascribed to GaAs on the basis of the optical absorption data discussed later. It was not possible to isolate any particles from the 0.008M solution for examination in the TEM.

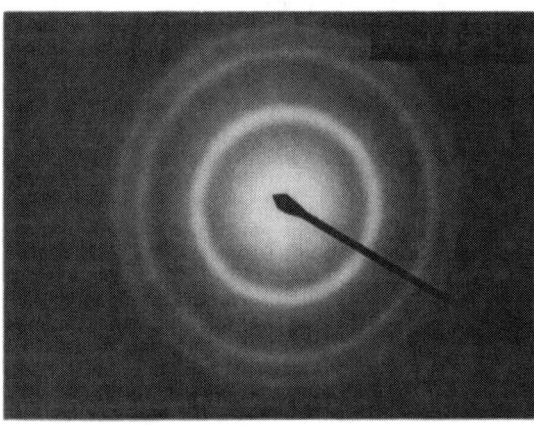

Figure 3. Electron Diffraction Pattern from InAs

The most uniform material was prepared in a 0.008M aqueous solution of $In(NO_3)_3$ as illustrated in Figure 2. From this it is evident that the spread in particle size is small and there is no evidence for clustering. The particles, whose mean diameter was 620Å, produced an electron diffraction pattern (Figure 3) from which the measured lattice spacings were in excellent agreement with those in the standard reference. At higher magnifications, lattice fringes were observed indicating that the particles are crystalline. However, the particles are not single crystals but are comprised of a very thin amorphous layer which surrounds a solid core of crystalline domains, each of which occupies an area of approximately 100Å x 100Å. The complete particle is occupied by such domains which are randomly aligned with respect to their neighbours and between which there are no voids. The precipitate from a 0.1M aqueous In^{3+} solution, also examined in the TEM, was characterised by aggregates of very fine clustered particles. The electron diffraction pattern from this material indicated that a predominance of the impurity phase although an X-ray diffractometer powder pattern from the bulk material was indexed to InAs. This suggests that the semiconductor may have aggregated and that the TEM sample was not therefore representative of the bulk material.

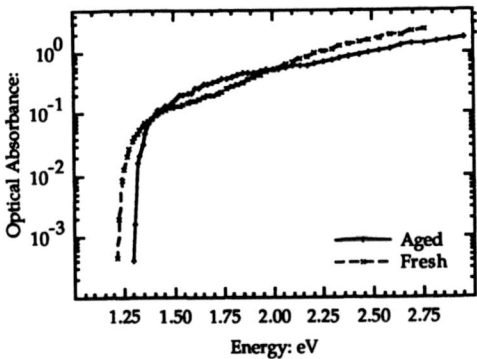

Figure 4. Optical Absorption Plot for GaAs

The optical properties of these materials bear an inverse relationship to their crystallinity. Optical absorption spectra from thin films of GaAs, electrophoretically deposited on to tin oxide coated glass substrates, exhibit a well defined absorption edge as illustrated in Figure 4. The fresh material was deposited from the 0.008M solution whilst still in suspension and the aged material from a more concentrated (0.02M) solution after it had been allowed to stand overnight. These spectra exhibit sharper edges than those reported for hydrogenated amorphous GaAs deposited by vacuum techniques [9, 10]. This is surprising in view of the lack of long range order suggested by the X-ray diffraction spectrum. The corresponding Tauc plots shown in Figure 5 reveal an extremely good fit for an amorphous material with intercepts on the energy axis ($\approx$1.4eV and $\approx$1.5eV) which lie close to the bandgap of crystalline material (1.42eV). The bandgap of hydrogenated amorphous GaAs is strongly dependent on the hydrogen content, with values only approaching those of crystalline material when deposited in a high partial pressure of H_2. This suggests that either this material does not contain a high concentration of dangling bonds or that the concentration of shallow defects is low.

The difference in behaviour between the fresh and aged colloids may be related to the particle size. An increase in the magnitude of the bandgap for semiconducting particles with very small dimensions has been reported previously [11]. If the aged particles have grown in solution, the effective bandgap would have changed correspondingly. Alternatively, this

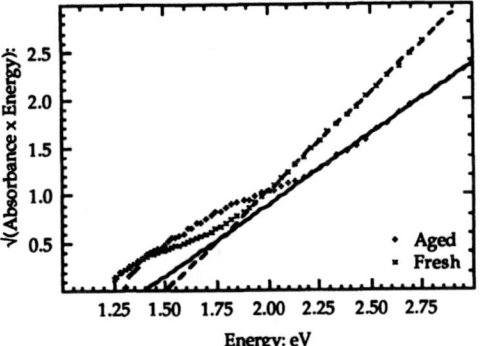

Figure 5. Tauc Plot for GaAs

aged material may have contained a very high concentration of shallow defects which reduced the measured value of the gap.

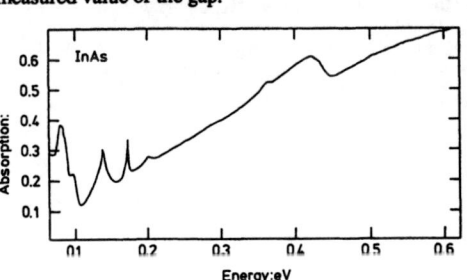

Figure 6. Absorption Spectrum from InAs/KBr

An optical absorption spectrum was also obtained from some InAs particles embedded in KBr (see Figure 6). The absence of an asymptotic value for the absorption may have been caused by light leakage through the sample due to a non-uniform dispersion of the InAs in the host medium. It was not therefore possible to obtain a value for the reflectivity which precluded calculation of the absorbance and an estimate for the value of the energy gap. The increasing optical absorption in this region of the spectrum corresponds to the value of the energy gap expected for crystalline InAs

(0.36eV); the size of these particles may lead to a larger value for the gap. The absorption peaks in this spectrum are not generally understood unless they are related to the size of the microcrystalline domains. The low energy peaks are too large for optical phonon activity but may be related to absorption by gap states.

CONCLUSIONS

This work has shown that the synthesis of the Group V metal arsenides in an aqueous solution of arsine is controlled by the ligand exchange rate of the solvated metal cation. The corresponding behaviour of the aqueous AsH_3 is also controlled by the ligand exchange rate. In concentrated In^{3+} solutions, a heavy precipitate is accompanied by an arsenic mirror whilst, for Ga^{3+} and Al^{3+} solutions under the same conditions, the reaction rate is barely observable. Attempts to reduce the contrasting behaviour between the different metals with mixed solvent systems were unsuccessful.

The preparation of AlAs in any solvent does not appear to be a possibility as is the incorporation of aluminium into a related ternary. Gallium solutions yield a colloidal material whose optical properties can be surprisingly well explained in terms of an amorphous semiconductor with very short range order. The most uniform material, prepared in dilute In^{3+} solutions, had a crystal structure which confirmed that InAs had been synthesised but for which the optical absorption spectrum did not exhibit a well defined band edge.

Further work will investigate the influence of different metal precursors and pH on the reaction rate in addition to obtaining more conclusive evidence for the synthesis of GaAs.

Acknowledgements

The authors would like to thank Dr.B.Guttler (Department of Earth Sciences) for the IR absorption measurements on InAs, Dr.E.C.Constable (Department of Chemistry) for useful discussions and Mr.M.D.Spencer (STC Ltd) for some of the early experimental work. This work was partially funded by Prudential Venture Managers Ltd., London.

References

1.   F.A.Cotton and G.Wilkinson, Advanced Inorganic Chemistry, 5th ed. (Wiley Interscience, New York, 1988) p. 391.
2.   A.Brukl Z. Anorg. Chem. 131, 236 (1923)
3.   J.Naud and P.Priest Mat. Res. Bull. 9, 337 (1974)
4.   S.M.Kulifay J. Am. Chem. Soc. 83, 4916 (1961)
5.   H.A.Stahl U.S. Patent No. 3 925 698 (1975)
6.   R.C.Weast, CRC Handbook of Chemistry and Physics, 68th ed. (CRC Press Inc., Boca Raton, 1987) p. D-163.
7.   J.Burgess, Metal Ions in Solution, (Ellis Horwood, Chichester, 1978) p. 317.
8.   L.Moser and A.Brukl Monatsh. 45, 25 (1924)
9.   K.K.Mahavadi, Ph.D. Thesis, University of Cambridge, 1985.
10.   W.Paul, T.D.Moustakas, D.A.Anderson and E.Freeman, in 7th Int. Conf. on Liquid and Amorphous Semiconductors, ed. by W.E.Spicer (CICL Univ., Edinburgh, 1977) p. 467
11.   R.Rosetti and L.Brus J. Phys. Chem. 86, 4470 (1982)

ROOM TEMPERATURE EXCITONIC ABSORPTION IN SMALL CdS CRYSTALLITES

D.K. RAI* AND BINOD KUMAR
University of Dayton Research Institute, Dayton, OH 45469
*Permanent address: Department of Physics, Banaras Hindu University,
Varanasi-221005, India

ABSTRACT

The absorption characteristics of commercial CdS-containing yellow glass which shows constant transmitted intensity over a range of incident CW laser intensity have been studied at room temperature. Although the thick specimen (t>0.6 mm) shows only a broad step-like feature near λ>460 nm, a thin (t~0.09 mm) specimen shows two absorption features which can be interpreted as the first two quantum-confined exciton absorption features corresponding to a crystallite size of ~45 Å. The absorption spectrum of a sample (t~0.6 mm) heated for 15 min. at 700°C shows two new absorption features at 450 nm and 380 nm, which correspond to a much smaller crystallite size of ~25 Å. This reduction in size is not inconsistent with estimates made from a well-known model for crystallite growth. Some consequences of these changes in the absorption features on the optical nonlinearities of the glass will be discussed.

INTRODUCTION

Materials with large nonlinear optical parameters are eagerly sought because of their proven and perceived capabilities as essential components in modern opto-electronics. Such applications include generation of coherent and non-coherent radiation at desired wavelengths, image modification, removal of optical aberration, optical signal processing, and optical computing [1]. In addition to the significant value of the nonlinear optical coefficients, these materials are expected to have certain additional properties such as the use of materials at low laser power and low cost with a reasonable life span in an unprotected environment.

The first observation that ordinary filter glasses containing CdS_xSe_{1-x} microcrystals have optical nonlinearities was made by Bret and Gires [2] in 1964 when they used the filter for Q-switching of ruby lasers. However, the physical origin and related mechanisms of the nonlinear effect were not explained. Interest in these glasses was revived after the work of Jain and Lind [3] who investigated degenerate four-wave mixing and reported a fairly strong phase conjugate signal, thereby implying a reasonably large third-order susceptibility. The large nonlinearity is believed to originate from the incident radiation created electron-hole plasma which causes a "band filling effect."

We have conducted an experimental study to investigate the nonlinear effects and room temperature excitonic absorptions in a commercial CdS-doped glass. This paper presents results and analyses of the study.

EXPERIMENTAL

Several 10x4 mm disc-shaped specimens were obtained from a bulk glass of composition SiO_2=76.66%, K_2O=4.88%, CaO=3.51%, ZnO=5.75%, BaO=0.32%, and CdS=0.88% commercially produced by Holophane Co., France. The transmission spectrum and optical density of the specimens were measured using a Beckman

5270 spectrophotometer. The laser transmission measurements as a function of intensity were conducted on 15 specimens using two lasers: the copper vapor laser emitting both 510 and 578 nm lines and a single line argon ion laser operating at 488 nm. A spot size of approximately 3 mm was used for both of the lasers. The transmission was measured as a function of incident laser power using a broadband detector and power meter. The power was progressively increased until the specimens fractured. The maximum power levels used were $\sim$50 watts/cm^2 and 10 watts/cm^2 for copper vapor and argon ion lasers respectively.

RESULTS AND DISCUSSION

Figure 1, the transmission spectrum of the specimen, shows the location of the absorption edge around 500 nm. The band gap calculated from the spectrum is 2.54 eV which is different from the band gap of the bulk CdS, 2.42 eV. The difference may have resulted from the confinement effect of CdS microcrystals in the glass. Figure 1 also shows the location of the laser lines employed for the laser transmission measurement.

A plot of incident versus transmitted intensity for the copper vapor laser is shown in Figure 2. At low energy levels, up to about 22 W/cm^2, the material shows a linear behavior. As the incident intensity increases from 22 watts/cm^2 up to 38 watts/cm^2, a remarkably wide plateau region is observed. With further increase in the incident intensity beyond 38 watts/cm^2, the transmitted intensity is enhanced and the specimen fractures around 50 watts/cm^2.

The argon ion laser transmittance as a function of intensity is shown in Figure 3. As expected, the transmittance of the specimen is much lower for this wavelength. Nonetheless, the data show behavior similar to the copper vapor laser experiment; again three regions are noted. Even at very low intensities, a slight nonlinearity is observed, followed by a plateau region and an enhanced transmittance region. It should be noted that the plateau region in this case begins at a much lower intensity, that is, at 3 watts/cm^2, and continues up to about 5.5 watts/cm^2. The fracture of specimens took place at 11 w/cm^2.

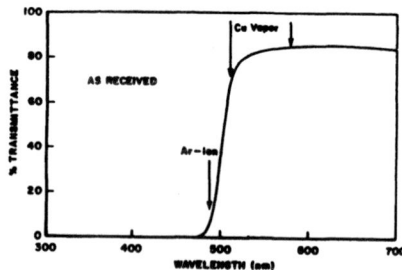

Figure 1. Transmission spectrum of CdS-containing glass.

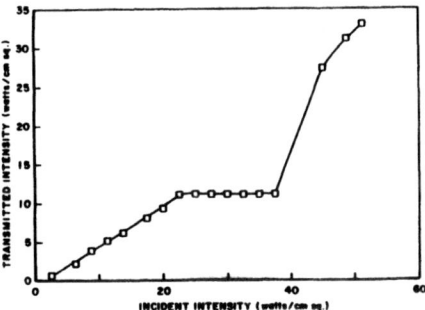

Figure 2. Transmission as a function of intensity of CdS-containing glass using multiline copper vapor laser (511-578 nm).

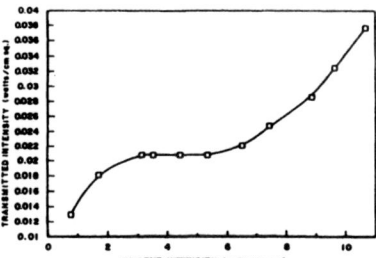

Figure 3. Transmission as a function of intensity of the CdS-containing glass using an argon ion laser (488 nm).

The observed data as shown in Figures 2 and 3 show an unusual region of constant transmitted power before the transmittance starts to increase superlinearly implying absorption saturation. The intensity limiting transmission has been reported for CW laser irradiation in semiconductors [4] as well as in waveguides [5] formed from microcrystallite containing glasses. Theoretical arguments which led to observation of transmitted intensity being independent of incident laser intensity have also been presented by Jungk [6] and Kochelap and co-workers [7]. The glass used in this study also showed significant optical phase conjugate signal [8] in a degenerate four-wave mixing experiment using the second harmonic of Nd-YAG 1.06 μm radiation.

These observations prompted us to make a systematic study of the absorption spectrum of these glasses. At first an attempt was made to modify the size distribution of CdS crystallites in the glass by heat treating the glass at various temperatures, e.g., 500°C, 600°C, 700°C for approximately 15 min. It was noted that the effect of heat treatment up to 650°C was rather minimal and the transmittance spectrum remained almost unchanged, but when the treatment temperature was increased to 700°C and 750°C, a large blue shift of the absorption edge occurs as shown in Figure 4. The blue shift may result from the quantum confinement effect and a possible formation of $Cd_{1-x}Zn_xS$ solid solution as reported by Borelli, et al.[9] on similar glass compositions. The evidence of quantum confinement and resulting excitonic absorption features have been investigated and presented in Figures 5 and 6. The possibility of $Cd_{1-x}Zn_xS$ solid solution formation at these heat treatment temperatures are currently investigated.

Figure 5 shows the absorption spectrum of a thin specimen (0.09 mm) of the as-received glass. Two absorption bands located around 450 and 480 nm are noted. Converting these energies to eV and making use of the Ekimov and Efros [10,11] analysis (identifying them as the j=1 and j=2 transitions) ΔE_1 and ΔE_2 are calculated to be 0.14 eV and 0.25 eV, respectively. The ratio $\Delta E_1/\Delta E_2$ is equal to 0.56 which is somewhat larger than the theoretical value of $X_1^2/X_2^2=0.49$. If one assumes that the reduced mass of the electron-hole pair in the CdS microcrystallite is ~0.2 m_e then the average size of the microcrystallites is estimated as 40-45 Å.

Figure 6 shows the absorption spectra of the CdS-doped glass heat treated at 700°C for 15 min. and then rapidly cooled. Two absorption bands are again observed but their positions have shifted toward the blue as compared to the band positions in Figure 5. The ΔE_1 and ΔE_2 were calculated to be 0.40 eV and 0.84 eV, respectively and the ratio $\Delta E_1/\Delta E_2$ to be $\simeq 0.50$

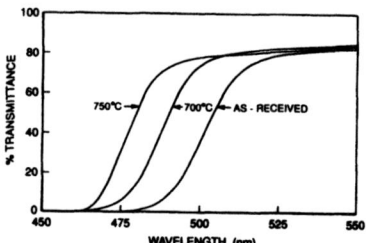

Figure 4. Transmittance spectrum of as-received and heat treated CdS-doped glass specimens.

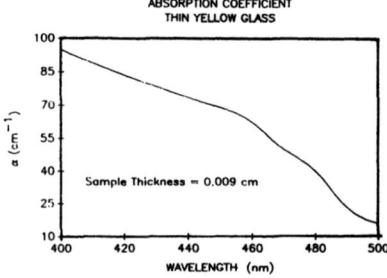

Figure 5. Absorption spectrum of a thin ($\simeq$ 0.09 mm) CdS-doped as-received specimen.

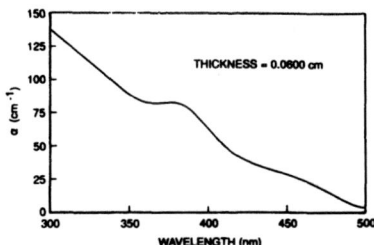

Figure 6. Absorption spectrum of a CdS-doped glass specimen heat treated at 700°C for 15 minutes.

which is much closer to the expected value of 0.49. The size of microcrystallite (with $\mu \sim 0.2\ m_e$) is estimated to be $\sim$20-25 A.

The reduction of particle size of CdS in the heat-treated specimen (Figure 6) is believed to result from the combined effect of dissolution of CdS particles at 700°C and the rapid cooling that prevents the growth of the CdS crystallites. The reduction of the particle size suggests that the heat treatment techniques may be tailored to develop crystallites of desired size and excitonic absorptions.

The remarkable low threshold for the nonlinear behavior of the CdS doped glass as shown in Figures 2 and 3 may be attributed to the following.

(i) The glass in the present case is believed to be saturated with CdS crystallites as the color was spontaneously obtained during the forming process. This is in contrast to most of the other investigations carried out on semiconductor-doped glasses, particularly $CdS_{1-x}Se_x$-containing materials in which case the glass remains relatively colorless after forming and the $CdS_{1-x}Se_x$ particles are nucleated and grown during a secondary thermal treatment process. These $CdS_{1-x}Se_x$-doped glasses are believed to be less saturated than that of the CdS-doped glasses studied in this investigation. The increased concentration of CdS is expected to increase the absorption coefficient of the composite material [12].

(ii) The size of the crystallites in the present case is estimated to be around 45 Å. This decreased size of the crystallites causes [9,13] a considerable shift ($\geq$0.1 eV) of the effective band gap as compared to the bulk CdS band gap. Further, it is known [12] that generation of electron-hole pairs due to light may cause, under suitable conditions, an incident intensity dependent increase of the absorption of the glass which will reduce the transmittance. The reduced size of the crystallites suppresses the formation of excitons and favors free electron-hole pairs, which further enhances the above-mentioned intensity dependent absorption.

(iii) The contribution of thermal effects to the observed nonlinearity as reported by Han, et al. [5] is probable. The interaction of the laser beam with the CdS-doped material will lead to a significant temperature increase of the specimen, and the absorption edge would shift towards lower energy (longer wavelength) resulting in enhanced absorption with increasing intensity of the laser beam.

(iv) The room temperature excitonic absorption features are present in the as-received thin specimens. A blue shift of these absorption features occurs for the specimen heat treated at 700°C and rapidly quenched.

Further experimentation and analysis are in progress to determine the contributions of various mechanisms to the origin of the observed nonlinear effect.

ACKNOWLEDGEMENTS

The authors thank Mr. John Detrio, Dr. Nils Fernelius, and Mr. T. Pottinger for useful discussions. The assistance of Drs. D. Ballal and R. Becker for the laser transmittance measurements and Mr. Ron Shimovetz for specimen preparation is sincerely acknowledged.

REFERENCES

1. D.H. Auston and A.A. Ballman, P. Bhattacharya, et al., App. Opt. 26, 211 (1987).

2. G. Bret and F. Gires, App. Phys. Letts. 4, 175 (1964).

3. R.K. Jain and R.C. Lind, J. Opt. Soc. Amer. 73, 647 (1983).

4. F. Yang, D. Li, N. Tian, G. Xiong, and X. Xu, J. Lumin 40&41, 527 (1988).

5. S.K. Han, Z. Huo, R. Srivastava, and R.V. Ramaswamy, J. Opt. Soc. Am. B6 663-668 (1989).

6. G. Jungk, Phys. Stat. Sol. (b) 150, 483 (1988).

7. V.A. Kochelap, A.V. Kuznetsor, and V.N. Sokoler, Phys. Stat. Sol. (b) 150 489 (1988).

8. T. Pottinger (private communication)

9. N.F. Borelli, D.W. Hall, H.J. Holland, and D.W. Smith, J. Appl. Phys. 61, 5399 (1987).

10. A.K. Ekimov and Al.I. Efros, in Laser Optics of Condensed Matter, edited by J.L. Bironan, H.Z. Cummins, and A.A. Kaplyansku (Plenum, New York, 1988) p. 199.

11. A.I. Ekimov and Al.I. Efros, Phys. Stat. Sol. (b) 150, 627 (1988).

12. K.C. Rustagi and C. Flytzanis, Opt. Letts. 9, 344 (1984).

13. H. Jerominek, M. Pigeon, S. Patela, Z. Jakubczyk, D. Delisle, and R. Tremblay, J. Appl. Phys. 63, 957 (1988).

PARTICLE-SIZE DISTRIBUTION OF CdSe QUANTUM DOTS
DETERMINED BY PHOTOLUMINESCENCE SPECTROSCOPY

E.N. PRABHAKAR,* C.A. HUBER,** and D. HEIMAN, Francis Bitter National Magnet Laboratory, MIT, Cambridge, MA 02139

ABSTRACT

Particle-size distribution effects on the energy levels of semiconductor quantum dots are investigated. By examining the low temperature photoluminescence spectra of microcrystals of the binary semiconductor CdSe embedded in a glass matrix, the distribution of energy levels due to three-dimensional confinement is determined. Calculations of the electron-hole pair ground state energy provide a relation between confinement energy and particle diameter. This allows conversion of the photoluminescence lineshape directly into a distribution of particle radii and facilitates analysis of the observed properties of the material. With extension to other systems the technique can become a valuable tool in the study of semiconductor microparticle composites.

INTRODUCTION

Probably the simplest and best known system in quantum mechanics is the 'particle in a box,' or infinite square well potential. While of great pedagogical interest, there are few observable systems which correspond to it. This is beginning to change — in recent years, technological advances have led to the fabrication of *quantum-dot semiconductors* (QDS). For instance, submicron lithography techniques have been used to make QDS of quantum layered GaAs in (Al,Ga)As with diameters less that 1000 Å. Semiconductor particles of CdS and CdSe less than 100 Å in diameter have been made from colloidal crystallization [1] and crystallization in optical glass matrices [2] and in zeolites [3].

These glass matrix materials and multiple quantum well structures with QDS are of increasing interest to researchers in photonics and solid state physics. They may become useful in nonlinear optical devices for controlling light flow, as well as in nonlinear electronic devices [4]. The physics of zero-dimensional systems is an infant field, it spans the gap between the chemistry of large molecules and the bulk properties of macroscopic materials. This field of investigation is relatively new, and there are as yet only a few models of the physical interactions involved. [5].

The process of photoluminescence (PL) is well understood in the case of bulk semiconductors. Incoming light creates electron-hole pairs with excess energy, which rapidly decay to the minima of the energy bands. These then form excitons which eventually recombine, giving off energy in the form of light. The quantum confinement in QDS affects both the kinetic energy of the particles and potential energy of the exciton system, and hence the recombination energy. The finite size also leads to surface polarization effects, which further increase the energy. Previous studies have been forced to rely on counting particles in electron micrographs to determine the distribution of particle sizes [2]. More recently, optical absorption measurements in combination with photoluminescence data at room temperature have been employed to deduce the particle-size distribution of semiconductor microcrystals in commercial color glass filters. [6]

We here report on low-temperature photoluminescence studies of experimental CdSe composite glasses. A simple model connecting the photoluminescence spectra with the distribution of particle radii provides us with a quick, reliable means of determining the size-distribution of particles in the material.

THEORY

Photoluminescence results from radiative recombination of photogenerated electrons and holes in the semiconductor. When a macroscopic sample of CdSe is illuminated by light above the band gap, an electron is excited into an empty state in the conduction band, leaving behind a hole in the valence band. Both of them loose kinetic energy as they relax toward the band minima (at the center of the Brillouin zone in CdSe), where they recombine. The recombination energy is slightly lower than the band gap energy E_g (1.841 eV in CdSe at low temperatures) by an amount $E_x = R_y \mu/m_o \varepsilon_1^2$, the exciton energy, due to the Coulomb interaction between the electron and the hole. Here $R_y = 13.6$ eV is the atomic Rydberg, $\varepsilon_1 = 10.6$ is the static dielectric constant of CdSe. $\mu = 0.11\, m_o$ is the reduced electron-hole effective mass calculated using $m_e = 0.13\, m_o$, $m_h = 0.56\, m_o$ for the hole effective mass appropriately averaged over the crystal directions parallel and perpendicular to the hexagonal c-axis. A value $E_x = 12.8$ meV is obtained in this way.

In a quantum dot, the situation is somewhat different. The energy gap is the same, since there are still enough atoms to form continuous bands. However, the ground state energy of the particle is slightly above the energy gap. The confinement in space results in a finite momentum, and hence a kinetic energy term. The difference in the optical band gaps of the semiconductor and the glass being about 2 eV, the dot is usually modelled as an infinite spherical potential well for both electrons and holes. For the ground state the kinetic energy term due to confinement is $E_o = \hbar^2\pi^2/2\mu R^2$, where R is the radius of the particle. In addition, the electron-hole Coulomb energy changes to $E_c = \beta e^2/\varepsilon_1 R$, with $\beta = 1.8$ the scaling factor determined by confinement [7]. Finally, the energy due to the dielectric discontinuity between the semiconductor and the glass is given by $E_d = \gamma e^2/R$, with $\gamma = \langle \sum_{n=1}^{\infty} \alpha_n (\frac{r}{R})^{2n} \rangle$ and $\alpha_n = (\varepsilon-1)(n+1)/\varepsilon_1(\varepsilon n+n+1)$ where $\varepsilon = \varepsilon_1/\varepsilon_2$, $\varepsilon_2 \approx 1.5$ the dielectric constant of the glass, and the brackets denoting an average over the ground-state wavefunction. For CdSe we find $\gamma = 0.060$ [8]. The lowest energy of an electron-hole pair in a CdSe microcrystallite is then:

$$E = E_g + E_o - E_c + E_d = E_g + \frac{\hbar^2\pi^2}{2\mu R^2} - \frac{\beta e^2}{\varepsilon_1 R} + \frac{\gamma e^2}{R} \tag{1}$$

The above equation can be rewritten as:

$$\frac{E-E_g}{E_x} = \frac{\pi^2}{(R/a_x)^2} - \frac{2\beta}{(R/a_x)} + \frac{2\varepsilon_1\gamma}{(R/a_x)} \tag{2}$$

with $a_x = a_o \varepsilon_1 m_o/\mu$ (a_o the Bohr radius) the exciton radius, calculated to be 53.2 Å by using the parameter values of bulk CdSe. Equation (2) represents a general scaled equation for quantum dot semiconductors with different E_g, μ and ε_1 parameters. It is accurate for cases where the kinetic energy is larger than the electrostatic energy, or analogously, when the particle size is smaller than the bulk exciton radius.

Introducing the variables $y = (E-E_g)/E_x$ and $x = R/a_x$, a dimensionless equation is obtained from (2):

$$y = \frac{\pi^2}{x^2} - \frac{2\phi}{x} \tag{3}$$

where $\phi = \beta-\varepsilon_1\gamma = 1.16$ for CdSe. Equation (3) represents a quadratic equation for x which is solved to give:

$$x = \frac{\phi}{y}\left[\left(1 + \frac{\pi^2}{\phi^2}y\right)^{1/2} - 1\right] \tag{4}$$

The above equation connects the ground state energy of an electron-hole pair with the size of the semiconductor crystallite.

EXPERIMENTAL

The experimental CdSe composite glasses were fabricated at Corning Glass Works. The microcrystalline CdSe phase was obtained by heat treatments at 600°C for 2 hours, sample A, and at 675°C for 0.5 hour, sample B. The samples have an orangish-red color. Transmission electron micrographs of samples prepared under the same conditions as sample A indicate this sample contains particles with diameters in the range from 25 to 35 A [9]. Sample B, which was annealed at a higher temperature, is expected to contain larger particles. X-ray diffraction analysis shows the microcrystals to have the wurtzite structure of bulk CdSe [9].

The samples were excited with the 4579 Å (2.71 eV) line of an argon-ion laser, typical excitation intensities were 30 μW over an area of 10^{-3} cm². The luminescence spectra were analyzed with a 0.86 m double monochromator and a photomultiplier tube in the photon-counting mode. Figure 1 shows the photoluminescence spectra of both samples at T = 2 K. The spectra have been corrected for the spectrometer efficiency. The photoluminescence spectrum of sample A showed a broad low-energy peak centered at 1.595 eV, 86% of the height of the main peak at 2.20 eV. This low-energy peak has also been seen by previous authors in both commercial and experimental semiconductor-doped glasses, it is presumably due to impurities or defects but its origin remains uncertain. We modelled the low-energy peak as a gaussian with a FWHM of 0.600 eV and subtracted it from the spectrum of sample A. Figure 1 shows the corrected spectrum. Sample B did not show a clearly discernible peak at low energies, but rather the long low-energy tail exhibited in Figure 1.

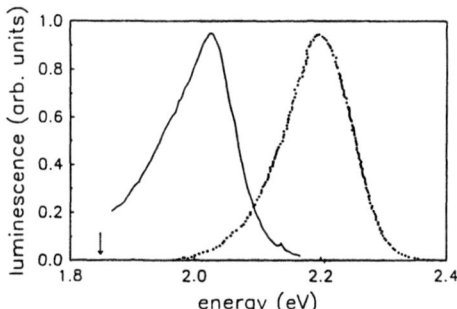

Figure 1. Low-temperature photoluminescence of CdSe crystallite-doped glasses, samples A (dotted line) and B (solid line). The arrow indicates the position of the energy band gap of bulk CdSe.

ANALYSIS AND DISCUSSION

The thermalization time of the photogenerated band electrons and holes, estimated to be of the order of 1 psec or less, is smaller than the 20-100 psec carrier lifetime reported for semiconductor-doped glasses [10]. Assuming the quasi-Fermi energy of electrons and holes to lie below the respective band edges by at least kT (~0.2 meV), at low temperatures the recombination radiation originates from electrons and holes in their lowest energy state, the thermal width of the energy distribution of the rapidly thermalized excess carriers being negligible in comparison to the spectral linewidths in Figure 1. Therefore, photons emitted at a given energy essentially originate from crystallites whose lowest excited state corresponds to that energy. Each photoluminescence spectrum in Figure 1 approximately represents v(E), the volume fraction of the crystallites with an energy band-gap E. The particle-numer distribution being given by n(E) ~ v(E)/R³. Through Equation (4) it is possible to convert the photoluminescence spectrum to a function of radius v(R) rather than energy. The distribution of particles over radii is then given by:

$$n(R) \sim \frac{v(R)}{R^3}\frac{dE}{dR} \sim \frac{v(R)}{R^5}\left[\frac{\pi^2 a_x}{R} - \phi\right]$$ (5)

The results of Equation (5) are shown in Figure 2 for each sample. Sample A shows a distribution with a maximum at 29 Å and approximately 4 Å wide consistently with the electron micrograph analysis. For sample B the distribution peaks at 39 Å and has an approximate width of 8 Å. Due to the presence of the additional low-energy peak in the photoluminescence spectrum of sample A and to the low-energy tail in the spectrum from sample B, the uncertainty is higher on the large-particle end of the distributions.

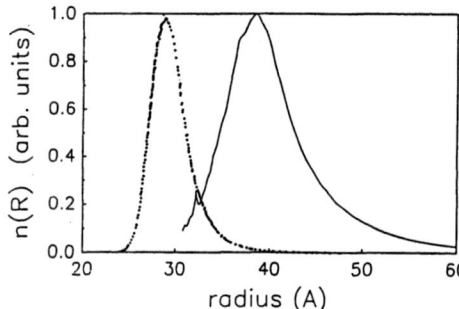

Figure 2. Particle-size distributions for sample A (dotted line) and sample B (solid line).

In the process of deconvoluting the particle-size distribution from the photoluminescence spectrum, the intrinsic low-temperature linewidth of the photoluminescence from a single crystallite has been neglected. Preliminary reports suggest that it may be as much as one-sixth of the total linewidth [9]. Transient spectral hole-burning experiments on colloidal preparations of 45 Å diameter CdSe crystallites reveal a homogeneous (single-particle) linewidth for the lowest allowed optical transition of 17 meV at low temperatures [11]. In that case, the dependence of the single-particle linewidth on the sample preparation procedure suggests that the broadening mechanisms are not intrinsic to the particle size.

SUMMARY

The spectroscopic method outlined provides a straightforward and reliable way of determining the distribution of crystallite sizes in the composite. With extension to other systems the technique can be a valuable aid in determining the distribution of particle sizes, facilitating analysis of its relationship to the observed properties of the material. This in turn would aid in providing composites with specific desired properties.

ACKNOWLEDGEMENTS

We thank N.F. Borrelli for supplying the excellent samples. The facilities at the Francis Bitter National Magnet Laboratory are supported by the National Science Foundation.

REFERENCES

*Present address: California Institute of Technology, Pasadena, CA 92215.

**On leave from the University of Puerto Rico, Rio Piedras, PR 00931.

1. R. Rossetti, R. Hull, J.M. Gibson, and L.E. Brus, J. Chem. Phys. **82**, 552 (1985); C.J. Sandroff, D.M. Hwang, and W.M. Chung, Phys. Rev. B33, 5953 (1986).

2. N.F. Borrelli, D.W. Hall, H.J. Holland, and D.W. Smith, J. Appl. Phys. **62**, 5299 (1987).

3. V.N. Bogomolov, V.V. Poborchii, S.V. Kholodkievich, and S.I. Shagin, Pis'ma Zh. Eksp. Teor. Fiz. **38**, 439 (1983) [Sov. Phys. JETP Lett. **38**, 533 (1983)]; Y. Wang and N. Herron, J. Phys. Chem. **91**, 257 (1987).

4. R.K. Jain and R.C. Lind, J. Opt. Soc. Am. **73**, 647 (1983).

5. Al. L. Efros and A.L. Efros, Fiz. Tekh. Poluprovodn. **16**, 1209 (1982) [Sov. Phys. Semicond. **16**, 772 (1982)]; G.W. Bryant, Phys. Rev. Lett. **59**, 1140 (1987).

6. P.D. Persans, A. Tu, Y.-J. Wu, and M. Lewis, J. Opt. Soc. Am. B6, 818 (1989).

7. L.E. Brus, J. Chem. Phys. **80**, 4403 (1984); **79**, 5566 (1983).

8. E.N. Prabhakar, B.S. Thesis, MIT, 1988 (unpublished).

9. N.F. Borrelli (private communication).

10. M.C. Nuss, W. Zinth, and W. Kaiser, Appl. Phys. Lett. **49**, 1717 (1986), and references therein.

11. A.P. Alivisatos, A.L. Harris, N.J. Levinos, M.L. Steigerwald, and L.E. Brus, J. Chem. Phys. **89**, 4001 (1988).

THERMAL ANNEALING OF AMORPHOUS CoMnNiO
FILM ON OXIDIZED Si SUBSTRATE

TAN HUI, QIN DONG AND TAO MINGDE
Xinjiang Institute of Physics, Academia Sinica
Wulumuqi 830011, Xinjiang, China
LIN CHENGLU AND ZOU SHICHANG
Ion Beam Laboratory,
Shanghai Institute of Metallurgy, Academia Sinica
Shanghai 200050, China

ABSTRACT

Amorphous CoMnNiO film is doposited on oxidized Si
substrate by RF sputtering equipment. Structure relaxation
occurs in the amorphous CoMnNiO film when it is annealed below
550°C. Annealed in the range from 600°C to 1000°C, the amorphous film is converted into the polycrystal. After annealing
in rich oxygen atmosphere, the amorphous film is transformed
into spinel solid solution with stable structure and good
electrical properties. The electrical conductivity will be
reduced due to formation of low valence oxides when annealed
without oxygen. As annealing temperature is higher than 1000°C,
some spinel solid solutions will be resolved into low valence
oxides CoO and NiO, reducing the conductivity of the CoMnNiO
film.

INTRODUCTION

Amorphous CoMnNiO film deposited by RF sputtering is a
practical material for fabrication negative temperature coefficient (NTC) thermistors [1,2], which exhibits a wide
operating temperature, a rapid response and a good stability.
Amorphous thin film is a substable structure, having
obvious thermal annealing characteristics. H. Matsunami et al.
[3] examined the influence of high and low temperature thermal
annealing on the structure and electrical properties of amorphous SiC thin film. But the thermal annealing behaviour on
amorphous CoMnNiO thin film have not been reported. In this
paper, the thermal annealing behaviour of amorphous CoMnNiO
thin film is investigated.

EXPERIMENTAL

The target for sputtering is made of the mixture of Co, Mn,
Ni by ratio for 3:2:1, which is sintered in oxygen atmosphere.
Amorphous CoMnNiO thin film is deposited by RF sputtering on
oxidized silicon substrate. The sputtering is performed in
argon or oxygen atmosphere for 4 hours. RF power is 200W,
substrate temperature is kept below 200°C. The thickness of
amorphous CoMnNiO film is about 1.5µm.
Thermal annealing of the samples in O_2 atmosphere is completed by a quartz tube stove and annealing temperature is from
500°C to 1000°C at the interval of 50°C to 100°C. Conventional
thermal annealing (CTA) in Ar atmosphere is carried at 550°C,
750°C and 850°C respectively for 4 hours and rapid thermal
annealing (RTA) in Ar atmosphere was performed at 1000°C,

$1100^{\circ}C$ and $1200^{\circ}C$ respectively for 30 sec. in a RF graphite heater.

Using Auger electron spectrum, x-ray diffraction, Raman spectroscopy, and resistance measurements, we have studied the compositions, structures and electrical properties of the CoMnNiO thin films.

RESULTS AND DISCUSSION

A. Conventional Thermal Annealing in Oxygen Atmosphere

The Auger electron peak intensity obtained by Auger electron spectrum measurements for the CoMnNiO samples before annealing is listed in table 1. The results indicate that the atomic ratio of Co, Mn, Ni and O in the CoMnNiO thin film is about 7:6:2:10.

Table 1. Auger electron peak intensity in CoMnNiO thin film before annealing

Sample	Content of O_2 in atmosphere (%)	Relative intensity				Normalized intensity			
		O	Mn	Co	Ni	O	Mn	Co	Ni
1	15	122	80.0	90.9	25.5	1.0	0.66	0.75	0.21
2	60	100	57.8	72.7	21.8	1.0	0.58	0.73	0.22

X-ray diffraction patterns of samples annealed in O_2 atmosphere are shown in Fig.1. Fig.2 shows the dependence of resistance on annealing temperature.

From Fig.1 and Fig.2, it is discovered that the CoMnNiO thin film before annealing is typical amorphous structure. As is known that the atoms from target are deposited on substrate at random. The composition of deposited films may be deviated from stoichiometric ratio of target material. And in the case

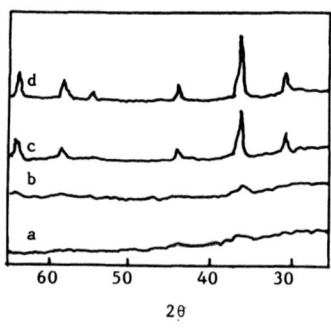

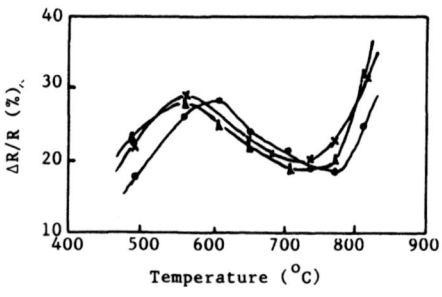

Fig.1 X-ray diffraction patterns of the CoMnNiO thin film before and after annealing in O_2 atmosphere
a. As-deposited   b. $550^{\circ}C$,4hr.
c. $750^{\circ}C$,4hr.   d. $850^{\circ}C$,4hr.

Fig.2 The dependence of resistance in thin film on annealing temperature
o, sample-1,   Δ, sample-2,
x, sample-3.

of low substrate temperature, deposited atoms can not arrange
in order. As a consequence, there are various defects in thin
film. During annealing at low temperature, such as 550°C, the
atoms in the thin film can not get sufficient energy to rearrange. Only bond angle and bond length in the thin film are
changed and adjusted, as well as some dangling bonds are compensated and over-coordination and disbounds are removed [4,5],
which results in reduction of the conductivity. Above phenomenon is called as the structure relaxation.

Annealed above 550°C, amorphous structure of the thin film
is crystallized and it is enhanced with increasing annealing
temperature. During crystallization, the atoms in the thin
film can draw thermal energy to rearrange, then amorphous structure is transformed gradually into polycrystal. In high temperature annealing, Co, Mn, Ni in the thin film may form a
serial cubic spinel structures or solid solutions which possess
near equal crystalline constants, higher solubility and stable
structure. When promoted annealing temperature further, the
crystal structure are perfected. Annealed in rich oxygen atmosphere, sufficient oxygen is provided for formation of cubic
spinel solid solutions. From results of x-ray diffraction
spectra and compositions of the target, the chemical structure
is written as

$$Co [MnCo_xNi_{1-x}] O_4$$

In crystallization and transforming, the changes of electrical properties in the thin film are due to different properties of spinel solid solutions. As is known, the conductivi-
ty σ of material can be represented as $\sigma=ne\mu$. Where n, e and μ
are electron concentration, a electron charge and mobility respectively. The carrier concentration in spinel solutions is of
variable valence ions in the material. Annealing in temperature range from 550°C to 750°C, the reduction of resistivity
indicates increasing the compositions which contribute to
variable valence ions. As annealed 800°C, increase of resistivity shows decreasing compositions that supply carriers. If
annealing temperature is higher than 1000°C, the resistivity
rises rapidly, which shows that spinel solid solutions are
resolved into low valence oxides that cann't contribute to
conduction.

B. Conventional Thermal Annealing in Ar Atmosphere

Fig.3 shows x-ray diffraction patterns of the CoMnNiO
samples after annealing in Ar atmosphere. Spreading resistance
vs depth of the samples same as Fig.3 is shown in Fig.4. We
find that when annealed in Ar atmosphere, owing to lack of
oxygen, the spinel solid solutions in the CoMnNiO thin film
reduce greatly, forming a solid solution $NiCoO_2$ made of low
valence oxides NiO and CoO, which possess less conductivity.
Therefore the resistivity of samples annealed in Ar atmosphere
is higher than that of samples annealed in O_2.

C. Rapid Thermal Annealing in Ar Atmosphere

Characteristics of RTA are high temperature and short time
during annealing. High temperature gives great kinetic energy

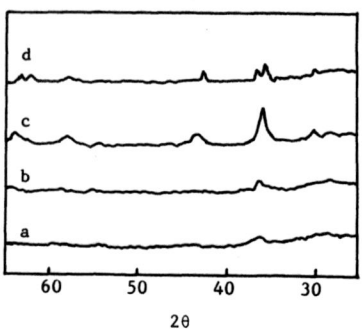

Fig.3 X-ray diffraction patterns of samples
before and after annealing in Ar atmosphere
a. As-deposited   b. 550°C, 4hr.
c. 750°C, 4hr.   d. 850°C, 4hr.

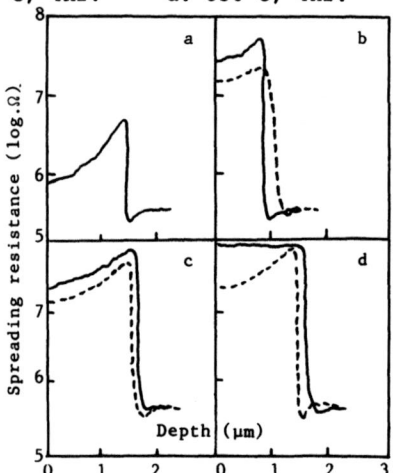

Fig.4 Spreading resistance vs depth of samples
annealed in different atmosphere
———, Ar;   ----, O_2
a. As-deposited   b. 550°C, 4hr.
c. 750°C, 4hr.   d. 850°C, 4hr.

for order arrangement of the atoms in the CoMnNiO thin film,
accomplishing crystallization. Fig.5 shows x-ray diffraction
patterns of the CoMnNiO samples after RTA in Ar atmosphere. As
compared Fig.5 with Fig.3, it is discovered that after annealing at 750°C in Ar atmosphere, the structure of the CoMnNiO
thin film is the imperfect spinel solutions. Tetrahedral distortion solid solutions $(Mn_{1-x}Co_x)_3O_4$ and low valence solid
solutions $(Ni_{1-x}Co_y)O$ are produced after annealing at 850°C.
The result after RTA at 1000°C is similar to that after conventional thermal annealing at 850°C. The higher the temperature
is, the larger the amount of oxygen released out from thin film.

Annealing in Ar atmosphere results in low conductivity.
Raman spectra for the CoMnNiO thin film are shown in Fig.6.
It is indicated that Raman peak is not found for the sample
before annealing, but after annealing at 550°C a wide and flat
peak appears in 650cm^{-1}∿750cm^{-1}, which suggests formation of
microcrystal oxides. A sharp Raman peak appears with increasing annealing temperature, which means crystallization of the
CoMnNiO thin film. The results of Raman spectra are in accor-
dance with that of x-ray diffraction analysis.

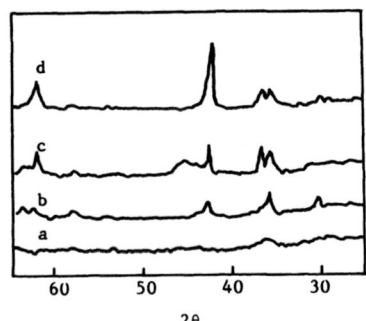

Fig.5 X-ray diffraction patterns of the samples
before and after RTA in Ar atmosphere
a. As-deposited   b. 1000°C, 30sec
c. 1100°C, 30sec   d. 1200°C, 30sec

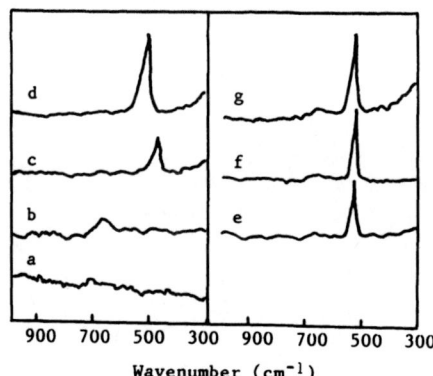

Fig.6 Raman spectra of the samples before and
after annealing in Ar atmosphere
a. As-deposited   b. 550°C, 4hr.
c. 750°C, 4hr.   d. 850°C, 4hr.
e. 1000°C, 30sec   f. 1100°C, 30sec
g. 1200°C, 30sec

SUMMARY

1. Conventional thermal annealing of the amorphous CoMnNiO thin films in O_2 atmosphere is profitable for formation of cubic spinel solid solution with stable structure and excellent electrical properties.
2. High temperature annealing makes spinel solid solution to be resolved into low valence oxides, reducing the conductivity of the thin films.
3. As a kind of material made thermistors, it is would be avoided that amorphous CoMnNiO thin film is thermally treated without oxygen.

REFERENCES

1. F.J. Hyde, D. Sc., M. Sc., B. Sc., Thermistors p10-33 (1969).
2. H. Tan, M.D. Tao and Y. Han, Chinese Journal of Semiconductors, Vol.10, No.11 (1989).
3. H. Matsunami, H. Masahiro and T. Tanaka, J. Electro. Mater., 8, No.3, 249 (May 1979).
4. Y. Hamakawa, Amorphous Semiconductor, p62-64 (1983).
5. M.H. Brodsky, Amorphous Semiconductors, p234-245 (1979).

PREPARATION AND CHARACTERIZATION OF MOLYBDENUM DISULFIDE
MICROCRYSTALS IN COLLOIDAL DISPERSION

E. Lu*, P. D. Persans*, A. F. Ruppert** and R. R. Chianelli**

* Physics Department and Center for Integrated Electronics,
Rensselaer Polytechnic Institute, Troy, NY 12180-3590

** Exxon Research and Engineering Company,
Annandale, NJ 08801

ABSTRACT

We report the preparation and general, optical and resonant Raman scattering characterization of MoS_2 platelets suspended in an insulating organic liquid. Structural results indicate that the platelets are predominantly 2H in structure and are about 500Å in diameter and 50-90 Å thick. The optical constants of the composite are close to those expected from the simple Maxwell-Garnett effective medium theory. We have observed resonant enhancement of the Raman cross-section for the MoS_2 phonon mode at 408 cm^{-1} as the laser excitation energy is scanned through the "B" exciton.

INTRODUCTION

Molybdenum disulfide consists of two-dimensional S-Mo-S layers which are weakly coupled to one another in the third direction by weak Van der Waals forces. It can be regarded as a nearly 2-D semiconductor system and posseses extreme anisotropy in both electronic and structural properties [1]. The electronic and optical properties of bulk crystal [1], thin film [2,3], amorphous and poorly ordered [4,5] MoS_2 have been reported in the literature. It has found a wide variety of applications including lubricants, batteries and catalysts [5]. We suggest that MoS_2 may also be a promising non-linear optical material because of sharp structure in the dielectric functions in the visible which is due to excitons near the direct band-gap. It has also been recently suggested that semiconductor microparticle composites may have enhanced nonlinear optical coefficients over those of bulk semiconductors due to local field effects [6] and quantum and other size-effects [7].

Nanoparticles are also of general interest because they may possess properties distinctly different from bulk crystals due to: i.) the breakdown in crystal momentum as a good quantum number [8]; ii.) changes in electron-phonon interactions [7,9]; and iii.) changes in electron-hole overlap [7].

In this paper, we report the preparation and structural and optical characterization of MoS_2 microparticles. In particular we focus on general and optical characterization to establish the nature of the crystallites in the dispersion. We also report the results of optical absorption measurements on dispersions and the comparison of those measurements to predictions from simple effective medium analysis. Finally we report preliminary resonant Raman scattering measurements for excitation energies near the upper exciton peak in the absorption spectrum.

MATERIALS PREPARATION AND GENERAL CHARACTERIZATION

MoS_2 crystals were prepared from the elements in a sealed silica tube at 1100°C with I_2 as a mineralizer. Preliminary dispersions were prepared by adding 75 mg of these crystals to 5 ml of propylene carbonate in a test tube and placing in an ultrasonic bath at room temperature for 16 hours. The resulting dispersion consisted of large (1μm) and small particles. This dispersion was then centrifuged at 1000g for 2 hours. The resulting supernatant fluid was then decanted. The final suspension is clear, with a green color and is stable for months against precipitation. It can be diluted by the addition of more propylene carbonate and remains stable and uniform. The colloid can be flocculated by the addition of ionic salts and redispersed by washing.

We determined the hydrodynamic radius of the particles in similarly prepared dilute suspensions using dynamic light scattering. The largest radius was found to be 250 Å. Other, faster, auto-correlation speeds were observed which suggested that the particles were anisotropic. TEM examination of flocculated materials gave similar results; the particles were platelet-shaped with 500Å as the largest diameter. In TEM we also observed that the particles tended to be "clumped" together. We believe that this clumping is characteristic of the flocculated material, not the suspension.

The c-axis crystallite size was determined to be 50-90 Å from line broadening of the (022) x-ray diffraction peak of flocculated material. We note that this size may be smaller than the platelet thickness.

EXPERIMENTAL DETAILS

Optical transmission spectra were measured on a Cary 17 dual-beam spectrophotometer over a wavelength range from 2600 nm to 330 nm. Silica cuvettes with path lengths from 1 mm to 10 mm were used and we performed preliminary investigations using different dilutions in order to check for multiple scattering effects. A small nonlinearity in absorptance versus concentration was observed which appeared to wavelength dependent. We did not correct for this effect in our present analysis.

Raman scattering measurements were carried out in the near-backscattering geometry. Typically, the 5145 Å line of an argon ion laser was used as the excitation source. Scattered light was focused onto the entrance slit of an ISA HG-2000 double monochromator which was coupled to a cooled RCA 310034A low dark count photomultiplier and photon counting electronics. Instrument resolution was set to 4 cm^{-1}. A Spectra-Physics 375B dye laser with R6G dye and optics was used as the exitation source for resonant Raman scattering. Spurious dye emission and plasma lines were excluded from the excitation beam by use of a Pellin-Broca prism monochromator after the laser. The wavelength tuning range is 5700-6500 Å.

RESULTS AND DISCUSSION

OPTICAL ABSORPTION

Fig. 1 shows the the room temperature absorption coefficient, α, plotted against photon energy. The absorption coefficient begins increasing sharply at 1.7 eV. Below this energy, α is small and is not measurable by our present transmission analysis. Bulk 2H-MoS$_2$ exhibits a direct gap at 1.7 eV with two excitons (denoted A and B) at 1.84 and 2.03 respectively [1] and an indirect gap at about 1.2 eV [3]. We observe two absorptance peaks at about 1.86±.01 eV and 2.04±.01 eV respectively. These positions are close to those of the A and B exciton peaks of natural bulk 2H-MoS$_2$. Therefore, we conclude that most particles in the colloid have this structure. We note that the exciton peaks for 3R MoS$_2$ have a smaller splitting than the peaks in 2H-MoS$_2$.

With some simplifying assumptions, the fraction of colloid volume occupied by crystals can be estimated. We assume that multiple scattering is negligible, and that the composite dielectric function can be modelled by a suspension of isolated, isotropic spherical particles, we also assume that the particles are much smaller than the wavelength of light. The most problematic assumption is that of spherical particles with isotropic optical constants, however, a complete solution to the general problem of anisotropic optical constants and anisotropic particles is not presently available. We are currently working on a model for platelet-type suspensions. The effect of this assumption on the current experiments can be checked by checking the shape of the effective absorption curve over a wavelength range where the optical constants change significantly.

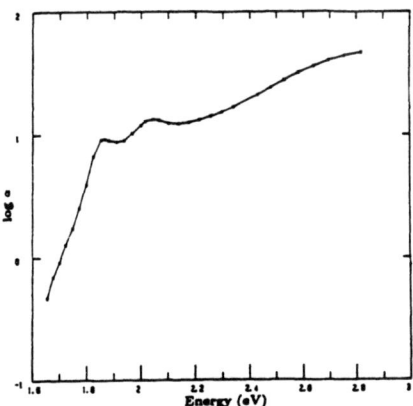

Figure 1.- Optical absorption coefficient, α, for a typical MoS$_2$ dispersion in propylene carbonate.

Under these assumptions we estimate the volume fraction f_c from the observed absorption coefficient α_{obs} and the real and imaginary parts of the dielectric constant ϵ of bulk MoS$_2$. In the limit that the crystallite dimensions are much smaller than the wavelength of light λ, the optical extinction cross-section for a spherical particle of radius a in a non-absorbing matrix is given by Mie scattering theory under the assumption that the dielectric constant of crystallites is the same as that of bulk:

$$\sigma_e = \frac{8\pi^2 a^3}{\lambda} Im \frac{\epsilon - \epsilon_M}{\epsilon + 2\epsilon_M} \tag{1}$$

where ϵ_M is the dielectric constant of the matrix. The observed absorption coefficient is thus:

$$\alpha_{obs} = \frac{4.5\epsilon_M}{(\epsilon_1 + 2\epsilon_M)^2 + (\epsilon_2)^2} f_c n\alpha \tag{2}$$

where $n = \epsilon_1^{1/2}$ and $\alpha = 2\pi\epsilon_2/n\lambda$

The values of the real and imaginary parts of ϵ for bulk 2H-MoS$_2$ were obtained from results of Kramers-Kronig analysis of reflectivity performed by Beal and Hughes [10], ϵ_M is taken to be 1.3. The volume fraction is thus calculated from Eq. (2). We performed the calculation at several wavelengths and found an average value of $f_c = 1.4 \times 10^{-3}$ for undiluted suspension. This value varied by about ten percent with wavelength and therefore we

156

know that either there are some small corrections which might result from the spherical, isotropic assumptions or that the dielectric function of crystals in suspension differs from bulk material. We are presently working to extend our optical analysis to anisotropic particles.

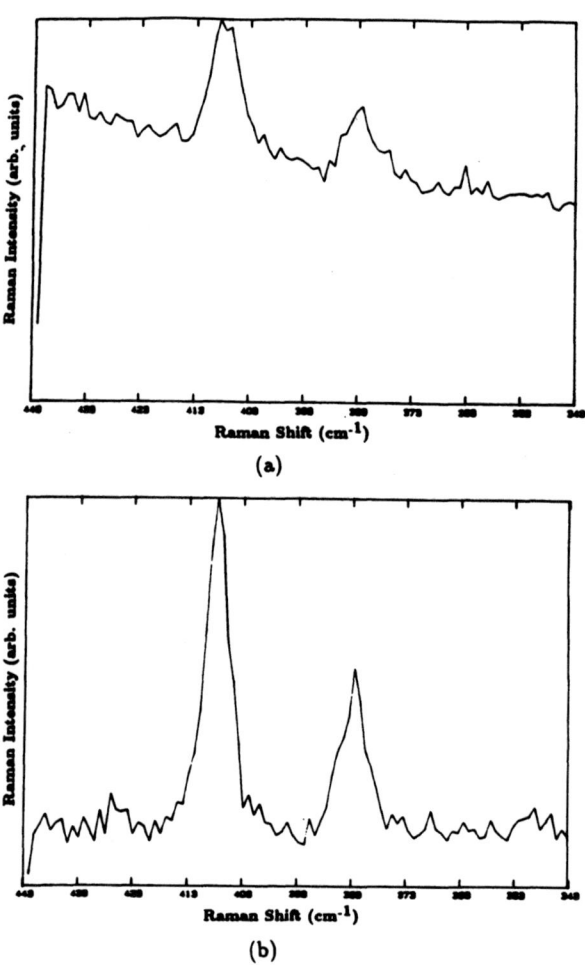

(a)

(b)

Figure 2.- Raman spectra of colloidal (a) and bulk (b) MoS_2. Excitation energy was 2.41 eV.

RESONANT RAMAN SCATTERING

The Raman spectra of colloidal and bulk MoS_2 are shown in Fig. 2. The Raman lines of colloid are at the same positions as and slightly broader than the corresponding lines of bulk. There are many more Raman lines in the colloid which arise from the propylene carbonate. They could be accounted for by measuring the propylene dicarbonate alone. The rising background in (a) is due to luminescence from the propylene carbonate solution. Two first order Raman lines at 383 cm^{-1} and 408 cm^{-1} were observed for both colloid and bulk 2H-MoS_2. These modes are assigned $E_{2g}^1(\Gamma)$ and $A_{1g}(\Gamma)$ respectively for bulk crystals [12,13] and they are due to optic-mode vibrations of atoms within a layer. The Raman peak positions from the dispersion are within one wavenumber of the bulk values; we do not observe size shifts such as those reported for nanocrystalline Si [8] and CdS [14] materials. This may be because the particles are relatively large (500Å) along the layer direction.

For resonant Raman scattering, the laser excitation energy was tuned from 2.0 to 2.1 eV since the B exciton is in this range. The range was also limited by the dye laser configuration. The Raman intensity of the 849 cm^{-1} mode of propylene carbonate was also measured as a calibration, and divided into the intensity of the MoS_2 mode. This procedure removes the ω^4, laser power dependence and sample volume effects automatically. We assume that propylene carbonate has no resonant structure in the wavelength range. The intensity ratio for both MoS_2 modes against excitation photon energy is shown in Fig. 3. A resonant enhancement in Raman intensity of the $A_{1g}(\Gamma)$ mode was observed, centered at 2.04 eV. The maximum resonance occurs at photon energy of 2.043 eV; close to the center position of B-exciton absorption peak. The scaled Raman intensity at this energy is about twice of those at the two ends. The maximum enhancement in Raman intensity of $E_{2g}^1(\Gamma)$ mode was observed for the lowest excitation energy. This suggests that the $E_{2g}^1(\Gamma)$ mode resonates below the B exciton but we have only one point in the present data. Our measurements will be extended to lower energy in the near future. For energies scanning through the B exciton we observe no resonance in the $E_{2g}^1(\Gamma)$ mode. This result is consistent with limited resonance data published on bulk 2H-MoS_2 in the literature [12,13].

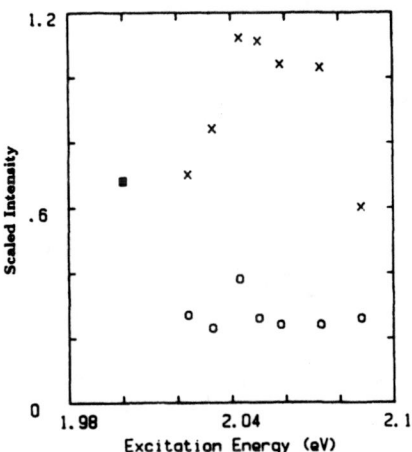

Figure 3.- Intensity of A_{1g} (X) and E_{2g}^1 (O) MoS_2 Raman modes plotted versus laser excitation energy. Intensities are scaled by the propylene carbonate mode at 849 cm^{-1}.

SUMMARY

We have prepared stable MoS_2 colloidal dispersions in propylene carbonate by a straighforward ultrasonic and centrifuge technique. We are able to isolate small particles with radius of gyration of $250\mathring{A}$ and thickness of $50\text{-}90\mathring{A}$. The optical absorption coefficient of the composite can be used to estimate the concentration of particles. In the present studies it is about 0.14%. We find that most particles are $2H\text{-}MoS_2$ by optical absorption and Raman scattering measurements.

We observe resonant Raman scattering which indicates that the A_{1g} phonon mode resonates weakly with the B exciton. We believe that polarized resonant Raman scattering and excitation spectroscopy will prove to be an extremely useful tool for the study of electronic states in anisotropic nanocrystalline materials. It should be possible to explore changes in wavefunction symmetries using electron-phonon coupling as the probe.

ACKNOWLEDGEMENTS

Work at Rensselaer was supported in part by the NSF under grant DMR-8714634 and by Rensselaer Polytechnic Institute.

REFERENCES

1. J. A. Wilson and A. D. Yoffe, Adv. Phys. 18, 193, (1969).
2. L. F. Mattheis, Phys. Rev. B, 8, 3719, (1973).
3. C. D. Roxlo, M. Daage, A. F. Ruppert and R. R. Chianelli, J. Catal., 100, 176, 1986.
4. R. R. Chianelli, Int. Rev. Phys. Chem., 2, 127, 1982.
5. C. H. Chang and S. S. Chan, J. Catal., 72, 139, 1981.
6. J. Haus, R. Inguva and C. M. Bowden, Phys. Rev. B, submitted.
7. S. Schmitt-Rink, D. A. B. Miller, and D. S. Chemla, Phys. Rev. B, 35, 8113, (1987).
8. H. Richter, Z. P. Wang, and L. Ley, Sol. State Commun., 39, 625, (1981).
9. A. P. Alivisatos, T. D. Harris, P. J. Carroll, M. L. Steigerwald, and L. E. Brus, J. Chem. Phys. in press.
10. A. R. Beal and H. P. Hughes, J. Phys. C, 12, 881, 1979.
11. A. M. Stacy and D. T. Hodul, J. Phys. Chem. Sol. 46, 405, (1985).
12. T. Sekine, T. Nakashizu, M. Izumi, K. Toyoda, K. Uchinokura, and E. Matsura, J. Phys. Soc. Japan, 49, (Suppl. A), 551, (1980).
13. E. Lu, P. D. Persans and K. Rajan, Phys. Rev. B, submitted.

PART III

Microcrystalline Silicon: Properties

THE ROLE OF HYDROGEN IN SILICON MICROCRYSTALLIZATION

S. WAGNER,* S.H. WOLFF* AND J.M. GIBSON**
*Department of Electrical Engineering, Princeton University, Princeton, New Jersey 08544
**AT&T Bell Laboratories, Murray Hill, New Jersey 07974

ABSTRACT

We report and interpret two groups of experiments on the role that hydrogen plays in the formation of silicon microcrystals. We show that the growth of single-crystal Si by molecular beam epitaxy at 475 °C is disrupted by H_2, which induces the formation of microcrystals. In crystallization experiments of non-hydrogenated a-Si and of hydrogenated a-Si:H on a hot stage in a transmission electron microscope, hydrogen facilitates the nucleation of crystallites. We explain our observations with a substantial reduction of the grain boundary energy by hydrogen.

INTRODUCTION

How does microcrystalline silicon form? How can we describe the structure and stability of microcrystalline silicon once it has been grown? These two questions have dominated our discussion of the "structure" of microcrystalline silicon over the past years. They have been answered in two ways: either by considering the thermodynamic stability of microcrystals, or by studying the growth kinetics that favor the formation of microcrystals. Veprek and coworkers [1,2] have used a thermodynamic stability argument for explaining the observation of a minimum microcrystal size. They suggest that the energy released on the growth surface by recombing hydrogen radicals allows to overcome the activation barrier to nucleation. Tanaka and Matsuda [3,4] and Tsai and coworkers [5,6] have interpreted microcrystal formation as a result of growth kinetics. Tanaka and Matsuda suggest that microcrystal growth is favored by a sufficiently large surface diffusion range of the growth species, the large range being permitted by high hydrogen coverage of the growing surface. Tsai emphasizes the importance of the reverse reaction to the growth, namely etching: hydrogen radicals preferentially etch amorphous tissue, to the benefit of the more stable microcrystals. Kinetic control is also espoused by Shimizu and coworkers [7], who discuss amorphous or crystalline network formation in terms of a subsurface reaction driven by energetic particles that arrive from the glow discharge.

In this paper we present two new experimental approaches to the study of microcrystallite formation in silicon, and interpret their results by analyzing the role of hydrogen. In one group of experiments we ask for the effect of hydrogen on the growth of single-crystal silicon. We will see that hydrogen can disrupt the growth of epitaxial silicon under certain conditions, to form microcrystalline material. In the second group of experiments we observe the recrystallization of amorphous silicon on a hot stage in the electron microscope. These experiments also show that hydrogen promotes the formation of microcrystals.

When silicon is reacted with hydrogen, three stable or metastable solids can be formed. These are hydrogen-coated single-crystal silicon, microcrystals with hydrogen-decorated boundaries, or hydrogen dispersed in an amorphous matrix. Our experiments deal with hydrogen breaking up a hydrogen-covered epitaxial growth front into independently growing, hydrogen-covered crystallites, and with hydrogen segregating from an initially amorphous matrix into the tissue between crystallites.

Mat. Res. Soc. Symp. Proc. Vol. 164. ©1990 Materials Research Society

We now proceed to describing our experiments. Then we interpret the results of these experiments by focusing on the role which hydrogen plays by reducing the grain boundary energy.

EXPERIMENTS

The MBE-Growth of Silicon in the Presence of Hydrogen

Temperatures around 500°C are of particular interest in the growth of silicon films, and in the interaction of hydrogen with silicon surfaces. Single-crystalline epitaxial layers can be grown by low-pressure chemical vapor deposition at temperatures down to 550°C [8]. Hydrogenated amorphous silicon has been grown by CVD from disilane at temperatures up to 475°C [9]. At 520°C the peak rate of hydrogen desorption from the (100) Si monohydride surfaces has been measured [10]. We have observed the slow chemisorption of molecular H_2 onto Si (111) 7x7 surfaces at 500°C [11] by transmission electron diffraction. Fig. 1 shows one result of this work, i.e., the occupancy of the center adatom site as a function of exposure to hydrogen gas at 4×10^{-6} Torr pressure. Molecular hydrogen is seen to reduce the occupancy under these conditions, though slowly.

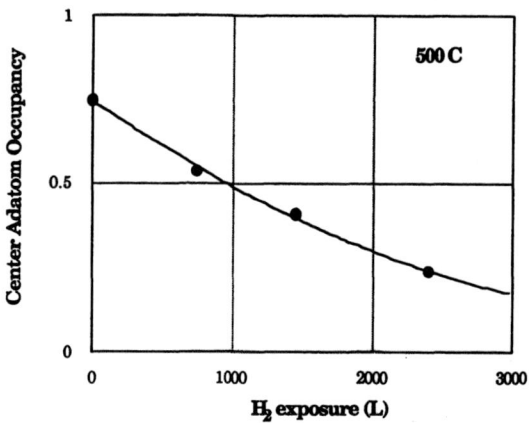

Figure 1: Occupancy of the center adatom site on the Si(111) 7×7 surface as a function of H_2 exposure at 500°C.

Taken together, these observations mean that over a small temperature range near 500°C, silicon may grow either crystalline or amorphous in the presence of H_2, and that H_2 may either desorb or chemisorb. Perhaps these two groups of observation are related, so that hydrogen may be employed to control the growth mechanisms of silicon?

We tested this question by growing silicon and silicon-germanium alloy layers by MBE in the presence or absence of hydrogen gas. We used Si(100) substrates, grew over a range of temperature, and at two growth rates [12]. The MBE system and the standard substrate preparation and film growth procedure have been described earlier [13]. The 4-inch diameter wafers were chemically cleaned, loaded into the growth chamber, argon-ion sputtered at room temperature and annealed at 775°C. After cooling to the growth temperature, H_2 was admitted. The H_2 pressure took 30 sec to stabilize. The H_2 flow was stopped at the termination of film growth. We varied the substrate temperature from 475 to 750°C, the hydrogen pressure was held at 0, 1×10^{-6} or 1×10^{-5} Torr and the growth rate was set at either 5 or 1Å/sec.

Cross sections of the samples were prepared, and were studied by transmission electron microscopy (TEM) and transmission electron diffraction (TED) in a JEOL 2000FX transmission electron microscope at 200 keV electron energy. The cross-sectional micrographs were taken in diffraction contrast with the electron beam close to the (011) direction such that the (400) planes were strongly diffracting. The hydrogen concentration profile was determined by secondary ion mass spectrometry (SIMS) coupled with sputter-profiling.

Figure 2 shows the key sequence of our growth results, as analyzed by TEM/TED. A length scale is given in Fig. 2e. The three silicon films of Fig. 2 all were grown at a rate of 5Å/sec. A layer grown at 475°C in the absence of H_2 (Figs. 2a,b) demonstrates the good crystal quality achievable by MBE at a low substrate temperature. Addition of 1×10^{-5} Torr of H_2 at 475°C (Figs. 2c,d) produces a layered film (Fig. 2c): the film begins to grow as a single crystal, but approximately midway converts to polycrystals along a ragged front. TED proves the polycrystallinity (Fig. 2d). Thus we observe that hydrogen coverage induces a transition from perfect crystal growth to microcrystalline growth, by promoting the nucleation of microcrystallites which then crowd out the single crystal. When the rate of growth was reduced to 1Å/sec, a completely epitaxial layer grew again, even in the presence of hydrogen (data not shown). Raising the substrate temperature to 550°C or higher restored complete crystallinity even at the growth rate of 5Å/sec (Figs. 2e,f). Table 1 summarizes the entire series of experiments and their results.

Table 1. Experimental results of MBE growth [12].

Film composition	Substrate temp.	H_2 pressure	Dept. rate	Crystal quality
Si	475°C	0	5Å/s	good
Si	475°C	1×10^{-5}Torr	5Å/s	poor
Si	475°C	1×10^{-5}Torr	1Å/s	good
Si	550°C	1×10^{-5}Torr	5Å/s	good
Si	625°C	1×10^{-5}Torr	5Å/s	good
Si	750°C	1×10^{-5}Torr	5Å/s	good
$Si_{0.8}Ge_{0.2}$	450°C	1×10^{-5}Torr	5Å/s	good
$Si_{0.8}Ge_{0.2}$	550°C	1×10^{-6}Torr	5Å/s	good

The observations of Fig. 2 and Table 1 clearly illustrate an effect of hydrogen on silicon growth. Fig. 2c shows that hydrogen is plowed ahead of the growth front until it induces the epitaxial layer to break into many small crystals. The return to completely epitaxial growth, when the growth rate is reduced, demonstrates that hydrogen affects the kinetic barriers to single or microcrystalline growth.

The impurity profiles determined by SIMS verify this link of hydrogen with the film structure (Fig. 3a,b). The SIMS profile of Fig. 3a was taken on the sample of Fig. 2a (growth rate 5Å/sec, substrate temperature 475°C, hydrogen pressure 1×10^{-5} Torr). The origin of the depth scale of Fig. 3a lies at the top of the grown film. The hydrogen content of the substrate (at depth $> 0.4 \mu m$) is approximately $10^{18} cm^{-3}$. The substrate/film interface is marked by the frequently observed large excursions of the nitrogen, carbon and oxygen concentrations, but the hydrogen concentration is not affected. Between 0.25 and $0.37 \mu m$, where the growth was found to

164

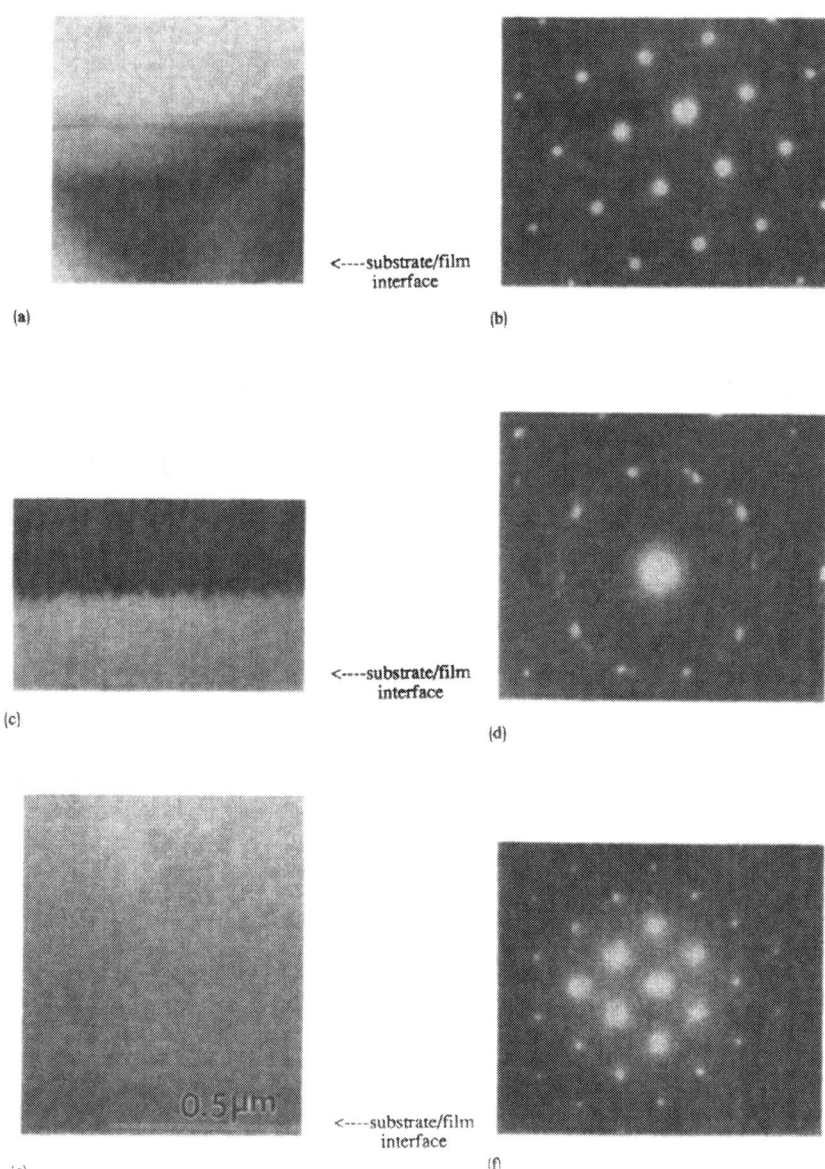

(a)

(b)

<----substrate/film
interface

(c)

(d)

<----substrate/film
interface

(e)

(f)

0.5μm

<----substrate/film
interface

Figure 2: TEM cross section and diffraction photographs for MBE-grown Si films at: (a)(b) 475°C without H_2; (c)(d) 475°C at 1×10^{-5} Torr of H_2, and (e)(f) 550°C with 1×10^{-5} Torr of H_2.

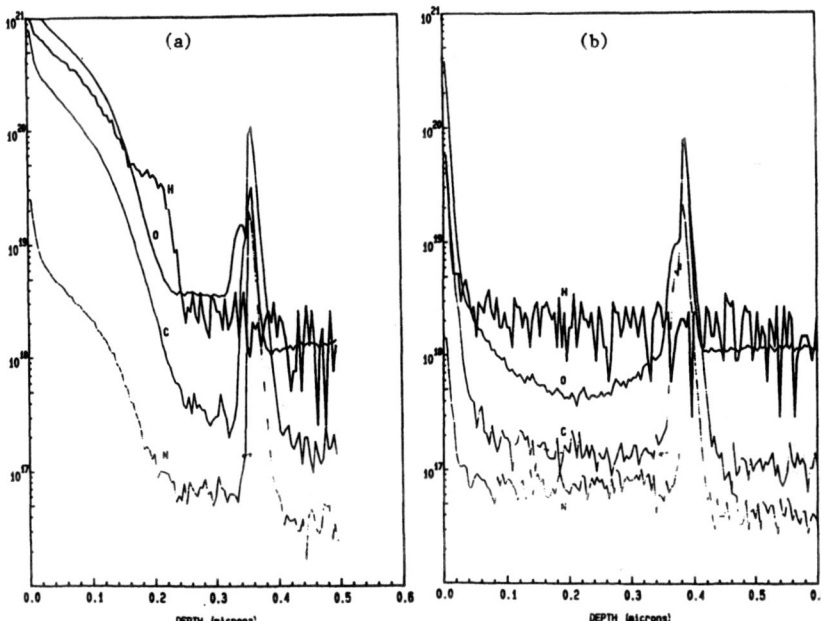

Figure 3: SIMS profiles for H,O,C, and N of MBE films grown at (a) 475°C with
1×10^{-5} Torr of H_2, and (b) 625°C with 1×10^{-5} Torr of H_2.

be epitaxial by TED, the hydrogen concentration retains the value it has in the bulk
crystal. At $0.25\mu m$, where single-crystal growth was disrupted, the SIMS profile
shows a sharp increase of the hydrogen concentration to the mid-10^{19}cm^{-3} range.
This reflects the ability of microcrystalline silicon to hold (in its grain boundaries) a
much higher concentration of hydrogen than single crystal Si. The hydrogen profile
near the epitaxial/microcrystalline interface clearly differs from the concentration
profiles of the atmospheric contaminants. We surmise that the N, C and O profiles
(and probably the hydrogen profile close to the surface) were established by in-
diffusion from the atmosphere after growth. The SIMS profile of a completely epitaxial layer, grown at 625°C, is shown in Fig. 3b. The hydrogen concentration
remains constant at a value which is characteristic for a single crystal. The absence
of grain boundaries also is suggested by the lack of extensive in-diffusion of the
atmospheric contaminants.

The results of these MBE growth experiments say that, regardless of hydrogen
exposure, epitaxial silicon can be grown above 550°C. But at 475°C, hydrogen can
disrupt epitaxial growth in a process that is rate-controlled.

The Crystallization of a-Si and of a-Si:H

The MBE experiments showed that hydrogen can fragment a crystallization
front, by making small crystallites more stable than a large single crystal when
hydrogen cannot be rejected fast enough from the growing surface. One way of
viewing these MBE results is as the segregation of hydrogen over ever shorter

distance until hydrogen is incorporated uniformly in a-Si:H. Does hydrogen also promote the formation of small crystallites when one starts with amorphous silicon? This is the question we wanted to answer in the crystallization experiment.

We studied the crystallization of two kinds of amorphous silicon. One was nominally pure a-Si produced by self-implantation of a single-crystal wafer. Si^{28} ions with 500 keV energy were implanted at 77K substrate temperature to a dose of $5\times10^{15}cm^{-2}$. This procedure amorphizes a layer approximately $1\mu m$ thick. The second kind of sample was hydrogenated a-Si:H produced by the glow-discharge deposition of SiH_4 on a single-crystal Si wafer at a substrate temperature of 250 °C.

The samples were recrystallized on a hot stage mounted within the JEOL 2000FX electron microscope. They were thinned from the back mechanically first, then chemically, to open a hole surrounded by a thin (100 to 300Å) film. The recrystallization was observed in transmission electron diffraction by dark-field imaging the thin region of the film with 200 keV electron energy. The samples were heated in an oven from room temperature to 600 °C within two minutes, or to 800 °C within three minutes. The thermocouple was calibrated on the Curie point of a series of NiCo alloys to an estimated accuracy of ±25 °C. The crystallization took place within a few seconds. Therefore, the entire process was recorded on videotape. The figures accompanying this paper are translations of video frames to gray-scale hard copy.

The self-implanted a-Si samples were observed to crystallize at 800 °C within one second. A frame taken during this period is shown in Fig. 4a.

(a)   (b)

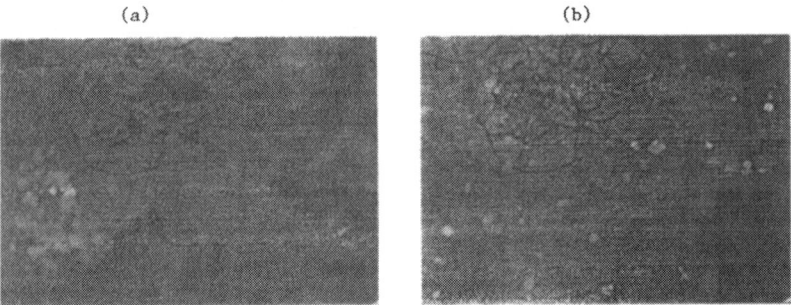

Figure 4: TEM micrographics taken during crystallization of (a) a-Si at 800 °C, and (b) a-Si:H at 600 °C.

The light gray area in the center is the amorphous material. It is on its way to crystallization. Two crystallites are growing; one is the light region on the lower left, the other is the black region encroaching from the bottom and the right. The figure is one μm wide. The tangle of fibers near the upper edge is imaged from the electron optics, not the sample.

We see that non-hydrogenated a-Si crystallizes at a relatively high temperature. The crystallites reach about $0.5\mu m$ in-plane diameter. The number of nucleation sites is so small as to suggest heterogeneous nucleation, either on the c-Si substrate or on surface oxide particles. We estimate that the crystal growth rate lies between 0.1 and $1\mu m/sec$.

Crystallization of hydrogenated a-Si:H produces a very different result, as is evident from Fig. 4b. The black region on the right of Fig. 4b is the etched hole. Again, this frame was taken during crystallization, which took place at 600 °C and was completed within about three seconds. The white spots are the Si crystallites. They become visible with diameters of around 100 Å and grow to 200 or 300 Å. The frame of Fig. 4a also is 1μm wide.

Thus hydrogenated a-Si:H crystallizes at a comparatively low temperature. Nucleation is homogeneous and small crystallites are formed. The growth rate of their diameter is about 100Å /sec, and they grow to about 0.03μm in-plane diameter.

The main lesson to be drawn from these experiments is that hydrogen promotes the crystallization of amorphous silicon. Hydrogen aids homogeneous nucleation at a relatively low temperature. How can we formulate the role that hydrogen plays? The MBE experiment suggested an interplay between the kinetics of silicon incorporation, hydrogen elimination, and grain stabilization by hydrogen. The crystallization experiments highlight the role of hydrogen in nucleation, a process in which grain stabilization plays a key role. Absent sufficient kinetic data, we focus on the stabilization of nuclei, once formed, by hydrogen.

THE NUCLEATION OF SILICON CRYSTALS FROM AN AMORPHOUS MATRIX

We discuss the effect of hydrogen on nucleation in the context of the Gibbs free energy of the nucleus. Our proposal is that in a-Si:H hydrogen, swept out of the recrystallized volume, stabilizes the nucleus by decorating its surface.

The nucleation of c-Si in an a-Si matrix costs much energy if the interface between the nucleus and the surrounding matrix remains coherent, that is, if the bonds across the interface remain intact. The energy of the coherently-strained interface can be reduced by breaking the strained bonds, i.e., by the production of dangling bonds. The interfacial energy can be reduced further by passivating these dangling bonds with hydrogen. Thus, in non-hydrogenated a-Si, the nucleus/matrix interface consists either of strained or of dangling bonds. In hydrogenated a-Si:H, the strained or dangling bonds can be replaced by Si-H terminations. We now formulate the Gibbs free energy of a c-Si nucleus, in a classical picture but using individual bond energies. This formulation will allow us to extract the effects of the energies and concentrations of hydrogen (and of dangling bonds) contained in the initial a-Si(H), on the Gibbs free energy - radius relation for the c-Si nucleus.

The gain in Gibbs free energy upon nucleation from an amorphous matrix is defined as:

$$\Delta G \equiv G(c-Si \text{ nucleus}) - G(a-Si \text{ matrix}) . \tag{1}$$

ΔG is made up from a bulk (crystal) and a surface contribution:

$$\Delta G = \Delta G(\text{bulk}) + \Delta G(\text{interface}) . \tag{2}$$

The bulk crystal contribution for a spherical nucleus which contains N_{Si}^b silicon atoms is

$$\Delta G(\text{bulk}) = N_{Si}^b \Delta g_{Si} = \frac{4\pi}{3}(\rho_{Si} N_L / A_{Si}) \Delta g_{Si} r^3 , \tag{3}$$

where Δg_{Si} is the free energy of crystallization per Si atom, ρ_{Si} is the specific weight of silicon, N_L the Avogadro number, and and A_{Si} the atomic weight of silicon. r is the radius of the nucleus. We approximate $\Delta G(\text{interface})$ by $\Delta E(\text{interface})$ and express the contributions from coherent bonds, dangling bonds and Si-H bonds as
$\Delta G(\text{interface}) = \Delta E(\text{interface}) =$

$$\Delta E (\text{coherent bonds}) + E (D.B.) + \Delta E (Si-H) , \tag{4}$$

and express the latter in terms of the number of sites on the interface N_{Si}^i, the atomic fraction of a given type of bond x^i, and the individual bond energy E:

$$\Delta G \text{ (interface)} \simeq$$

$$N_{Si}^i(x_{CB}^i \epsilon_{CB} + x_{DB}^i \epsilon_{DB} + x_{Si-H}^i \epsilon_{Si-H}) =$$

$$4\pi(\rho_{Si}N_L/A_{Si})^{2/3}(x_{CB}^i \epsilon_{CB} + x_{DB}^i \epsilon_{DB} + x_{Si-H}^i \epsilon_{Si-H})r^2 \quad . \tag{5}$$

Now we assume that all dangling bonds and Si-H bonds in the interface were swept out from the amorphous volume that the nucleus now occupies. Their atomic fractions in the amorphous material were x_{DB}^b and x_{Si-H}^b. Expressing ΔG(interface) in terms of x_{DB}^b and x_{Si-H}^b, and adding ΔG(bulk) results in

$$\Delta G = (4\pi\rho_{Si} N_L/3A_{Si})[\Delta g_{Si} + x_{DB}^b(\epsilon_{DB} - \epsilon_{CB}) + x_{Si-H}^b(\epsilon_{Si-H} - \epsilon_{CB})] r^3 \tag{6}$$

$$+ 4\pi(\rho_{Si} N_L/A_{Si})^{2/3}\epsilon_{CB} r^2 \quad .$$

Note that the contributions by the dangling and the Si-H bonds enter with r^3. The critical radius r^* of the nucleus, determined from the condition $\partial\Delta G/\partial r = 0$, is given by

$$r^* = [2(A_{Si}/\rho_{Si} N_L)^{1/3}\epsilon_{CB}]/[(g(a-Si) - g(c-Si)$$

$$+ x_{DB}^b(\epsilon_{CB} - \epsilon_{DB}) + x_{Si-H}^b(\epsilon_{CB} - \epsilon_{Si-H})] \quad . \tag{7}$$

We define a radius of incoherence r^i at which the entire interface is decorated with hydrogen (the energy of the dangling bond being so large and its concentration so small that its contribution may be neglected). For a fully H-covered interface,

$$x_{Si-H}^i = 1 = \frac{1}{3}(\rho_{Si}N_L/A_{Si})^{\frac{1}{3}} x_{Si-H}^b r^i \quad . \tag{8}$$

$$r^i \simeq 3(A_{Si}/\rho_{Si} N_L)^{\frac{1}{3}}/x_{Si-H}^b \quad . \tag{9}$$

For $x_{Si-H}^b = 0.1$, $r^i \simeq 75\text{Å}$.

The schematic Gibbs free energy - radius diagram of Fig. 5 illustrates the effect of hydrogen. The curve near the bottom marked "Crystal" represents ΔG(bulk), the gain in bulk free energy upon crystallization. The top curve marked "Coherent Interface" is the energy cost of maintaining bonds across the entire interface. The two contributions sum to the total Δ G, marked a-Si. The reduction in interface energy by hydrogen is shown on the lower left as a—Si:H$_x$, using the "Crystal" value as the baseline. The new ΔG for a nucleus with a partially or completely incoherent interface is denoted in the center as a—Si:H$_x$. The lowering of ΔG by decoration of the interface with hydrogen, and the concurrent reduction of the radius of the critical nucleus are evident.

Critical Fluctuations

The participation of hydrogen in the nucleation process raises a fascinating question: Will the nuclei in a-Si:H only grow but not decay? The latter process is implied to occur by our resorting to a thermodynamic treatment (which presumes finite and nearly equal rates of growth and decay). We discuss the question in terms of the ability of hydrogen to enter and to leave the Si-Si network, and of its attempt frequency for doing so, because it must be the motion of hydrogen which controls the rate of nucleation.

As mentioned earlier, H_2 chemisorbs on and desorbs from c-Si surfaces around $500°$ C. H_2 molecules are occluded in a-Si:H at high pressure [14]. Nuclear magnetic resonance experiments suggest that H_2 can leave or enter the Si-Si network [15].

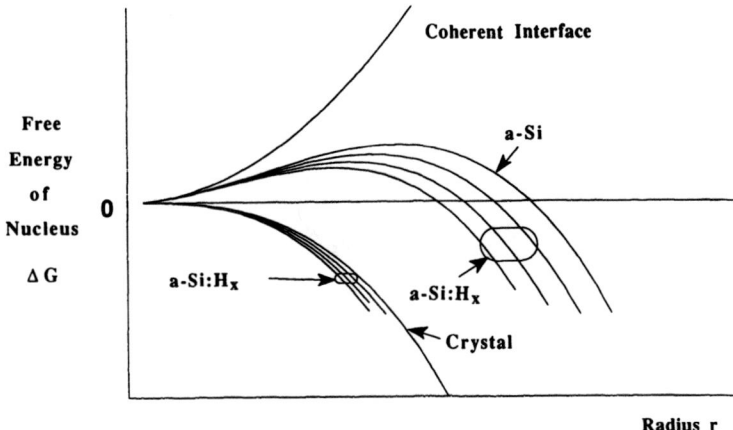

Figure 5: Schematic diagram of the Gibbs free energy of a nucleus formed in amorphous silicon vs. its radius. The effect of hydrogen is illustrated by the curves marked a—Si:H$_x$.

Therefore, fluctuations of nucleus size could occur by Si-H bonds breaking and reforming in exchange with H$_2$ molecules. The rate of such fluctuations, gauged by the rate of H$_2$ chemisorption or desorption at 500°C, may be slow even at 600°C. Thus, exchange of hydrogen between Si-H bonds and occluded H$_2$ is unlikely to control nucleation. (We exclude the extreme case of dehydrogenation of the sample).

The diffusion coefficient of hydrogen in a-Si:H,D$_H$, is approximately 1.66×10^{-3} exp $[-1.45$ eV/kT]cm^{2}s^1[16]. Taking the jump frequency of H, $\Gamma = 6D/a^2$, where a is the jump distance (~ 2.5Å). Γ at 600°C, the temperature of crystallization of a-Si:H, is 10^5sec^{-1}. The diffusion mechanism has not yet been identified with certainty, but is likely to involve bound hydrogen atoms in the initial and final states. One may conceive of hydrogen leaving Si-H sites in the crystallizing network ahead of the crystallization front. This process can proceed in reverse when a subcritical nucleus shrinks. If hydrogen diffusion is the rate-controlling step in nucleation, fairly frequent fluctuations of the nucleus size are seen to be possible at 600°C. At the typical growth temperature for a-Si:H of 250°C, however, $\Gamma \simeq 1$, which renders reversible nucleation unlikely. This means that a thermodynamic treatment of nucleation with participation of H is realistic at 600°C, but that kinetic factors dominate at 250°C.

SUMMARY

We have studied the growth of silicon layers by molecular beam epitaxy in the absence and presence of hydrogen gas. At a substrate temperature of 475°C hydrogen breaks up the growing single crystal surface into many small crystallites. We also have observed the recrystallization of non-hydrogenated and of hydrogenated amorphous silicon on a hot stage in the transmission electron microscope. Hydrogenated material crystallizes homogeneously and at lower temperature than non-hydrogenated a-Si. We interpret our observations in terms of a reduction of the grain boundary energy by hydrogenation. This reduction facilitates the break-up of a single-crystalline growth front into small crystallites, and the homogeneous nucleation of crystals in an amorphous matrix.

Acknowledgements

We acknowledge stimulating discussions with J.C. Bean, R. Hull, D. Loretto, P.M. Fauchet, and F. Spaepen. The work at Princeton University is carried out within a program on amorphous silicon-germanium alloys sponsored by the Electric Power Research Institute under Contract No. RP 2824-2.

List of References

1. S. Veprek, Z. Iqbal, and F.A. Sarott, Phil. Mag., B,45, 127 (1982).

2. S. Veprek, Mat. Res. Soc., Symp. Proc. Vol. 164.

3. K. Tanaka and A. Matsuda, Mat. Sci. Repts., 2, 139 (1987).

4. A. Matsuda and T. Goto, Mat. Res. Soc. Symp. Proc. Vol. 164.

5. C.C. Tsai, J.C. Knights, C. Chang, and B. Wacker, J. Appl. Phys., 59, 2998 (1983).

6. C.C. Tsai, G.B. Anderson, R. Thompson, and B. Wacker, 13th 1CALS, to be published in J. Non-Cryst. Solids.

7. I. Shimizu, 13th 1CALS, to be published in J. Non-Cryst. Solids.

8. T.N. Nguyen, D.L. Harame, J.M.C. Stork, F.K. LeGomes, and B.S. Meyerson, Proc. Int. Electron Devices Meeting, Los Angeles, Dec. 7-10, 1986, p. 304.

9. M. Akhtar, V.L. Dalal, K.R. Ramaprasad, S. Gau, and A.J. Cambridge, Appl. Phys. Lett., 41, 1146 (1982).

10. G. Schulze and M. Henzler, Surf. Sci., 134, 336 (1983).

11. S.H. Wolff, S. Wagner, J.M. Gibson, D. Loretto, I.K. Robinson, and J.C. Bean, unpublished data.

12. S.H. Wolff, S. Wagner, J.C. Bean, R. Hull, and J.M. Gibson, Appl. Phys. Lett., 55, 2017 (1989).

13. J.C. Bean and E.A. Sadowski, J. Vac. Sci. Technol., 20, 137 (1982).

14. Y.J. Chabal and C.K.N. Patel, Phys. Rev. Lett., 53, 210 (1984) and 53, 1771 (1984).

15. J.B. Boyce, S.E. Ready, M. Stutzmann, and R.E. Norberg, 13th 1CALS, to be published in J. Non-Cryst. Solids.

16. S.F. Chou, R. Schwarz, Y. Okada, D. Slobodin, and S. Wagner, Mat. Res. Soc. Symp. Proc., 95, 165 (1987).

17. F. Buda, G.L. Chiarotti, R. Car, and M. Parinello, Phys. Rev. Lett., 63, 294 (1989).

18. G.L. Chiarotti, F. Buda, R. Car, and M. Parinello, Mat. Res. Soc. Symp. Proc. Vol. 163.

MICROCRYSTAL SI FILMS PREPARED BY REMOTE PLASMA CVD

SUNG CHUL KIM, JUNG TAE HWANG, SEUNG KYU LEE, CHANG YOUNG JUNG, SUNG MOO SOE,
SUNG OK KOH, KWAN SOO CHUNG, JIN JANG, Dept. of Physics and Dept. of
Electronics, Kyung Hee University, Dongdaemoon-ku, Seoul 130-701, Korea

ABSTRACT

The effects of deposition temperature, rf power and hydrogen dilution
ratio on the growth, structure and transport of p-type microcrystal(μc-) Si
films deposited by remote plasma CVD have been investigated. While low
substrate temperature and low rf power yield small grain sizes, high
temperature and high rf power tend to supress the growth of grains. The
etching of Si by hydrogen radicals plays an important role to grow μc-Si, but
excess etching supresses the growth of crystallites. We obtained 400 Å of
grain size and 3.5 S/cm of room temperature conductivity for p-type μc-Si.

INTRODUCTION

There has been much work on microcrystal silicon(μc-Si) since it can be
used as window material in amorphous silicon(a-Si:H) solar cells, photosensors
and as ohmic contact materials in a-Si:H solar cells, thin film transistors,
and diodes. Hydrogen dilution in the glow discharge silane plasma is known to
produce μc-Si films at relatively low temperatures. The growth of μc-Si is
enhanced by increasing the hydrogen dilution ratio(H_2/SiH_4), and it is
attributed to the etching of amorphous silicon phase by hydrogen radicals
[1,2].

There are two kinds of remote plasma enhanced chemical vapor deposition
(RPCVD) methods. Shimizu's group developed a hydrogen radical CVD, which
utilizes the hydrogen radicals produced by remote microwave plasma. The
hydrogen radicals are mixed downstream with silane or SiF_4 molecules[3].
Another method is the use of He plasma to generate metastable He for the
downstream excitation of SiH_4[4].

We used the metastable He for the dissociation of H_2 and SiH_4. In the
present work, the influences of the deposition temperature, rf power and
hydrogen dilution ratio on the growth and the transport properties of p-type
Si films by RPCVD has been investigated.

EXPERIMENTAL

P-type microcrystalline Si films were prepared by a remote plasma CVD
technique, where plasma generating region is connected upstream of a remote
deposition chamber. Silane and hydrogen mixture are introduced into the
downstream reactor. While helium atoms pass through the plasma generating
region, inside of the cylindrical quartz tube of diameter 3.8 cm, and some of
them are excited to metastable state. The substrate temperature was varied
over a range of 50 to 350 °C, and the rf power was applied to the induction
coil, turned on the outside of the cylindrical quartz tube. While the hydrogen
dilution ratio (H_2/SiH_4) was fixed at 50. Other deposition conditions are: a
total gas flow rate of 2000 sccm, a pressure of 2 torr, and a various rf power
levels applied to the coil. Films of about 0.6 μm in thickness were deposited
on both Corning 7059 for x-ray, Raman and conductivity measurements, and high
resistivity Si wafer for infrared vibrational absorption experiment. The
details of the preparation conditions are shown in Table I.

The Si films have been characterized by optical absorption, Raman
scattering, dc conductivity and x-ray diffraction. Coplanar electrode
configuration was used in the conductivity measurement. Raman study was
carried out utilizing the 5145 Å argon laser radiation.

Mat. Res. Soc. Symp. Proc. Vol. 164. ©1990 Materials Research Society

Table I Preparation conditions of Si thin films

Gas mixing ratio	$SiH_4/H_2=1/50$, $B_2H_6/SiH_4=1/100$
Total pressure	2 Torr
Flow rate	$SiH_4+H_2 = 100$, He = 2000 sccm
Diameter of discharing tube	3.8 cm
RF power	7 - 50 W
Substrate temperature	80 - 330 °C

RESULTS AND DISCUSSIONS

The growth rate of μc-Si films gives the information about the growing mechanism. The dependence of the deposition rate on the rf power is shown in Fig. 1. With increasing rf power, the deposition rate increases until 23 W and then it decreases. The dependence of the deposition rate on the substrate temperature is shown in Fig. 2. The deposition rate increases with the growing temperature. This behavior is quite different from the case in hydrogenated amorphous silicon(a-Si:H). In glow-discharged a-Si:H films, the deposition rate is nearly independent of the substrate temperature and increases with rf power.

Raman scattering identifies the local structural order of the materials. Raman spectra of the deposited Si films as a function of substrate temperature are shown in Fig. 3. All films were fabricated with rf power of 35 W. The deposited Si film at 230 °C is characterized by the sharp peak near 518 cm^{-1} in the spectrum. This peak at 518 cm^{-1} is due to the Si-Si TO modes and it determines the degree of crystallity of the Si films. The peaks of the Si films deposited at 180 °C and 330 °C are broader than that deposited at 230° C. The Si film deposited at 80 °C show a broad peak near 480 cm^{-1} as well as the peak at 520 cm^{-1}, indicating the structure of this film is a mixture of crystalline and amorphous phases.

Figure 4 shows the Raman spectra of the Si films as a function of rf power. All the films were deposited at 230 °C. With increasing rf power level, the degree of the crystallinity reaches a maximum at 15 W and then the crystalline quality degrades. The Si film deposited with 7 W shows amorphous structure.

The size of crystallite can be determined from the full width at half maximum[FWHM] using Scherrer's formula[5]. The grain size was determined from the FWHM of the (111) peak of the x-ray data. The results are shown in Fig. 5 as a function of rf power. With increasing rf power, the peak becomes sharper until 15 W, but it gets broader at higher rf levels.

Device-quality amorphous Si films show only Si-H stretching mode absorption at 2000 cm^{-1}, but it shifts to 2080 cm^{-1} in microcrystalline Si, which is considered as the bonded hydrogens in the interconnecting, boundary regions. As shown in Fig. 6, the Si film deposited with 7 W at 230 °C shows only Si-H modes, but the films deposited with rf power of > 15 W show only 2080 cm^{-1} peak.

We measured the temperature dependence of the conductivity for the deposited Si films. With increasing rf power, the room temperature conductivity reaches a maximum at 15 W and then decreases as shown in Fig. 7. The room temperature conductivity deposited with 15 W of rf power is 3.5 S/cm. The activation energy obtained from the temperature dependence of the conductivity is shown in Fig. 8. The film deposited with 15 W of rf power at 230 °C shows the smallest value of 25 meV. At higher or lower rf power levels the activation energy increases.

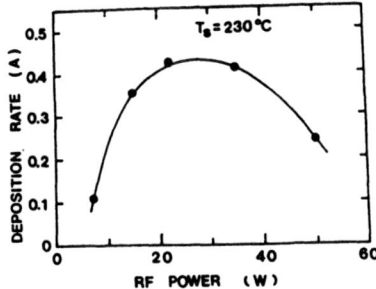

Fig. 1. The dependence of the growth rate on rf power.

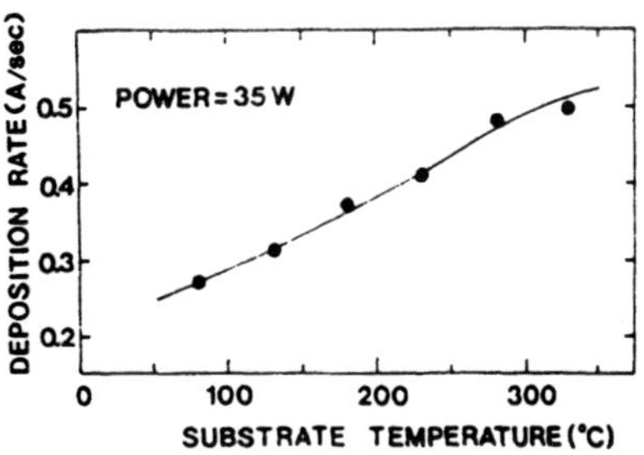

Fig. 2. The dependence of the growth rate on the substrate temperature.

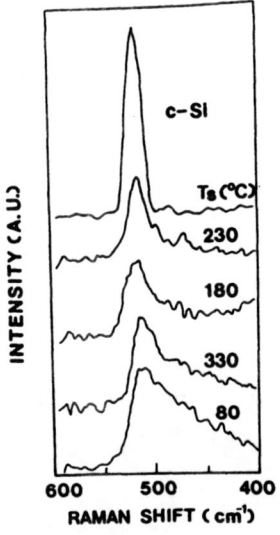

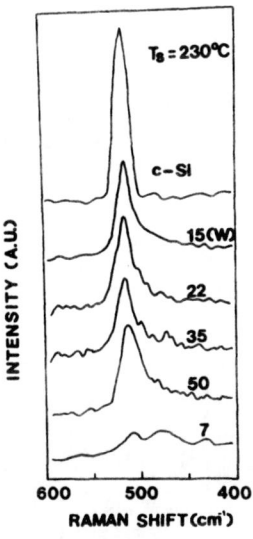

Fig. 3. Raman spectra(of excitation of 5145 A) vs. the growth temperature. The films were deposited with rf power of 35 W.

Fig. 4. Raman spectra vs. rf power when the growth temperature is 230° C.

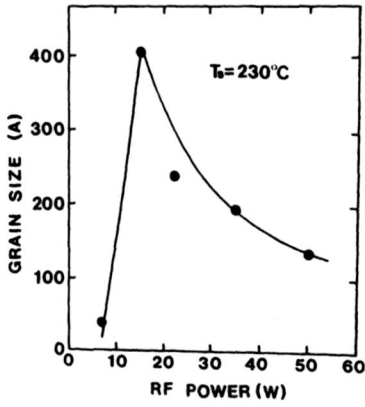

Fig. 5. Grain size of the Si films as a function of rf power.

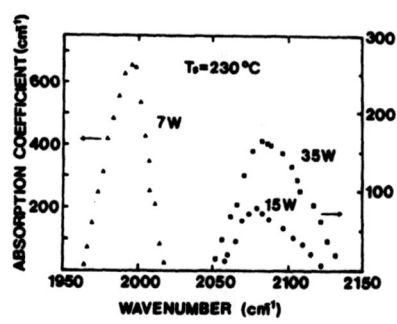

Fig. 6. The effect of rf power on Si-H bonding configuration.

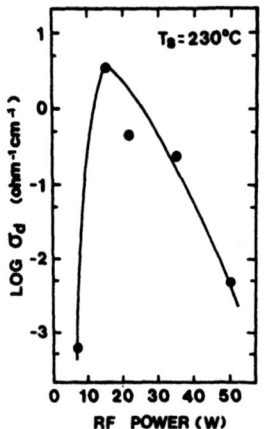

Fig. 7. The room temperature conductivity of p-type Si films vs. rf power.

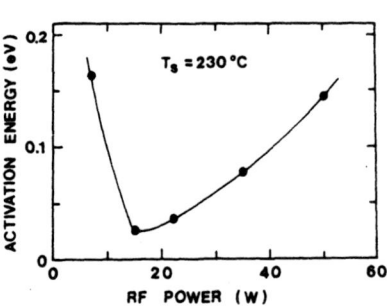

Fig. 8. Activation energy of the p-type Si films vs. rf power

So far, we studied the Si films deposited by RPCVD using the preparation conditions shown in Table I. We also studied the effects of hydrogen dilution and He flow rate on the structural and electrical properties of the Si thin films.

The deposition rate decreases with increasing the hydrogen dilution ratio and is zero at H2/SiH4=100. With increasing the hydrogen dilution ratio above 20, the μc-Si films are formed and the grain size increases until H2/SiH4 = 50, but the grain size start to decrease for further increase of hydrogen dilution ratio, as shown in Fig. 9. The effect of He flow rate on the deposition rate is shown in Fig. 10. With increasing He flow rate, the deposition rate first increases, but it decreases slightly at Fr[He] > 20 Fr [SiH4+ H2].

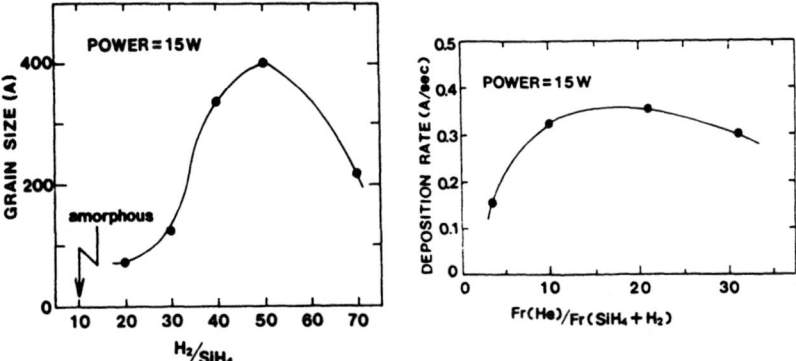

Fig. 9. The effect of hydrogen dilution ratio on the grain size of the μc-Si films.

Fig. 10. The effect of He flow rate on the growth rate.

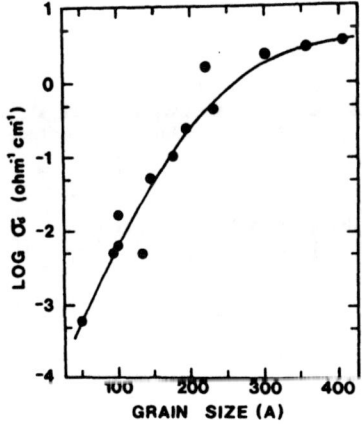

Fig. 11. Room temperature conductivity of Si films vs. the grain size.

The room temperature conductivity of the μc-Si films as a function of the grain size is shown in Fig. 11. The conductivity increases with the grain size, but it tends to saturate at higher than 300 A.

The deposition of Si films by a RP-CVD can be considered as follows. The helium is excited into the metastable states by passing through the plasma generating region and then it is transported toward the substrate. Above the substrate, the feeding H_2 and SiH_4 are dissociated by the metastable heliums. The energy difference(20 eV) between the ground and metastable states is sufficient to decompose SiH_4 and H_2. As a result, the SiH_3 and/or SiH_2 precursers can be formed. The adsorption of silicon containing precures and dehydrogenation will build up of the silicon network. In this case, absorption of hydrogen and/or hydrogen coverage of the growing surface can give rise to the desorption of silicon hydrides [SiHx]. This is the etching of silicon surface by hydrogen radicals.

The structural, electrical, and optical properties change with growth temperature, rf power level and hydrogen dilution ratio. With increasing temperature, the relaxation of Si atoms and the evolution of hydrogen from the material accelerate. The growth of microcrystal Si is considered as a result of preferential etching of the amorphous phase. During the growth, the amorphous and microcrystal regions are formed. The etching of the Si by atomic hydrogen atoms can be considered as a result of the hydrogen attachment to the growing surface and desorbing of SiHx from the growing surface. The fast evolution of the hydrogen from the growing surface can suppress the formation of microcrystal Si at high temperatures above 230 °C.

With increasing rf power level, the size of grains reaches a maximum and and then decreases. The increase of the hydrogen radical with rf power enhances the formation of the microcrystal Si up to 15 W, but the excess etching of the amorphous and microcrystal regions suppress the formation of the grain, resulting in the decrease of the grain size and slow down of the deposition rate with an increase of rf power.

CONCLUSION

Deposition temperature, rf power, hydrogen dilution ratio, and rf power were found to have a profound effect on the grain size and transport properties of the microcrystal Si films prepared by remote plasma CVD. The etching of Si by hydrogen radicals enhances the growth of grains, but excess etching suppresses it. The p-type μc-Si film deposited at near the optimum deposition condition shows 400 A of grain size and >3S/cm of room temperature conductivity.

ACKNOWLEDGEMENTS

This work was supported in part by the Education Ministry, Korea through the free collect subject(1987) and in part by Inter-University Research Center.

REFERENCES

1. C.C. Tsai, in Amorphous Silicon and Related Materials, edited by H. Fritzsche(World Scientific Publishing, Singapore, 1989), vol.1, p. 123.
2. R.C. van Oort, M.J. Geerts, J.C. van den Heuvel, J.W. Metselaar, Electronics Letters 23, 967(1987).
3. N. Shibata, A. Miyauchi, A. Tanabe, J- Hanna, S. Oda and I. Shimizu, Jpn J. Appl. Phys. 25, 1783(1986).
4. G.N. Parsons, D.V. Tsu and G. Lucovsky, J. Non-Cryst. Solids 97&98, 1375(1987).
5. B.D. Cullity, "Elements of x-ray diffraction", Addison-Wesley. Reading, MA. 1978, p. 284.

FRACTAL-LIKE STRUCTURES PRESENT IN HYDROGENATED AMORPHOUS AND
MICROCRYSTALLINE SILICON

M.J. Geerts, R.C. van Oort and J.C. van den Heuvel, Delft University
of Technology, Faculty of Electrical Engineering, Mekelweg 4, 2628 CD
DELFT, The Netherlands.

Abstract

In hydrogenated amorphous silicon NMR studies indicate a two-
phase structural inhomogenity. Hydrogenated microcrystalline films
consist of small crystals with a typical size of 100 Å, embedded in an
amorphous web.　A hydrogen rf plasma is able to etch both type of
films, but with different etch rates. The more crystalline parts of a
film are etched more slowly, which makes hydrogen plasma etching a
technique that can reveal structural inhomogenities and differences in
structural disorder as present, both in amorphous and in
microcrystalline silicon films.
Amorphous and microcrystalline films were etched and fractal-like
structures were visible when using a SEM at a magnification of 20000
times. In microcrystalline films the fractals form a closed network.
The number and typical size of the fractals present in amorphous films
can be influenced by the conditions during the deposition.

INTRODUCTION

Hydrogenated amorphous silicon (a-Si:H) has become an important
material in the fabrication of electronic devices such as solar cells
and thin film transistors. It is generally accepted that hydrogenated
amorphous silicon is indeed amorphous. Evidence for this comes from
transmission electron microscopy (TEM), x-ray and neutron diffraction,
extended x-ray absorption fine structure spectroscopy (EXAFS) 29Si-
nuclear magnetic resonance (NMR) and Raman scattering [1].
The properties of a-Si:H, structural, electrical and optical are
strongly dependent on the deposition technique and on the deposition
conditions used. Under specific conditions hydrogenated
microcrystalline (μc-Si:H) films are formed. These films contain
crystallites with a diameter in the range of 30 - 300ÅÅ. In a-Si:H the
experimental evidence seems to indicate important structural
inhomogenities. Nuclear magnetic resonance (NMR) studies indicate a
two-phase structural inhomogenity in which the hydrogen atoms are
found in two kinds of environment [2]. The material can be described
by means of a two-phase model: a phase, (a) , of low defect density
with hydrogen bonded mainly in the form of SiHx (x=1) groups. This
phase is imbedded in a second phase, (b) , of poor quality, with a
high defect density. In phase (a) a fraction of the monohydride is
atomically dispersed, the remainder is clustered. Phase (b) contains
large amounts of hydrogen, which is bonded not only in the form of
SiHx (x=1), but also as SiHx (x>1) and as $(SiH_2)n$ (n>1) chains.
In good quality material phase (a) dominates. With TEM no
structure can be resolved in as grown films at a scale exceeding 10 -

30 Å for films with a thickness not exceeding 1000 Å [1]. Both types of film, a-Si:H and μc-Si:H, can be etched in a hydrogen plasma [3]. The etch rate of the amorphous films is about ten times higher than the etch rate of microcrystalline films. The difference in etch rate is due to a smaller etch rate of the microcrystals (material with a higher structural order) present in the μc-Si:H films. Thus, by using a hydrogen rf plasma one is able to reveal differences in structural order between a-Si:H and μc-Si:H and between various a-Si:H films have been etched using a hydrogen rf plasma to study the different phases: (a) and (b) of the a-Si:H films and the crystals and the network embedding the crystals of the μc-Si:H film.

EXPERIMENTAL SETUP

The intrinsic a-Si:H and μc-Si:H films used for the etching experiments were grown by the rf glow-discharge method. The deposition conditions are given in table I. Corning 7059 glass was used as a substrate. The same reactor, model SAMCO PD 10, was applied for both growing and etching. Before the etching experiments were performed the reactor was mechanically cleaned to obtain reproducible, well-defined etching conditions in the reactor. The μc-Si:H and a-Si:H films were etched in the same run. The etched films were examined using scanning electron microscopy.

Table I: Deposition conditions for a-Si:H and μc-Si:H and some properties of the a-Si:H films.

	a-Si:H		μc-Si:H
Film number	704	701	289
silane flow rate (sccm)	100	55	2
hydrogen flow rate (sccm)	-	45	98
substrate temperature (K)	523	523	523
pressure (torr)	0.3	0.3	0.3
rf power (mW/cm^2)	62	36	300
DOS (cm^{-3}, eV^{-1}, 10^{16})	5.8	3.4	-
2070 cm^{-1} I.R. signal	present	absent	-

The a-Si:H films no. 701 and 704 differ in their electrical and structural properties as can be seen from table I. In this table the density of states (DOS), as measured by the space-charge-limited current method [4], is given. The DOS near the Fermi-level is smaller for film 701, compared to film 704, indicating a better electrical quality and a higher structural order. The latter is also reflected by infrared spectroscopy analysis of these films as presented in [5]. The presence of an infrared peak near 2070 cm^{-1} is a measure for the amount of microstructure present in the film. With increasing order the amount of microstructure increases which leads to a more pronounced 2070 cm^{-1} signal. From infrared spectroscopy analysis it is clear that film no. 704 exhibits a more pronounced microstructure based on the presence of the 2070 cm^{-1} signal while the signal was absent for film no. 701.

RESULTS AND DISCUSSION

In Figures 1, 2, 3, and 4 scanning electron microscopy
photographs (magnification 20.000x) are shown of an as-grown a-Si:H,
of two a-Si:H film surfaces, originally deposited under different
conditions, and one μc-Si:H film surface etched in a hydrogen rf
plasma. The etched surface clearly is roughened, as can be expected
for a material consisting of more than one phase as a-Si:H and μc-Si:H
are generally believed to be.
For both the amorphous and microcrystalline films the SEM pictures
clearly show that in the etched films one phase (white part in
picture) if of a higher structural order and consequently is etched
less strongly which leads to the coarsened surface. The structures are
fractal-like.

The fractals in the μc-Si:H film form a closed network. This
network is a strong indication that the fractals are of higher
structural order. The fractals probably do consist of microcrystals.
In general, doped μc-Si:H films have a small electrical resistance
compared to doped a-Si:H films. The small resistance is possible only
if the crystals in μc-Si:H are in close contact and form a network.
The fractals in the a-Si:H film do not form a closed network, which
leads to a high electrical resistance.
Referring to the description of the structure of a-Si:H in the
introduction, we suggest that the fractals present in the a-Si:H film
of Figure 2 and 3 represent phase (a). The fractals are embedded in
phase b. The electrical properties of the a-Si:H film are governed by
phase (b) and by the interfase of phase (a) and (b).
From figures 2 and 3 it is clear that in a-Si:H films, which differ in
their structural and electrical properties, differences are present in
their fractal-like structures. The number of the fractals is larger
for the a-Si:H film which is of better quality while the size is
almost equal. Thus with increasing structural order the amount of
crystalline parts present in the film is increased.
An important question remains. Can the etch proces be responsible
for the structures visible? The fact cannot be ruled out but for films
prepared under different conditions, such as films no. 701 and 704, we
find a difference between the etched film surfaces, which means that
hydrogen plasma etching is a technique capable of revealing
differences in film structures of various kinds of films. Thus, the
suggestion to ascribe the presence of the fractals solely to the etch
process can be ruled out. On the other hand, the etch process can have
an influence on the fractals, for example by increasing their size due
to co-deposition occuring preferentially on the surface of the
fractals.
Currently more detailed investigations are being carried out to
quantisize the observed differences in typical fractal numbers and
structures and to apply the percolation theorie to determine the the
deposition conditions where amorphous material changes to
microcrystalline material.

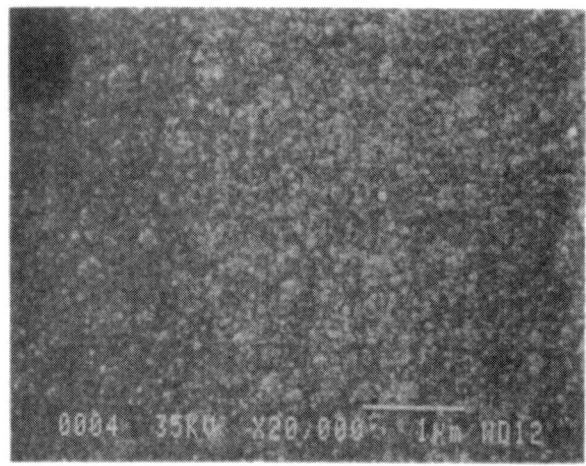

Figure 1: SEM picture of an as-grown a-Si:H film surface.

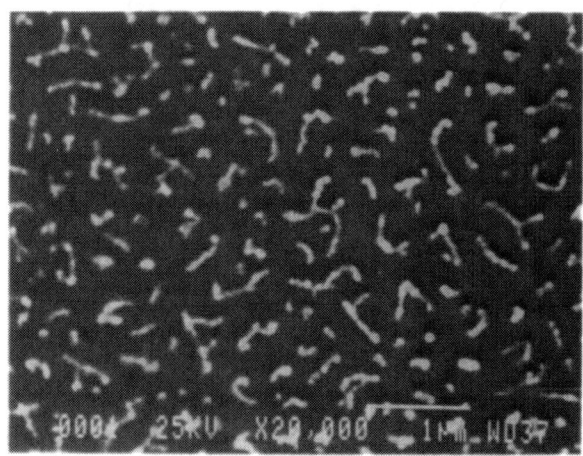

Figure 2: SEM picture of an a-Si:H film surface (film no. 704) etched in a hydrogen plasma. Etching conditions were an rf power of 825 mW/cm² and a substrate temperature of 523 K. Approximately 1.2 μm of the film was removed by etching. The thickness of the as-grown film was 1.8 μm.

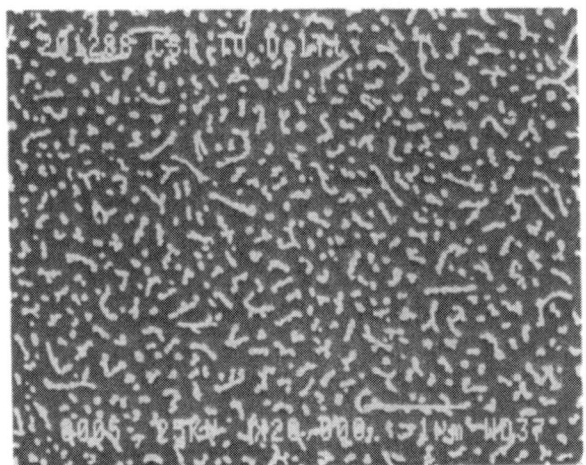

Figure 3: SEM picture of an a-Si:H film surface (film no. 701) etched in a hydrogen plasma. Etching conditions were an rf power of 825 mW/cm² and a substrate temperature of 523 K. Approximately 1.2 μm of the film was removed by etching. The thickness of the as-grown film was 1.8 μm.

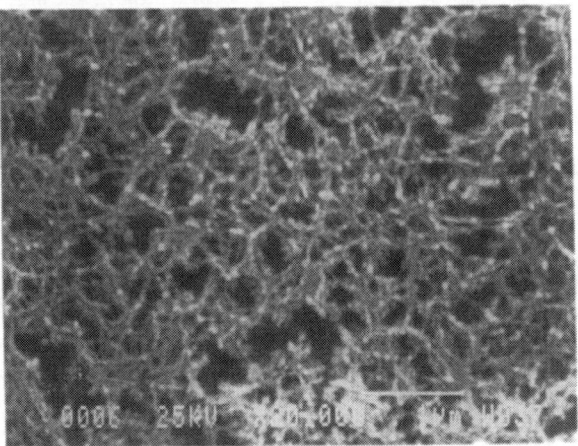

Figure 4: SEM picture of an uc-Si:H film surface (film no. 289) etched in a hydrogen plasma. The etching conditions were an rf power of 482 mW/cm² and a substrate temperature of 523 K. Approximately 0.3 μm of the as-grown film was removed by etching. The Thickness of the as-grown film was 0.8 μm.

CONCLUSIONS

By means of rf hydrogen plasma etching one is able to reveal fractal-like structures present in a-Si:H films. The fractals are crystalline-like and seem to govern the electrical properties of uc-Si:H films, in which the fractals form a closed network. The fractals in a-Si:H are embedded in an amorphous network. Between a-Si:H films deposited under different conditions there are differences in the number of fractal-like structures. With increasing structural order in the film, the number of the crystalline-like structures is raised. The influence of co-deposition on the size of the crystals during the etch proces is still not clear.

AKNOWLEDGEMENT

This work was supported by NWO, the Netherlands Foundation for Scientific Research. The authors thank M. Kleefstra and JB vanStaden for valuable discussions and JW. Metselaar for his encouragement.

REFERENCES

[1]   J.C. Knights, in Topics in Applied Physics, vol. 55, ed. J. D. Joannopoulus, G. Lucovsky (Springer Verlag, Berlin, 1983), chapter 2.
[2].   H. Fritzsche, Thin Solid Films, 90, 119, (1982).
[3].   R.C. van Oort, M.J. Geerts, J.C. van den Heuvel, J.W. Metselaar Elec. Lett., 23, (8), 967, (1987).
[4].   R.C. van Oort, M.J. Geerts, J.C. van den Heuvel, in Photon. Beam. and Plasma-enhanced Processing, edited by E.F. Krimml and V.T. Nguyen (Proc. E-MRS, vol XV, Strassbourg, France, 1987) pp. 449-453.
[5].   R.C. van Oort, J.C. van den Heuvel, M.J. Geerts, in Proc. of the 1988 Int. Topical Conf. on Hydrogenated Amorphous Silicon Devices an Technology", edited by J. Kanicki (Yorktown Heights, USA, 1988) pp. 27-30.

FAST-PULSE EXCIMER-LASER-INDUCED PROCESSES IN a-Si:H

K. WINER, R.Z. BACHRACH, R.I. JOHNSON, S.E. READY, G.B. ANDERSON, and J.B. BOYCE, Xerox Palo Alto Research Center, Palo Alto, CA 94304

ABSTRACT

The effects of fast-pulse excimer laser annealing of a-Si:H were investigated by measurements of electronic transport properties and impurity concentration depth profiles as a function of incident laser energy density. The dc dark conductivity of laser-annealed, highly-doped a-Si:H increases by a factor of ~350 above a sharp laser energy density threshold whose magnitude increases with decreasing impurity concentration and which correlates with the onset of hydrogen evolution from and crystallization of the near-surface layer. The similarities between the preparation and properties of laser-crystallized a-Si:H and μc-Si:H are discussed.

INTRODUCTION

The electronic conductivity and mobility of hydrogenated microcrystalline silicon (μc-Si:H) can be orders of magnitude higher than those of hydrogenated amorphous silicon (a-Si:H), like a-Si:H, a promising candidate for large-area display and sensor applications. In addition, the *in-situ* growth of μc-Si:H is compatible with existing a-Si:H processing techniques, which allows for hybrid amorphous/microcrystalline device fabrication. On the other hand, the low thermal growth rates of μc-Si:H are undesirable for efficient large-scale device fabrication. Also, growth of μc-Si:H alloyed with Ge or C, important for many sensor applications, is not yet completely reliable. Fast-pulse excimer laser annealing of a-Si:H shares all of the important advantages of μc-Si:H, but offers the further advantage of spatial selectivity, the capability of post-deposition processing, and efficient crystallization of a-Si:H alloys. Although improved device performance utilizing laser-crystallized a-Si:H has been demonstrated [1,2], further progress toward the application of such materials requires a better understanding of the underlying laser-induced a-Si:H crystallization process [3,4].

EXPERIMENTAL CONSIDERATIONS

In order to investigate this process, two sets of samples were employed. Firstly, 1000 Å thick high-quality a-Si:H films were deposited by the rf glow discharge decomposition (2 W, 230 °C) of either PH_3 or B_2H_6 diluted in silane for n- or p-type a-Si:H, respectively, onto 3000-Å-thick a-SiN$_x$ buffer layers supported by glass substrates. Chrome contacts in the gap cell configuration were evaporated onto the resulting films for dc dark conductivity measurements. These laser-annealed a-Si:H films were also characterized by Raman spectroscopy, X-ray diffraction, transmission electron microscopy (TEM), and Hall effect measurements. Secondly, a 500 Å thick 10^{-2} P-doped a-Si:H layer was deposited onto a 1 μm thick undoped a-Si:H layer supported on a similar nitride buffer on a crystalline Si (c-Si) substrate in order to examine P and H concentration depth profiles after laser annealing. The concentration depth profiles were measured by secondary ion mass spectrometry (SIMS) using a 14.5 keV Cs^+ beam. Both types of films were scanned at room temperature under $\leq 1 \times 10^{-6}$ Torr vacuum with 2560 pulses at 40 Hz from an XeCl excimer laser (308 nm, 17 ns) at laser energy densities between 60 and 420 mJ/cm^2.

RESULTS

The intense uv laser light is absorbed in the first ~170 Å of the a-Si:H surface. As the resulting heat pulse near the surface is propagated into the bulk, the

Mat. Res. Soc. Symp. Proc. Vol. 164. ©1990 Materials Research Society

temperature of the pulse rapidly decays. The accompanying crystallite nucleation and growth processes can be investigated by monitoring impurity activation and diffusion as the laser beam or starting a-Si:H material properties are varied.

DC Dark Conductivity

The most dramatic effect of laser crystallization of a-Si:H is observed in the room temperature dc dark conductivity which increases from ~4 x 10^{-2} $\Omega^{-1}cm^{-1}$ for as-deposited 10^{-2} P-doped a-Si:H to ~15 $\Omega^{-1}cm^{-1}$ over a very narrow laser energy density range (80-90 mJ/cm²) as shown in Fig. 1a. The electron (Hall) mobility similarly increases from ~0.1 to ~10 cm²V⁻¹s⁻¹ upon laser-induced crystallization [3]. The laser energy density threshold (≈85 mJ/cm² in 10^{-2} P-doped a-Si:H) increases and broadens as the doping level of the a-Si:H film decreases. In undoped a-Si:H, the threshold is completely washed out and the conductivity increases approximately exponentially with incident laser energy density. Similar results are observed for B-doped a-Si:H (Fig. 1b) and for both n- and p-type a-SiC:H alloys. In the latter case, the threshold increases and broadens as the carbon content of the alloys increases.

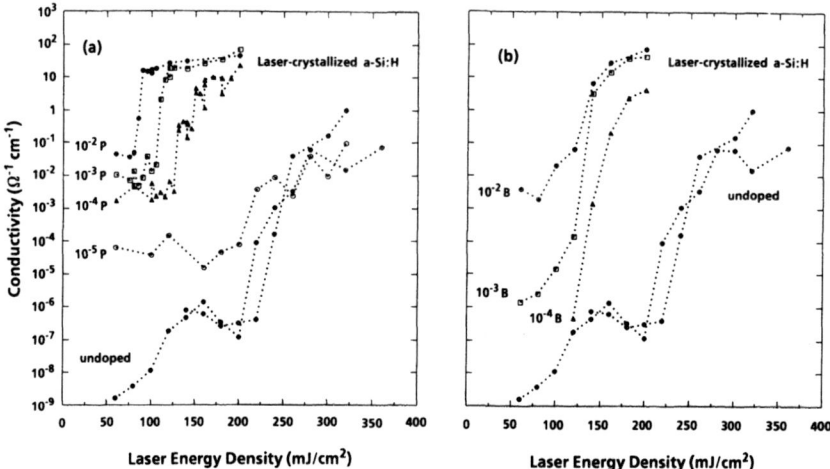

Fig. 1   Dc dark conductivity of (a) P-doped and (b) B-doped a-Si:H following excimer-laser annealing as a function of incident laser energy density.

The structure of the 10^{-2} P-doped a-Si:H films has been investigated just below, at, and just above the ≈85 mJ/cm² threshold by Raman, X-ray diffraction, and TEM measurements. The Raman spectra exhibit the sharp peak near 520 cm⁻¹ characteristic of crystalline Si only in films just above threshold (≥90 mJ/cm²). X-ray diffraction results show essentially the same behavior, with the average grain size increasing from the amorphous silicon background at and below threshold to ≈ 30 Å just above threshold. TEM micrographs show the formation of small (10-20 Å diameter) crystallites distributed discontinuously within the first ~50 Å of the a-Si:H surface just at threshold. Just above threshold, a continuous surface layer of crystallized a-Si:H is formed whose thickness increases with increasing laser energy density. This crystallized layer continues to grow until the entire film is transformed into a continuous, large-grained (~1000 Å diameter) polycrystalline layer near 400 mJ/cm². At intermediate energy densities, heterogeneous layers form composed of large polycrystals near the surface separated from the remaining amorphous material by a layer of smaller microcrystallites and microvoids characteristic of explosive

explosive crystallization. Similar results have been observed upon laser crystallization of amorphized Si [5].

The increase in the dc dark conductivity near threshold is, therefore, a percolation phenomenon with the largest single increase in conductivity arising from the formation of a continuous conduction path and correspondingly enhanced effective mobility. This is seen clearly in the temperature dependence of the dc dark conductivity shown in Fig. 2.

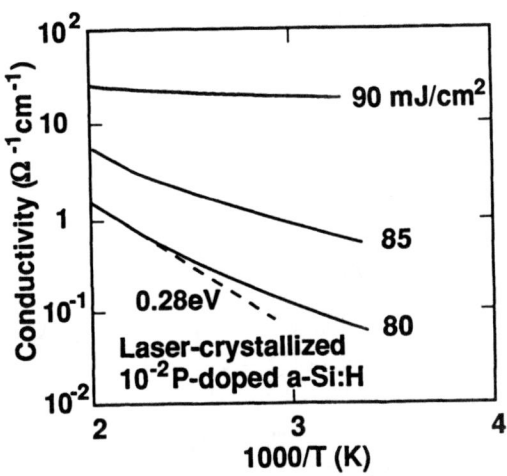

Fig. 2   Dc dark conductivity of P-doped a-Si:H as a function of inverse tempera-ture just below, at, and just above threshold.

Below threshold, the temperature dependence of the conductivity is identical to that of as-deposited 10^{-2} P-doped a-Si:H with an equilibrium activation energy E_A of 0.28 eV. At the 85 mJ/cm^2 threshold, the shape of the conductivity curve is still characteristic of an amorphous film and E_A has not changed even though the conductivity itself has increased by a factor of ten. Just above threshold, the conductivity increases by another factor of 30 and the small activation energy ($\ll 0.1$ eV) is characteristic of conduction in a nearly degenerately-doped semiconductor.

The further slow increase in the conductivity above threshold (Fig. 1) is due to the increasing thickness and, correspondingly, increasing average grain size, of the crystallized a-Si:H films. In fact, at or above threshold the conductivity of the dominating crystallized layer is actually as much as ten times higher than that given in Fig. 1 due to the use of the nominal a-Si:H thickness (1000 Å) rather than the less-well-known thickness of the crystallized layer in the conversion of the measured current to conductivity units.

Lastly, the conductivity of laser crystallized P-doped a-Si:H near 200 mJ/cm^2 is essentially independent of doping between 10^{-4} and 10^{-2} mole fraction of phosphine in the gas. This is most likely due to the $\sim 1 \times 10^{20}$ cm^{-3} P solubility limit in c-Si which limits the conductivity of P-doped laser crystallized a-Si:H films to $\sigma = n\mu e \sim 1 \times 10^{20}$ cm$^{-3} \times 10$ cm^{2}V^{-1}s$^{-1} \times 1.6 \times 10^{19}$ Vs$\Omega^{-1} \sim 500$ Ω^{-1}cm^{-1}. This is about the same value observed in Fig. 1 at 200 mJ/cm^2 when the actual thickness (~ 200 Å) of the polycrystalline layers is taken into account. Similar behavior is also observed for B doping.

Impurity Concentration Depth Profiles-Phosphorus

The P concentration depth profiles measured with SIMS are shown in Fig. 3. The P-doped surface layer is well-defined; the long P tail into the as-deposited undoped a-Si:H layer is due to the low spatial resolution of the high energy Cs$^+$ probe beam. Up to a laser energy density of 100 mJ/cm^2, no change in the P concentration depth profile is observed. At 140 mJ/cm^2, a slight enhancement in the P concentration near the P-doped/undoped interface is observed, which increases with respect to the remaining portions of the P-doped layer with increasing laser energy density. Above 200 mJ/cm^2, the near-surface P concentration begins to decrease and the P tail into the undoped region begins to broaden. At the highest energy density of 420 mJ/cm^2, the P has diffused nearly 500 Å from the original P-doped layer into the initially undoped a-Si:H layer.

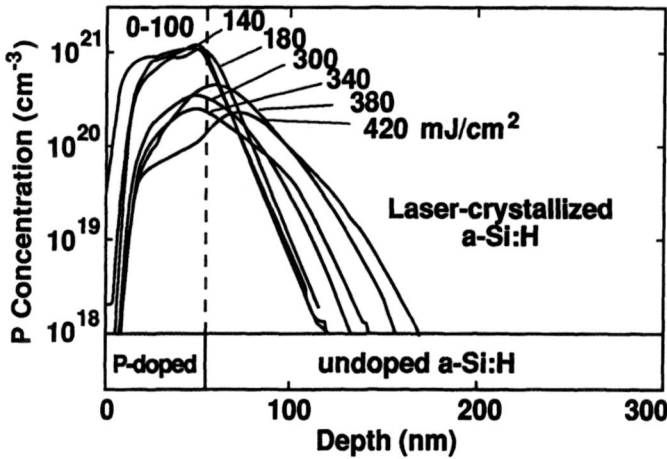

Fig. 3   P concentration depth profiles as a function of laser energy density.

The laser pulse width is 17 ns. The time the sample is strongly heated is estimated to be about 100 ns per pulse from transient reflectivity measurements. For P to diffuse over a distance of L = 500 Å during the integrated heating time τ = 2560 x 100 ns would require a diffusion constant D of $L^2/2\tau$ = 1 x 10^{-4} cm^2/s, which corresponds to a temperature for P diffusion in c-Si of 2400 K [6], well above the 1700 K c-Si melting temperature. This is a rough estimate of the average temperature required to account for the observed P diffusion assuming that the P diffusion constant in laser-crystallized a-Si:H is the same as that in c-Si. Performing the same calculation for a P diffusion distance of 10 Å corresponding to the near threshold condition gives an average temperature of 1630 K. The P diffusion data clearly indicate that laser-induced crystallization of a-Si:H results from solidification from the melt at laser energy densities above 100 mJ/cm^2. Such a conclusion agrees with interpretations of laser-induced crystallization of amorphized Si [5] and provides a reasonable explanation for the large grain sizes obtained at higher laser energy densities.

Impurity Concentration Depth Profiles-Hydrogen

The H concentration depth profiles are shown in Fig. 4. The H concentration is slightly higher in the near-surface P-doped a-Si:H layer than in the undoped a-Si:H bulk, where $[H] \approx 5 \times 10^{21}$ cm^{-3} = 0.1[Si]. As observed for P, no change in the H concentration depth profile is observed below an energy density of 140 mJ/cm^2. At 140 mJ/cm^2, the H concentration near the P-doped/undoped interface is enhanced relative to that in the remaining P-doped layer. At still higher laser energy densities the P-doped layer becomes increasingly depleted of H as crystallite formation proceeds. Interestingly, H appears to become trapped at the surface and P-doped/undoped interfaces even after H depletion, and presumably crystallization, extends deep into the undoped layer.

H depletion of the near-surface clearly accompanies large grain polycrystallite formation. TEM micrographs show that the microcrystalline layer between the poly-crystalline near-surface and the amorphous bulk contains a large concentration of microvoids, which most likely result from H$_2$ gas bubble formation. The polycrystalline layer can only increase in thickness if H in this defective microcrystalline layer can diffuse through the polycrystalline/melt region and escape to the vacuum during laser annealing.

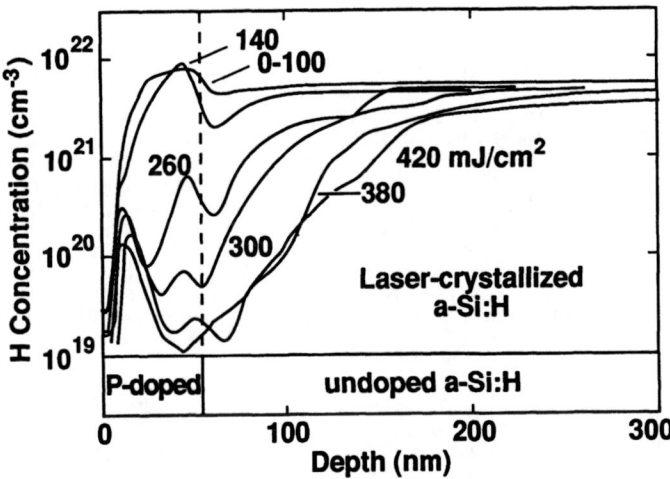

Fig. 4   H concentration depth profiles as a function of laser energy density.

The changes in the H and P concentration depth profiles with increasing incident laser energy density correspond exactly; the initial changes in both begin at 140 mJ/cm^2 and P diffuses into the undoped a-Si:H layer a distance equal to the depth of H depletion. H diffusion out of the heated near-surface region and P diffusion into the undoped a-Si:H layer are both driven by chemical potential gradients and enabled by the high temperature for short times of the near-surface melt zone.

DISCUSSION

The solid-state chemical reactions that control defect formation and dopant activation in a-Si:H are mediated by H diffusion; the reactions can be accelerated by increasing the carrier concentration which increases the H diffusion rate [7]. While it is unlikely that similar hydrogen-mediated chemical reactions lead to crystallite nucleation, the presence of 10% H in a-Si:H does have some effect on the crystallization processes. It has been shown [4] that undoped LPCVD a-Si containing less than 0.1% H completely (poly-)crystallizes near 50 mJ/cm^2 (under identical laser conditions) without the formation of an intermediate defective microcrystalline layer, in contrast to the behavior of undoped a-Si:H observed here and of amorphized Si under 693 nm illumination [5]. This suggests that H impedes large polycrystalline grain growth in a-Si:H, perhaps by frustrating crystal organization. The defective microcrystalline layer, which consists of small (<100 Å diameter) grains in a H-rich matrix, should have properties similar to those of microcrystalline Si:H grown by plasma-enhanced CVD.

The decrease in the crystallization threshold with increasing doping might be due to the increased diffusion rate of H in doped a-Si:H, which would decrease the energy cost of crystallite nucleation. Evidence against this interpretation lies in the higher (≈ 120 mJ/cm^2), broader laser energy density threshold in 10^{-2} B-doped a-Si:H compared to that in 10^{-2} P-doped a-Si:H, even though the H diffusion constant of the former is a factor of ten higher than that of the latter. However, the mobility of holes is much smaller than that of electrons in a-Si:H. Therefore, the threshold dependence on doping suggests that the nucleation threshold in a-Si:H depends directly on the charge carrier concentration, and may proceed via some as yet unidentified solid-phase mechanism. Of course, the impurity atoms themselves may be acting as nucleation sites, but the large change in threshold behavior between 10^{-4} and 10^{-5} P-doping levels, for example, is more consistent with the factor of ~twenty decrease in carrier concentration rather than the factor of three decrease in total P concentration. Given that in Si at the melt temperature the intrinsic carrier concentration is of order 10^{20} cm^{-3}, it is unlikely that extrinsic dopants can greatly affect the crystallite growth or H depletion processes, so that their effect should be limited to nucleation.

The process of laser-induced crystallization of a-Si:H, which takes place at high temperatures and short times, is completely different from the low-temperature, quasi-equilibrium growth of μc-Si:H. However, both processes can result in material with large, polycrystalline grains and with improved transport properties. In general though, the conductivity and mobility of laser-crystallized a-Si:H can be more than ten times higher than those in μc-Si:H. This difference in material properties is due to the large concentration of H incorporated in μc-Si:H, which inhibits very large grain growth and leads to increased defect formation. One possibility for improved μc-Si:H properties might be to increase the temperature of growth in order to decrease the H content and promote defect annealing.

In summary, we have measured the electronic transport and impurity diffusion behavior in a-Si:H upon excimer-laser-induced annealing. We have interpreted the observed doping dependence of the conductivity threshold in terms of a solid-phase nucleation process whose rate depends on the carrier concentration in the initial a-Si:H film. Crystallization at high laser energy densities is determined to result from regrowth from the melt.

We thank Bob Street for many informative discussions. This work supported by the Solar Energy Research Institute (Golden, CO).

REFERENCES

[1]   T. Sameshima and S. Usui, Mat. Res. Soc. Symp. Proc. **71** (1986) 435.
[2]   P.H. Fang, Solar Cells **25** (1988) 27.
[3]   S.E. Ready, J.B. Boyce, R.Z. Bachrach, R.I. Johnson, K. Winer, G. Anderson, and C.C. Tsai, Mat. Res. Soc. Symp. Proc. **149** (1989) 345.
[4]   R.Z. Bachrach, K. Winer, J.B. Boyce, S.E. Ready, R.I. Johnson, and G.B. Anderson, J. Electronic Materials (to be published).
[5]   J. Nayaran, C.W. White, M.J. Aziz, B. Stritzker, and A. Walthuis, J. Appl. Phys. **57** (1985) 564.
[6]   S.M. Sze, *Physics of Semiconductor Devices* (Wiley, New York, 1981).
[7]   K. Winer and W.B. Jackson, Phys. Rev. B**40** (1989) 12558.

HYDROGEN PASSIVATION OF DOPED AND UNDOPED MICROCRYSTALLINE SILICON

M. STUTZMANN, C.P. HERRERO*, M. INGELS AND A. BREITSCHWERDT
Max-Planck-Institut für Festkörperforschung, D-7000 Stuttgart 80, FRG
*Instituto de Ciencia de Materiales, Serrano, 115 dpdo, 28006 Madrid, Spain.

ABSTRACT

The effect of hydrogen plasma treatment on the properties of undoped and doped microcrystalline silicon are investigated by Raman and infrared spectroscopy, conductivity measurements, hydrogen effusion, and subgap absorption. It is found that hydrogen causes a passivation of dopant atoms similar to that seen in crystalline silicon. However, the efficiency of the dopant passivation appears to be smaller in μc-Si, and the vibrational spectra of dopants and hydrogen show distinct broadening due to local disorder. It is suggested that intergrain regions may efficiently getter atomic hydrogen introduced from the plasma and also provide the origin for a weakly bound hydrogen phase observed in the effusion experiments.

INTRODUCTION

Passivation of point defects in semiconductors by atomic hydrogen has been widely used in crystalline silicon and III-V compounds, and also in amorphous thin film semiconductors. In the specific case of silicon, special attention has been paid to acceptor and donor passivation in crystals [1], and to passivation of deep defects (dangling bonds) in the amorphous phase, which contains typically 10 at.% H as a major chemical constituent (a-Si:H). Also microcrystalline silicon (μc-Si) is usually prepared under conditions which favor hydrogen incorporation (e.g. plasma-enhanced CVD of SiH_4/H_2 mixtures) [2], and it is known that hydrogen incorporated into μc-Si during growth helps to passivate coordination defects at the grain boundaries. This fact has also been exploited for the passivation of grain boundaries in polycrystalline Si [3,4]. In the present paper, we study the effect of hydrogen plasma treatment on μc-Si (doped and undoped) which does not contain bonded hydrogen in the as-prepared state. This allows us to distinguish hydrogen passivation effects similar to those observed in poly- or single crystalline Si from H-related effects due to sample deposition in a hydrogen-rich environment.

EXPERIMENT

Intrinsic, p-type, and n-type μc-Si samples were obtained by recrystallizing amorphous silicon samples (undoped or doped with ≈ 0.1 at.% P or B) for one hour at T = 800°C. Samples prepared in this way had crystallite sizes of 100 Å or more and were free of bonded hydrogen. The thickness of the μc-Si films was app. 1.5 μm in all cases. Hydrogen plasma passivation was performed in a remote glow discharge system. Before and after passivation, all samples were characterized by Raman scattering (phonon and local modes), spectroscopic ellipsometry (crystallinity and surface oxide), infrared reflection (free carrier concentration), and subgap absorption (free carriers, deep defects). Deuterium effusion measurements were used to determine the bonding state of the incorporated H (D) and the total hydrogen content.

RESULTS AND DISCUSSION

Infrared reflection spectra of n-type and p-type μc-Si are shown in Figs. 1 and 2, respectively. Both figures show the plasma edge caused by the free carrier in the 1000 cm^{-1} wavenumber region and the thin film interference fringes at higher wavenumbers. From the position of the plasma frequency, $\omega_p^2 = (Ne^2/m^*\varepsilon)$, the charge carrier density, N, can be estimated, assuming the same effective masses as in crystalline silicon. For the as-prepared

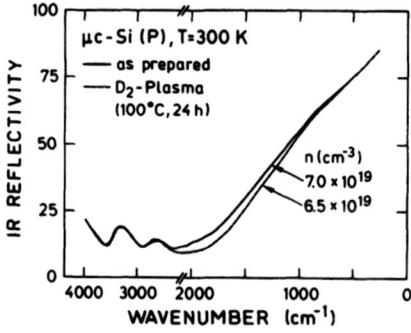

Fig. 1 Infrared reflection spectra of phosphorus-doped μc-Si before (solid curve) and after (dashed curve) D_2-plasma passivation for 24 h at 100°C. n indicates the free-electron density calculated from the position of the plasma edge.

state of the sample in Fig. 1 and Fig. 2 we then obtain an electron density of 7×10^{19} cm^{-3} and a hole density of 5×10^{19} cm^{-3}. These densities compare favorably with the expected active dopant concentration of 1000 ppm $\cong 5 \times 10^{19}$ cm^{-3} estimated from the volume fraction of PH_3 and B_2H_6 dopant gases introduced during the film deposition (0.1 vol.%), indicating that almost all dopant atoms have become electronically active during the crystallization. Exposure to a hydrogen plasma reduces the free carrier densities again. According to Fig. 1, the reduction is very small in n-type μc-Si (–10%, dashed curve). In crystalline silicon, passivation at the same temperature and under identical conditions leads to a reduction of the free electron density by more than 30%, in agreement with earlier studies [5]. A similar inhibition of shallow level passivation in μc-Si compared to c-Si is also observed in p-type samples (Fig. 2). H_2-plasma treatment at 180°C for 48 h only results in a reduction of the free hole density by a factor of two, whereas a similar treatment for crystalline Si would be sufficient to passivate more than 95% of the acceptor atoms [6]. The reason for the low passivation efficiency for dopants in μc-Si is not yet understood. Since it is known that a surface oxide layer can hinder in-diffusion of hydrogen [7], we have determined the oxide layer thickness on our recrystallized samples by spectroscopic ellipsometry. Typical values were in the range 20 – 40Å, and thus similar to native oxides on the crystalline samples used for comparison. Also, HF etching of the

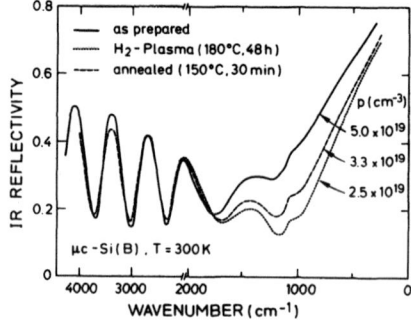

Fig. 2 IR reflectiviy of boron-doped μc-Si at room temperature in the as-prepared state (solid curve), after hydrogen passivation (dotted curve), and after partial anneal (dashed curve).

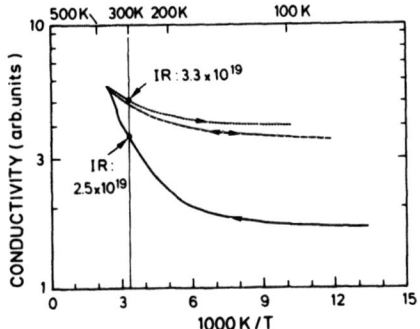

Fig. 3   Temperature dependence of the dc-conductivity for the sample in Fig. 2. See text for details.

microcrystalline samples just before H_2-plasma treatment did not improve the passivation efficiency. Therefore, we believe that the lower efficiency is related to the sample structure rather than to secondary effects. We will come back to this point in the discussion of the Raman spectra below.

As in crystalline silicon, the dopant passivation in μc-Si is reversible by annealing at temperatures between 150 and 200°C for a short time. As an example, the dashed curve in Fig. 2 shows the recovery of the free hole density in B-doped samples after a 30 min. anneal at 150°C to an intermediate value of 3.3×10^{19} cm^{-3}. The thermally induced reactivation of the dopants can also be observed in the conductivity of the samples. In Fig. 3, we show the development of the conductivity as a function of temperature for a number of heating and cooling cycles. (A relative conductivity of 1 in Fig. 3 corresponds to $\sigma \approx 0.1\Omega^{-1}$cm^{-1}.) The sample is the same one which has been used to obtain the IR reflectance spectra in Fig. 2, and the measurement starts after H-passivation to a hole density of 2.5×10^{19} cm^{-3}. The sample is heated to 150°C (solid line in Fig. 3) and kept at this temperature for 25 min. After that, the sample is rapidly quenched with liquid nitrogen to $\approx$80 K and again heated up slowly to 150°C (dashed curve). The conductivity is measured both during the rapid quench and the following heating period, in order to check for possible metastable configurations similar to those observed in doped hydrogenated amorphous silicon [8]. However, within the experimental error $\sigma(T)$ was identical during rapid quenching and slow heating. The dotted curve in Fig. 3 shows $\sigma(T)$ after an additional 5 min anneal at 150°C and corresponds to a slow cooling process. According to Fig. 3, the change in the room temperature conductivity between the solid and dotted curve is about a factor of 1.4. This compares favorably with a factor of 1.3 deduced from the infrared data in Fig. 2.

In order to determine the concentration and the bonding state of hydrogen in the passivated μc-Si samples, we have used hydrogen effusion and Raman scattering. The results of the effusion measurements are summarized in Fig. 4 and are also compared to similar measurements performed on crystalline silicon samples. Effusion spectra were obtained by heating the samples with a constant rate (25°C/min) from room temperature to 900°C while recording the H_2 (D_2) partial pressure in the constantly pumped effusion system with a mass spectrometer. (Deuterium instead of hydrogen was used because of the much lower D_2 background). In the case of boron-doped μc-Si (Fig. 4(a)), the effusion spectrum consists of a dominant peak at 300°C and a smaller shoulder at 500°C. Compared to crystalline silicon passivated under identical conditions (dashed curve), the effusion of deuterium in μc-Si shows a definite shift towards lower temperatures. (Note that in Fig. 4(a) the total amount of D-atoms incorporated into the μc-Si and c-Si samples is roughly equal, despite of the factor of five difference in boron concentration. This is compensated by a much larger passivation depth in the case of the crystalline specimen (5 μm vs 1.5 μm for the μc-Si sample).) A similar trend is also observed in n-type samples (Fig. 4(b)).

However, in this case the effusion spectrum follows more closely the structure observed in crystalline silicon (small peak at 500°C, stronger narrow peak at app. 350°C), with the exception of an additional, low-temperature effusion peak at 200°C. One can speculate that the effusion of hydrogen in μc-Si is enhanced by the intergrain regions, where H_2 molecules can be trapped and diffuse relatively easily. More details concerning the hydrogen effusion in μc-Si and c-Si will be given in a future publication [9].

The vibrational properties of boron and hydrogen in passivated μc-Si can be investigated with the help of Raman scattering. This technique has been used quite successfully for a study of boron-doped c-Si [6,10], and some pertinent results obtained for μc-Si are described below. Figure 5 shows the spectral region of the [11]B and [10]B local vibrational modes in the Si matrix for a total boron concentration of 5×10^{19} cm^{-3}. Crystalline silicon (dotted curve) shows in this region two distinct peaks for the two natural boron isotopes at 620 cm^{-1} ([11]B) and 643 cm^{-1} ([10]B) with an intensity ratio of 4:1. In unpassivated μc-Si (B) (solid curve), only the [11]B-mode is observed, whereas the smaller [10]B-peak is no longer resolvable, due to the inhomogeneous disorder-broadening of the larger [11]B-line. Hydrogen passivation leads to a reduction of the [11]B-peak intensity by about a factor of two due to B–H complex formation. This reduction is in quantitative agreement with the changes in the free hole density deduced from Fig. 2. As in c-Si, the decrease in the hole density can also be detected via the less pronounced free hole scattering [6]. In Fig. 5, this can be seen as an overall decrease of the background signal. The new vibrational mode due to [11]B–H complexes at 650 cm^{-1} observed in crystalline samples is not clearly observed in Fig. 5, because of the low complex concentration and of the disorder broadening in μc-Si.

In addition to the hydrogen-induced changes in the B-related local modes, also the vibrations of the incorporated hydrogen atoms can be studied by Raman scattering. Figure 6 shows a comparison of the low temperature (T = 20 K) Raman scattering in the spectral region around 2000 cm^{-1} for c-Si and μc-Si samples doped with similar boron concentrations and subjected to the same plasma passivation. In the c-Si sample, only the sharp line of H-atoms in B–H complexes is visible [6]. In μc-Si, the same vibrational mode can be detected, however, broadened by disorder-induced local strain. In addition,

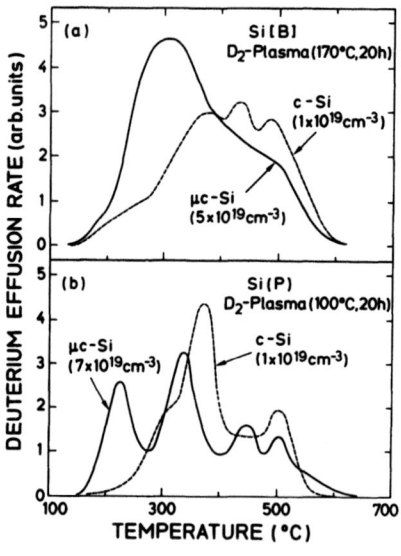

Fig. 4 Deuterium effusion of doped, D_2-plasma passivated μc-Si samples. (a) Boron-doped, (b) Phosphorus-doped. Dashed curves indicate similar effusion spectra of doped crystalline Si samples subjected to the same plasma treatment.

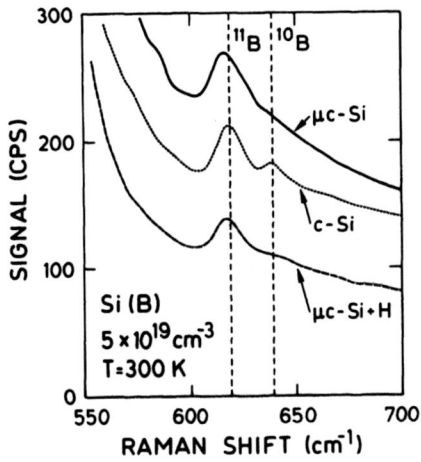

Fig. 5 Raman spectra of boron local vibrational modes in p-type microcrystalline Si before (solid curve) and after H₂-passivation (dashed curve). The dotted line is the corresponding spectrum of crystalline silicon with the same dopant concentration.

a stronger peak around 2150 cm⁻¹ is observed. The 2150 cm⁻¹ peak is also seen in the Raman spectra of H-passivated undoped and phosphorus-doped μc-Si, and is most likely due to Si–H bonds in the intergrain regions. These regions act as effective H-getters for the hydrogen atoms entering the μc-Si samples and are probably also the reason for the much lower dopant passivation efficiency mentioned above.

Finally, we would like to briefly mention some results obtained in undoped μc-Si. The effect of H-passivation in this case has been studied by subgap absorption measurements (photothermal deflection spectroscopy) and Raman scattering. From the Raman spectra (appearance of the 2150 cm⁻¹ peak) it was obvious that a considerable amount of hydrogen could be introduced by plasma treatment. Also, it is known from electron spin resonance investigation [e.g. 4] that plasma passivation leads to a strong (order of magnitude) reduction of the Si dangling bond related resonance signal. On the other hand, subgap absorption measurements show only a small reduction of the absorption coefficient in the subgap region ($\hbar\omega \leq 1$ eV [11]) by less than a factor of two. In fact, our data obtained in undoped μc-Si are almost identical to those observed earlier by Jackson et al. for polycrystalline Si [3]. A more systematic study, combining subgap absorption, spin density, IR, and hydrogen effusion measurements will be necessary to clarify the effect of hydrogen in undoped μc-Si.

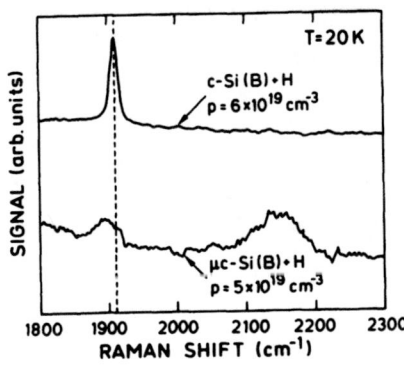

Fig. 6 Hydrogen vibrations at T = 20 K in passivated, boron-doped crystalline (upper trace) and microcrystalline silicon (lower trace).

REFERENCES

[1] S.J. Pearton, J.W. Corbett, T.S. Shi, Appl. Phys. A **43**, 153 (1987).

[2] N.M. Johnson, S.E. Ready, J.B. Boyce, C.D. Doland, S.H. Wolff, J. Walker, Appl. Phys. Lett. **53**, 1626 (1988).

[3] W.B. Jackson, N.M. Johnson, D.K. Biegelsen, Appl. Phys. Lett. **43**, 195 (1983).

[4] S. Hasegawa, S. Takenaka, Y. Kurata, J. Appl. Phys. **53**, 5022 (1982).

[5] K. Bergman, M. Stavola, S.J. Pearton, J. Lopata, Phys. Rev. B **37**, 2770 (1988).

[6] M. Stutzmann, Phys. Rev. B **35**, 5921 (1987).

[7] C.P. Herrero, M. Stutzmann, A. Breitschwerdt, this meeting, Symposium G.

[8] R.A. Street, J. Kakalios, C.C. Tsai, T.M. Hayes, Phys. Rev. B **35**, 1316 (1987).

[9] M. Stutzmann, M. S. Brandt (unpublished).

[10] C.P. Herrero, M. Stutzmann, Solid State Commun. **68**, 1085 (1988).

[11] M. Ingels, M. Stutzmann, S. Zollner, this volume.

CONTROL OF CHEMICAL REACTIONS FOR GROWTH OF CRYSTALLINE Si AT LOW SUBSTRATE TEMPERATURE

Isamu SHIMIZU, Jun-ichi HANNA*, and Hajime SHIRAI
The Graduate School, Imaging Sci. & Eng. Lab.* , Tokyo Institute of Technology,
4259 Nagatsuta, Midori-ku, Yokohama, Japan

ABSTRACT
A systematic study has been made on the formation of Si-network of amorphous(a-), microcrystalline(μc-) and epitaxial (epi)-Si prepared by Plasma-Enhaced (PE-) CVD under control of flow of atomic hydrogen. The control of the Si-network structures requires a deliberate selection of the precursor, i.e., SiHn (n$\leq$3) and SiFnHm (n+m$\leq$3), as well as an intentional acceleration of the chemical reactions for the propagation of Si-network in the vicinity of the growing surface by impinging of atomic hydrogen. A plausible interpretation was given to the growing mechanism of c-Si at low temperature.

INTRODUCTION
In recent years a novel field termed "Giant Micro-Electronics" has been enthusiastically opened up, which led to the development of the electronic devices requiring large sizes such as solar cells, display devices and sensors. Hydrogenated amorphous silicon (a-Si:H) and its related thin films prepared by PE-CVD played a major role in the early stage of this new field because of their attractive characteristics as follows:
(1) excellent photoconductivity and feasibility of the valence control by impurity doping,
  and
(2) ability to make thin homogeneous films with a large area on inexpensive substrates
  at a rather low substrate temperature (Ts).

A great attention has been focussed on μc-Si prepared by PE-CVD under a slightly different condition from that for a-Si:H because of its promising natures, namely, higher electric conductivity of the doped films arising from higher doping efficiency[1] or mobility[2], excellent optical transparency for visible light and rather high optical absorption in near ir region. In practice, the doped μc-Si is widely adopted to the a-Si:H solar cells as the thin layers at the contact with electrodes.[3] Some attempts have been made to fabricate μc-Si TFT[4], expecting increment of the ON-current. A higher collection efficiency was obtained in the solar cells combining c-Si one with the a-Si:H cells in tandem structure.[5]

First systematic studies were made on μc-Si by Veprek and his coworkers mainly with respect to the control of the plasma chemistry.[6] According to their results, μc-Si is preferentially deposited by PE-CVD from gaseous mixture of SiH$_4$ and H$_2$ under the silane-depletion condition and the longer dwell time of silane in plasma to provide "the partial chemical equilibrium" between deposition and etching.[7] Recent results obtained by Tsai and her coworkers gave a strong support to this etching model in making μc-Si.[8] On the contrary, the diffusion of the precursors adsorbed on the growing surface was taken into account by Matsuda as an important role played in making μc-Si.[9] According to their model, the diffusion length is mainly ruled by the chemical activity of the growing surface. In other words, the crystal growth is greatly promoted by diffusion of the precursors adsorbed on the surface covered with hydrogen.

Accordingly, dilution with hydrogen is thought to be effective on the passivation of the surface.

We found marked different behaviors in making μc-Si from fluorinated silyl radicals, SiHnFm (n+m≤3), compared with those from SiHn, as follows:[10]
(a) μc-Si was able to be deposited at a high growth rate of 30 A/ s or more,
(b) epi-Si was also grown at a high growth rate on the c-Si(100),[11] (110)[12] substrate at
Ts of 350 °C or below.

Specific characteristics were also found in the electric properties of the μc-Si prepared from SiHnFm under the control of atomic hydrogen as follows:[13] (1) high electric conductivity of --10^2 S/cm or more at room temperature in P-doped film with the aid of a high doping efficiency of 30 % or more, (2) high Hall mobilities of 11 cm^2/Vs (electron) in μc-Si and 150 cm^2/Vs or more in the epi-Si prepared at Ts=300 °C, (3) high drift mobility of holes of --1 cm^2/Vs, and (4) rather low electric conductivity of --10^{-7} S/cm and high photoconductivity due to the rather low dangling bond density (3x10^{16} cm^{-3}).

In this work, we performed a systematic study focussed on the chemical reactions in the vicinity of the growing surface in making crystalline Si at a rather low substrate temperature with the assistance of atomic hydrogen.

EXPERIMENTAL
The film deposition was performed with a conventional RF glow discharge apparatus equipped with a microwave (2.45 GHz) plasma generator of atomic hydrogen. In addition, a remote-plasma-type reactor[13] driven by microwave plasma and a ECR plasma apparatus were utilized for the deposition of Si-thin films under control of atomic hydrogen.

The measurements were made for ir absorption, Raman scattering, X-ray diffraction, RHEED and SIMS to investigate the structure of the chemical composition of Si-network.

CHEMICAL REACTIONS IN THE VICINITY OF SURFACE
Immediately after sticking to the growing surface, the precursors generated in the reactive plasma are inferred to undergo changes in making Si-network by evolving out an excessive amount of monovalent elements, i.e., H or F. The sticking reaction is mostly independent of Ts, whereas the succeeding reactions for propagation of Si-network are efficiently accelerated by rising Ts. Consequently, structures of Si-network are determined by both of sticking and propagation reactions taking place in the vicinity of the growing surface. Sticking coefficients (β),almost independent of Ts were quantitatively measured for various silyl radicals, SiHn (n≤3) by Schmitt and Perrin-Allain by means of the grid method[14] and by Tsai et al. and A.Yuuki et al. by measurement of the step coverage in a trench [15),16)] , i.e., β(SiH3)=0.1, β(SiH2)=0.7--1.0. Okada et al. measured the β value by the step coverage method for fluorosilylanes generated by RF glow discharge of SiF$_4$ + H$_2$ and obtained β=0.2.[17]

Table 1 Kinetic parameters for dehydrogenation

precursors	activation energy (ΔE eV)	pre-factor (A (at.%)$^{-2}$)
SiHn	0.15	8
SiHnFm	0.15---0.23	14
SiHnClm	0.41	19

With regard to evolution of the monovalent elements, H and F, resulting in the formation of Si-Si bonds in endothermic reactions in the vicinity of the growing surface, their reaction kinetics were characterized by the chemical species used as the precursors. Table 1[18] shows the kinetic parameters, activation energy (ΔE eV) and pre-factor A obtained from the slope and the intercept with the ordinate of Arrhenius plots, respectively are shown for the dehydrogenation reactions with various precursors, i.e., SiHn, SiHnFm and SiHnClm. The reactions with the halogenized silyl radicals, SiHnFm and SiHnClm, are characterized by large activation energies and large A values compared with those of SiHn. It should be noted that the activation energies of the reactions with SiHnFm tend to be reduced by impinging of atomic hydrogen. Apparent differences in the kinetics of these dehydrogenation reactions depending on the precursors led us to a conclusion that both the sticking and the propagation reaction are markedly dependent on the chemical species used as the precursors. In addition, impinging of atomic hydrogen is a powerful tool of control over the kinetics of both reactions taking place in the vicinity of the growing surface for the construction of the main frame of Si-network.

CHEMICAL ANNEALING WITH ATOMIC HYDROGEN

We pointed out in our previous study[19] that µc-Si was preferentially made under the PVD condition defined by Tsai et al.[15] but hardly under the CVD condition when the film was deposited by RF glow discharge of SiH$_4$ with the assistance of atomic hydrogen. This is consistent with the condition in making µc-Si by RF glow discharge from a gaseous mixture of silane and hydrogen presented first by Veprek et al described above.[7] There remained some puzzles in the roles of atomic hydrogen because we could not observe a marked "chemical etching" by impinging of atomic hydrogen under these conditions.

First, we investigated "the chemical etching" by exposing the surface of a- and µc-Si to atomic hydrogen flow of the same intensity as preparation of films. Changes in the shape of surfaces were observed by SEM and SIMS for an a-Si:H and a µc-Si after exposure to the flow of atomic hydrogen for long time (more than 1.5 hr). Although no marked changes in the thickness were observed in both films, the flat surfaces were considerably roughed up by exposure to atomic hydrogen for a long period,

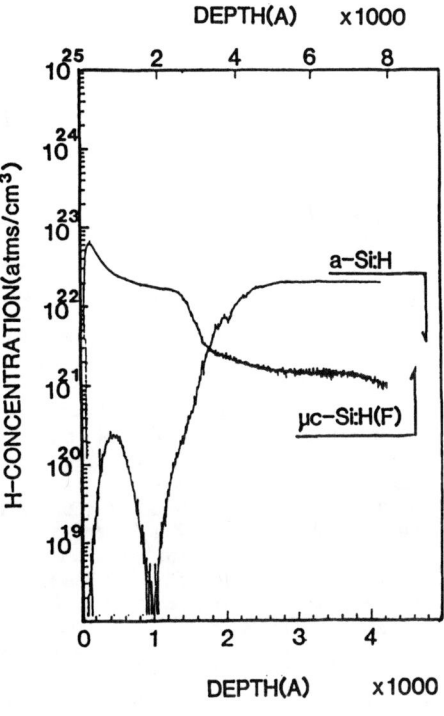

Fig.1 Depth profiles of hydrogen after exposure to atomic hydrogen

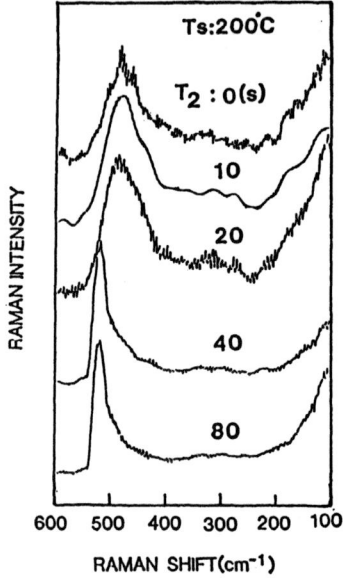

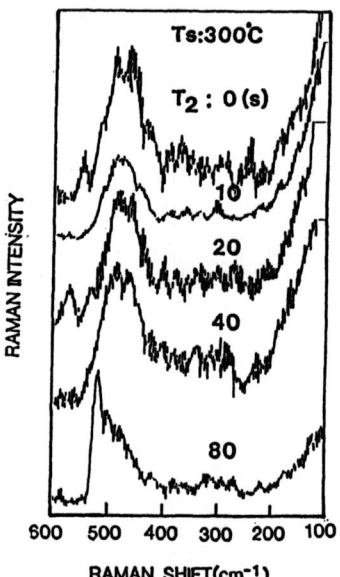

RAMAN INTENSITY

RAMAN SHIFT(cm⁻¹)

RAMAN INTENSITY

RAMAN SHIFT(cm⁻¹)

Fig.2 Raman scattering spectra of a-Si:H prepared under CVD condition by varying T_2

The numbers in the figure indicate the annealing period (T_2s)

exhibiting simultaneous, apparent occurrence of morphological changes in a certain depth from the surface of both the a- and μc-Si.

Fig.1 shows the depth profiles of hydrogen in the films (C_H) measured by SIMS for both a- and μc-Si thin films treated with atomic hydrogen. A great reduction in the C_H was found within the depth of ~20 nm from the surface of a-Si:H. On the contrary, a slight increase in the C_H was observed in a depth of 30 nm or more from the surface of μc-Si. These results suggest us that both hydrogenation and dehydrogenation are caused by absorption of atomic hydrogen in a certain depth in the surface region, which accompanies structural changes in Si-network including chemical etching. The chemical etching took place in the localized sites inhomogeneously with a fairly lower efficiency compared with those of deposition of precursors or dehydrogenation, which resulted in desorption of Si-related species .

Secondly, an attempt was made at forming films by cyclic interruption of the deposition for certain period (T_2 s) to expose the growing surface to a shower of atomic hydrogen in a repeated deposition of a-Si:H layer during a constant period (T_1=10 s corresponding to 2 nm in thickness) under the CVD-like condition. The similar experiment was first made by Asano[20] by means of the multiplasma-zone reactor equipped with a rotatory susceptor. He found a marked transition from amorphous to crystalline caused by exposure to atomic hydrogen of a-Si:H thin layers repeatedly deposited under the CVD condition. In Fig.2, Raman spectra are illustrated for films prepared for the different period T_2 at the two levels of Ts, 200 °C(left) and 300 °C(right). Marked changes are seen in the TO bands' indicating the transition from amorphous to crystalline for the periods (T_2) of 80 s and 40 s at the Ts of 300 °C and 200 °C, respectively. No marked chemical etching was observed during these periods and on the contrary, a slight increase in the thickness of the deposited layer per a cycle experiment. This was considered to be due to a small amount of silane molecules remaining in the reactor during T_2.

Asano interpreted this treatment with atomic hydrogen in terms of an enhanced the coverage with hydrogen responsible for promotion of the surface diffusion of the precursors in the deposition process. We believe, however, that impinging of atomic hydrogen promotes effectively the propagation reaction in a solid state within a certain thickness from the surface with the aid of a strong ability of the atomic hydrogen to penetrate through Si-network. This elucidation is strongly supported by the evidence that the transition from amorphous to crystalline was caused for the shorter T_2 at lower Ts. Either penetration of atomic hydrogen or evolution of molecular hydrogen, which leads to the propagation of Si-network, is inferred to be more difficult in a rigid Si-network prepared at higher Ts.

Since these changes in the structure of Si-network caused by impinging of atomic hydrogen are phenomenologically similar to the thermal annealing, we gave a term of "chemical annealing" to the structural changes caused intentionally by atomic hydrogen. We still believe that this chemical annealing with atomic hydrogen is more likely than the balance of the chemical etching and the deposition as the main phenomenon responsible for the crystallization when μc-Si is made from SiHn.

With regard to the chemical etching by atomic hydrogen, monovalent elements such as H and F in Si-network are preferentially evolved out compared with the Si-related species. Consequently, it is a plausible interpretation that movement of the Si-segments responsible for promotion of propagation reactions is efficiently enhanced in the growing region by impinging of atomic hydrogen resulting in reduction in crosslinking density of Si-network. Evidently, a monotonous decrease is caused by increasing T_2 in the width of Raman TO band attributed to the disorder of the bond angles ($\Delta \emptyset$) and the content of hydrogen remaining in films. This gives support to the idea of "chemical annealing".

The similar experiments were performed by ECR plasma for preparation of a-Si:H or a-Si:H(F) from SiH_4 or SiF_4 by excitation by low pressure (-- 1 mtorr) hydrogen plasma. As for the low pressure plasma, crystallization is thought to be suppressed by bombardment with ionic species.[7] Fig.3 illustrates the band width (FWHM cm^{-1}) of Raman TO band for films prepared from SiHn (a) and SiHnFm (b) plotted as a function of T_2 for exposure to atomic hydrogen. The transition from amorphous to crystalline is observed in the films prepared from SiHnFm but not in those from SiHn. The C_H in the films from SiHnFm is smoothly reduced down to 4 at.% or below with an increase in T2, and tends to be saturated at around C_H--7 at % in the films from SiHn, as shown in the inset. These evidences suggest that fluorine remaining in films is preferentially evolved out by impinging of atomic hydrogen, resulting in crystallization during the chemical annealing. With reducing C_H, optical gaps (Eo eV) obtained by Tauc plots became commonly narrow from Eo=2.1 eV to 1.9 eV. They are still rather wider than that a-Si:H or a-Si:H(F) of (Eo=1.7 eV) prepared by RF glow discharge at the same Ts, implying that the main frame of Si-network prepared by ECR plasma may be loose compared with that prepared by RF glow discharge.

Consequently, the long-range structural relaxation in making highly ordered crystalline structure did not take place during T_2 in the films prepared from SiHn despite the evidence that μc-Si was made from SiHnFm by the chemical annealing. Direct chemical etching with atomic hydrogen is unlikely as a decisive process in the chemical annealing because no marked reductions were seen in the thickness of the deposited layer during exposure to the flow of atomic hydrogen.

PREPARATION OF CRYSTALLINE FILM FROM SiHnFm

Some specific features of the fluorinated precursors, SiHnFm (n+m≤3), are observed in the chemical reactions in the vicinity of the growing surface when the surface is exposed to the flow of atomic hydrogen. In the similar manner to the behaviors in the chemical annealing, the C_H is efficiently reduced down to 2--3 at% with increasing atomic hydrogen flow. Taking into account the fact that a fairly large amount of fluorine, 10 at % or more, was included in films prepared at room temperature, the fluorine of the precursors was preferentially evolved out by heating the substrate to accelerate propagation reactions. At Ts=300 °C, the C_F was reduced by the level of 10^{19} cm^{-3} which is less than the C_H by two orders of magnitudes. Accordingly, it is said that fluorine is an efficient mediator in making rigid Si-network.

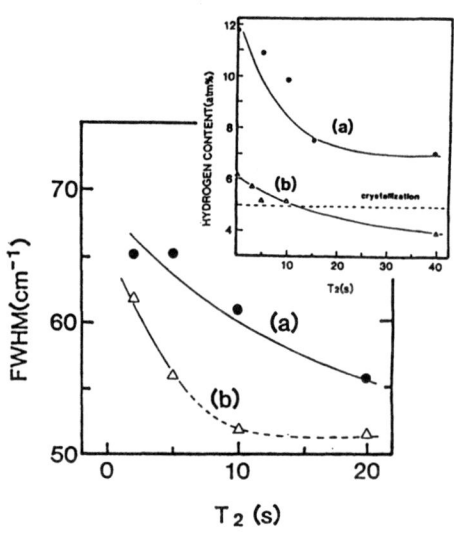

Fig. 3 TO band width (FWHM) plotted as a function of annealing period (T_2s)

In the remote type of plasma used in this experiment, direct bombardment of the growing surface with ionic species can be avoided, which is thought to be an advantage in making crystalline films. According to the systematic study on the gas-phase chemistry of the mixture of SiF_4 and H_2 by the mass spectrometry, SiF_4 was successively reduced with atomic hydrogen in plasma through some intermediates, SiF_2H and $SiFH_2$, which are considered to be the precursors.[21] These intermediates are more stable than SiHn because of the strong electron negativity of F, which was supported by the evidence that these intermediates were efficiently transported in a space toward the susceptor apart from the end of plasma by a several cm. Rather low sticking coefficients, ß=0.2 or less[17] give another support to this idea.

Table 2 Typical conditions to make c-Si

conditions	A	B
pressure(mtorr)	400	100
flow rate SiF$_4$ (sccm)	76	76
H$_2$	30	30
A	76	76

The rate of propagation reaction accompanied with evolution of terminators are greatly accelerated by rising Ts and also impinging of atomic hydrogen as described in the previous section. We chose two typical conditions as shown in Table 2 and prepared the films by varying Ts. Under condition (A) where an excessive amount of atomic hydrogen was added, both C_H and C_F were markedly reduced at Ts of 160 °C or below and an onset of the transition from amorphous to crystalline was observed at this temperature. The density depend on crystalline grains was efficiently increased under this condition by rising Ts though the sizes of the grains tended to be limited within about 50 nm in dia. Highly ordered epi-Si-like film was grown on c-Si(110) under the (A) condition at Ts of 300 °C.

On the other hand, a rather large amount of H and F remained in films when films were prepared under the condition (B) at Ts of 300 °C or below. In Fig.4, Raman spectra are shown for films deposited on c-Si(100) and (110) substrate at various Ts under the (B) condition. Crystallization was observed at Ts of 200 °C or higher and simultaneously C_H was reduced down to 5 at %. Under (B) condition, the grain size became larger compared with that under (A) condition, leading us to an idea that the growth of the grain is over nucleation under the (B).

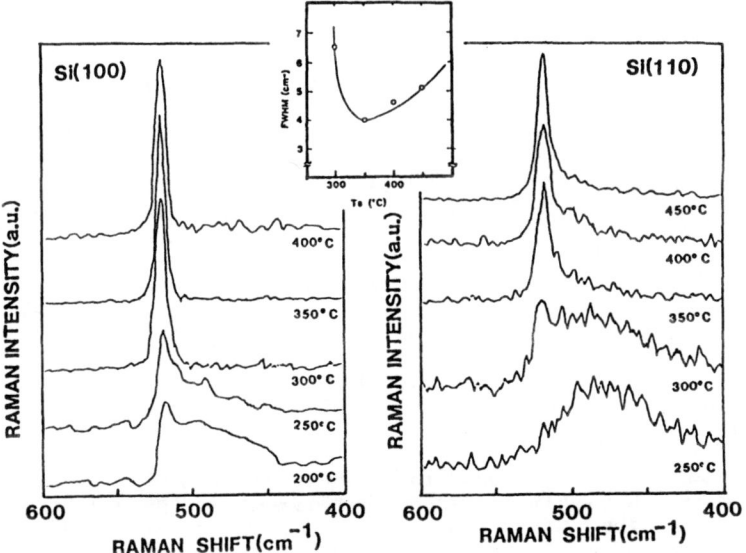

Fig.4 Raman scattering spectra for Si thin films

Remarkable changes happened in the structure of Si-network when films were made on c-Si (100) substrate under the (B) condition. The TO band became very sharp at the Ts of 300 °C or over, indicating an onset of epitaxial growth. A minimum band width in FWHM of about 4 cm^{-1} was obtained in c-Si deposited at Ts of 350 °C, which is the same result as that of c-Si substrate. A highly ordered structure supported by fine streaks and Kikuch lines in RHEED pattern was made in the epi-Si grown under the optimum condition, resulting from enhancement of the surface diffusion even at low Ts of around 350 °C. The epi-Si was able to be grown up to the thickness of 4 μm or more without deterioration in the structure. The substrate temperature used in this method is fairly low compared with the Ts of about 700 °C in making epi-Si by PE-CVD from silane.[22] On the other hand, the similar epitaxial growth of c-Si at low Ts was independently performed by either PE-CVD and Photo-CVD from SiH_2F_2[23] and the spontaneous chemical deposition.[24]

Under the (A) condition, the deposition rate was almost constant in a wide range of Ts from 160 °C to 400 °C. On the other hand, a marked reduction in the deposition rate was observed under the (B) condition at around 300 °C to 400 °C, where the transition from amorphous to crystalline took place. These behaviors were independent of the substrates used for the deposition.

MEASUREMENT OF H AND F BY MEANS OF SIMS

The content of terminators, H and F, remaining in Si-network is the key to the kinetics of the chemical reactions in making Si-network in the vicinity of the growing surface. Both C_H and C_F measured by SIMS for films deposited under the (A) and the (B) conditions on either c-Si(110) or (100) substrate are plotted as a function of Ts. (see Fig.5) Here closed and open symbols indicate C_H and C_F, respectively. Under the (A) condition, C_H was smoothly decreased by 3×10^{20} cm^{-3} (1.6 at%) at Ts of 300 °C, whereas C_F was reduced down to 10^{19} cm^{-3} or less.

On the contrary, remarkable reductions in both C_H and C_F by 5×10^{19} and 1.5×10^{18} cm^{-3}, respectively, were observed at Ts=350 °C when films were grown epitaxially on the c-Si(100) substrate. In the epi-Si thin film prepared under an optimized condition, both C_H and C_F were efficiently lowered in Si-network down to ---10^{17} cm^{-3}, which was not observed in μc-Si deposited on the substrates such as c-Si(110), (111) or glass. These results lead us to a conclusion that the selective chemical reactions favorable to crystal growth take place preferentially by heating substrates in the vicinity of the growing surface under the (B) condition . Evidently, on the other hand, we could obtains an epi-Si exhibiting highly ordered structure with thickness of 4 μm or more under the (B) condition. This suggests us that diffusion of precursor is efficiently activated on the surface of the c-Si(100) at Ts of 300 °C or more , which is responsible for the epitaxial growth.

PLAUSIBLE MECHANISM OF CRYSTAL GROWTH AT LOW Ts

In PE-CVD, control of chemical reactions in the vicinity of growing surface is considered to be a decisive treatment in making Si-network, though the network is made through complicated chemical reactions either in a gas-phase or on the surface. There are two chemical processes at least distinguished clearly by their temperature dependence, -- the sticking and the propagation of Si-network accompanied with release of terminators such as H and F. In the former the efficiencies are mostly independent of Ts , whereas the reactions in the latter is endothermic and dependent clearly on Ts. The rate of the propagation reaction resulting in dehydrogenation is promoted efficiently not only by heating substrates but also impinging of atomic hydrogen. There is, however, still puzzles in their microscopic processes in making the ordered structure at such a low substrate temperature. Two processes, namely, the surface diffusion and the desorption of Si-related species due to the chemical etching with atomic hydrogen are plausible causes for promotion of a long range relaxation of the species for crystallization.

The conspicuous reduction in the deposition rate corresponding to the transition from amorphous to crystalline was found under the (B) condition, implying the occurrence of reduction in the sticking coefficient of the precursors on the rigid crystalline surface.

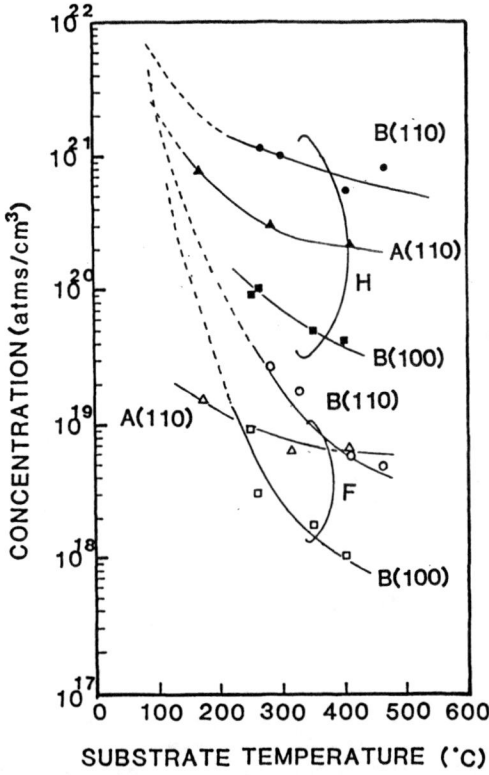

Fig.5 Concentration of H and F in films prepared under A and B condition on various c-Si substrates

The etching mechanism is unlikely to elucidate this reduction in the deposition rate with increasing Ts because the etching rate is far slow than the deposition rate. In fact, the process in making nuclei was overcome by the growth under this condition at higher Ts. This gives a strong support to the idea that the diffusion of precursor is dominant factor promoting the growth of crystal. Either the chemical activity of precursors themselves or the intentional promotion of the propagation reaction with atomic hydrogen is greatly decreased under the (B) condition compared with the (A) condition, giving advantage to the crystal growth caused at a rather low substrate temperature.

ACKNOWLEDGEMENTS
The authors thanks M.Nakata,H.Azuma,T.Uematsu,Y.Shinagawa ,and A.Sakai for their enthsiastic experimental efforts and helpful discussion. This work was supported in part by the Sunshine Project under MITI and by Grant-in-Aid for Scientific Research on Control of Reactive Plasma in Priority Area.

REFERENCES
1. N.Shibata, K.Fukuda,H.Ohtoshi,J.Hanna,S.Oda, andI.Shimizu:Jpn.J.Appl. Phys.,26(1987)L10
2. W.Spear,G.Weilleke,P.G.LeComber,and G.Fitzgerrald:J.Phys.(Paris),Suppl. 10(1981)C4-257
3. Y.Ucjida,T.Ichimura,M.Ueno, and M.Ohsawa:J.Phys.(Paris)42(1981)C4-265
4. A.Sakri,A.Le Glaunec,Y.Colin, and O.Bonnaud:Springer Proc.Phys.,35(1989) 365
5. Y.Hamakawa,H.Okamoto,and H.Takakura: Proc.18th IEEE Photovoltaic Specialist Conf.(Las Vegas 1985)813
6. S.Veprek,Z.Iqbal, and F.A.Sarott:Phil.Mag.,B45(1982)137
7. S.Veprek,M.Heintze,F.A.Sarott,M.Juricik-Rajman, and P.Willmott: Mat.Res. Soc.Symp.Proc.118(1988)49
8. C.C.Tsai,R.Thompson,C.Doland,F.A.Ponce,G.B.Anderson, and B.Wacker: Mat.Res.Soc.Symp.Proc.118(1988)49
9. A.Matsuda:J.Non-cryst.Solids,59/60(1983)767
10. I.Shimizu: Proc. 13th ICALS (Asheville,NC, 1989) in press
11. J.Hanna,S.Oda,H.Shibata,A.Miyauchi,A.Tanabe,K.Fukuda,T.Ohtoshi, H.Nguyen, and I.Shimizu:Mat.Res.Soc.Symp.Proc.,70(1986)11
12. I.Shimizu:J.Non-cryst.Solids,97/98(1987)257
13. J.Hanna,N.Shibata,K.Fukuda,H.Ohtoshi,S.Oda,and I.Shimizu:"Dis-ordered Semiconductors" ed. by M.Kastner,G.Thomas,S.Ovshinsky (Plenum,1987)pp345
14. J.P.M. Schmitt:J.Non-cryst.Solids,59/60(1983)649
15. C.C.Tsai,J.C.Knights,G.Cheng, and B.Wacker:J.Appl.Phys.,59(1986)2998
16. A.Yuuki,Y.Matsui, and K.Tachibana: Jpn.J.Appl.Phys.,28(1989)212
17. Y.Okada,J.Chen,I.H.Campbell,P.M.Fauchet, and S.Wagner:Proc.13th ICALS(Asheville,NC,1989) in press
18. A.Azuma,H.Shirai,J.Hanna, and I.Shimizu:Mat.Res.Soc.Symp.Proc.,(San Diego,CA,1989) in press
19. H.Tanabe,M.Azuma,T.Uematsu,H.Shirai,J.Hanna and I.Shimizu:Mat.Res. Soc.Symp.Proc.,(San Diego,CA,1989)in press
20. A.Asano:Proc.Workshop on Sunshine Project(Tokyo,1989)126(Japanese)
21. N.Shibata,K.Fukuda,H.Ohtoshi,J.Hanna,S.Oda,and I.Shimizu:Mat.Res.Soc. Symp.Proc.,95(1987)225
22. R.Reif:J.Electrochem.Soc.,131(1984)2430
23. S.Nishida,T.Shiimoto,A.Yamada,S.Karasawa,M.Konagai,and K.Takahashi: Appl.Phys.Lett.,49(1986)79
24. J.Hanna,A.Kamo,M.Azuma,N.Shibata,H.Shirai and I.Shimizu:Mat.Res.Soc. Symp.Proc.,118(1988)79

EFFECTS OF HYDROGEN ATOMS ON PASSIVATION AND GROWTH OF MICROCRYSTALLINE SI

TOSHIMICHI ITO*, TATSURO YASUMATSU*, HIROKUNI WATABE**, MOTOHIRO IWAMI+ AND AKIO HIRAKI*
*Department of Electrical Engineering, Osaka University, Suita, Osaka 565, Japan
**Matsushita Electric Industrial Co. Ltd., Shiromi, Osaka, Osaka, Japan
+Research Institute for Surface Science, Faculty of Science, Okayama University, Okayama, Okayama, Japan

ABSTRACT

Effects of hydrogen atoms bonded to silicon atoms on microcrystalline surfaces in μ c-Si:H and anodized porous silicon have been investigated using infrared spectroscopic technique and a semi-empirical molecular-orbital calculation (AM1 method). Experimental results on thermal stability of the Si-H bonds and oxidation of H-covered Si can be explained consistently in terms of system total energies and heats of formation of various clusters constructed by 13-18 Si and 18-24 H atoms (with an additional O atom). Possible role of H on the growth of μ c-Si:H is discussed in relation to stable Si-H bonds.

INTRODUCTION

Hydrogen bonded to silicon plays an important role on stabilization and passivation of Si dangling bonds in a-Si:H, μ c-Si:H and anodized porous Si (po-Si).[1-3] In a-Si, hydrogen-silicon is incorporated in bulk to kill gap states. In μ c-Si, H atoms cover microcrystalline Si surfaces and boundaries. In po-Si, which is anodically produced from monocrystalline Si, H atoms terminate Si dangling bonds on surfaces ($200-800$ m^2/cm^3) of a great number of micro pores (typical size of $\sim$ 10nm) present in the monocrystalline Si.[2] Thus, these materials contain more than atomic percents of hydrogen, meaning importance of the Si-H bonds. Therefore, many works have been reported on the Si-H system in these materials.[1-10] Our previous studies have revealed structures of these materials by using infrared spectroscopy, ion scattering spectrometory, electron spectroscopy and transmission electron microscopic techniques.[7-10] However, quantitative energy considerations were insufficient to explain our experimental data. In order to solve these problems, we have performed electronic energy-state calculations for various Si-H(-O) clusters. For saving computing times, we employ a semiempirical molecular-orbital calculation, called as AM1 method.[11] Another reason for the usage of the AM1 method is its relatively good calculation accuracy for Si-related molecules,[12] compared to other semiempirical methods such as MINDO/3 and MNDO methods.[13] In the present paper, we report on AM1 calculation results as well as experimental evidences on thermal stability and initial oxidation of μ c-Si:H and po-Si in relation to passivation of H-covered Si surfaces and growth mechanism of these materials. Charge distribution in the Si-H couples will also be discussed with observed fine structures of the Si-H stretching vibration absorption band.

EXPERIMENTAL AND CALCULATION

Details of preparation methods of μ c-Si[6,8] and po-Si[9] have been described elsewhere, so that a very brief description is given here. Thin films of μ c-Si:H were prepared at ~ 100℃ in pure hydrogen (~ 0.5Torr) by using a conventional rf-sputtering apparatus while po-Si films were produced by anodization of p-Si(111) wafers with very low resistivity in a HF solution.

In AM1 calculations, we used a program supplied by QCPE which has been originally developed by Dewar et al. Therefore, details of the AM1 calculation procedure should be referred to the original[11] and its related reports.[12,13] In the present calculations, we also employed a convenient supporting program named as 'MOL-GRAPH' supplied by Daikin Ltd. AM1 calculations were done for various clusters constructed by 13-18 Si and 18-24 H atoms, additionally with an O atoms if necessary. Molecular orbitals with a minimum total energy (E_T) or a minimum heat of formation (H_F) were selfconsistently determined for each cluster. (Therefore, a lower H_F means a more stable state.) Typical calculation times ranged from 30 to 130 min.

RESULTS AND DISCUSSION

Desorption of bonded H from microcrystalline surfaces

Infrared spectra can give information on H atoms bonded to Si atoms through vibration analysis of Si-H oscillators. Figure 1 shows spectral changes of Si-H stretching vibration band for specimens of μ c-Si(a) and po-Si(b) during heat treatments at low temperatures (300-400 ℃) in vacuum.[7] A common modification for both specimens appears as clear decreases in absorption intensity located at ~ 2150cm⁻¹, meaning the presence of

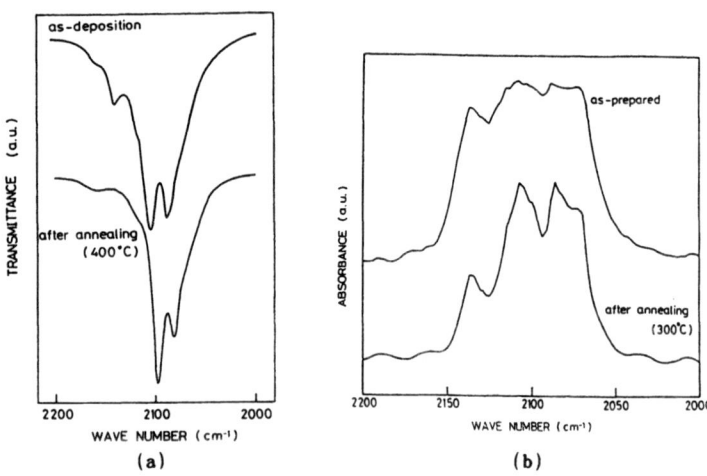

Fig.1.   Transmittance of μ c-Si:H (a) and absorbance of anodized porous Si in the stretching vibration band of Si-H bonds with and without one-hour annealing in vacuum at low temperatures of (a) 400 and (b) 300 ℃.

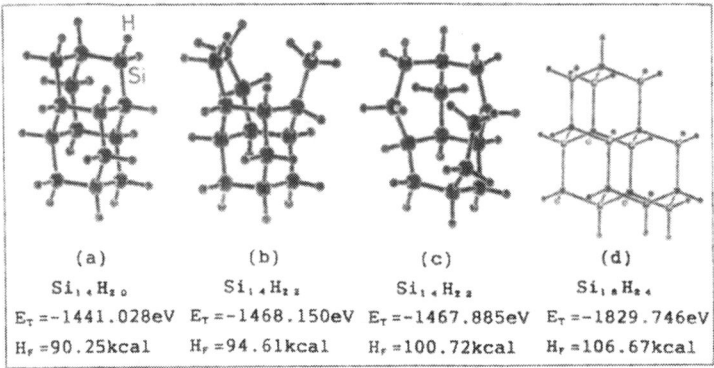

Fig.2.   Optimized cluster structures of Si_nH_m with a minimum
total energy (E_T) and a minimum heat of formation (H_F) calculated from the AM1 method: (a)$Si_{14}H_{20}$, (b)$Si_{14}H_{22}$, (c)$Si_{14}H_{22}$
and (d)$Si_{18}H_{24}$.

metastable H sites to be decomposed at the low temperatures in
as-deposited μ c-Si and as-prepared po-Si. H_2 desorption can
also be directly observed as a function of specimen temperature.[9] This metastable peak has been assigned as absorption by
SiH_3 vibrators. However, because of a relatively strong
bond between Si and H atom, it is not straightforward to
explain the decomposition process of the Si-H bonds.

Thermal phenomena should be related to free energies of
the system concerned. In the present case, electronic energy of
the microcrystalline surfaces may be specially important in the
free energy because the surface area is very large for both
cases. Thus we calculated energy states of various Si_nH_m
clusters as shown in Fig.2. The results indicate that a SiH_3
site (b) is not energetically preferable so that desorption of
two H atoms (or finally a H_2 molecule) from adjacent Si-H
couplings and Si-Si bonding between their Si atoms can
simultaneously take place ((b) → (a)) . It is also suggested
that increase in number of SiH_2 structure make the system
metastable (c).

Oxidation of H-covered Si microsurfaces

Figure 3 shows a time (t_{ox}) dependence of an integrated
absorption intensity with typical spectra of the Si-H

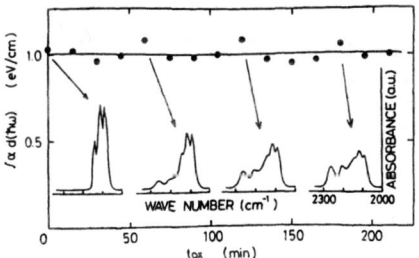

Fig.3.   Time (t_{ox}) evolution of integrated absorption coefficient in the
region of the Si-H stretching vibration band. Porous Si specimens measured were kept at 180℃ in air.
Insets show typical spectra.

stretching band measured for po-Si during its partial oxidation
in air at 180 ℃.[10] One should note that no change observed in
the total intensity means no change in number of the Si-H
vibrators during the low temperature oxidation while a spectral
change is observed with appearance of a new structure in the
high energy side, indicating the partial oxidation of Si atoms
bonded with H atoms. In other words, the low temperature
oxidation occurs selectively for Si-Si backbonds of the Si-H
configurations.[10]

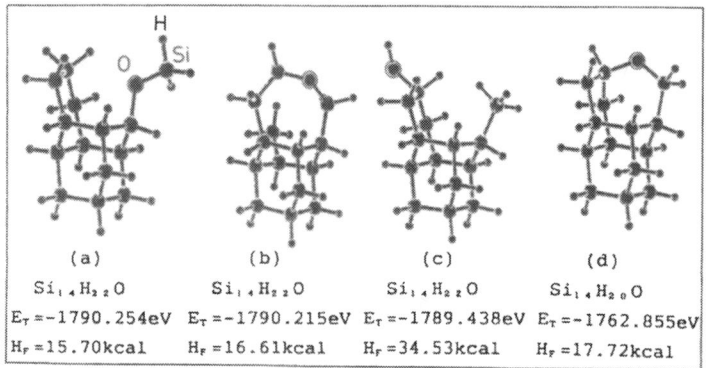

Fig.4. Optimized cluster structures of Si_nH_mO with a minimum
total energy (E_T) and a minimum heat of formation (H_F) calculated from the AM1 method: (a)-(c)$Si_{14}H_{22}O$ and (d)$Si_{14}H_{20}O$.

        AM1 calculation results consistently explain the above
experimental results. In this case, clusters shown in Fig.4
were used. The results indicate as follows: (1) The Si-Si
backbonds can be oxidized more easily than the Si-H bonds
(compare (a),(b),(d) to (c)). (2) The selective backbond
oxidation is more preferable than a simultaneous H desorption
and Si-O-Si formation (compare (a),(b) to (d)).

Charge distribution of Si-H bonds

        In an infrared spectrum of the Si-H vibration absorption,
one usually observes a fine structure.[7,9] Such a phenomenon
appears more clearly in (second) derivative of the stretching
absorption band as is typically shown in Fig.5. The origin of
the fine structure has been studied experimentally[1,3] and
theoretically,[5] and is considered to be due to charge
redistribution of the Si-H vibrators[3] and/or to various modes
of the Si-H vibrators with different symmetries.[1,5] In the
former model, the oscillation frequency (ν) of Si-H vibrators
is well represented by a following relation:

$$\nu = a + b \sum_j SR(R_j),$$

where SR is the stability ratio electronegativity of atom (or
group) R_j and is relatd to the charge state of the Si-H
vibrator concerned. The present AM1 calculation can give a net
electronic charge for each atom.[13] The results are shown in
Fig.6, indicating that the (formal) charge distribution is

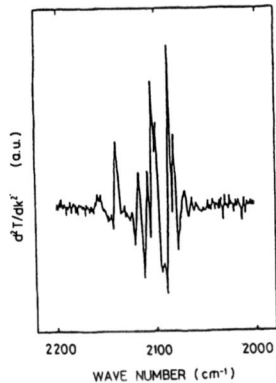

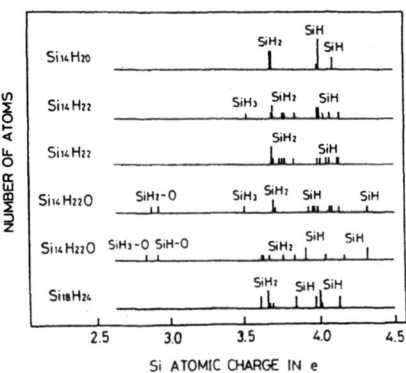

Fig.5. Fine structures of Si-H
stretching absorption band observed in second derivative of
transmittance for μ c-Si:H.

Fig.6. Charge distribution of H-
bonded Si atoms in various clusters
calculated by the AM1 method.

strongly dependent on the configuration of Si and H atoms (and
O atom if any). This suggests that the fine structures observed
are originated from different charge states of the SiH_n
vibrators. When we compare the IR spectra in Fig.1 with and
without the heat treatment, minor modifications can be seen in
the low energy side as well. This is also the case when the low
temperature oxidation proceeds as is shown in the insets of
Fig.3. These observations are well correlated with the fact
that the AM1 calculation can predict modification of the charge
distribution of the Si-H vibrators induced by the structural
modification.

Possible role of H atoms on the Si growth

Recent studies show an importance of SiH_3 radicals on
the growth of the Si:H materials.[6] From the present AM1
calculations, possible role of H atoms on the growth can be
discussed. When we compare $Si_{14}H_{22}$ to $Si_{14}H_{20}$ (Fig.2), the
latter is more preferable in energetics, meaning that a reaction
of SiH_3 radical to the H-covered Si surface leads to a Si
deposition on microcrystalline Si through automatical
desorption of H_2. Furthermore, a more symmetrical cluster has
a lower energy, suggesting a continuous growth of the micro-
monocrystalline silicon. However, when a microcrystallite
becomes large in size to meet an adjacent grain or microcrystallite, it is more difficult for all of the radicals to
find their most stable states specially at the grain
boundary. In this case, such continuos growing of micro-mono-
crystallines is terminated and a different microcrystalline
growth starts near a site with a metastable state at the grain
boundary.

CONCLUSIONS

We have studied H-covered Si microcrystalline surfaces in μ c-Si:H and porous Si (po-Si) using infrared spectroscopy and semiempirical AM1 calculations. The following conclusions can be deduced:
(1) Weak bondings of SiH₃ are present on H-covered surfaces in μ c-Si:H and po-Si,and can be decomposed during low-temperature annealing in vacuum, leading to desorption of hydrogen gas and formation of Si-Si bond between adjacent SiH$_n$ configurations.
(2) At low temperatures oxidation of the H-covered Si surfaces can selectively take place at backbond Si-Si sites without oxidation of the Si-H bonds or any H desorption.
(3) Observed fine structure of the absorption band of the Si-H stretching vibrations are originated from charge redistributions of the Si-H couples in different configurations.

ACKNOWLEDGEMENTS

The authors would like to thank Mr. T.Minamino of Daikin Ltd. for his help on AM1 calculations.

REFERENCES

1. G.Lucovsky, J.Non-cryst. Solids 76, 173(1985) and references therein.
2. G.Bomchil, R.Herino and K.Barla, in MRS Europe 1985,(1986), pp.463 and references therein.
3. G.Lucovsky, J.Vac.Sci.Technol. 16, 1225(1979).
4. A.Matsuda, T.Yoshida, S.Yamasaki and K.Tanaka, Jpn.J.Appl. Phys. 20, L439(1981).
5. W.B.Pollard and G.Lucovsky, Phys.Rev. B26, 3172(1982).
6. A.Matsuda and K.Tanaka, J.Appl.Phys. 60, 2351(1986).
7. T.Satoh and A.Hiraki, Jpn.J.Appl.Phys. 24, L491(1985).
8. A.Hiraki, in Structure and Bonding in Noncrystalline Solids (Plenum, 1986) pp.219-236 and references therein.
9. T.Ito, Y.Kato and A.Hiraki, in The Structure of Surfaces, edited by J.F.van der Veen and M.A.Van Hove (Springer-Verlag, Berlin, 1988) pp.378-383.
10. Y.Kato, T.Ito and A.Hiraki, Jpn.J.Appl.Phys. 27, L1466(1988).
11. M.J.S.Dewar, E.G.Zoebisch, E.F.Hearly and J.J.P.Stewart, J. Amer.Chem.Soc. 107, 3902(1985).
12. M.J.S.Dewar and C.Jie, Organometallics 6, 1486(1987).
13. M.J.S.Dewar and W.Thiel, J.Amer.Chem.Soc. 99, 4899(1977).

A DISCUSSION OF ELECTRONIC OPTICAL ABSORPTION SPECTRA OF NANOCRYSTALLINE SILICON THIN FILMS

ETIENNE BUSTARRET* AND M.A. HACHICHA**
*Max-Planck-Institute für Festkörperforschung, D-7000 Stuttgart 80, FRG
**Lab. Etudes de Propriétés Electroniques des Solides, CNRS, BP 166x, F-38042, Grenoble, France (associated to Univ. J. Fourier de Grenoble).

ABSTRACT

Both fully crystallized nanocrystalline as-deposited silicon layers with an average grain size ranging between 6 and 80 nm and mixed-phase hydrogenated doped and undoped silicon films are studied at room temperature by Photothermal Deflection Spectroscopy (PDS), Transmission Spectroscopy and Spectroscopic Ellipsometry. The differences with regard to similar data obtained on monocrystalline and hydrogenated amorphous silicon are discussed, with an emphasis on the low-energy part of the 0.6–5.6 eV explored range.

RESULTS AND DISCUSSION

Even though part of the recent interest in nanocrystalline silicon (nc-Si) layers deposited below 400°C by chemical transport (CT) in a H_2 plasma [1], by plasma enhanced chemical vapor deposition (PECVD) of SiH_4-based mixtures [2], or by sputtering (SP) of a monocrystalline silicon (c-Si) target in H_2 [3] stems from their remarkable combination of optical absorption and electronic transport features [4], there are only a few published investigations of their optical properties, when compared to the wealth of data [5] available on their higher temperature atmospheric or low pressure chemical vapor deposition counterparts (APCVD and LPCVD). This is noteworthy, since for the latter materials, the main motivation was the quantitative study of the surface roughness (a critical parameter for integrated circuits manufacturing), and one would expect the solar cells applications of the former to be associated with thorough studies and a tentative engineering of the optical absorption profile, as in the case of amorphous hydrogenated silicon (a-Si:H) layers.

The most exhaustive reports on the optical properties of low-temperature films [6,7] concentrate on the enhanced effective absorption measured between 0.5 and 2 eV in most of the samples obtained by the CT method, as exemplified by Fig. 1 where the absorption coefficient spectra $\alpha(E)$ of CT [7] and LPCVD nc-Si layers are compared to those of a-Si:H and of c-Si [8]. It is now understood that this enhancement is related to that of the inelastic (Raman) and diffuse elastic scattering cross sections [9] as in many APCVD and some LPCVD films [5]. The effects are attributed to an enhanced coupling of the electromagnetic field of the incident light to the charge-density fluctuations at the grain boundaries [9].

Maybe because such peculiarities dominated the early studies, the optical investigations of other sets of nanocrystalline silicon thin films deposited at low temperature either by PECVD [10] or by SP [3,11] are restricted to limited ranges. While contributing to the available stock of experimental data, it is one aim of this paper to distinguish preparation-dependent spectral features from more universal profiles. For this purpose, we preferred to use three different measurement techniques on the same samples rather than to apply the classical transmission/reflection procedure to films of varying thicknesses deposited under the same conditions [6,7]. This choice was prompted by the experimental observation [7,12] that films of different thicknesses did not have the same optical surface, and by the low sensitivity to diffuse scattering of the PDS method.

Unlike other authors we did not try to derive the absorption coefficient α at energies greater than 2.5 eV from spectroscopic reflectance [6] or ellipsometry measurements, since in that range the probed region is much thinner than the actual film, so that the results depend strongly on the surface roughness and on the presence and nature of superficial overlayers [13,14]: all results given below have been obtained on as-deposited samples, often after a prolonged exposure to air. We shall therefore give more emphasis to the low and medium absorption region, i.e. to the absorption spectra measured below 2.8 eV, than to the high energy range, where the imaginary part ε_2 of the effective pseudo-dielectric function is nonetheless sensitive (see Fig. 2) to the degree of crystallinity [12–15].

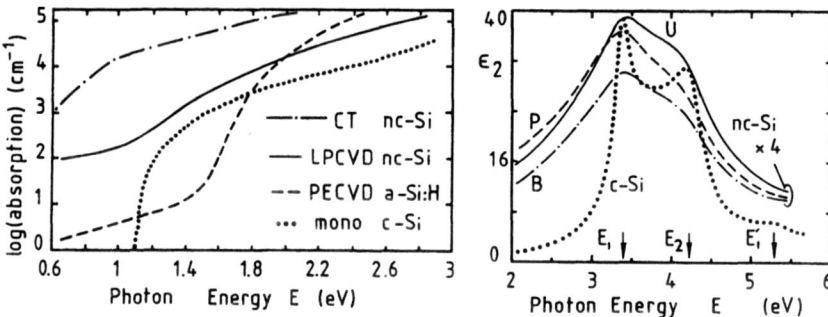

Fig. 1   Typical absorption spectra of CT and LPCVD-deposited nc-Si films compared to those of c-Si [8] and a-Si:H.

Fig. 2   Imaginary part of the pseudo-dielectric functions of as-deposited PECVD doped and undoped films exposed to air. The spectra of a-Si:H and c-Si conserved under the same conditions are given for comparison.

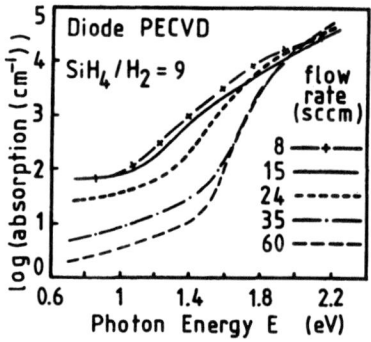

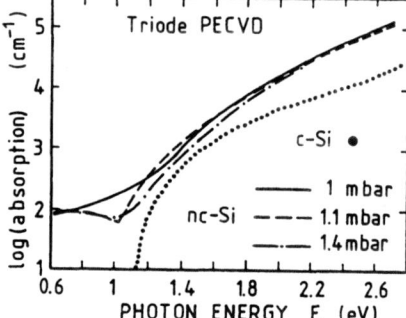

Fig. 3   Absorption profiles of our diode PECVD layers deposited at various total gas flows.

Fig. 4   Absorption profiles of our triode PECVD layers deposited at various total pressures, and comparison to that of c-Si [8].

The PDS apparatus was used in the 0.6 to 2.2 eV range, and a rotating analyzer spectroscopic ellipsometer from 1.7 to 5.6 eV. The transmission spectra (0.4–2.8 eV) were recorded with a Beckman 5240 double beam spectrometer. All resolutions were better than 25 meV. The mean crystallite sizes were deduced from grazing angle (typically 0.5°) x-ray diffraction [18].

Diode configuration: amorphous and mixed-phase undoped hydrogenated silicon

Because of the low exciting frequency of 50 kHz, our classical capacitive PECVD reactor had to be operated at a total gas pressure P_{tot} greater than 2 mbar for the discharge to be stable. The films studied below have been deposited at 350° from a $SiH_4:H_2$ (1:9) mixture under a total pressure of 3 mbar, with a power density of 300 mWcm^{-2}. The preparation conditions of all undoped 500 nm-thick films described below have been detailed elsewhere [16], along with their physicochemical and transport properties.

It has been shown [17] that the fundamental parameter for the deposition of a microcrystalline silicon film near equilibrium is the residence time τ of SiH_x species in the reactor. This is illustrated by Fig. 3 where the experimental $\alpha(E)$ spectra vary from the typical a-Si:H profile (Tauč gap 1.8 eV, Urbach parameter 55 meV, deep defect density 3×10^{-6}cm^{-3}) of the sample obtained for a total gas flow of 60 sccm ($\tau \simeq \tau_e/4$) to those of 50% crystallized [18] nc-Si layers deposited at 15 and 8 sccm (resp $\tau \simeq \tau_e$ and $\tau \simeq 2\tau_e$), where the mean grain size is 6 nm. for $E < 1.2$ eV, the absorption coefficient of the latter is lower than that of the LPCVD sample for Fig. 1 (probably because the 6% H atomic contents [16] ensure a relatively low deep defect density) but more than one order of magnitude higher than that of the amorphous film. The low crystalline volume fraction f_c in the diode PECVD nc-Si layers leads to an enhanced absorption in the 1.6 to 2.2 eV energy range. Other deposition temperatures at $\tau = \tau_e$ yield similar trends: the Urbach edge of a-Si:H disappears as soon as $f_c > 10\%$, and an inflection region associated with the indirect gap of silicon develops around 1.2 eV.

Triode configuration: as-deposited fully crystallized undoped materials

We have shown elsewhere that in the presence of a d.c. magnetic field and of a mesh between the two electrodes, it was possible to lower the total pressure, thus increasing the grain size by a factor of three for a total flow of 15 sccm [16]. The present undoped layers and the doped samples of the next section were prepared with the mesh and the substrate holder electrode held at the same potential. Under these conditions, f_c as deduced from Raman spectra [18] is found to depend critically on P_{tot}: f_c is greater than 90% in samples deposited at 1 mb and below; it is equal to 60% at 1.4 mb and above, while the average grain size remains at 16 ± 2 nm in the whole pressure range. The absorption spectrum of the 1 mb film represented on Fig. 4 is identical within experimental error to that of the larger grain (80 nm) LPCVD sample of Fig. 1, confirming the trends of the diode nc-Si set but at lower Hydrogen contents (below 2% [16]). The spectra of films deposited at 1.1 and 1.4 mb also given on Fig. 4 differ markedly from this "standard" nc-Si $\alpha(E)$ profile between 1.6 and 0.8 eV, where they are parallel to c-Si $\alpha(E)$ variations [8] down to 1 eV.

In the case of the 1 mb and 1.1 mb samples, the observed spectral differences near the indirect gap edge are particularly puzzling, since both films have similar Hydrogen concentration and bonding [16], a similar dilation of the bonds (3×10^{-3}, i.e. ca. 10 kbar) and the same texture. Apart from a 15% difference in f_c, these two films have widely distinct transport properties: the 1 mb layer yields room temperature d.c. conductivity ($3 \times 10^{-2} \Omega$cm^{-1}) and Hall mobility (0.8 cm^{2}V^{-1}s^{-1}) which are one order of magnitude superior to those of the 1.1 mb sample, while the 1.4 mb material has intermediate features. Since the PDS signal has the same phase over the whole spectrum of all studied samples, the common profile between 0.6 and 0.8 eV represents the bulk of the material. The plateau below 0.9 eV present in the data on the 1.1 and 1.4 mb films and which becomes a shoulder in the case of the layer deposited at 1 mb has features similar to those of photoinduced absorption spectra measured at lower temperatures in microcrystalline silicon with a comparable grain size [19]. These involved both the enhanced free carrier absorption [20] of the grains and the

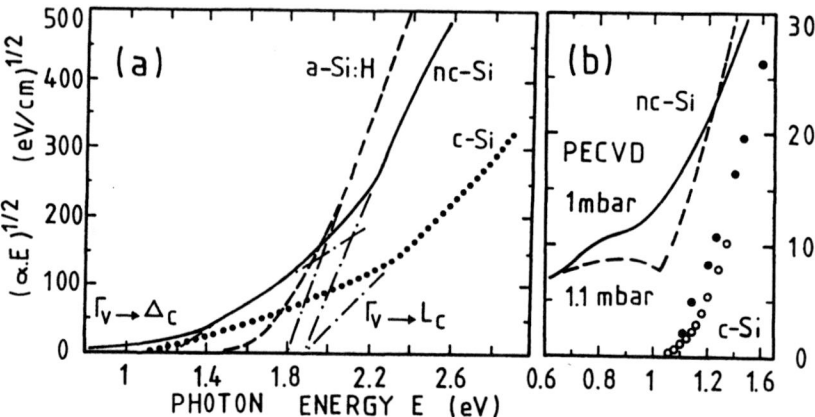

Fig. 5   Tauč plots below the direct absorption edge: (a) determination of the indirect optical gaps; (b) the low energy region.

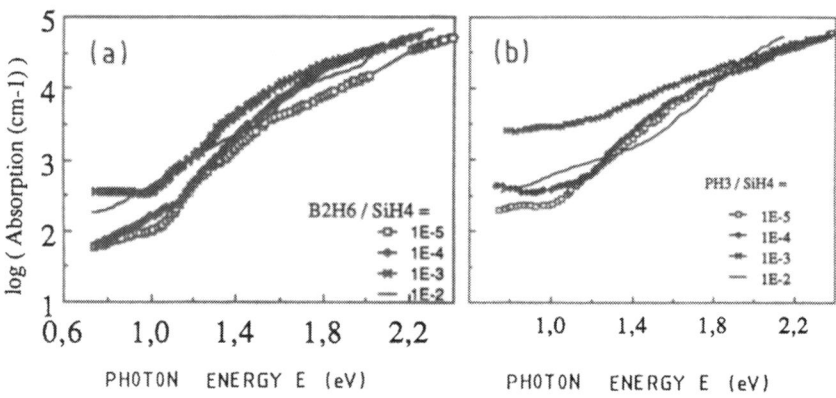

Fig. 6   Absorption spectra of triode PECVD samples. (a) B-doped; (b) P-doped.

a-Si:H-like recombination mechanism in the inter grain region. Other possibilities, such as the presence in the grain boundary of our stressed films of intrinsic deep defects have to be considered before we can explain this original low energy $\alpha(E)$ profile.

The $(\alpha E)^{1/2}$ vs E plot has already been applied to monocrystalline silicon [8,21,22] below the direct gap. As seen in Fig. 5a, its extrapolation (in the $E > 2.2$ eV range) to zero absorption yields an optical gap of 1.9 eV both for c-Si and nc-Si, slightly in excess of the expected value for the $\Gamma_v \rightarrow L_c$ transition (around 1.7 eV), but in good agreement with the Tauč gap of a-Si:H [22]. Between 1.8 and 1.4 eV (i.e. for $\alpha E \gtrsim 10^4$ eVcm^{-1}), the linear plots extrapolate to the indirect $\Gamma_v \rightarrow \Delta_c$ gap values at 1.1 and 1.2 eV for c-Si and nc-Si, respectively. This difference is too large to be explained by the 10 kbar tension of the films mentioned earlier. In both energy ranges, the slope of the straight line associated to nc-Si is about twice that of c-Si. The intermediate behavior observed between 2.2 and 1.8 eV disappears when a $\alpha^{1/2}$ vs E plot is drawn.

The low energy data for c-Si [8,21] and for two triode nc-Si samples are blown-up on Fig. 5b. In c-Si, the curve deviates from a linear edge below 1.17 eV [21]. Down to 1 eV, the $(\alpha E)^{1/2}$ profile of the nc-Si layer grown at 1.1 mb is almost parallel to that of c-Si (the shift is vertical rather than horizontal). Finally, the linear tail extrapolating to 0.4 eV which has been detected on CT nc-Si layers was not observed in our material [7].

Triode PECVD B and P-doped layers

The introduction of B_2H_6 or PH_3 in the chamber at volumic ratios r (to SiH_4) ranging from 10 ppm to 1% leads to the deposition of doped films which are described in more detail elsewhere in this volume [24]. As seen on the ε_2 spectra given in Fig. 2, the films remain nanocrystalline upon doping, even though the grain size of all doped layers is 9 nm (except for $PH_3/SiH_4 = 1\%$ where it drops to 6 nm), about half the value of the undoped material deposited under similar conditions ($P_{tot} = 1$ mb, 350°C, triode configuration). Taking into account these grain sizes [18] for the analysis of Raman results [24], we determine $f_c \simeq 0.8$ for all layers doped at $r \leq 1^o/_{oo}$. For $B_2H_6/SiH_4 = 1\%$, the f_c decreases to approximately 0.7, while it drops to 0.5 in the film prepared with $PH_3/SiH_4 = 1\%$. In general, the P doped samples have lower f_c than the B doped material deposited at the same nominal (gas phase) doping ratio, as illustrated by the $\varepsilon_2(E)$ curves of Fig. 2.

The absorption profiles of the B and P doped films are given in Figs. 6a and 6b. In the case of Boron doping, all spectra remain similar to those of the undoped material studied above, with a slightly higher absorption coefficient at low energies for the two higher doping ratios. The extrapolated $\Gamma_v \rightarrow \Delta_c$ decreases from 1.3 to 1.2 eV as r increases from 10 ppm to 1%. On the other hand (see Fig. 6b), this indirect gap increases from 1.25 to 1.3 eV with Phosphine doping ratios between 10 ppm and $1^o/_{oo}$. The spectrum of the heavily P-doped samples ($r = 1\%$) has some of the features of the absorption profile of a-Si:H, with only the wider optical gap at 1.6 eV. Finally, the layer obtained at $PH_3/SiH_4 = 1^o/_{oo}$ displays an enhanced absorption much weaker than, and spectrally different of that reported [7] for un-doped CT films (see Fig. 1), while the low-energy part of the spectrum of the 10^{-4} P-doped samples has been tentatively associated [24] to free carrier absorption [20] When applied to the undoped sample grown at 1.1 mb, such analysis yields a carrier concentration of 2×10^{18} cm^{-3} and a mobility of 8 cm^2 V^{-1}s^{-1}, a factor 8 higher than the Hall effect values.

CONCLUSION

The shape of the $\varepsilon_2(E)$ peak near the direct E_1 gap of silicon depends strongly on the average grain size. For well crystallized films with grain sizes in the 5 to 15 nm range, this can be explained by finite size effects.

At intermediate photon energies (1.4 eV $< E < 2.8$ eV), the α spectrum lineshape is determined by the amorphous volume fraction f_c. The extrapolated indirect $\Gamma_v \rightarrow L_c$ gap has the same value in our triode PECVD nc-Si films as in a-Si:H.

At still lower frequencies, two different $\alpha(E)$ profiles were observed on otherwise very similar films, and a subgap absorption feature was identified. In well crystallized samples (PECVD and LPCVD) the spectrum does not seem to depend on the mean grain size, so that finite size effects are thought to be negligible. Finally, we did not detect the enhanced optical absorption reported by other authors [6,7], even for doped or highly stressed films. We conclude rather than apart from its energy features ($E < 1.4$ eV), the spectrum of well crystallized nc-Si films with grain sizes in excess of 15 nm is now relatively independent from the preparation procedure.

REFERENCES

[1] S. Vepřek, V. Mařeček, Solid State Electron. **11**, 683 (1968).

[2] M. Morel, F. Morin, Appl. Phys. Lett. **35**, 686 (1979).

[3] H. Ishida, M. Noda, H. Shimizu, Jap. J. Appl. Phys. **22**, L73 (1983).

[4] Y. Ushida, T. Ishimura, M. Ueno, M. Ohasawa, J. Phys.(Paris) **42**, C4-265 (1981).

[5] G. Harbeke, in "Polycrystalline Semiconductors", Springer Series on Solid St. Sciences, **57**, 156 (1985), and references therein.

[6] H. Richter, L. Ley, J. Appl. Phys. **52**, 7281 (1981).

[7] Z. Iqbal, F.A. Sarott, S. Vepřek, J. Phys. C **16**, 2005 (1983).

[8] W.C. Dash, R. Newman, Phys. Rev. **99**, 1151 (1955).

[9] S. Vepřek, F.A. Sarott, Z. Iqbal, Phys. Rev. B **36**, 3344 (1987).

[10] M. Matsuda, M. Matsumura, S. Yamasaki, H. Yamamoto, T. Imura, H. Okushi, S. Iizima, K. Tanaka, Jap. J. Appl. Physics **20**, L183 (1981).

[11] T. Hata, T. Hatsuda, T. Komatsu, Jap. J. Appl. Physics **24**, 1463 (1985).

[12] S. Kumar, B. Drevillon, C. Godet, J. Appl. Physics **60**, 1542 (1986).

[13] D.E. Aspnes, Thin Solid Films **89**, 249 (1982).

[14] S. Logothetidis, J. Appl. Phys. **65**, 2416 (1989).

[15] B.G. Bagley, D.E. Aspnes, A.C.A. Adams, C.J. Mogab, Appl. Phys. Lett. **38**, 56 (1981).

[16] M.A. Hachicha, J.C. Bruyere, E. Bustarret, A. Deneuville, M. Brunel, "Ion and Plasma Assisted Techniques 87", Int. Conf. Proc. published by Conferences Exhibitions Publications Consultants Ldt, Edinburgh (UK), p. 360 (1987).

[17] S. Vepřek, M. Heintze, F.A. Sarott, M. Jurcik-Rajman, P. Willmott, Mater. Res. Soc. Symp. Proc. **118**, 3 (1988).

[18] E. Bustarret, M.A. Hachicha, M. Brunel, Appl. Phys. Lett. **52**, 1675 (1988).

[19] Hsiang-Na Liu, D. Pfost, J. Tauč, Solid St. Comm. **50**, 987 (1984).

[20] Y. Mishima, M. Hirose, Y. Osaka, J. Appl. Phys. **51**, 1157 (1980).

[21] G.G. Macfarlane, T.P. McLean, J.E. Quarrington, V. Roberts, Phys. Rev. **111**, 1245 (1958).

[22] R. Tsu, in "Disordered Semiconductors", M.A. Kastner, G.A. Thomas, S.R. Ovshinksy, Eds., Plenum Press, New York (1987), p. 479.

[23] B. Welber, C.K. Kim, M. Cardona, S. Rodriguez, Solid St. Comm. **17**, 1201 (1975).

[24] M.A. Hachicha, E. Bustarret, this volume.

GROWTH OF MICROCRYSTALLINE SILICON
IN ULTRATHIN LAYERS

Y.-J. WU*, P. D. PERSANS*, B. ABELES** AND S.-L. WANG***

* Physics Department and Center for Integrated Electronics, Rensselaer Polytechnic Institute, Troy NY 12180
** Exxon Research and Engineering Company, Annandale, NJ 08801
*** James Franck Institute, University of Chicago, Chicago 60637

ABSTRACT

We report recent studies of the crystallization of ultrathin (< 10 nm) amorphous silicon layers clad by silicon dioxide. We observe changes in Raman scattering spectra which indicate that nanocrystalline silicon layers formed by crystallization of amorphous silicon continue to evolve structurally with further annealing. There is evidence for significant strain in crystallized material.

INTRODUCTION

The ability to produce ultrathin layers of hydrogenated amorphous silicon (a-Si:H) and related materials affords us the opportunity to study the effects of interfaces on the structure and stability of amorphous materials [1,2]. In previous work we have reported on hindered crystallization in ultrathin a-Si:H layers clad by a-SiO$_2$ [3,4]. In the present paper we extend these studies to include detailed Raman characterization of the amorphous silicon/ microcrystalline silicon/ oxide composite as the materials is annealed beyond the first appearance of crystallization. These studies are particularly relevant to our understanding of microcrystalline silicon because it should be possible to control crystallite size in the direction parallel to layer growth by varying layer thickness.

An understanding of crystallization, diffusion and relaxation of ultrathin amorphous silicon, germanium, oxide and nitride layers is important for technological and fundamental reasons. For example, it is important to understand nucleation and growth of crystallites near Si/SiO$_2$ interfaces in order to better control low temperature epitaxy [5]. This may have applications for three dimensional intergated circuits. Ultrathin amorphous multilayers are especially attractive for the study of heterojunctions between deposited layers because a high density of interfaces can be produced reproducibly [2] .

EXPERIMENTAL DETAILS

Periodic amorphous multilayers were grown in two different deposition systems at Exxon and at the University of Chicago. Both systems employed low pressure rf plasma assisted chemical vapor deposition using SiH$_4$ and N$_2$O as the source gases. Results on similar samples from different laboratories were consistent with one another. Typically, a-Si:H layers were grown from pure SiH$_4$ and oxide layers were grown from a mixture of SiH$_4$:N$_2$O in the ratio 1:50. Fused silica and borosilicate glass substrates were clamped to the heated (200-300°C) anode of either reactor. Layer thicknesses were controlled by varying the layer growth time. Average multilayer growth rates and the average optical index of refraction were used to check layer thicknesses inferred from separate growth rate measurements of the sublayer components [6].

Samples were isochronally annealed in a dry flowing N$_2$ ambient with temperature steps of 15°C to 50 °C, depending on the sample and temperature. Typically, smaller steps were taken in the vicinity of the crystallization temperature for a particular sample after it was located with other samples and larger steps [3,4].

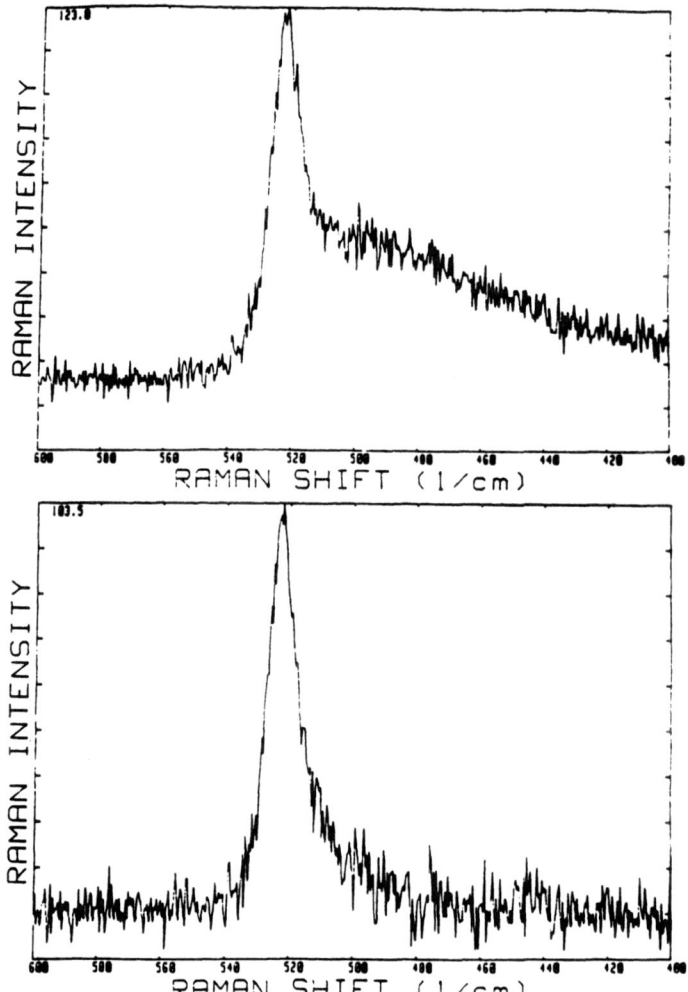

Fig. 1 : upper: Raman scattering spectrum for 10 nm sample at annealing temperature Ta=715°C; lower: the same spectrum after the subtraction of background level and the amorphous contribution.

After each anneal the Raman and optical transmission spectra were measured at room temperature. Raman spectra were performed in the near backscattering configuration with 5145Å excitation. The scattered light was collected into a J - Y 2000M double 1 m monochromator and detected with a low dark count RCA 31034A cooled photomultiplier with photon counting electronics. The Raman spectra were fit by first subtracting a flat background and then decomposing the spectrum into an a-Si component centered at 480 cm^{-1} and a Gaussian peak near the crystalline Si TO mode position. The peak was fit by least squares to the high frequency side of the sharp crystalline mode. A typical spectrum before and after a-Si subtraction is shown in Fig. 1. The Raman spectrum of a crystalline Si wafer was frequently measured to check position calibration and resolution of the system. The ratio in area under the microcrystalline Si peak to the crystalline plus amorphous peaks was used to characterize crystallized volume fraction. We have assumed that the volume-corrected scattering cross sections for a-Si and crystalline Si are the same, consistent with the literature [8].

Optical transmission spectra were measured on a Cary 17 spectrophotometer. The absorption coefficient and index of refraction were extracted by standard thin film analysis with the assumption that the multilayer film could be approximated by a uniform effective medium [9].

RESULTS AND DISCUSSION

Our present annealing studies quantitatively confirm previous reports of hindered crystallization in a-Si:H/a-SiO$_2$ multilayers [3,4]. In Fig. 2 we plot the crystallization temperature for several films with varying Si sublayer thickness. The crystallization temperature T_C is defined as the temperature at which the film achieves 50% of its high temperature anneal crystal fraction. We note that near the crystallization temperature we found variations in the crystalline volume fraction from site to site across a single sample and from piece to piece of the same sample. This caused the crystallization temperature to vary by about 25°C from sample to sample. The data in Fig. 2 represent an average of several samples for each point. In Fig. 3 we plot the crystallized volume fraction against annealing temperature for two samples with silicon sub-layer thicknesses of 4 nm and 10 nm respectively. We see from this figure that the crystallization temperature can be quite well defined. We note that x-ray and TEM studies of similar samples indicate that the average crystal size perpendicular to the layers is close to the layer thickness, that the crystals are randomly oriented and that the lateral dimensions of the crystals are close to the vertical dimensions [3,4].

The Si:H/SiO$_2$ system is similar in behavior to the Ge/GeN$_x$ [10], Ge:H/GeN$_x$ [10] and Ge:H/SiN$_x$ [11] systems. Crystallization appears to be hindered in all four systems, with significant increases in crystallization temperature for layers thinner than 10 nm. This effect has been discussed in terms of the relative interface energy for amorphous semiconductor/ insulator and crystalline semiconductor/ insulator heterojunctions [3,4,10,11].

Raman scattering studies of microcrystalline silicon (μc-Si) have demonstrated a correlation between crystallite size, Raman peak width and Raman peak position [12,13]. This correlation is explained in terms of the breakdown in crystal momentum as a good quantum number and the consequent coupling of lower-energy non-zone-center phonons to the optical wave [13]. We might therefore expect that Raman scattering peak position and width would be useful measures of the crystal structure.

In Fig. 4 we show crystallite Raman peak width plotted against annealing temperature for samples with 4, 7, and 10 nm Si nominal sublayer thicknesses d_S. In Fig. 5 we plot the Raman peak positions for the same set of films. The data show a general trend toward narrower peak width for thicker sublayers and higher annealing temperatures. The final peak width for the $d_S= 4$ nm is about 11 cm^{-1} whereas the peak width for thicker layers is close to 7 cm^{-1}. For comparison the bulk crystalline width is about 4 cm^{-1}. Decreasing

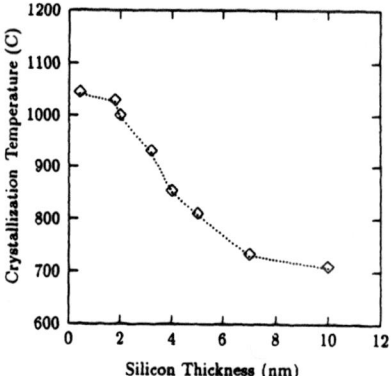

Fig. 2: Crystallization temperature Tc plotted against silicon layer thickness in nm; Tc is the temperature at which the crystallized volume fraction is greater than .5

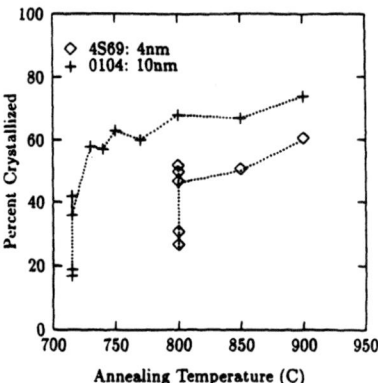

Fig. 3: Volume fraction crystallized from the ratio of the integrated area under micro-crystalline silicon peak to the total area under both the amorphous and microcrystalline peaks of the Raman spectra plotted against the annealing temperature Ta for 4 nm (diamond) and 10 nm (+) multilayers.

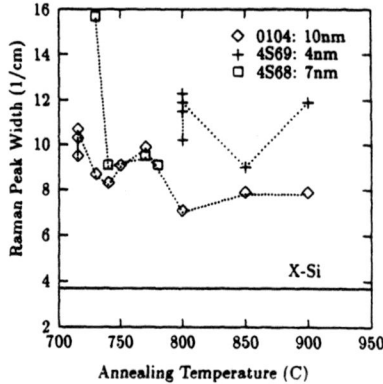

Fig. 4: Raman peak width (TO mode) plotted against annealing temperature for samples with 10 nm (diamond), 7 nm (square), 4 nm (+) silicon layer thickness; the peak width for crystal silicon is denoted as the horizontal line with width 3.7 wavenumbers.

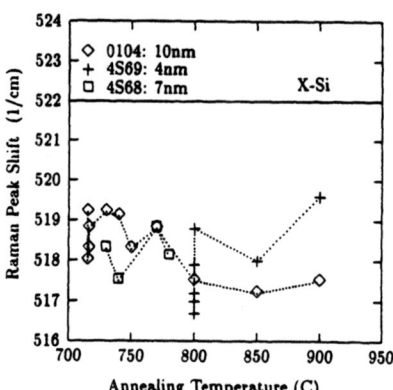

Fig. 5 : Raman peak shift (TO mode) plotted against annealing temperature for samples with 10 nm (diamond), 7 nm (square), 4 nm (+) silicon layer thickness; the peak shift for crystal silicon is denoted as the horizontal line with 522 wavenumbers of shift.

peak width with increasing T_A is consistent with the hypothesis the for T_A near T_C the sample is composed of a broad size distribution of crystallites with interstitial disordered material. As annealing proceeds, we would expect the average crystallite size to increase , consuming amorphous material and smaller crystallites. Larger crystallites are expected to have narrower Raman peak width [12,13] and peak positions closer to the bulk value.

If average particle size were the only consideration then we would expect the peak positions to correlate with peak width and with sublayer thickness. This is not the case for the present experiments. The layered crystal data are all clustered about 518 cm^{-1} and the bulk position is at 522 cm^{-1}. For the $d_S =$ 7 nm and 10 nm samples the peak even appears to move down with increasing annealing. The fact that the peak position is not dependent on sample or annealing conditions suggests that some other effect is controlling the position and that the size-interpretation of the width may be questionable. The shift in position could be due either to tensile strain or to oxygen impurity in the Si layers.

SUMMARY

Research at Rensselaer was supported in part by the NSF under grant DMR-8714634.

REFERENCES

1. B. Abeles and T. Tiedje, Phys. Rev Lett. 51,2003 , (1983).
2. J. Kakalios, H. Fritzsche, N. Ibaraki and S. R. Ovshinsky, J. Non- Cryst. Sol. 66, 339 , (1983).
3. P. D. Persans, A. F. Ruppert and B. Abeles, J. Non-Cryst. Sol. 102, 130, (1988).
4. P.D. Persans, A.F. Ruppert and B. Abeles, in: Multilayers: Synthesis, Properties and Non-electronic Applications, eds, T. Barbee, L. Greer and F. Spaepen, (Mat. Res. Soc., Pittsburgh, 1988)
5. J. A. Roth, S. A. Kokorwoski, G. Olson, L. Hess, in Laser and Electron Beam Interactions with Solids, ed. B. Appleton and G. Celler, (Elsevier, Amsterdam, 1982), p. 169.
6. C. Roxlo and B.Abeles, Phys. Rev.B, 34, 2422, 1986
7. R. Johanson, Ph.D. Thesis, University of Chicago.
8. J. Gonzales-Hernandez and R. Tsu, Appl. Phys. Lett., Vol. 42, No. 1, 90 1983
9. H. Ugur, R. Johanson, and H. Fritzsche, in Tetrahedrally Bonded Amorphous Semiconductors, ed. D. Adler and H. Fritzsche, (Plenum, New York, 1985), p. 425.
10. I. Honma, H. Komiyama and K. Tanaka, J. Appl. Phys, 66, 1170, (1989).
11. C.V.M. Williams, A. Bittar, H. J. Trodahl, J. Appl. Phys., in press.
12. Z. Iqbal and S. Veprek, J. Phys. C, 15, 377, (1982).
13. H. Richter, Z. Wang and L. Ley, Sol. St. Commun., 39, 625, (1981).

PICOSECOND PHOTOMODULATION STUDY OF
NANOCRYSTALLINE HYDROGENATED SILICON

M. Wraback, Lingrong Chen, J. Tauc and Z. Vardeny*
Brown University, Department of Physics and Division of Engineering,
Providence, Rhode Island 02912
*University of Utah, Department of Physics, Salt Lake City, Utah 84112

ABSTRACT

We have extended our photomodulation studies of nc-Si:H to the picosecond time domain. We measured the decays of photoinduced reflectivity with 100fs temporal resolution as a function of light intensity. Comparison with the data obtained on a-Si:H and c-Si indicates that ultrafast trapping and recombination processes are mainly the properties of the amorphous phase. It has also been observed that nc-Si:H is unstable under high illumination.

Introduction

We have applied the photomodulation (PM) spectroscopy to the study of electron states and carrier dynamics in nanocrystalline hydrogenated silicon (nc-Si:H). This technique employs two light sources: a pump beam for photogeneration of carriers, and a probe beam for measuring photoinduced changes in transmission and reflectivity. We have measured steady-state and time-resolved PM spectra in the time domain from 10^{-7} to 10^{-2}s, and have found that the data are compatible with the presence of two phases, amorphous and crystalline [1]. More recently we have employed the picosecond pump and probe technique to study ultrafast decays in reflectivity with a temporal resolution of 100fs. In this paper, we summarize the results on long time scales and present our new work in the picosecond regime, which provides information about ultrafast carrier trapping and recombination as well as photoinduced defects in nc-Si:H.

Experimental

The samples were nc-Si:H films $3-5\mu m$ thick deposited on crystalline Si substrates. They were prepared by the glow-discharge process in low-pressure hydrogen plasma in the floating potential condition with the substrate held at $300\,^{\circ}C[1]$. For the steady-state measurements the pump was a mechanically chopped cw Ar$^+$ laser beam with an intensity of $100mW/cm^2$ and the probe beam was derived from an incandescent lamp dispersed by a monochromator. The transient spectra with nanosecond resolution were measured on the system developed by Stoddart [2,3]. The pump pulse was obtained from a frequency-doubled Nd:YAG laser, and had a duration of 10ns, energy of $100\mu J$, and repetition rate of $20s^{-1}$. The diameter of the illuminated spot was about 1cm. The probe beam was produced with 0.1eV spectral resolution by an incandescent lamp and a set of interference filters of known transmission characteristics [2]. The transmission T and its photomodulation ΔT were recorded in the spectral range 0.25-1.8eV. In this range photoinduced changes in reflectivity can be neglected [4], and $-\Delta T/T = d\Delta\alpha$, where d is the film thickness and $\Delta\alpha$ is the change in the absorption coefficient α.

The light source for the low intensity picosecond experiments was a colliding pulse modelocked ring laser (CPM) [5], which produces 2eV optical pulses of 100fs duration and 0.1nJ pulse energy. These pulses were amplified to $2\mu J$ with a copper vapor laser pumped amplification stage [6] for the high intensity experiments. The beam was split into pump and probe pulses. Under high excitation conditions the pump pulse was focused to a diameter of $80\mu m$

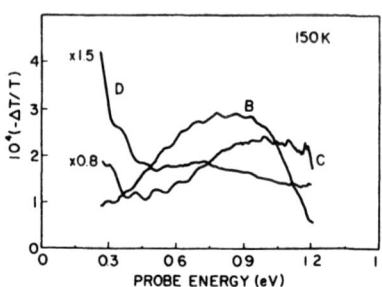

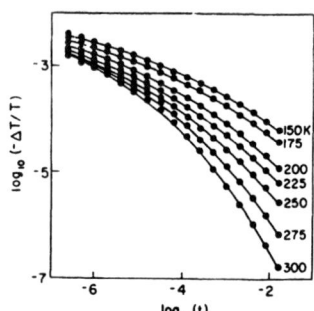

Fig. 1. Steady-state PM spectra of three samples--D, C, and B--at 150K: grain sizes are 830Å (sample D), 420Å (sample C), and 70Å (sample B).

Fig. 2 Measured PFCA decays of sample D using a probe energy of 0.25 eV at several temperatures. The fits (solid lines) are stretched exponentials.

on the sample. The probe pulse was delayed using a translation stage and focused inside the pump area to ensure the probing of a uniform carrier density. For the low excitation measurements both pump and probe pulses were focused to 25μm. In both cases the pump-induced change in reflectivity ΔR as measured by the modulation of the probe pulse was recorded as a function of the delay time between the probe and the pump pulses.

Steady-State and Transient Spectra

The steady-state PM spectra of three nc-Si:H films with crystallite grain sizes of 70Å (sample B), 420Å (sample C), and 830Å (sample D) at temperature θ = 150K are shown in Fig. 1 [1,7]. The spectrum obtained for sample B is similar to that of a-Si:H [8], with an onset of photoinduced absorption (PA) at about 0.5eV followed by photoinduced bleaching (PB) at about 1.0eV. In sample D this PM band is very weak, but there is a very strong onset of photoinduced absorption at low energies not found in sample B or a-Si:H. This long wavelength absorption has a frequency dependence of approximately ω^{-2} and is ascribed to photoinduced free-carrier absorption (PFCA) [7]. The spectrum of sample C shows both characteristics of samples B and D; with respect to sample B the PA band is shifted toward higher energies. This data provides strong evidence for the existence of amorphous and crystalline phases in nanocrystalline samples.

This two-phase model is corroborated by the transient spectra. It was found [1] that the PM band of the amorphous phase in nc-Si:H has spectral features and time dependencies similar to those observed in a-Si:H [2,3], but with differences which can be understood as being due to increased disorder. Fig. 2 shows the decay of PFCA at 0.25eV as a function of delay time. The data can be described using stretched exponentials, which extend into much longer times than the exponential decays due to recombination in crystalline Si. They also show that the rate of recombination of free carriers in the crystalline phase is determined by recombination in the amorphous phase. The interpretation proposed in Ref. 1 assumes that the carriers generated in the grains have high mobilities and reach the grain boundaries very quickly, but only the holes become directly trapped in the amorphous phase. The electrons encounter a barrier in the conduction band due to the gap mismatch between the two phases and this prevents them from entering the higher lying conduction band or bandtail states in the amorphous phase. Thus, recombination is slower than in c-Si because the hole density in the grain has

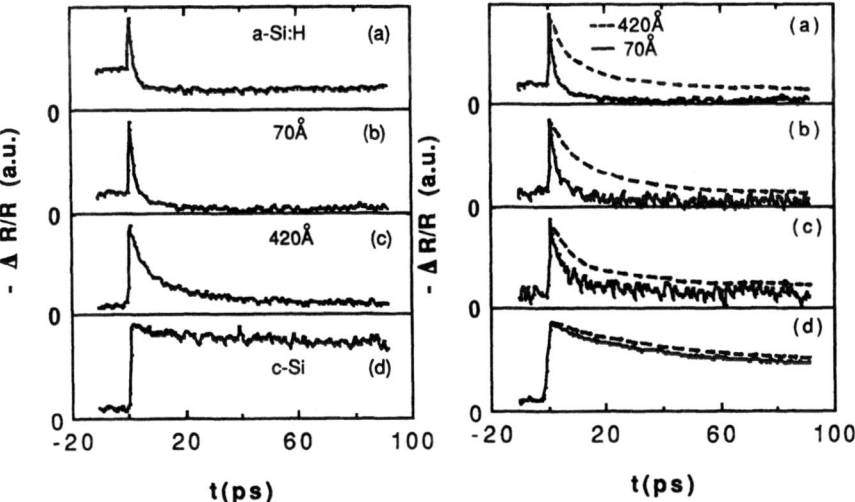

Fig. 3. Decays in photoinduced reflectivity -ΔR at 4.5mJ/cm² pump intensity for four samples: a) a-Si:H; b) sample B; c) sample C; d) c-Si.

Fig. 4. Comparison of decays in ΔR for samples B (solid lines) and C (dashed lines) at four carrier densities (a) N=7×10²⁰ (b) 3×10²⁰ (c) 1×10²⁰ (d) 2×10¹⁸cm⁻³.

been reduced. The dominant recombination rate will be determined by how fast the electrons from the grain can reach a hole. Using this model and the transient decays measured in Ref. 1 one obtains stretched exponentials as in Fig. 2 which are consistent with the existence of amorphous and crystalline phases.

Results in the Picosecond Time Range

Previous studies [9,10] of a-Si:H and c-Si [10] under low intensity 2eV excitation (~10¹⁷ - 10¹⁸ carriers/cm³) have shown striking differences in the lifetime of carriers in extended states. Carriers in a-Si:H are trapped in low mobility states in tens of picoseconds, while those in c-Si live longer than a nanosecond. Recently, it has been found [11-13] that under high excitation conditions (10²⁰-10²¹ carriers/cm³) the lifetime of carriers in extended states for both of these materials is sharply reduced, coinciding with heat production on a picosecond time scale. In a-Si:H [11,12] the lifetime τ is approximately proportional to N^{-1}, whereas in c-Si [13] it is proportional to N^{-2}, where N is the density of photoexcited carriers. The former has been attributed to a fast nonradiative recombination of highly spatially correlated e-h pairs [14] due to either an Auger [11] or bimolecular [12] process. The latter is due to the normal Auger recombination. Since nc-Si:H may be viewed as a two-phase material, it is interesting to examine the picosecond trapping and recombination processes to determine under what conditions the amorphous or crystalline phase dominates on very short times scales.

Fig. 3 compares the transient decays in −ΔR for a-Si:H, c-Si, sample B and sample C under high illumination. The response in a-Si:H (Fig. 3a) is characterized by an initial decrease in reflectivity due to the excitation of free carriers by the pump pulse. This is followed by a subpicosecond decay to a constant positive value in ~10ps. This decay is attributed to the nonradiative recombination of free carriers and subsequent production of heat [11,12], which

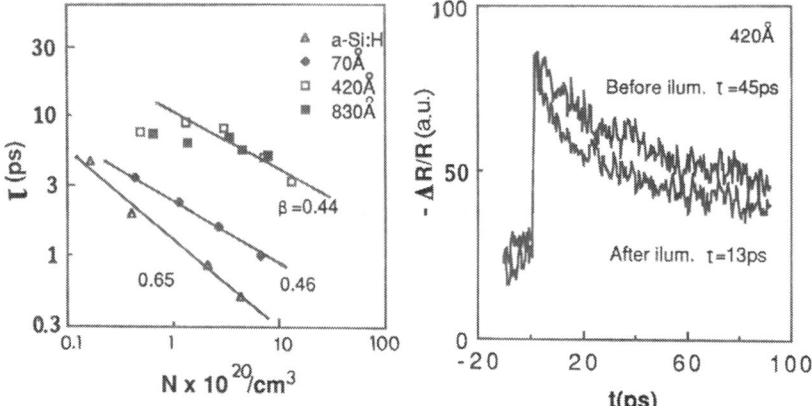

Fig. 5. Power law behavior of the decay time τ for a-Si:H and the three nc-Si:H samples. β is the slope of the straight line fits.

Fig. 6. Comparison of the decays in Sample C before and after high intensity illumination ($>5mJ/cm^3$).

causes a positive change in reflectivity. A similar behavior is observed in sample B, but the decay is somewhat slower, and ΔR decays to a less positive value than in a-Si:H. In sample C there is a fast decay component in the first 10ps followed by a much slower one over the duration of the trace, but no crossing of the zero line. The data in sample D are nearly identical to those on sample C, except for a slightly slower decay. In c-Si there is almost no decay on this time scale.

Further insight can be gained from Fig. 4, in which the decay curves for samples B and C are shown for a number of carrier densities between 10^{18} and $10^{21}/cm^3$. For all carrier densities $-\Delta R$ decays more slowly in the sample with the larger grain size. Moreover, for both samples the decays become faster with increasing carrier density. The data for sample B exhibit no positive ΔR below $2 \times 10^{20}/cm^3$ and are qualitatively similar to those of a-Si:H over the entire range of carrier densities studied. Finally, we note that for all the nc-Si:H samples there is a striking difference in the decay time between the data at high excitation densities ($>10^{19}/cm^3$) and those at low densities ($\sim 10^{18}/cm^3$). In Fig. 4a through 4c, both samples exhibit a rapid decay in $-\Delta R$ in the first 10ps; in Fig. 4d the decays are clearly much slower--on the order of hundreds of picoseconds. This indicates that we are in a different regime of carrier dynamics.

The data for all three grain sizes and a-Si:H were quantitatively analyzed by defining the decay time τ as the time at which $-\Delta R$ reaches half its initial value. Log-log plots of τ versus carrier density N for each sample are shown in Fig. 5. For a-Si:H, $\tau \alpha N^{-\beta}$, where $\beta=0.65$. The data for samples B and C also exhibit power law behavior with exponent $\beta \cong 0.45$. It is important to note, however, that as the grain size increases one needs larger carrier densities to reproduce the same decay times and decay curves as are found in a-Si:H. In sample C, τ becomes insensitive to N below $10^{20}/cm^3$, and in sample D it is relatively independent of N except for the highest carrier densities studied.

Discussion

These observations lead us to suggest the following model to explain the data. At low intensities the carriers generated in the amorphous phase thermalize quickly ($<1ps$) and are trapped, while the electrons generated in the grains live in extended states for a long time, in agreement with the stretched exponentials from the transient PM spectra shown in Fig. 2. As the grain size increases, a larger percentage of the total number of photogenerated carriers is excited in the grains. This is evident from the steady-state spectra in Fig. 1, in which the relative strength of PFCA increases with respect to the PM band characteristic of the amorphous phase as the grain size becomes larger. One therefore sees an initial fast contribution to the decay in $-\Delta R$ due to trapping, and a very slow contribution is probably due to free carriers in the crystalline phase. Typical trapping times in good quality a-Si:H are 20-30ps [11]. On the other hand, $\tau = 408$ps for c-Si. The results for nc-Si:H for the same carrier density lie in between these values, with $\tau = 81$ps for sample B, 207ps for sample C, and 303ps for sample D. This indicates that one sees an average of these two contributions weighted by the grain size.

As noted earlier, the decay times drop dramatically for high excitation densities. These results strongly suggest that carrier trapping is replaced by an ultrafast nonradiative carrier recombination process as the dominant decay mechanism. This is most clearly evident in sample B, which exhibits (Fig. 4a,b) a decay to a positive ΔR that cannot be explained by trapping, but only by heat production. Since the power law exponent β for nc-Si:H is closer to that of a-Si:H than the value reported for c-Si [13], we conclude that this recombination process is primarily a characteristic of the amorphous phase. At high intensities spatially correlated e-h pairs are generated in both phases at t = 0. However, the electrons and holes diffuse rapidly away from each other as they thermalize in the crystalline phase, with the holes becoming trapped in the amorphous phase while the electrons remain behind in the grains. Therefore, only those carriers generated in the amorphous phase remain spatially correlated long enough to participate in a fast recombination process. It then follows that this process would be less efficient in nc-Si:H than in a-Si:H and would become even more so with increasing grain size. Sample C, for example, requires more carriers to obtain the same decay time measured in sample B because a greater percentage of carriers go into the crystalline phase, where they become ineffective in the recombination process. When the effective carrier density is rendered low enough, τ becomes nearly insensitive to the excitation density, as is exhibited in sample D and in sample C at intermediate intensities. It is surprising that $\beta < 1$ for all the samples measured. This indicates that the fast recombination process observed is not entirely the same as the one described in Refs. [11] and [14]. It is also clear that dispersive bimolecular recombination of trapped electrons and holes [15] is not applicable, since this would yield $\beta > 1$. In our case, the excitation densities under high illumination are large enough ($>10^{20}$/cm^3) to saturate the shallow traps in the conduction band tail, thus indicating that at very short times (<1ps) we may be in a different regime of recombination kinetics than the one described in Ref. 15.

We have also found that the nc-Si:H samples are unstable under high illumination independent of grain size. Fig. 6 shows the decays in sample C at a given light intensity before and after the sample had been exposed to high illumination levels. The decay is more than three times as fast after the exposure. This is attributed to the creation of light-induced defects in the films, which increases the efficiency of the recombination process. Therefore all the measurements for this work were performed after the samples had been illuminated until they did not change anymore. It is important to note that under similar experimental conditions this effect was found to be negligible in a-Si:H.

In conclusion, steady-state and time-resolved spectral measurements have been performed which indicate that nc-Si:H can be described by a two-phase model with amorphous and crystalline phases. Ultrafast trapping and nonradiative recombination under high excitation conditions are primarily characteristics of the amorphous phase. It has also been shown that nc-Si:H is much more sensitive than a-Si:H to defect production under high illumination.

The authors thank T.R. Kirst for technical assistance. M.W. thanks IBM for a fellowship. This work was principally supported by NSF grant DMR

8706289.

References

1. Lingrong Chen, J. Tauc and Z. Vardeny, Phys. Rev. B $\underline{39}$, 5121 (1989).
2. H.A. Stoddart, Ph.D. Thesis, Brown University, 1987.
3. H.A. Stoddart, Z. Vardeny and J. Tauc, Phys. Rev. B $\underline{38}$, 1362 (1988).
4. H.T. Grahn, C. Thomsen and J. Tauc, Opt. Commun. $\underline{58}$, 226 (1986).
5. R.L. Fork, B.I. Greene and C.V. Shank, Appl. Phys. Lett. $\underline{38}$, 671 (1981).
6. W.H. Knox, M.C. Downer, R.L. Fork and C.V. Shank, Opt. Lett. $\underline{9}$, 552 (1984).
7. Hsiang-na Liu, D. Pfost and J. Tauc, Sol. State Commun. $\underline{50}$, 987 (1984).
8. Z. Vardeny, T.X. Zhou, H. Stoddart and J. Tauc, Sol. State Commun. $\underline{65}$, 1049 (1988).
9. Z. Vardeny and J. Tauc, in <u>Disordered Semiconductors</u>, eds. M.A. Kastner et al., (Plenum Press, N.Y., 1987) p. 339.
10. J. Kuhl, E.O. Gobel, Th. Pfeifer and A. Jonietz, Appl. Phys. A $\underline{34}$, 105 (1984).
11. A. Esser, K. Seibert, H. Kurz, G.N. Parsons, C. Wang, B.N. Davidson, G. Lucovsky and R.J. Nemanich, Proceedings of the 13th International Conference on Amorphous and Liquid Semiconductors, J. Noncryst. Sol. (to be published).
12. A. Mourchid, D. Hulin, C. Tanguy, R. Vanderhaghen and P.M. Fauchet, Proceedings of the 13th International Conference on Amorphous and Liquid Semiconductors, J. Noncryst. Sol. (to be published).
13. M.C. Downer and C.V. Shank, Phys. Rev. Lett. $\underline{56}$, 761 (1986).
14. W. Rehm and R. Fischer, Phys. Stat. Sol. (b) $\underline{94}$, 595 (1979).
15. J. Orenstein and M.A. Kastner, Sol. State Commun. $\underline{40}$, 85 (1981).

OPTICAL PROPERTIES OF MICROCRYSTALLINE SILICON

MARTIN INGELS, MARTIN STUTZMANN, AND STEFAN ZOLLNER
Max-Planck-Institut für Festkörperforschung, D-7000 Stuttgart 80, FRG

ABSTRACT

Optical properties of undoped, microcrystalline silicon are investigated by photothermal deflection spectroscopy, spectroscopic ellipsometry and Raman scattering. Samples are prepared by recrystallization of hydrogenated amorphous silicon in the temperature range 680 – 900°C. The increase of grain sizes with increasing annealing temperature and the disappearance of amorphous tissue lead to noticeable changes in the observed spectra. It is argued that much of the pertinent structural information of μc-Si can be obtained by a suitable combination of optical measurements alone.

INTRODUCTION

Like hydrogenated amorphous silicon (a-Si:H) microcrystalline silicon (μc-Si) appears to be a promising material for use in large area electronic devices. Both types of materials can be deposited at relatively low temperatures with the same deposition methods, e.g. plasma-enhanced chemical vapor deposition, but have quite different electronic, optical, and transport properties. A detailed understanding of these properties in μc-Si is complicated by the fact that the amorphous-to-microcrystalline transition occurs gradually: both, the volume fractions of amorphous versus microcrystalline material as well as the crystallite size in the μc-fraction are known to vary almost continuously as a function of deposition parameters such as substrate temperature, hydrogen dilution, gas residence times in the deposition reactor, etc. [1–3]. Therefore, in general many different experimental techniques probing structural and electronic properties are necessary for a sufficient characterization of any given μc-Si sample. Since such a complete characterization is in many cases not practical, it is useful to try to establish correlations between structural (Raman, x-ray, TEM) and electronic properties of μc-Si [4,5]. In the following, we will describe our attempts to test the applicability of such correlations to recrystallized a-Si:H samples.

EXPERIMENTAL DETAILS

μc-Si has been obtained by annealing a-Si:H samples at different temperatures between 300 and 900°C. Amorphous silicon samples were deposited by glow-discharge decomposition of silane at 230°C onto quartz substrates. Two different films thicknesses (0.5 and 1.5 μm) were used to check for possible influences of this parameter on the results. All samples were nominally undoped. Optical absorption data in the energy range 0.4 – 5.6 eV have been recorded by a combination of photothermal deflection spectroscopy [6], standard transmission measurements, and spectroscopic ellipsometry [7]. Raman spectra were obtained with an Ar$^+$ ion laser ($\lambda = 5145$ Å) and a double monochromator, using a photon-counting system.

RESULTS AND DISCUSSION

Figure 1 shows a summary of the optical absorption spectra of a-Si:H, a-Si (after hydrogen evolution at 650°C), μc-Si just after thermal crystallization (685°C), and polycrystalline silicon on sapphire (SOS). In the case of high-quality a-Si:H ([H] $\approx$ 10 at.%), the absorption coefficient, α, is high for $\hbar\omega > 3$ eV ($\alpha \approx 10^6$ cm^{-1}) and connects smoothly with the exponential Urbach tail region (1.5 eV $\leq \hbar\omega \leq 2$ eV). For $\hbar\omega < 1.5$ eV, the defect absorption band of the midgap dangling bond defect levels is visible ($N_D \approx 10^{16}$ cm^{-3}). Upon desorption of the bonded hydrogen, the defect absorption increases by about two

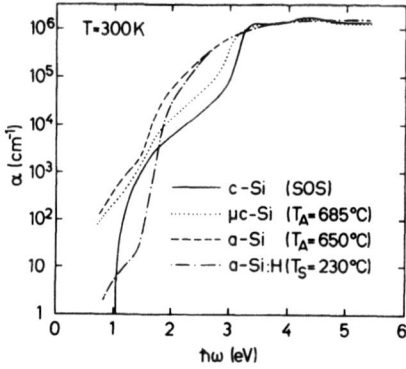

Fig. 1 Optical absorption coefficient as a function of energy for crystalline silicon on sapphire (solid curve), μc-Si (dotted curve), amorphous silicon (dashed curve) and hydrogenated amorphous silicon (dash-dotted curve).

orders of magnitude (i.e. 10^{18} cm^{-3}). In addition, the optical absorption in the Urbach tail region and the visible part of the optical spectrum increases (dashed line in Fig. 1). After crystallization of the sample at $T_A = 685°C$, the absorption in the visible region of the spectrum decreases by about a factor of four as the indirect character of the optical transitions in this energy range is reestablished. At higher photon energies, the absorption spectra of microcrystalline Si develop the two-peak structure caused by the direct transitions across the E_1 and E_2 gaps in the Si band structure. However, in this logarithmic scale there is little difference between the absorption coefficient for $\hbar\omega > 3.5$ eV in all four samples shown in Fig. 1.

The more interesting spectral region below $\hbar\omega \approx 3$ eV is shown in detail in Fig. 2. The dotted curves indicate the two extreme cases of pure amorphous silicon obtained by annealing at 675°C, and of crystalline Si. Absorption coefficients for μc-Si always fall between these two limits. As the annealing temperature for the μc-Si is raised from 685 to 800°C, the absorption above 1.7 eV decreases slightly, whereas that at lower photon energies increases again. We attribute the annealing-induced decrease of the absorption coefficient in the region 1.7 eV $\leq \hbar\omega \leq 3$ eV mainly to a decrease of the volume fraction of amorphous inter-grain tissue, which occurs when the microcrystallites grow from typically 80 Å diameter just after crystallization to about 130 Å diameter after annealing at 900°C (see below). k-uncertainty due to the finite size of the crystallites appears to be less important in this energy range [8].

An interesting point is that the absorption in the subgap region ($\hbar\omega \leq 1.3$ eV) in-

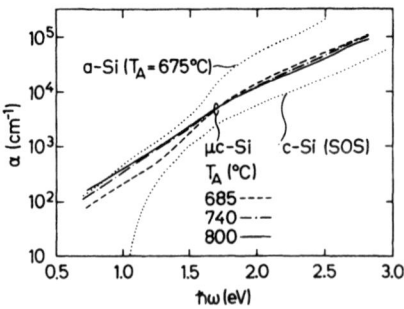

Fig. 2 Absorption coefficient in the range 0.6 eV to 3 eV for microcrystalline silicon annealed at different temperatures. The absorption coefficients for amorphous and crystalline silicon are included for comparison (dotted lines).

creases with increasing annealing temperature. Based on a study of polycrystalline silicon, it was suggested by Jackson et al. [9], that the subgap absorption, as in amorphous silicon, is a measure for the density of grain boundary defect states (dangling bonds). However, it is known that the dangling bond density as determined by electron spin resonance actually decreases upon annealing of undoped μc-Si. Also, hydrogen passivation is known to strongly reduce the density of ESR centers in μc-Si [10], but is causes only small changes in the subgap absorption [9,11]. These results indicate that the connection between subgap absorption and dangling bond defect density in μc-Si appears to be less straightforward than originally believed. Apparently there is a large, approximately constant density of spinless defects which contribute to the subgap absorption.

For a structural characterization of our samples (amorphous fraction, crystallite size) we have used Raman scattering, and some characteristic spectra obtained before and after crystallization are depicted in Fig. 3. Up to an annealing temperature of 650°C, all samples remain amorphous, with little change in the Si vibrational density of states. The main structural change in this temperature range is the loss of bonded hydrogen, which gives rise to the broad Raman peak around 2000 cm^{-1}. For the sample in Fig. 3, crystallization occurs at 675°C. However, the silicon optical phonon peak at $\approx$ 520 cm^{-1} is quite broad (FWHM 10 cm^{-1}), and there exists still a noticeable fraction of amorphous tissue as indicated by the low energy shoulder in the Raman spectrum around 480 cm^{-1}. The grain size can be estimated from the width of the phonon peak and the peak position to about 70 Å $\pm$ 20 Å just after crystallization [12]. Upon annealing at 900°C, the amorphous fraction disappears almost completely and the phonon peak narrows and shifts towards higher energies, indicating further growth of microcrystallites to a size of $\approx$ 120 Å $\pm$ 20 Å, in agreement with x-ray diffraction spectra obtained for the same samples. However, the Raman peak width is still considerably larger than that of the silicon-on-sapphire (SOS) sample used as a reference for c-Si.

It is known from the literature that the crystallite size in μc-Si can also be estimated from the broadening of the E$_1$-optical transition at $\hbar\omega \approx 3.4$ eV [5]. In order to test

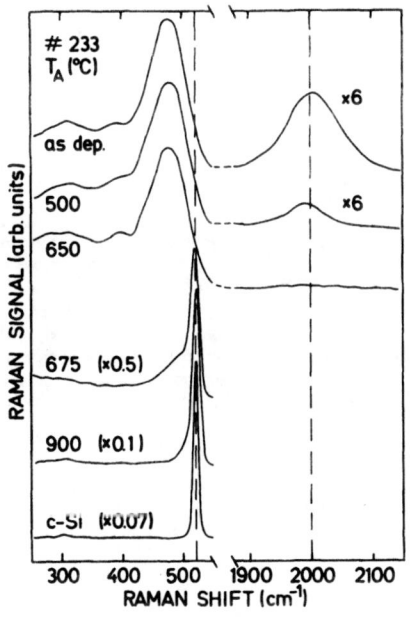

Fig. 3 Raman spectra of sample #233 after anneal at different temperatures.

232

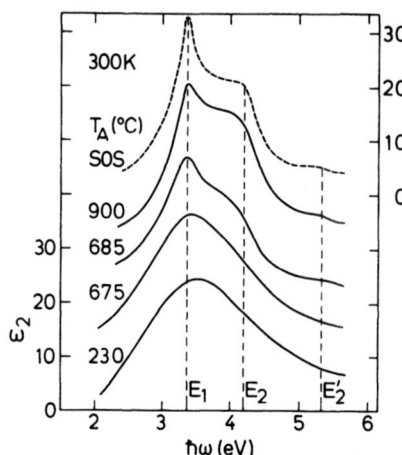

Fig. 4 Room temperature spectra of the imaginary part, ϵ_2, of the dielectric constant. Spectra are determined by spectroscopic ellipsometry and are shown for different annealing stages. The spectrum of SOS is included for comparison (dashed curve).

this for the present samples, we have measured the pseudo-dielectric constants ϵ_1 and ϵ_2 by spectroscopic ellipsometry in the energy range 2 eV $< \hbar\omega <$ 5.6 eV. The results obtained for the imaginary part $\epsilon_2(\hbar\omega)$ of the dielectric constant as a function of energy are shown in Fig. 4. As long as the sample remains amorphous, the ϵ_2-spectra consist of a single broad peak around $\hbar\omega \approx 3.5$ eV. Crystallization of the sample becomes noticeable by a change to a more structured spectrum between $T_A = 675°C$ (amorphous) and $T_A = 685°C$ (μc-Si). In particular, the peaks corresponding to direct optical transitions between the highest valence band and the lowest conduction band become visible: the $E_1(E_0')$ peak caused by the almost parallel bands between the Γ- and the L-point, and the E_2 peak at $\hbar\omega \approx 4.2$ eV due to optical transitions at the X-point of the Brillouin zone. With increasing crystallite size one observes a decreasing width, Γ, of the E_1-transition. For a more quantitative analysis, we have extracted Γ by fitting the second derivative of the dielectric function, $d^2\epsilon_2/d(\hbar\omega)^2$, to the theoretically expected form for a two-dimensional critical point in the c-Si band structure [13]

$$\frac{d^2\epsilon_2}{d(\hbar\omega)^2} \propto \mathrm{Im}\left\{(\hbar\omega - E_1 + i\Gamma)^{-2}\right\}. \qquad (1)$$

The obtained values for the peak width $\Gamma(E_1)$ are plotted in Fig. 5 versus the optical phonon linewidth, Γ_{TO}, deduced from the Raman spectra. Over the limited range of crystallite sizes investigated in this study, we find a roughly linear dependence between the two quantities in Fig. 5, with $\Gamma(E_1) \approx 160$ meV for the smallest crystallites just after crystallization (≈ 80 Å diameter) and $\Gamma(E_1) \approx 70$ meV for SOS, in agreement with room temperature values for bulk crystalline silicon [13]. With a suitable calibration, it should therefore be possible to estimate crystallite sizes in μc-Si solely from optical data obtained in the spectral region of the E_1-transition, without the necessity for more direct structural measurements (Raman, x-ray diffraction, or electron microscopy). However, to this end a more extensive data base would be desirable.

We would like to thank R. Arce for his help with sample preparation and M. Brunel (CNRS, Grenoble) for the x-ray diffraction measurements. Valuable discussions with E. Bustarret are gratefully acknowledged. We also thank M. Brandt and C. Herrero for a critical reading of the manuscript.

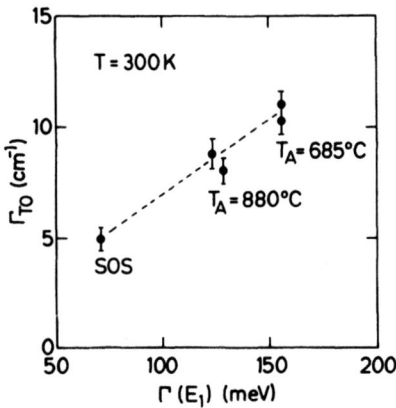

Fig. 5 Correlation between the width Γ_{TO} of the optical phonon Raman peak and the width parameter $\Gamma(E_1)$ of the E_1-transition at 3.4 eV.

REFERENCES

[1] Z. Iqbal, S. Vepřek, J. Phys. C **15**, 377 (1982).

[2] Y. Mishima, S. Miyazaki, M. Hirose, Y. Osaka, Phil. Mag. B **46**, 1 (1982).

[3] N.M. Johnson, S.E. Ready, J.B. Boyce, C.D. Doland, S.H. Wolff, J. Walker, Appl. Phys. Lett. **53**, 1626 (1988).

[4] S. Kumar, D.K. Pandya, K.L. Chopra, J. Appl. Phys. **63**, 1497 (1988).

[5] S. Logothetidis, J. Appl. Phys. **65**, 2416 (1989).

[6] W.B. Jackson, N.M. Amer, A.C. Boccara, D. Fournier, Appl. Optics **20**, 1333 (1981).

[7] D.E. Aspnes, A.A. Studna, Appl. Optics **14**, 220 (1975).

[8] M.A. Hachicha, E. Bustarret, this volume.

[9] W.B. Jackson, N.M. Johnson, D.K. Biegelsen, Appl. Phys. Lett. **43**, 195 (1983).

[10] S. Hasegawa, S. Takenaka, Y. Kurata, J. Appl. Phys. **53**, 5022 (1982).

[11] M. Stutzmann, C.P. Herrero, M. Ingels, A. Breitschwerdt, this volume.

[12] I.H. Campbell, P.M. Fauchet, Solid State Commun. **58**, 739 (1986).

[13] P. Lautenschlager, M. Garriga, L. Viña, M. Cardona, Phys. Rev. B **36**, 4821 (1987).

TRANSPORT PROPERTIES OF B-, P-DOPED AND UNDOPED 50 kHz PECVD MICROCRYSTALLINE SILICON

M.A. HACHICHA* AND ETIENNE BUSTARRET**
*LEPES-CNRS, BP 166x, F-38042, Grenoble, France (associated to Univ. J. Fourier de Grenoble).
**Max-Planck-Institute für Festkörperforschung, Heisenbergstr. 1, D-7000 Stuttgart 80, FRG

ABSTRACT

Undoped 500 nm-thick silicon layers with a crystalline fraction around 95% and an average grain size of 20 nm have been deposited at 350°C by 50 kHz triode PECVD in a H_2/SiH_4 mixture, in the presence of a magnetic field. Their room temperature (rt) dc conductivity σ_{rt} is 0.03 $\Omega^{-1}cm^{-1}$ for a Hall mobility of 0.8 $cm^2V^{-1}s^{-1}$.

The study by SIMS, infrared absorption, grazing angle x-ray diffraction and Raman scattering spectroscopies of the doped samples shows how the crystalline fraction and the grain size drop as the B_2H_6/SiH_4 and PH_3/SiH_4 volumic ratios increase from 10 ppm to 1%.

The rt dc conductivity reaches 2 $\Omega^{-1}cm^{-1}$ (Hall mobility: 15 $cm^2V^{-1}s^{-1}$) for a solid phase density of 10^{19} cm^{-3} boron atoms, and 30 $\Omega^{-1}cm^{-1}$ (Hall mobility: 55 $cm^2V^{-1}s^{-1}$) at the maximum P incorporation of $8 \times 10^{20}cm^{-3}$.

TRANSPORT PROPERTIES OF DOPED AND UNDOPED nc-Si

Because of the very distinct microstructures associated with the generic "microcrystalline" or "nanocrystalline" silicon (nc-Si) term, the reported data on transport properties of doped and undoped nc-Si layers show a strong scatter associated to various preparation and film conservation procedures [1]. The general picture that a minimum crystalline fraction is needed to ensure percolation of the electronic transport path in the grains hold both for undoped [2] and doped [3] nc-Si materials. Most of the resulting enhancement in conductivity (over 8 orders of magnitude) results from the crystalline character [1-5], while heavy doping leads to a more limited gain of three [6] or two [7] orders of magnitude, respectively, for P- and B-doping. This is generally attributed to dopant segregation in the grain boundaries. Finally, the general decrease of the conductivity with grain size in well crystallized films [1,5] has been recently linked to quantum size effects [8].

In this work, we shall first report the transport properties of our undoped and doped 50 kHz triode PECVD samples [5] deposited at 350°C under a total pressure of 1 mbar, with a 300 $mWcm^{-2}$ power density and in the presence of a dc magnetic field which confines the plasma [5,7,9]. The dependence of the solid phase concentration (deduced from SIMS profiles) of B and P on the volumic diborane and phosphine to silane flow ratios given in Fig. 1 does not follow the power-law observed for a-Si:H grown from the same $SiH_4:H_2$ (1:9) mixture in the same reactor [10], and the incorporation of Boron is less efficient than expected [7]. We shall then give a detailed physico-chemical description of the doped samples, to our view a prerequisite to any quantitative interpretation of the optical [11] and transport properties of nc-films grown at low temperature.

Our nc-Si films have been characterized by the rt Hall effect (Van der Pauw method) and dc dark conductivity (gap cell geometry) as a function of temperature between 270 and 500 K. The straight Arrhenius plots observed for all layers on the whole range yield activation energies E_a of 150 ± 50meV for all samples save those deposited with $PH_3/SiH_4 \geq$ $1^0/_{00}$ or $B_2H_6/SiH_4 = 1\%$ where E_a drops to 30 ± 10 meV. Similar values obtained by others [1,7,9] lead us to consider the oxygen contamination as negligible [1], as confirmed by our infrared transmission data. In most samples, the grains are completely depleted so that the Hall mobility is that of the dc conductivity characterized by σ_{rt} and by the prefactor σ_0 (see Fig. 2). The maximum σ_{rt} values of 0.03, 2 and 30 $\Omega^{-1}cm^{-1}$ obtained respectively on the undoped, 1% B and 1%P-doped materials are associated with Hall mobilities of 0.8, 15 and 55 $cm^2V^{-1}s^{-1}$ and carrier densities (see Fig. 1) of 2, 19 and 100 $\times 10^{17}cm^{-3}$, respectively. If one takes into account the upper limits of our solid state dopant incorporation, these results

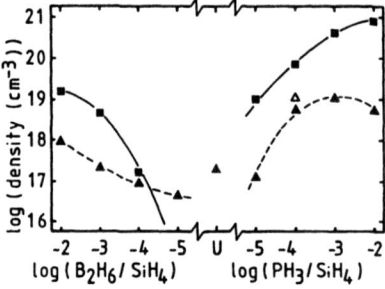

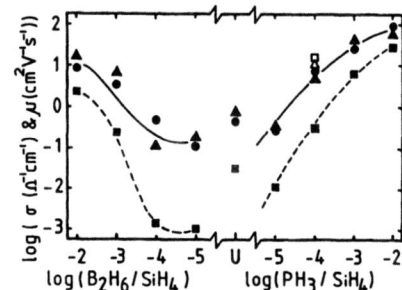

Fig. 1   Solid phase (SIMS) density of dopant atoms (■), Hall carrier density (▲), optically determined free carrier concentration (△) as a function of the gas phase nominal doping ratio.

Fig. 2   Room temperature Hall mobility (▲), dc conductivity σ_{rt} (■), and prefactor σ_0 as a function of the gas phase ratio. Open symbols correspond to values determined optically.

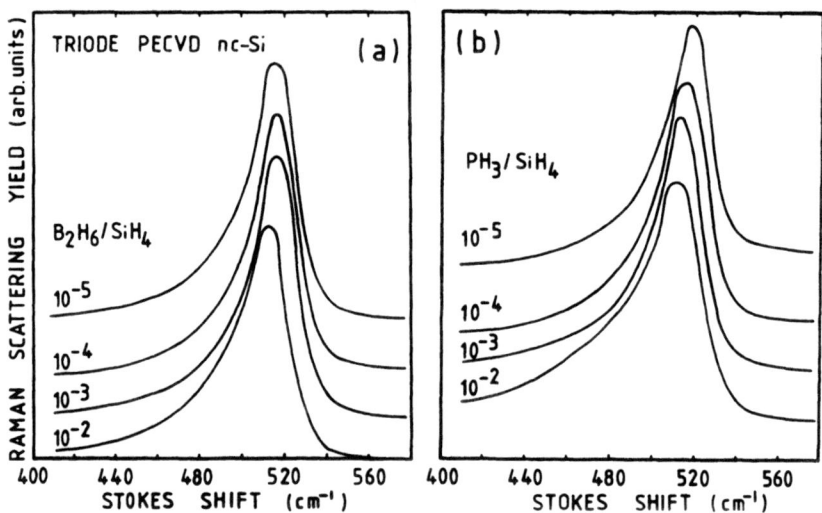

Fig. 3   Raman scattering spectra of B (a) and P (b) doped nc-Si layers.

are quite similar to those published for comparable materials [1,3–7,9]. Finally, the optical absorption profile observed [11] on the layer prepared at $PH_3/SiH_4 = 10^{-4}$ in the 0.65 to 0.95 eV range has been tentatively treated as free carrier absorption [12]. This yields a band mobility of 10 $cm^2V^{-1}s^{-1}$ and a carrier concentration of 1×10^{19} cm^{-3}, within a factor 2 of the Hall results. The dc conductivity deduced from the optical data is, however, at least one order of magnitude higher than that measured directly (see Figs. 1,2). In order to discuss the possible microscopic transport mechanisms responsible for both the present results and temperature-dependent Hall effect measurements under progress, we shall now determine the structural parameters of our doped films.

STRUCTURE AND HYDROGENATION OF THE DOPED LAYERS

X-ray diffraction studies at a grazing angle of 0.5° chosen to yield the pattern of the bulk of our 500 nm-thick film show that the average grain size is 9 nm in the (111) direction and 7 ± 1 nm in the (220) and (311) directions for all films except the $PH_3/SiH_4 = 1\%$ sample where it was 6 nm in all directions. This means a reduction by a factor two of the grain size upon doping. The ratios of the intensity of the (220) and (311) diffraction peaks to that of the (111) line rise with increasing B_2H_6/SiH_4 gas ratio to reach those of the c-Si powder pattern at $1^0/_{00}$. In the case of P-doping, the (111) preferential orientation of the undoped material is preserved, but this texture is less marked at intermediate than at high phosphine to silane ratios.

A preliminary Raman study of the layers deposited on the polished high purity c-Si substrate also used for the infrared transmission was performed at a 514.5 nm excitation in the Vertical Unpolarized configuration. The spectra given in Fig. 3 and their decomposition into a Gaussian and a Lorentzian component, together with a previous calibration of the Raman scattering cross-section of the TO-like phonon as a function of the grain size [13], enabled us to determine the crystalline volume fraction f_c. From 0.93 in the undoped layer, f_c decreases to 0.8 in all doped layers, except for the 1% diborane and phosphine flow ratios which yield samples with $f_c = 0.7$ and 0.5, respectively.

Finally, we performed optical transmission measurements in the 200 to 4000 cm^{-1} range in a N_2 purged double beam commercial spectrometer. We estimated the total bonded Hydrogen density C_H from the integrated normalized absorption I_W ($C_H/I_W = 1.6 \times 10^{-19}cm^2$) around 630 cm^{-1} attributed to Si-H wagging motions, neglecting the B-H bonds of the 1% doped layer which give rise to other lines at 475 and 2450 cm^{-1}, but subtracting the contribution of a nearby peak at 700 cm^{-1} associated to Si-P stretching vibrations. C_H corresponds to 1.8 at.% in the undoped sample, to a constant 4 at.% in the B-doped films and to 5 ± 1% in the P-doped layers.

Some of the absorption spectra in the 1900 to 2200 cm^{-1} range are given in Fig. 4. Between $B_2H_6/SiH_4 = 10^{-5}$ and 10^{-3}, the profiles do not vary significantly. They are dominated by the line at 2100 cm^{-1} with a shoulder at 2000 cm^{-1}. These two spectral components are well separated in the case of the heavily B-doped (1%) sample, where the 2000 cm^{-1} peak ascribed to isolated Si-H bonds in a disordered matrix is stronger. The assignment of the 2100 cm^{-1} peak is not straightforward since interacting Si-H units and isolated or linked SiH_2 groups in a disordered matrix can contribute to it, as well as SiH units on (111) or (220) crystalline surface or SiH_2 groups on a (100) surface [5,14]. In the case of P-doping, the strong 2100 cm^{-1} component observed at low doping ratios decreases steadily with increasing PH_3/SiH_4 ratios. When compared to the spectrum of the undoped samples [5], most of the stretching mode absorption profiles of the doped layers are stronger than expected on the basis of their relative C_H values. This enhanced absorption confirms that both the hydrogen and the doping atoms are segregated in the amorphous region and in the grain boundaries.

In conclusion, the physico-chemical characteristics of the most conductive B and P doped nc-Si films are quite different from those of the moderately doped samples. A quantitative study of the latter is possible with transport models derived from polycrystalline case.

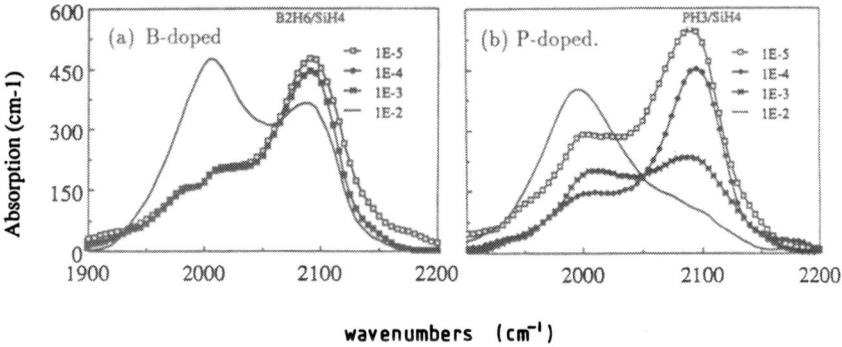

Fig. 4   Infrared absorption spectra of B (a) and P (b) doped nc-Si layers.

REFERENCES

[1] S. Veprek, Z. Iqbal, R.O. Kuhne, P. Capezzuto, F.A. Sarott, J.K. Gimzewski, J. Phys. C **16**, 6241 (1983).

[2] M. Konuma, H. Curtins, F.A. Sarott, S. Veprek, Phil. Mag. B **55**, 377 (1987).

[3] R.Tsu, J. Gonzalez-Hernandez, S.S. Chao, S.C. Lee, K. Tanaka, Appl. Phys. Lett. **40**, 534 (1982).

[4] W.E. Spear, G. Willeke, P.G. Lecomber, A.G. Fitzgerald, J. de Physique (France) **42**, C4-257 (1981).

[5] M.A. Hachicha, J.C. Bruyere, E. Bustarret, A. Deneuville, M. Brunel, "Ion and Plasma Assisted Techniques 87", Int. Conf. Proc. published by Conferences Exhibitions Publications Consultants Ltd., Edinburgh (UK), p. 360 (1987).

[6] G. Willeke, W.E. Spear, D.I. Jones, P.G. Lecomber, Phil. Mag. B **46**, 177 (1982).

[7] K. Mori, M. Kitagawa, T. Hirao, S. Ishihara, M. Ohno, Jpn. J. Appl. Phys. **20**, 2431 (1981).

[8] N. Lifshitz, S. Luryi, T.T. Sheng, Appl. Phys. Lett. **51**, 1824 (1987).

[9] T. Hamasaki, H. Kurata, M. Hirose, Y. Osaka, Appl. Phys. Lett. **37**, 1084 (1980).

[10] E. Bustarret, F. Vaillant, B. Hepp, Mater. Res. Soc. Symp. Proc. **118**, 123 (1988).

[11] E. Bustarret, M.A. Hachicha, this volume.

[12] Y. Mishima, M. Hirose, Y. Osaka, J. Appl. Phys. **51**, 1157 (1980).

[13] E. Bustarret, M.A. Hachicha, M. Brunel, Appl. Phys. Lett. **52**, 1675 (1988).

[14] T. Satoh, A. Hiraki, Jap. J. Appl. Phys. **24**, L491 (1985).

SUPPRESSION OF ACCEPTOR DEACTIVATION IN SILICON BY DISORDERED SURFACE REGIONS

K. SRIKANTH AND S. ASHOK
Center for Electronic Materials & Processing and Department of Engineering Science & Mechanics, The Pennsylvania State University, University Park, PA 16802

ABSTRACT

Permeation of atomic hydrogen in p-type Si damaged with ion implantation or deposited with polycrystalline or amorphous Si has been studied. Following ion implantation or film deposition, atomic hydrogen was introduced by low energy H ion implantation or from an electron cyclotron resonance (ECR) plasma. Spreading resistance profiles indicate that deactivation of acceptor dopant boron atoms by atomic hydrogen is drastically reduced in silicon wafers with any of the above disordered surface layers, and secondary ion mass spectroscopy (SIMS) traces this reduction to the suppression of hydrogen movement into the crystalline Si substrate. Trapping of hydrogen or formation of molecular hydrogen at defect sites in the surface disordered regions apparently is responsible for this phenomenon.

INTRODUCTION

The role of atomic hydrogen in crystalline Si has been a subject of intense interest in recent years, and in particular dopant neutralization under hydrogen plasma or low-energy hydrogen implantation has been widely investigated [1-5]. Acceptor neutralization in p-type Si is a singularly dominant effect, with the hole concentration dropping by 3-4 orders of magnitude at the surface and the reduction extending over several microns from the surface. While the neutralization phenomenon has been studied in ion-implanted Si samples that have been *annealed for dopant activation* [6], there has been no study reported on the influence of atomic hydrogen in ion beam damaged Si. Such an investigation is also of interest in assessing the role of ion damage encountered in dry etching technology using hydrogen-containing gases (eg., hydrocarbons or hydrogen halides) where ion damage is present concurrently with atomic hydrogen. More generally, the role of disorder at Si surface on the permeation of H in Si has not been investigated. We present evidence for suppression of acceptor deactivation in crystalline Si (c-Si) when the hydrogenation process is preceded by ion bombardment or poly/amorphous Si deposition.

EXPERIMENTAL

In the present experiment, ion beam damage of Si surface was accomplished by implanting Ar in a commercial (Varian 350D) ion implanter. Such an approach to creating ion damage offers the advantages of minimal contamination as well as controlled ion dose and penetration depth. Boron-doped Si wafers, with 1-10 ohm-cm resistivity and (100) orientation were implanted with 20 keV Ar at two doses - 1×10^{13} and 1×10^{15} cm^{-2}. There was no intentional heating of the substrate and the temperature rise was estimated to be less than 50°C at the highest dose. The choice of Ar energy was dictated by the lowest possible value in the implanter that would yield a reasonaly high ion current, while the doses were chosen to straddle the amorphization threshold of about 1×10^{14} cm^{-2} for 20 keV Ar in Si. For samples where the disordered surface regions are to consist of deposited Si films, chemical vapor deposited poly-Si (200 nm thick) and sputtered amorphous Si (100 nm thick) were used. These wafers, together with unimplanted control samples, were subjected to the atomic hydrogen treatment using a low-energy H ion beam (5" dia.) in a Commonwealth Scientific ion beam sputtering system. Since the Si substrate temperature during hydrogenation is crucial in determining the influence of H in Si [7], the H-implant was performed with and without intentional heating corresponding to substrate temperatures of approximately 85 °C and 200 °C respectively. The H-beam energy, current density and implant time were held constant at 0.4 keV, 1mA/cm^2 and 10 min. respectively for all the runs. Electron

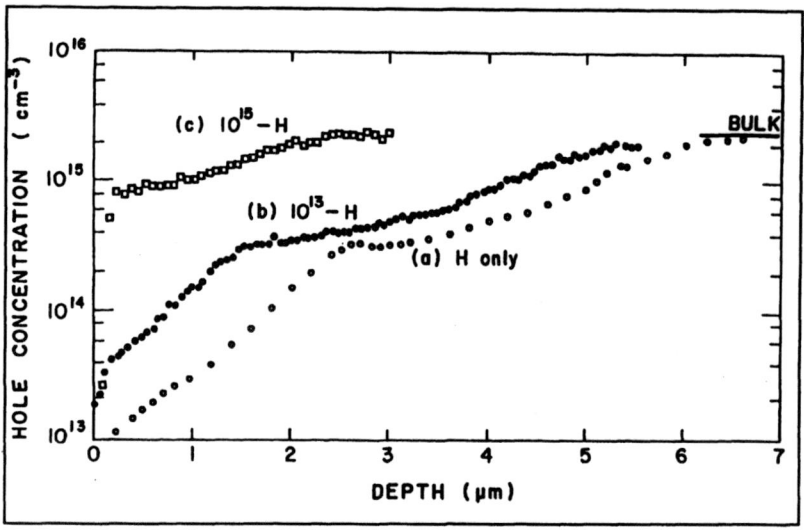

Fig. 1. Hole concentration profiles obtained from spreading resistance measurements for p-Si (a) with 0.4 keV H only, (b) implanted with 20 keV Ar at 10^{13} cm^{-2}, followed by 0.4 keV H and (c) implanted with 20 keV Ar at 10^{15} cm^{-2}, followed by 0.4 keV H.

cyclotron resonance (ECR) hydrogenation was performed at microwave power level of 500 W for 1 hour at 250 °C substrate temperature; the system pressure was 4×10^{-4} torr, and the hydrogen flow rate 11.7 sccm. Some of the experiments were repeated with deuterium replacing hydrogen in order to carry out secondary ion mass spectroscopy (SIMS) measurements. Following hydrogenation/deuteration, spreading resistance measurements were done on a Solid State Measurements automated spreading resistance profiler. Al Schottky dots, 1 mm in dia., were also evaporated on the samples for evaluating the current-voltage (I-V) and capacitatance-voltage (C-V) characteristics.

RESULTS

Fig. 1 gives the carrier concentration profiles of hydrogen implanted Si, without and with preceding Ar implantation. Curve (a) in Fig. 1 shows extensive deactivation of boron up to 6 µm into the bulk, with the surface concentration reduced by over 2 orders of magnitude. The profiles exhibit a characteristic knee corresponding to the change in diffusion coefficient. These are typical of most hydrogenation experiments and have been widely observed [4,8]. Under identical hydrogenation conditions, the p-Si sample that had earlier been implanted with Ar to a dose of 10^{13} cm^{-2} displays considerably reduced depth as well as extent of deactivation as seen in curve (b). A more drastic change is evident under a higher Ar dose of 10^{15} cm^{-2} (curve (c)). In fact the change due to hydrogenation is hardly noticeable for the latter case. For comparison, the hole concentration profiles of p-Si samples that were implanted with Ar only (no H-implant) reveal only a minor reduction in hole concentration near the surface, apparently due to residual hydrogen penetration caused by the various Si cleaning steps [9]. It is evident from Fig. 1 that the Ar implantation tends to suppress acceptor deactivation of atomic hydrogen, with the degree of suppression increasing with Ar dose.

Samples that were hydrogenated with in-situ heating (substrate temp. 200 °C) exhibit similar inhibition of acceptor deactivation by Ar implantation. However, they exhibit a distinctly different deactivation profile, with a marked absence of the two diffusion regimes, indicating that the diffusion kinetics have changed.

The $\log I$-V characteristics of Al/p-Si Schottky diodes fabricated on samples corresponding to Fig. 1

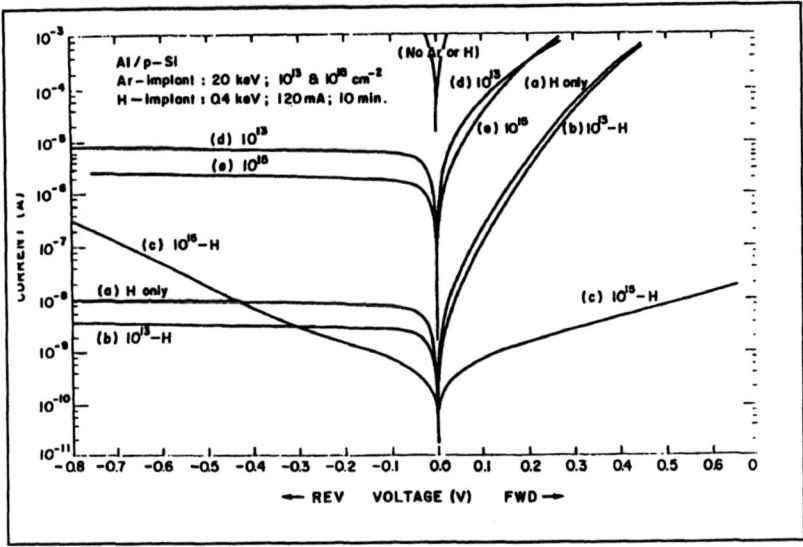

Fig. 2. Forward (FWD) and reverse (REV) logI-V characteristics of Al/p-Si Schottky diodes fabricated p-Si with Ar and H implants: (a) 0.4 keV H only, (b) 20 keV, 10^{13} cm^{-2} Ar followed by 0.4 keV H, (c) 20 keV, 10^{15} cm^{-2} Ar followed by 0.4 keV H, (d) 20 keV, 10^{13} cm^{-2} Ar only, and (e) 20 keV, 10^{15} cm^{-2} Ar only. The "ohmic" I-V characteristics of the low-barrier Al/p-Si control device with no Ar or H implant are also shown for comparision.

are shown in Fig. 2. It is seen from the I-V plots that improved rectification corresponding to Schottky barrier height increase occurs after Ar implantation (curves (d) and (e)) and much more so after subsequent hydrogenation (curves (a)-(c)). The increase in Schottky barrier height on p-Si due to ion damage is a universal phenomenon, and some of the highest barrier heights on p-Si have been found with sequential argon and hydrogen implantation [10]. We shall allude to these results in subsequent discussion. In passing , we note that the set of curves (c), for sample with 10^{15} cm^{-2} Ar implant followed by hydrogenation, do not display Schottky diode-like behavior but suggest considerable alteration of interfacial transport by the ion bombardment-modified surface layer. From the C-V characteristics, $1/C^2$ plots were obtained for all the samples of Fig. 2, and the variation in their slopes are consistent with the spreading resistance profiles of Fig. 1.

In order to elucidate the role of the Ar implanted layer in inhibiting acceptor deactivation by hydrogen, we replaced hydrogen with deuterium and carried out spreading resistance and SIMS measurements. The spreading resistance profiles were similar to those with H, and the SIMS profiles of deuterium (D) reveal that the Ar implantation damage prevents the entry of D into the crystalline Si bulk. As seen in Fig. 3, both the depth of penetration of D in Si and its magnitude in the bulk are progressively reduced with increasing Ar implantation dose. The D profiles are consistent with the spreading resistance profiles (not shown) of the corresponding samples. Thus it is clear that the Ar ion implant damage directly suppresses the penetration of deuterium into the bulk crystalline Si, and not merely affect the deactivation mechanism.

An additional experiment to ascertain the influence of surface disorder on permeation of H or D in Si was carried out by depositing thin layers of poly-Si or amorphous Si (a-Si). Fig. 4 displays the spreading resistance profiles following 0.4 keV D implant: The control sample (with no surface films) shows the expected acceptor deactivation to a depth of about 3 μm while almost complete absence of acceptor deactivation is seen with both 200 nm of poly-Si and 100 nm of a-Si. Thus even a very thin layer of disordered Si surface layer is adequate to act as a sink for D, and prevent its penetration into crystalline Si below. The suppression of D-penetration by the poly-Si layer is also corroborated by the SIMS profile

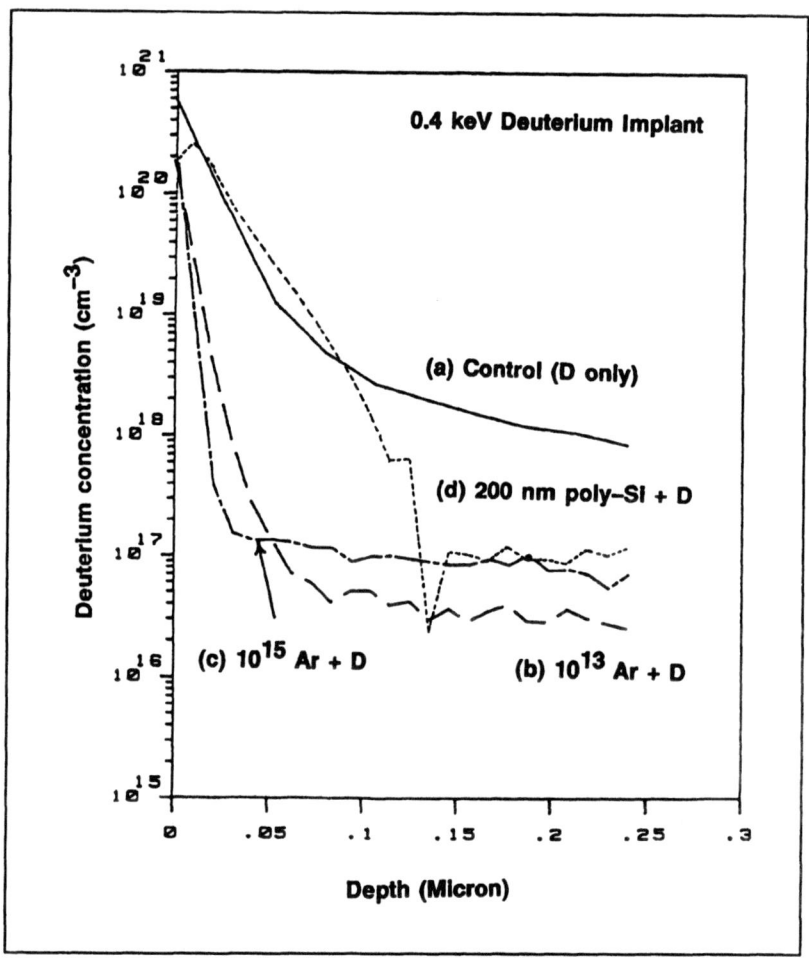

Fig. 3. SIMS profiles of 0.4 keV deuterium implanted samples: (a) Control (c-Si only), (b) with 20 keV, 10¹³ cm² Ar implant, (c) with 20 keV, 10¹⁵ cm² Ar implant, and (d) with 200 nm of poly-Si.

shown in Fig. 3.

Results similar to the above for low-energy hydrogen/deuterium implantation have also been observed with ECR hydrogenation, where the ion energy is lower but the flux level is very high.

DISCUSSION

The decreasing trend in acceptor deactivation with Ar implantation as seen in the spreading resistance data of Fig. I may be attributed to (1) Fermi level position near the surface, (2) hydrogen accumulation near the surface (3) defect site trapping of hydrogen. The movement of Fermi level near the surface upon Ar implantation has been seen in Fig. 2 in the form of a change in Schottky barrier

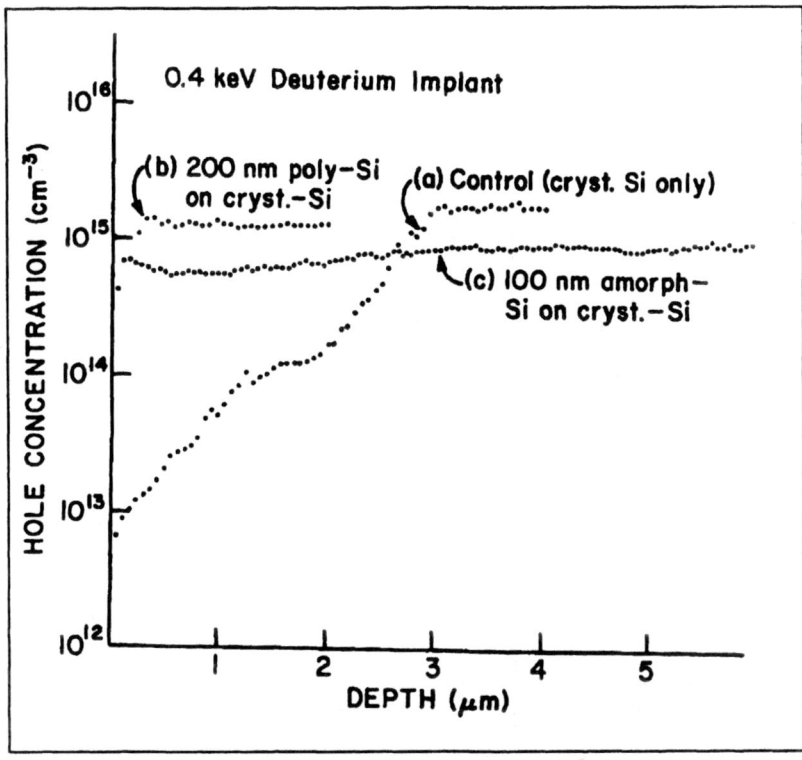

Fig. 4. Hole concentration profiles obtained from spreading resistance measurements for (a) control (c-Si only), (b) 200 nm of poly-Si on c-Si, and (c) 100 nm of a-Si on c-Si, following 0.4 keV deuterium implant.

height. This movement is mainly due to donor-like defect sites generated by the Ar bombardment damage [4,10]. Hydrogen diffusion into Si may be affected by the Fermi level position. However, as seen from the I-V plots (d) and (e) of Fig. 2 there is only a very small difference in the Schottky barrier heights of samples implanted with Ar at doses of 10^{13} and 10^{15} cm^{-2} respectively, indicating only a very small change in the Fermi level position between the two Ar implants. Hence this slight movement of Fermi level fails to explain the dramatic reduction in deactivation at the higher Ar dose.

From the trend in spreading resistance data, and the SIMS profiles of Fig. 3 it is clear that the hydrogen never gets to diffuse into Si in the heavily Ar-implanted case. It appears that hydrogen tends to accumulate at the surface damage region, even though SIMS may not be accurate in the near-surface region, which corresponds to a projected range of 21.8 nm and a straggle of 9.7 nm for 20 keV Ar in Si. It is also possible that a larger fraction of the incident H ions are reflected away from the Si surface when its surface is disordered. The form in which hydrogen exists and the reason for its accumulation are the basic questions. It has been reported that the formation of hydrogen molecules will be very significant in n-type Si [2,11]. It is well known that Ar ion bombardment damage of Si creates donor-like defects form an n-type surface layer. Thus, in our hydrogenation experiments on Ar-implanted p-Si, the "n- type" surface damaged layer may favor the formation of molecular hydrogen. This favored formation of H_2 molecules in the Ar-implanted surface layer hinders the permeation of atomic hydrogen deep into the bulk. Zundel et al. [8], in their recent report see this as a dominant mechanism and have attributed the absence of deep migration for long hydrogenation times(> 30 min.) to formation of large amount of molecular hydrogen beneath the surface, which strongly hinders the transfer of the implanted atomic hydrogen to the bulk. This would lead us to speculate that an n-type substrate should result in poor permeation of hydrogen in silicon. This is indeed the case, as observed by Pankove, Magee, and Wance [7] in their work on As implants in p-type Si. However, our results using poly-Si and a-Si surface layers [Figs. 4 & 3] cast doubt on this notion, since the deposited poly-Si and a-Si are undoped high resistivity materials. Thus it appears that the suppression of H penetration and acceptor deactivation in crystalline Si is a direct result of surface disorder caused by implantation or film deposition, rather than damage-induced donor levels.

In the scenario involving the higher Ar dose (10^{15} cm^{-2}), two other key effects work against hydrogen permeation. With increase in dose, more defect sites are generated and these defect sites are effective centers for trapping hydrogen. Pankove, Wance and Berkeyheiser [12] have speculated on similar trapping of hydrogen in defects formed by implantation process. It is also known that a silicon vacancy can accommodate upto four bonded hydrogen atoms. Furthermore it is known from low-temperature transport measurements of Ar-implanted Schottky diodes and high resolution electron microscopy that Si microcrystallites can appear in the surface region under Ar implantation [13]. It is conceivable that the hydrogen penetration could be reduced by the grain boundaries in a manner similar to that reported for Si ribbons [14]. Our results of Fig. 4 using poly-Si and a-Si surface layers clearly tend to support this model.

In summary, we have observed a dominant role played by surface disorder created by ion implantation or poly/amorphous Si deposition on crystalline Si in restricting hydrogen permeation. The introduced hydrogen apparently accumulates near the surface in the Ar-implanted sample and may be in the form of subsurface molecular hydrogen. It also appears that the total dose of hydrogen introduced into the Si sample is also reduced in the presence of surface disorder; the reason for this behavior is unclear at this time.

ACKNOWLEDGMENTS

This work was supported in part by the Penn State Center for Particle Beam Interactions with Solids, funded by IBM. We would also like to thank Solid State Measurements Inc., Pittsburg, for assistance with spreading resistance measurements, and C. Houser for the SIMS measurements. We would also like to thank the reviewer of the manuscript for useful comments.

REFERENCES

1. C. T. Sah, J. Y. Sun, and J. J. Tzan, *Appl. Phys. Lett.* 43, 203 (1983).
2. S. J. Pearton, J. W. Corbett, and T. S. Shi, *Appl. Phys. A* 43, 153 (1987).
3. N. M. Johnson, *Phys. Rev. B* 31, 5525 (1985).
4. M.W. Horn, J.M. Heddleson and S.J. Fonash, *Appl. Phys. Lett.* 51, 490 (1987).
5. A.E. Jaworowski, *Radiation Effects and Defects in Solids* 110, (1989).
6. J. I. Pankove, C. W. Magee and R. O. Wance, *Appl. Phys. Lett.* 47, 748 (1985).
7. H.-C. Chien, S. Ashok and M.-C. Chen, *Jap. J. of Appl. Phys.* 27, L1317 (1988).
8. T. Zundel, A. Mesli, J. C. Muller and P. Siffert, *Appl. Phys. A* 48, 31 (1989).
9. C.H. Seager, R.A. Anderson and J.K.G. Panitz, *J. Mat. Res.* 2, 96 (1987).
10. S. Ashok and K. Giewont, *Jap. J. of Appl. Phys.* 24, L533 (1985).
11. A. J. Tavendale, A. A. William, D. Alexiev and S. J. Pearton, in *Oxygen, Carbon, Hydrogen and Nitrogen in Silicon*, edited by J.C. Mikkelson,Jr., S.J. Pearton, J.W. Corbett and S.J. Pennycock (Materials Research Society, Pittsburgh, 1986), p.460.
12. J. I. Pankove, R. O. Wance and J. E. Berkeyheiser, *Appl. Phy. Lett.* 45, 1100(1984).
13. H.-C. Chien and S. Ashok, *J. Appl. Phys.* 60, 2886 (1986).
14. C. Dube, J. I. Hanoka and S. B. Sandstrom, *Appl. Phys. Lett.* 44, 425 (1984).

PART IV

Optical Properties

PROPERTIES OF BINARY Si:H MATERIALS
PREPARED BY HYDROGEN PLASMA SPUTTERING

SHOJI FURUKAWA   AND   TATSURO MIYASATO
Kyushu Institute of Technology, Faculty of Computer Science and
Systems Engineering, 680-4 Kawazu, Iizuka-shi, Fukuoka-ken 820,
Japan

ABSTRACT

Binary Si:H materials are prepared by means of the rf
sputtering technique in pure hydrogen atmosphere on low
temperature (about 100 K) and room temperature substrates. The
physical properties of the obtained materials are very much
affected by the rf power and substrate temperature during the
deposition. The material prepared at a low substrate temperature
with a low rf power has a wide optical gap, and shows a visible
photoluminescence at room temperature. On the other hand, the
material prepared at room temperature with a high rf power
contains many Si microcrystals, whose diameters are relatively
large, and its optical gap becomes very small. The latter
condition causes the dependence of the crystalline direction of
the material film on the substrate crystal even at the room
temperature. An rf power-modulated multi-layered structure
(superlattice) is also proposed, and an apparent diffraction
peak can be observed in the low-angle X-ray scattering
measurement.

INTRODUCTION

To date, many researchers have reported on topics related
to binary Si:H materials. Hydrogenated amorphous silicon (a-Si:H)
is the most popular material among them, and is expected to be
used for solar cells [1], thin film transistors [2], charge-
coupled devices [3], linear image sensors [4], electro-
photographies [5], image pickup tubes [6], optical recordings
[7], and light-emitting diodes [8], etc. Besides the a-Si:H,
hydrogenated microcrystalline silicon (μc-Si:H) [9,10] is also
expected to be used for those devices.
In this paper, we deal with binary Si:H materials prepared
by means of an rf sputtering technique in pure hydrogen
atmosphere. The physical properties of the obtained materials
are very much affected by deposition parameters, and the optical
gap can be controlled over the range from 1.5 to 2.4 eV. A wide
optical gap material can be obtained at a low substrate
temperature with a low rf power, and it shows a visible photoluminescence (red) at the room temperature. On the contrary, a
narrow optical gap material is obtained at the room temperature
with a high rf power, and it contains relatively large Si
microcrystals. The latter material shows the dependence of its
crystalline direction on the substrate crystal even at the room
temperature. By utilizing the rf power dependence of material
properties, we propose a technique that can prepare a superlattice only by changing rf power during the process of
deposition. The growth mechanism of such materials is also
discussed.

Mat. Res. Soc. Symp. Proc. Vol. 164. ©1990 Materials Research Society

248

EXPERIMENT

The Si:H materials used in this study were fabricated by means of an rf sputtering technique in a pure hydrogen atmosphere. Two different sputtering systems were used in the present study. One is a planar magnetron sputtering system [11-14] in which the substrate holder can be cooled using liquid nitrogen during the deposition process. The schematic diagram of this sputtering system is shown in Fig. 1. A substrate holder made of copper was used in order to keep the thermal conductivity between the substrate and the liquid nitrogen high. The target is a single-crystal Si of 7.5 cm diameter and 0.4 cm thickness. A planar magnet is buried under the target to increase the deposition rate of the material. The maximum intensity for the horizontal component of the magnetic field at the target surface was 200 Gauss, and that for the vertical one was 350 Gauss. The distance between the substrate and the target is 4.0 cm. The base pressure in the sputtering chamber was exhausted to less than 2×10^{-6} Torr by use of a rotary vacuum pump and an ion vacuum pump.

The other sputtering system employs a conventional substrate holder, in which the substrate is maintained at room temperature. The schematic diagram of the latter system is shown in Fig. 2. A single-crystal Si target of 10.0 cm diameter and 0.8 cm thickness is used in this case. The distance between the substrate and the target is 4.5 cm. The base pressure of this sputtering chamber was exhausted to less than 1×10^{-6} Torr by a rotary vacuum pump and a diffusion vacuum pump.

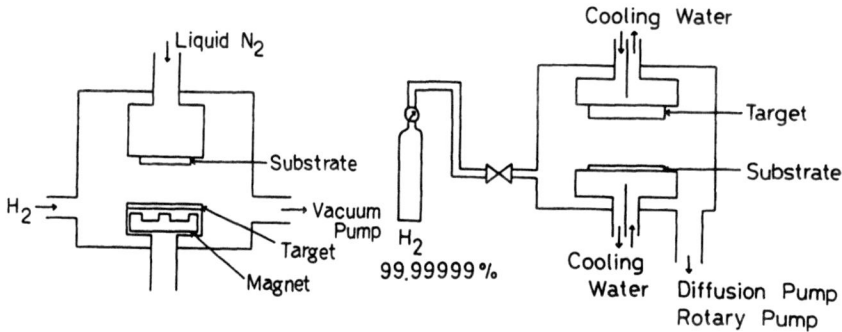

Fig. 1 Schematic diagram of the planar magnetron sputtering system.

Fig. 2 Schematic diagram of the second sputtering system.

After evacuation, pure H_2 gas was introduced into the chamber for both the cases. The purity of the H_2 gas was 99.99999 %. After 20 min. of the pre-sputtering process of the target, the Si:H material was deposited onto a single-crystal Si wafer and a quartz glass, which were set in the vacuum chamber at the same time. The rf power and hydrogen gas pressure were varied, respectively, from 10 to 300 W and from 0.2 to 20 Torr.

The materials on the quartz glass substrates were observed by an optical absorption (0.2-0.9 μm) and a Raman scattering (200-1000 cm^{-1}). The materials on the single-crystal Si substrates were observed utilizing infrared absorption (2.5-25 μm) and X-ray diffraction. In order to measure the electrical properties, aluminum was evaporated onto the surface of the materials on quartz glass substrates. The photoconductivity measurement was performed using a xenon arc lamp that provides an excitation light source. The visible photoluminescence was observed at 72 K and the room temperature, where an Ar ion laser (514.5 nm line) and an optical filter were used.

STRUCTURAL ANALYSIS

Raman scattering

Figure 3 shows the Raman spectrum for the specimen prepared at about 100 K with an rf power of 55 W. It shows a broad peak at about 480 cm^{-1}, indicating that the material is amorphous, or consists of very small Si microcrystals. Except for this peak, no apparent peaks were observed over the wave number range from 200

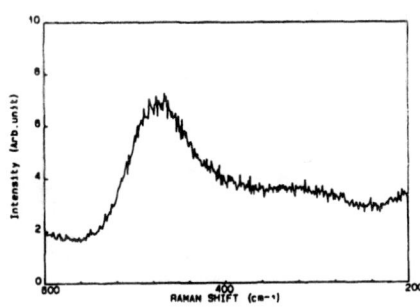

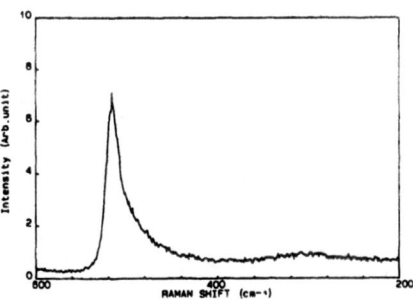

Fig. 3 Raman spectrum for the specimen prepared at about 100 K with an rf power of 55 W.

Fig. 4 Raman spectrum for the specimen prepared at the room temperature with an rf power of 300 W.

to 1000 cm^{-1}. The previously reported Raman spectra [11-14] are not correct, where the error had probably incurred due to laser annealed crystallization or some other effects.

Figure 4 shows the Raman spectrum for the specimen prepared at the room temperature with an rf power of 300 W. It shows a relatively sharp peak at 520 cm^{-1}, and also has a 480 cm^{-1} component, indicating that the material consists of large Si microcrystals and amorphous Si:H or small Si microcrystals.

X-ray diffraction

Figure 5 shows the X-ray diffraction pattern for the specimen prepared at about 100 K with an rf power of 55 W. The broad peak corresponds to Si (111) plane, which indicates the presence of amorphous or very small microcrystals. The center of the diffraction angle $2\theta=27.9 \pm 0.2°$ [(111) plane for CuKα] shifts towards the low-angle side with respect to that of single-crystal Si, indicating that the lattice constant of the material is about 2 % larger than that of the single-crystal.

Figure 6 shows the X-ray diffraction pattern for the specimen prepared at about 100 K with an rf power of 150 W. It shows an apparent diffraction peak at about $2\theta=28°$, indicating the presence of Si microcrystals.

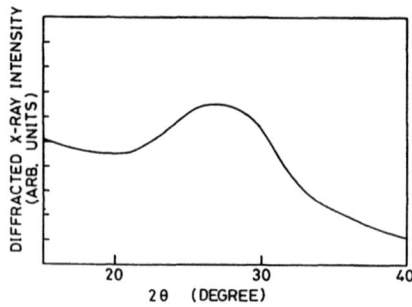

Fig. 5 X-ray diffraction pattern for the specimen prepared at about 100 K with an rf power of 55 W. (Ref. 11)

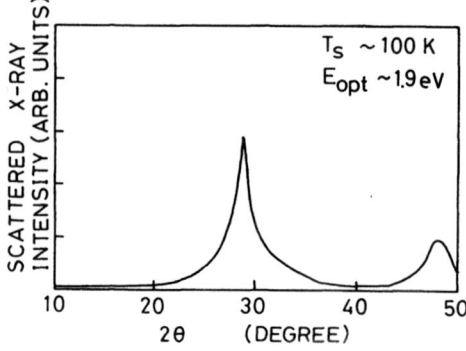

Fig. 6 X-ray diffraction pattern for the specimen prepared at about 100 K with an rf power of 150 W.

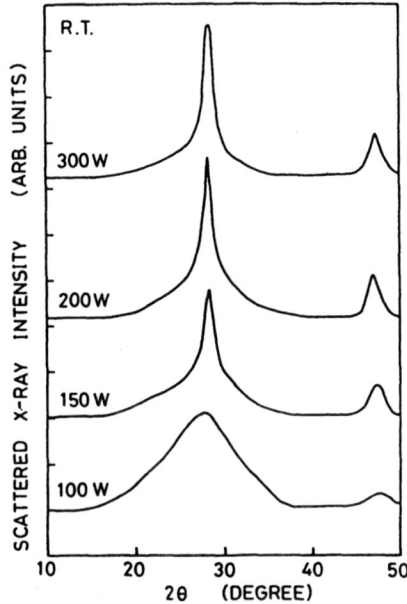

Fig. 7 X-ray diffraction
patterns for the
specimens prepared
at the room
temperature with
rf powers of 300,
200, 150, and 100 W.

Figure 7 shows X-ray diffraction patterns for specimens prepared at the room temperature with rf powers of 300, 200, 150, and 100 W. Both a sharp peak and a broad tail are observed in these cases. The sharp peak grows as the rf power during the deposition increases. A small peak corresponding to the Si (220) plane is also observed. The full-width at the half-maximum (FWHM) of the sharp peak does not seem to correlate with the rf power, whereas the FWHM of the broad tail seems to correlate with it. Concerning the optical gap, it increases with decreasing the rf power during the deposition.

Infrared absorption

Figure 8 shows the infrared absorption spectra for the specimens prepared at hydrogen gas pressures of 20, 7, and 2 Torr [15]. The substrate temperature and the rf power for those specimens were maintained at about 100 K and 55 W, respectively. The spectra show the absorption peaks at 640-660, 850-860, 890-910, and 2100-2140 cm^{-1}. Since the spectra of the materials have two peaks in the range from 850 to 910 cm^{-1}, $-(SiH_2)_n-$ $(n \geq 1)$ or $-SiH_3$ groups are considered to be contained in the materials. It has been reported that the wave numbers of rocking, wagging, bend-scissors, and stretching modes of $-(SiH_2)_n-$ groups are 630, 845, 890, and 2100 cm^{-1}, respectively [16-19]. The peaks at 628, 840, 902, and 2144 cm^{-1} are assigned to rocking, symmetric deformation, degenerate deformation, and stretching modes of $-SiH_3$ groups, respectively [16]. Therefore, in the case of spectrum (a), the peaks at 645, 850, 892, and 2100 cm^{-1} are responsible for rocking, wagging, bend-scissors, and stretching

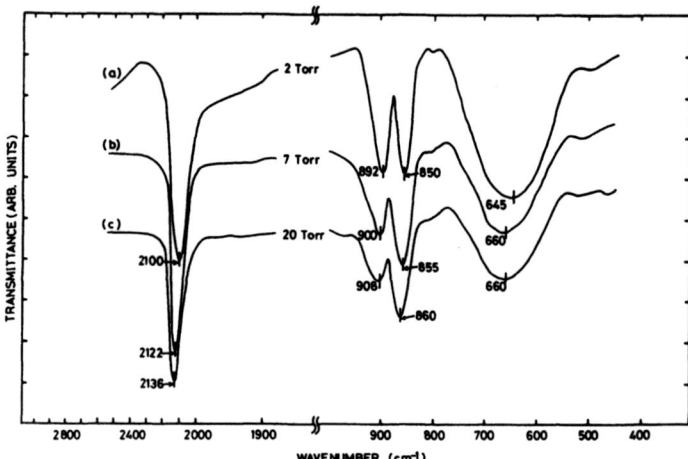

Fig. 8 Infrared absorption spectra for the specimens
prepared at hydrogen gas pressures of 20, 7,
and 2 Torr. (Ref. 15)

modes of $-(SiH_2)_n-$ ($n \geq 1$) groups, respectively. From the stretching absorption wave number (~ 2100 cm^{-1}), it is concluded that
$-(SiH_2)_n-$ groups are almost composed of $=SiH_2$ and $-(SiH_2)_2-$
groups [16-18]. In the case of spectrum (c), the peaks at 660,
860, 908, and 2136 cm^{-1} are responsible for rocking, symmetric
deformation, degenerate deformation, and stretching modes of
$-SiH_3$ groups, respectively. In the case of spectrum (b), both
modes of $-(SiH_2)_n-$ ($n \geq 1$) and $-SiH_3$ are considered to be contained
in the material. As shown in Fig. 8, $-(SiH_2)_n-$ ($n=1$ or 2) groups
in the material change to $-SiH_3$ groups as the hydrogen gas
pressure during the deposition increases.

Figure 9 shows the relation between the stretching
absorption wave numbers and ratios of $\alpha(\sim 850$ cm$^{-1})/\alpha(\sim 890$ cm$^{-1})$
for the Si:H materials. In Fig. 9, the data for the polysilane
alloys [17], which contain many $-(SiH_2)_n-$ groups, are also shown.
It has been reported that the stretching absorption wave number
increases slightly with the increase of the effective electronegativity [16-18], and the absorption strength for the peak at
~ 850 cm^{-1} shows a nonlinear behavior [18]. From the data in
Fig. 9, it is found that these values are systematically varied
with the variation from $\equiv SiH$, $=SiH_2$, $-(SiH_2)_n-$, ... to $-SiH_3$.

Figure 10 shows the relation between the stretching
absorption wave number (~ 2100 cm^{-1} peak) and the wagging or
symmetric deformation wave number (~ 850 cm^{-1} peak) for both the
present materials and the polysilane alloys. As shown in Fig. 10,
the wagging or symmetric deformation absorption wave numbers for
the present materials are slightly larger than those for the
polysilane alloys. This suggests that the bending force of Si-H
bonds in the present materials is larger than that in the alloys
containing long-chained polysilanes. Therefore, it is plausible
that the hydrogen atoms in the present materials are attached to
Si clusters, and do not form polysilanes.

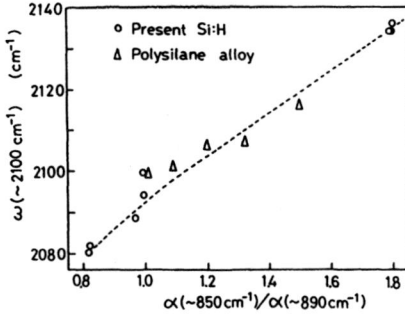

Fig. 9 Relation between stretching absorption wave numbers and ratios of α ($\sim$850 cm^{-1}) / α ($\sim$890 cm^{-1}).

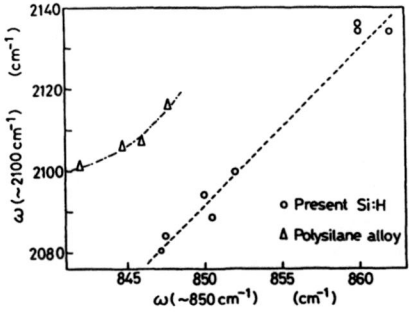

Fig. 10 Relation between stretching absorption wave number ($\sim$2100 cm^{-1} peak) and wagging or symmetric deformation one ($\sim$850 cm^{-1} peak).

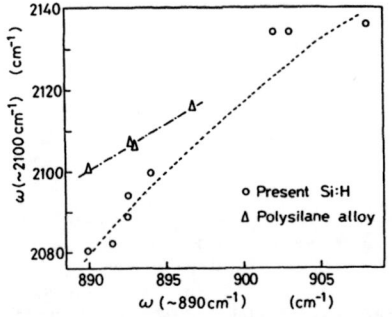

Fig. 11 Relation between stretching absorption wave number ($\sim$2100 cm^{-1} peak) and bend-scissors or degenerate deformation one ($\sim$890 cm^{-1} peak).

Figure 11 shows the similar relation between the stretching absorption wave number and the bend-scissors or the degenerate deformation wave number. In this case, the result also suggests an increase in the bending force in the present materials.

Figure 12 shows the infrared absorption spectra for the specimens prepared at the room temperature. As shown in Fig. 12, the Si-H stretching absorption wave number is about 2100 cm^{-1} (not $\sim$2000 cm^{-1}) for all the materials prepared with rf powers of 300, 200, and 150 W. This suggests that the hydrogen atoms are incorporated not in the form of $\equiv$SiH , but in the forms of $-(SiH_2)_n-$ $(n \geq 1)$ or $-SiH_3$. The absorption wave number is slightly increased as the rf power during the deposition increases.

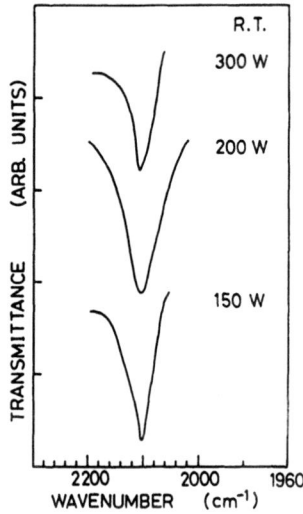

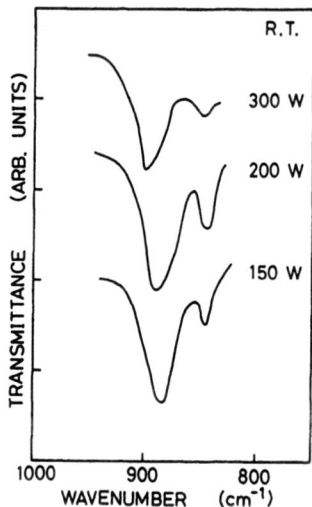

Fig. 12 Infrared absorption
spectra (stretching
region) for the
specimens prepared
at the room
temperature.

Fig. 13 Infrared absorption
spectra (bending
region) for the
specimens prepared
at the room
temperature.

Figure 13 shows the infrared absorption spectra in the bending region for the same specimens as in Fig. 12. There appear two absorption peaks in the range from about 850 to 910 cm^{-1}, and their wave numbers increase with increasing the rf power. From these results, it is concluded that the hydrogen atoms are incorporated in the form of $-(SiH_2)_n-$ ($n \geq 1$) for the materials prepared in a low rf power, and in the form of $-SiH_3$ for the materials prepared in a high rf power.

Structural model

From the structural analysis performed in the above, it is suggested that the material consists of amorphous or crystalline Si clusters or grains surrounded by H atoms, which are bonded in the dihydride or trihydride form. In fact, the investigation of the material prepared by the present method revealed the inclusion of polyhedral or sphere-like grains as also indicated in the transmission electron microscope photograph [12].

PHOTOLUMINESCENCE AND CONDUCTIVITY

Photoluminescence

The material prepared at about 100 K with an rf power of 55 W shows a visible (red) photoluminescence at room temperature,

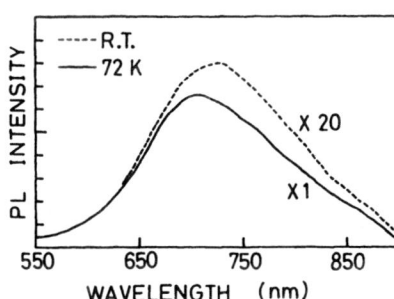

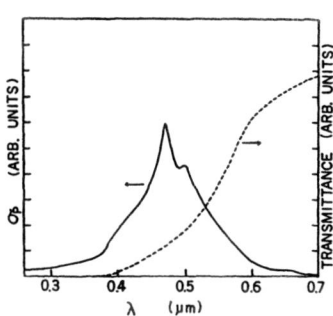

Fig. 14 Photoluminescence
spectra measured at
the room temperature
and 72 K.

Fig. 15 Photoconductivity
dependence on incident
photon wavelength for
the material prepared
at about 100 K. (Ref.13)

which arises from the widening of the optical gap. Figure 14
shows the photoluminescence spectra measured at the room
temperature and 72 K. There appears a broad peak near 700 nm for
both the cases, and its wavelength slightly decreases with a
decrease of the measurement temperature. The photoluminescence
intensity at 72 K is larger by about one order of magnitude than
that at the room temperature.

Conductivity

The dark-conductivity of the material showing the visible
photoluminescence is very small, and its magnitude is estimated
to be less than $\sim 10^{-11}$ S/cm. However, the material showed a
photoconductive property, although its value is very small.
Figure 15 shows the photoconductivity dependence on the incident
photon wavelength for the material prepared at about 100 K [13],
and its optical gap is about 2.3 eV. It shows a blue shift, and
has a peak at $\lambda < 0.5$ μm.

EFFECTS OF CRYSTALLINE SUBSTRATE ON MATERIAL PROPERTY

As shown in Figs. 4 and 7, the material prepared at room
temperature with a high rf power consists mostly of Si microcrystals, whose grain size is relatively large. This may cause
an epitaxial-like growth on the crystalline substrate.
Figure 16 (a) shows the X-ray diffraction pattern for the
material deposited onto the (511) cut single-crystal Si
substrate [20]. The Si:H material was obtained at the room
temperature with an rf power of 200 W. Figure 16 (b) is for the
substrate alone. The broad peak near $2\theta=60°$, which is probably
related to (311) diffraction, due to the substrate Si crystal.
However, the other peaks in Fig. 16 (a) have a different
character from those in Fig. 7. First, the scattered X-ray
intensity corresponding to (220) diffraction, which appears near
$2\theta=47°$, is relatively large compared with that in Fig. 7. Second,

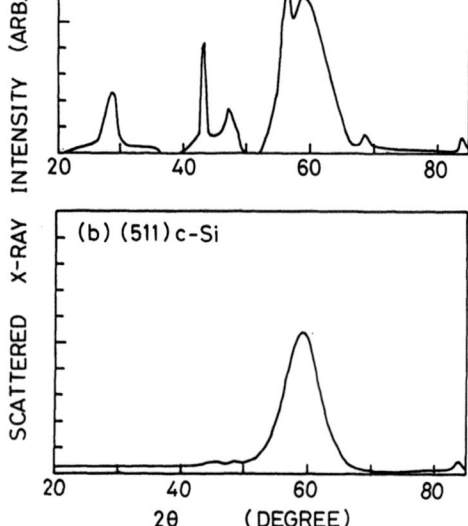

Fig. 16

(a) X-ray diffraction
pattern for the material
deposited onto a (511)
cut single-crystal Si
substrate.
(b) X-ray diffraction
pattern for the
substrate.

the sharp peak near 2θ=56° corresponding to (311) diffraction is
clearly acknowledged in Fig. 16 (a). Third, the peak corresponding to (400) diffraction also appears, although its intensity is
weak. The sharp peak near 2θ=43° is probably related to (211)
diffraction, which is restricted to the perfect crystal.

As shown in Fig. 16, the present Si:H film on a (511) Si
substrate is not consisted of a single-crystal structure, but
contained some defects. Nevertheless, it is clearly observed
that the crystalline direction of the microcrystals in the film
depends on the surface potential of the substrate crystal even
at the room temperature. By selecting the appropreate deposition
parameters, much better crystal may be obtained as to its
structure.

POWER-MODULATED MULTI-LAYERED STRUCTURE

The data in Figs. 3-7 indicate that the material structure
is very much affected by the rf power during the deposition. By
using these results, a power-modulated multi-layered structure
(superlattice) has been fabricated [20]. The layered structure
with 300 periods was prepared by changing the rf power during
the deposition. From the deposition rate of the two conditions
(rf powers of 300 and 110 W), the periodicity was estimated to
be 68 Å. As a result, an apparent diffraction peak was obtained
at about 2θ=1.3°, although its FWHM is as large as ~0.5°.

GROWTH MECHANISM

The present result showed that μc-Si:H can be prepared by selecting deposition parameters even at a low substrate temperature. On the other hand, it has been reported that μc-Si:H can be prepared by the plasma chemical vapor deposition method only with a substrate temperature higher than 400 K [9, 10]. Therefore, in the latter case, it is considered that μc-Si:H is formed on the growing surface, at which the SiH$_x$ species arrived there are reconstructed by obtaining thermal energy from the substrate. In the present case, μc-Si:H can be formed at a low temperature. Therefore, it is plausible that the microcrystals are formed on the growing surface (or in the gas phase) by obtaining the photon energy from the hydrogen plasma or by obtaining the energy from the hydrogen radicals[21].

SUMMARY AND CONCLUSIONS

Binary Si:H materials have been prepared by the hydrogen plasma sputtering technique. The properties of the materials are very much affected by the substrate temperature and the rf power during the deposition. The material prepared at a low substrate temperature with a low rf power has a wide optical gap of up to 2.4 eV, and shows a visible photoluminescence at the room temperature. On the contrary, the material prepared at a high substrate temperature with a high rf power has a narrow optical gap, and contains many Si microcrystals, whose grain size is relatively large. The crystalline direction of the microcrystals in the Si:H material prepared at the room temperature with an rf power of 200 W depends on the substrate crystal. An rf power-modulated multi-layered structure has also been proposed, and an clear diffraction peak corresponding to the period of the fabricated multi-layered structure has been observed in the low-angle X-ray scattering measurement.

ACKNOWLEDGEMENT

The authors would like to express their thanks to Professor Kiyoshi Takahashi and Professor Makoto Konagai of the Tokyo Institute of Technology, and the members of their laboratories, for their valuable discussion, and Raman and photoluminescence measurements. The authors would also like to express their thanks to Dr. Hiroyuki Fujishiro of the Kyushu Institute of Technology for his valuable discussion.

REFERENCES

1. D.E.Carlson, IEEE Trans. Electron Devices, ED-24, 449 (1977).

2. P.G.LeComber, W.E.Spear, and A.Ghaith, Electron. Lett. 15, 179 (1979).

3. S.Kishida, Y.Nara, O.Kobayashi, and M.Matsumura, Appl. Phys. Lett. 41, 1154 (1982).

4. S.Kaneko, M.Sakamoto, F.Okumura, T.Itano, H.Kataniwa, Y.Kajiwara, M.Kanamori, M.Yasumoto, T.Saito, and T.Ohkubo, Tech. Dig. Int. Electron Devices Meet., p.328, 1982.

5. S.Oda, S.Terazono, and I.Shimizu, Sol. Energy Mater. $\underline{8}$, 123 (1982).

6. S.Ishida, Y.Imamura, Y.Takasaki, C.Kusano, T.Hirai, and S.Nobutoki, Jpn. J. Appl. Phys. $\underline{22-1}$, 461 (1983).

7. M.A.Bosch, Appl. Phys. Lett. $\underline{40}$, 8 (1982).

8. H.Munekata and H.Kukimoto, Appl. Phys. Lett. $\underline{42}$, 432 (1983).

9. T.Hamasaki, H.Kurata, M.Hirose, and Y.Osaka, Appl. Phys. Lett. $\underline{37}$, 1084 (1980).

10. A.Matsuda, Proc. Int. Conf. Amorphous & Liquid Semicond. edited by K.Tanaka and T.Shimizu, p.767, 1983.

11. S.Furukawa, M.Komori, and T.Miyasato, Int. Conf. Phys. Semicond. Warsaw, 1988.

12. S.Furukawa and T.Miyasato, Phys. Rev. B $\underline{38}$, 5726 (1988).

13. S.Furukawa and T.Miyasato, Jpn. J. Appl. Phys. $\underline{27}$, L213 (1988).

14. S.Furukawa and T.Miyasato, Superlattices and Microstructures, $\underline{5}$, 317 (1989).

15. M.Komori, S.Furukawa, and T.Miyasato, Phys. Lett. A $\underline{135}$, 401 (1989).

16. G.Lucovsky, R.J.Nemanich, and J.C.Knight, Phys. Rev. B $\underline{19}$, 2064 (1979).

17. S.Furukawa and N.Matsumoto, Phys. Rev. B $\underline{31}$, 2114 (1985).

18. S.Furukawa, N.Matsumoto, T.Toriyama, and N.Yabumoto, J.Appl. Phys. $\underline{58}$, 4658 (1985).

19. S.Furukawa, M.Seki, and S.Maeyama, Phys. Rev. Lett. $\underline{57}$, 2029 (1986).

20. S.Furukawa and H.Fujishiro (unpublished).

21. H.Fujishiro, S.Furukawa, and T.Yamazaki (to be published).

CRITICAL REVIEW OF RAMAN SPECTROSCOPY AS A DIAGNOSTIC TOOL FOR SEMICONDUCTOR MICROCRYSTALS

P.M. Fauchet and I.H. Campbell
Department of Electrical Engineering
Princeton University
Princeton, NJ 08544

ABSTRACT

Raman scattering is becoming a widely used tool for the characterization of semiconductor microcrystals due to its sensitivity to crystal sizes below a few hundred angstroms. Through detailed analysis of the first order Raman spectrum it is possible to determine the size and shape of microcrystalline grains. First order spectra must be examined with care however, since they are sensitive to other factors including: stress/strain, surface vibrations, mixed amorphous/microcrystalline phases and intragrain defects. Second order Raman spectra are more sensitive to microcrystalline effects than first order spectra. They offer the potential to measure crystal sizes greater than a few hundred angstroms but much work remains to be done to quantify the size dependence of the second order spectra.

INTRODUCTION

The precise characterization of micro- and nano-crystalline semiconductors is difficult. This results from the small size of the grains, which is not easily measurable with conventional techniques. In addition, the nature and influence of grain boundaries and the uniformity of the grain sizes and properties are not easily measured. Raman scattering has been among the most useful non-destructive diagnostic tools for microcrystalline semiconductors[1]. In particular, the first order optical phonon line is used to measure grain size in microcrystalline silicon and related alloys, as can be seen from other papers in this volume[2].

In this paper, we first examine the information that can be obtained from the first order Raman spectrum of semiconductor nanograins. Emphasis is on size and shape effects, and the possible influence of surface phonons and electromagnetic resonance is also briefly discussed, in the spirit of our recent review[1]. Then, we turn our attention to the second order Raman spectrum, with special attention to the optical phonon contributions at high wavenumber.

FIRST ORDER OPTICAL PHONON RAMAN SPECTRUM

Around 1980, it was recognized that the Raman spectrum of microcrystals could differ from that of single crystals[3,4]. The difference in boundary conditions may affect the spectrum in different ways. First, consider an isolated grain. Phonons inside the grain can no longer be described by plane waves. The grain boundaries may be structurally or compositionally distinct from the bulk of the grain and thus have other vibrational properties. Therefore, one can expect a modification of the phonon spectrum of the single crystal and the presence of new phonon modes related to the interface. If the number of atoms in the grain becomes extremely small, it might be more appropriate to talk of a cluster. Clusters may have atomic arrange-

Mat. Res. Soc. Symp. Proc. Vol. 164. ©**1990 Materials Research Society**

ments which are not those of a crystal of the same material[5] and thus have new vibrational frequencies. Due to the large difference between the dielectric functions of the grain and of vacuum, special attention must also be paid to the electromagnetic field strength inside and at the surface of the grain. Second, consider a film made of many grains, either directly in contact with each other or embedded in a matrix. This matrix is typically the same material, in the amorphous phase. The boundary conditions are clearly different from those for the isolated grain and can usually not be described analytically. It seems intuitively clear however that the amplitude of the surface vibration are reduced. It is further assumed that an optical phonon cannot cross from one grain to another. The electromagnetic field strength can be assumed to be uniform. Experiments have been performed on many systems which fall in all these categories. Here, we discuss selected results that have a direct impact on the characterization of nanocrystalline semiconductors.

Consider a spherical grain, typically 10 nm in diameter. We neglect surface vibrations. The best description for the vibrations within the grain is a wave packet whose spatial dimensions are comparable to the crystallite size[4]. The restriction on the position of the wave packet leads to an uncertainty in momentum or wave vector. The one phonon Raman spectrum, which usually probes the center of the Brillouin zone, now probes a finite range of wave vectors. Since the dispersion curves for optical phonons are not entirely flat, the Raman line is broadened to include those new frequencies. The choice of the most appropriate function for the wave packet has

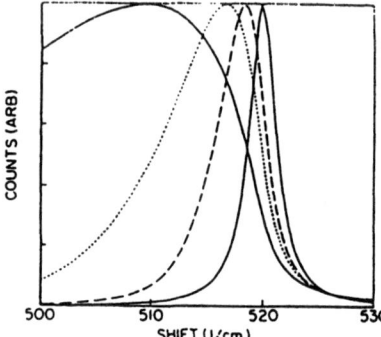

COUNTS (ARB)

500   510   520   530
SHIFT (1/cm)

Fig. 1. Calculated first order Raman spectra as a function of decreasing size for spherical Si microcrystallites (L= ∞, 100 Å, 60 Å and 30 Å). The peak shift, asymmetry and line width increase as the size decreases.

been discussed previously[6]. In Figure 1, we show the calculated first order optical phonon line shape for silicon as a function of decreasing size and in Figure 2, we show the relation between size L, line shift $\Delta\omega$, line width Γ, and asymmetry Γ_a/Γ_b for silicon at room temperature. In this idealized situation, when the size exceeds 20 nm, it is impossible to distinguish microcrystals from single crystals. Similar calculations have been performed for germanium[7].

The repeated success of such a simple model should however not lead us to use it without care. Several factors must be discussed : size inhomogeneity, shape of the grains, stress, defects inside the grain, mixed amorphous/crystalline phases, and surface vibrations.

1) Variations in the grain size lead to additional line broadening and decreased asymmetry. The Raman line becomes inhomogeneous and its exact shape now depends on the exact size distribution. If the three parameters that describe the line shape all fit on the curves of Figure 2, then the grain size is uniform.

2) Although to our knowledge, there has been no experimental verification yet, we have predicted that the line shape of grains containing the same number of atoms but having non-spherical shapes will be measurably different from that of spheres[7]. The difference is almost exclusively in the low wavenumber tail of the line, so that the line width and asymmetry are affected whereas the line shift remains the same. The possibility of non-spherical grains must always be kept in mind.

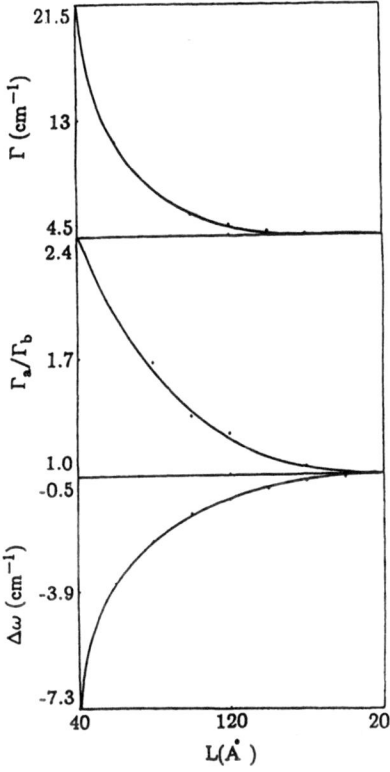

Fig. 2. Calculated shift $\Delta\omega$, asymmetry Γ_a/Γ_b, and line width Γ, in Si as a function of the diameter L.

3) Stress is another factor of importance for thin film applications[8,9]. For example, the line shifts with applied stress (see for example the spectrum of silicon-on-insulator). In addition, if the stress is non-uniform, the line will broaden. The magnitude of the line distortion depends on the magnitude of the stress. For silicon, germanium, gallium arsenide and other common semiconductors, a line shift by 1 wavenumber corresponds to a few times 10^9 dynes/cm^2 of stress or a few times 0.1 % of strain.

4) Defects within the grain may also modify the Raman spectrum. It has been reported that for defective grains, the Raman size could be less than the size determined from transmission electron microscopy (TEM)[9]. It is conceivable that some crystallographic imperfections effectively decouple vibrations from one region of the grain to another. The electrical conductivity has also been correlated to the Raman size, not to the TEM size[9].

5) Films containing nanograins often contain a volume fraction that is amorphous or at least strongly disordered[10]. In polycrystalline silicon films, the intergrain material often appears as a weak shoulder close to 495 wavenumbers[1,11]. When the grain size decreases well below 10 nm, the volume fraction of the intergrain material increases and the low wavenumber tail of the Raman line may overlap the 495 cm^{-1} line. The grain size can only be determined from a careful line shape analysis.

6) The atoms on the surface of the grain feel a different environment and that may lead to new vibrational frequencies. This effect has been observed only with isolated grains, where the surface vibrations are not damped and where under some conditions the strength of the electromagnetic field is large[12]. For example, isolated grains as large as a few tens of nanometers have a Raman line that is a combination of the crystalline and amorphous spectra and grains smaller than 10 nm display only the amorphous-like component, due to the large amplitude of the free surface vibrations and the concentration of the electromagnetic field in that shell region of intermediate composition. However, in films where surface vibrations are likely to be damped and the dielectric function of the film is rather uniform, this effect is not seen.

SECOND ORDER OPTICAL PHONON RAMAN SPECTRUM

In contrast to the extensively reported and interpreted line shape changes in the one phonon Raman spectra of microcrystals, two phonon spectra have not been widely discussed. Two phonon spectra are more difficult to quantify than first order spectra; they involve contributions from phonons throughout the Brillouin zone and peaks in the spectrum result from critical points in the phonon density of states modified by appropriate momentum and symmetry selection rules. The dominant contribution to the second order spectrum is from phonons near the edge of the Brillouin zone where the phonon dispersion relation is relatively flat. Early work with graphite implied that the primary effect of finite crystal size was to relax selection rules[3]. As the crystal size decreased the peak widths increased and new peaks appeared in the spectrum corresponding to relaxation of the momentum and symmetry selection rules, respectively.

Although first and second order spectra are both altered by finite crystal size effects the interpretation of the two phonon spectra is less clear. As an example, we consider the second order Raman spectrum of microcrystalline Si in the spectral region from 550 cm^{-1} to 1100 cm^{-1}. Figure 3 shows the evolution of the second order Raman spectrum of microcrystalline Si with decreasing crystal size and Figure 4 contains the corresponding first order spectra. In materials with the diamond structure the second order optic phonon spectrum is dominated by overtones e.g. 2TO(X) and 2LO(Γ) while many combinations e.g. TO(L)+LO(L) are symmetry forbidden[13,14]. The main peak from 900 cm^{-1} to 1000 cm^{-1} in the second order spectrum of crystalline Si is due to several overtones: 2TO(X), 2O(Q), 2O(S$_I$), 2TO(W) and 2TO(L)[13,14]. The dominant peaks in both the second order spectra and the first order spectra exhibit a down shift and broadening as the crystal size decreases. The effect is more pronounced in the two phonon spectra because the changes imposed by microcrystallinity affect each phonon which leads to a larger absolute shift and broadening. Of particular interest is the comparison between Figs. 3c,d and 4c,d. In this case the first order spectra of the microcrystals are virtually identical but the second order spectra are radically different.

We cannot apply the phonon confinement model to the second order spectra. Since the phonon energy is at a minimum at the zone edge, relaxing momentum conservation should shift the two phonon peaks to higher energy and not to lower energy as observed. This contradiction implies that the short wavelength phonons involved in the two phonon spectra do not experience significant confinement. Phonon confinement may be less significant because the amorphous matrix that surrounds the microcrystal has many vibrational modes that are at the same frequency as the zone edge optical phonons. This is in contrast to the first order optic phonons which have energies 4-5 meV above the dominant vibrations of the amorphous material.

In addition to the down shift and broadening of the second order spectrum there is relaxation of the symmetry selection rules that forbid general combinations of zone edge phonons. This is clearly seen in the spectra of crystalline and microcrystalline Si: the crystalline spectrum contains weak scattering from TO+TA combinations at $\sim$ 630 cm^{-1} but as the crystal size decreases the magnitude of the scattering in this spectral region generally increases.

Strain and electromagnetic resonance effects that are of concern in first order spectra may be even more important in second order spectra. The effect of strain is essentially the same as in first order scattering but the magnitude of the peak shift is twice as large because each phonon is shifted in energy[15]. Second order scattering in bulk crystals is sensitive to the exciting wavelength[16]; the relative weights of the different phonon overtones that make up the spectrum can be modified significantly. This sensitivity to strain and resonance effects implies that the changing electronic and atomic structure associated with microcrystalline material may be responsible

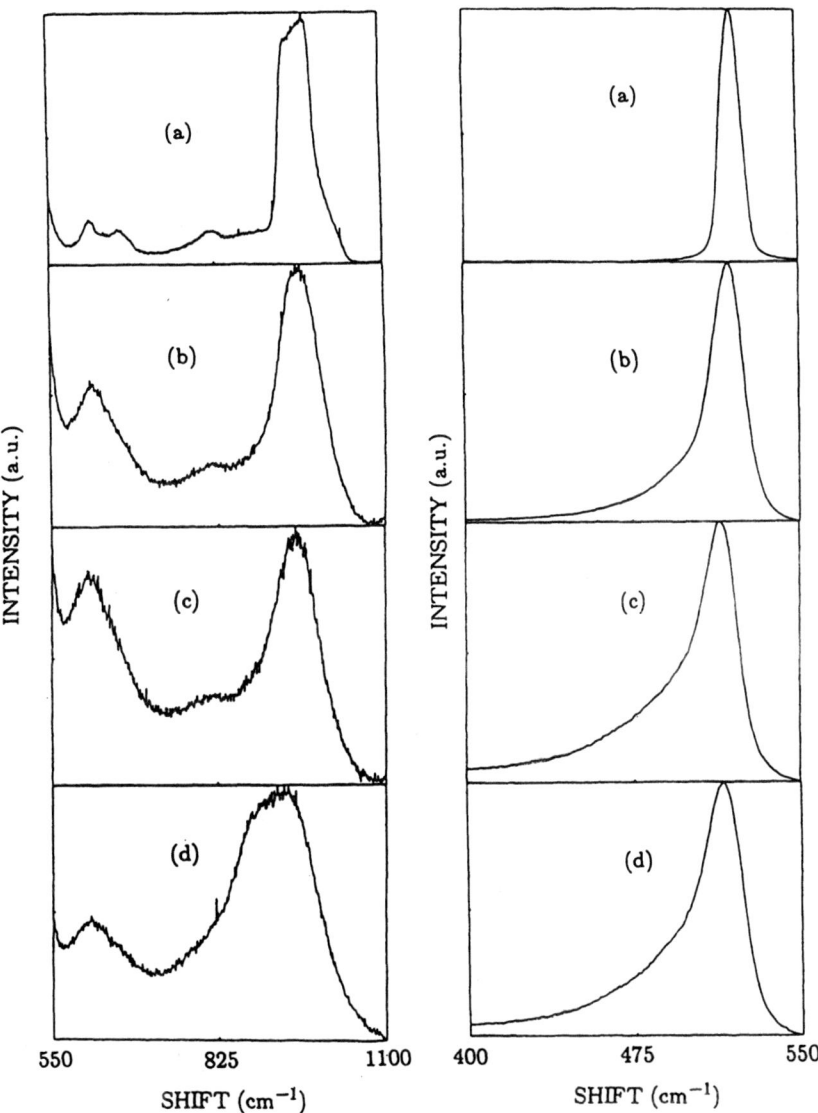

Fig. 3. Second order Raman spectra of silicon for three crystal sizes (a) Si wafer (b) ∼ 100 Å (c) and (d) ∼ 50 Å. As the crystal size decreases the dominant second order peak at ∼ 970 cm^{-1} broadens and down shifts and there is also an increase in symmetry forbidden combination scattering at ∼ 630 cm^{-1}.

Fig. 4. First order Raman spectra corresponding to the second order spectra shown in Fig. 3. Spectra (c) and (d) are virtually identical in contrast to their second order spectra.

for at least some of our observations. At present, we do not have a model that explains our results. It is clear however that the second order spectra are more dramatically altered by microcrystallinity, including grain sizes much larger than 10 nm.

CONCLUSION

Raman scattering is a very useful diagnostic tool for microcrystalline semiconductors. The first order Raman spectrum is widely used to measure grain size. We have pointed out several factors which, if overlooked, may lead to erroneous conclusions. The second order Raman spectrum is more sensitive to microcrystallinity than the first order spectrum. We have presented our recent results that demonstrate the point. At present, a self-consistent model that describes the dramatic changes we have observed is not available.

ACKNOWLEDGEMENTS

This work was supported in part by Coherent Inc., Newport Research Corporation and the National Science Foundation through the Presidential Young Investigator Program (ECS 86-57263). P.M.F is an Alfred P. Sloan Research Fellow.

REFERENCES

1. P.M. Fauchet and I.H. Campbell, Crit. Rev. Solid State Mater. Science **14**, 579 (1988).

2. Y. Okada et al.; R.J. Nemanich et al.; Y. Sasaki et al., this volume.

3. R.J. Nemanich and S.A. Solin, Phys. Rev. **20**, 392 (1979).

4. H. Richter, Z.P. Wang and L. Ley, Solid State Commun., **39**, 625 (1981).

5. C.J. Sandroff and L.A. Farrow, Chem. Phys. Lett., **130**, 458 (1986).

6. I.H. Campbell and P.M. Fauchet, Solid State Commun., **58**, 739 (1986).

7. I.H. Campbell and P.M. Fauchet, Proc. 18th Int. Conf. Physics Semiconductors, O. Engstrom ed., World Scientific, Singapore, 1987, 1357.

8. I.H. Campbell, P.M. Fauchet and F. Adar, Mat. Res. Soc. Symp. Proc. **53**, 311 (1986).

9. J. Gonzalez-Hernandez, G.H. Azarbayajani, R. Tsu and F.H. Pollak, Appl. Phys. Lett. **47**, 1350 (1985).

10. J. Gonzalez-Hernandez and R. Tsu, Appl. Phys. Lett., **42**, 90 (1983).

11. P.M. Fauchet, I.H. Campbell and F. Adar, Appl. Phys. Lett. **47**, 479 (1985).

12. S. Hayashi and K. Yamamoto, Superlattices and Microstructures **2**, 581 (1986).

13. P.A. Temple and C.E. Hathaway, Phys. Rev. **B7**, 3685 (1973).

14. J.L. Birman, Phys. Rev. **B131**, 1489 (1963).

15. B.A. Weinstein and G.J. Piermarini, Phys. Rev. **B12**, 1172 (1975).

16. P.B. Klein, H. Masui, J.-J. Song and R.K. Chang, Solid State Commun., **14**, 1163 (1974).

RAMAN SCATTERING FROM MICROCRYSTALLINE FILMS: CONSIDERATIONS OF COMPOSITE STRUCTURES WITH DIFFERENT OPTICAL ABSORPTION PROPERTIES

R.J. Nemanich[*], E.C. Buehler[*], Y.M. LeGrice[*], R.E. Shroder[*], G.N. Parsons[*], C. Wang[*], G. Lucovsky[*], and J.B. Boyce[**]

[*] Department of Physics, and Department of Materials Science and Engineering
North Carolina State University, Raleigh, NC 27695-8202 USA
[**] Xerox Palo Alto Research Center, Palo Alto, CA 94304 USA

ABSTRACT

Raman scattering measurements are used to characterize the components of microcrystalline Si and carbon films. A model is described which addresses the properties of Raman scattering from composites of materials of different optical absorption. The analysis shows that the observed spectra are dependent on both the percentage of the components and on the domain size of the more highly absorbing domains. Carbon films produced by different enhanced CVD techniques show both sp^2 (graphite) and sp^3 (diamond) regions. Silicon films prepared by excimer laser exposure of hydrogenated a-Si and by magnetron sputtering showed features due to both microcrystalline and amorphous regions. The experimental results reflect the length scales of the domains and vibrational excitations.

INTRODUCTION

CVD and other thin film deposition techniques often lead to the formation of metastable structures. In some cases the films are composites of different structures which are both stable and metastable. Characterization methods must then be sensitive to both materials, and also account for variations in crystalline domain size. Two examples are diamond films and microcrystalline Si films. This study describes the application of Raman spectroscopy to characterize the films.

Diamond films have been produced by various enhanced CVD techniques, and they generally show diamond regions and regions with graphite-like sp^2 bonding. Microcrystalline Si films are often produced by variations of the conditions for the deposition of hydrogenated amorphous Si (a-Si:H). The films exhibit a structure which is a combination of amorphous Si and crystalline Si regions. Raman spectroscopy is particularly useful for identifying diamond or graphite structures and whether a Si film exhibits amorphous or crystalline structure. A major issue in characterizing the films by optical techniques is that the optical absorption of the regions is quite different, and this will lead to differing sensitivities for the structures.

The Raman spectra of diamond and graphite structures have recently been summarized [1]. The feature due to diamond occurs at $1332 cm^{-1}$ and features due to the sp^2 structures exhibit peaks in the range of 1350 to 1620 cm^{-1}. The Raman spectrum of crystalline Si is dominated by a sharp feature at 520 cm^{-1} while the spectrum of a-Si:H displays features which resemble the broadened density of vibrational states of Si. There have been several detailed analysis of the evolution of the Raman spectra as a function of crystalline size[2-4]. For the particular case of Si, as the crystalline domain size decreases, the Raman peak broadens and shifts to lower frequency[3,4]. Similar effects have been proposed for microcrystalline diamond regions. The Raman spectra of graphitic structures also show distinct variations with crystal size. The goal of the study is to determine quantitative methods of characterizing the amount of diamond in the carbon films and the amorphous and microcrystalline components in the Si films.

RESULTS

A. Diamond

Several series of carbon films with diamond regions prepared by different deposition techniques have been examined by Raman spectroscopy. Two representative examples are shown in Fig. 1. The spectra show a sharp feature at 1332 cm⁻¹ which is due to diamond in addition to broader spectral features with peaks which range from ~1355 to ~1620cm⁻¹. These higher energy features are due sp² bonded or graphite-like structures. The sp² regions while complicated can be assigned to two different structures, disordered and microcrystalline. The disordered regions exhibit a broad peak centered at ~1490cm⁻¹ while the microcrystalline regions exhibit two features centered at ~1355 and ~1600 cm⁻¹.

To examine the aspects of Raman characterization of transparent diamond and absorbing graphite, composite samples were prepared. The samples were made from powders of diamond and graphite. The Raman spectra of several of the samples are displayed in Fig. 2. Each spectrum shows a feature at 1332 cm⁻¹ due to diamond and 1580 cm⁻¹ due to graphite. It has been reported that the Raman cross-section of graphite is 50 times stronger than that of diamond. Thus it is somewhat unexpected that the feature due to diamond is as intense as the feature due to graphite for the 1% diamond sample. This is due to the fact that the optical absorption strongly attenuates the light in the graphite regions, and the whole sample volume is not uniformly probed.

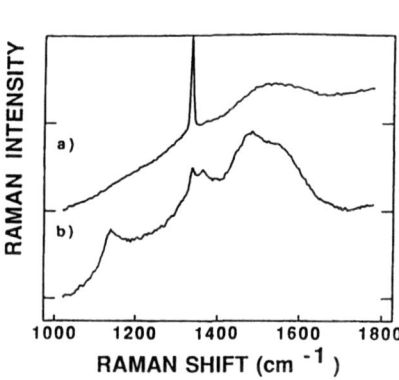
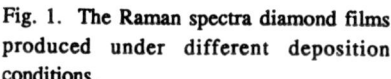
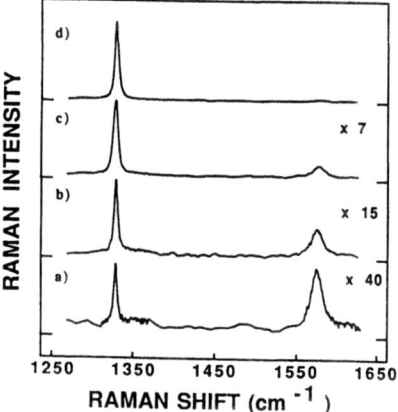

Fig. 1. The Raman spectra diamond films produced under different deposition conditions.

Fig. 2. The Raman spectra of composites of diamond-graphite powders. The relative concentrations of diamond in the samples are: (a) 1.3%, (b) 6.6%, (c) 21.5%, (d) 50.0%. The features due to diamond (D) and graphite (G) are identified.

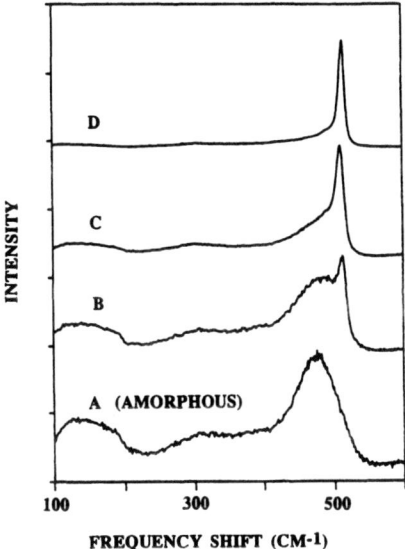

Fig. 3. The Raman spectra of excimer laser crystallized films (B-D) compared to that of amorphous Si (A).

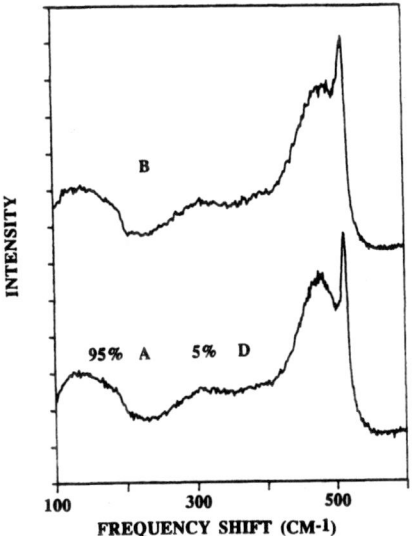

Fig. 4. The computer reconstructed Raman spectra of film B from Fig. 1.

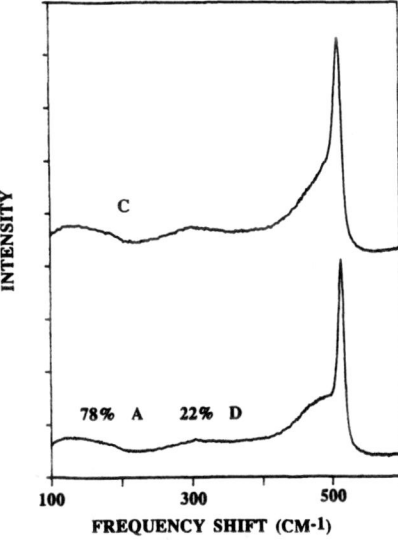

Fig. 5. The computer reconstructed Raman spectra of film C from Fig. 1.

B Silicon

Series of microcrystalline Si samples were prepared by excimer laser exposure of a-Si:H and by magnetron sputtering with high H concentrations. The Raman spectra of the series of films prepared by excimer laser exposure are shown in Fig. 3. The Raman spectra were excited with ~150 mW of 514.5nm radiation from an Ar ion laser, and the scattered light was dispersed with a triple grating monochromator. Essentially similar spectra were obtained from the magnetron sputtered films. The spectrum A is representative of the amorphous Si network vibrations while D represents predominantly crystalline regions with small domain sizes. Spectra B and C display spectral components which are attributed to both crystalline and amorphous regions in the sample. The goal of this study is to define the limits of using Raman spectroscopy to determine the relative amounts of these components.

To model the crystalline and amorphous components of the samples, the spectra A and D were added together to duplicate the results shown in spectra B and C. The results are shown in Fig. 4 and 5. The general aspects of the spectra are well described by this procedure. There is a small discrepancy at ~500cm^{-1} which could not be fit in either spectra. This aspect is discussed later. The results indicate that spectrum B can be described by 0.95 of A and 0.05 of D, and spectrum C was described by 0.78 of A and 0.22 of D. The following discussion describes the limitation of using this analysis to represent the relative concentrations of the amorphous and crystalline components.

THEORY

The optical absorption differences of the domains in the samples will affect the observed Raman results[5]. The ratio of the Raman intensities of the transparent and absorbing regions can be given by

$$\frac{I_a}{I_t} = \frac{A_a\, N_a\, V_a}{A_t\, N_t\, V_t} \qquad (1)$$

where I, A, N and V are the Raman intensity, Raman cross-section, atomic density, and illuminated volume respectively, and the subscripts a and t refer to the absorbing and transparent components. The illuminated volume will depend on the relative fraction of the absorbing, P_a, and transparent regions, $(1-P_a)$, and on the absorption constant, α, and size of the domains of the absorbing regions, l. Eq. (1) then becomes

$$\frac{I_a}{I_t} = \frac{A_a\, N_a\, P_a}{A_t\, N_t\, (1-P_a)} \left[\frac{1}{\alpha_a l}\right] \qquad (2)$$

This equation is valid for $\alpha_a^{-1} < l$. This model neglects effects due to feild enhancement and different shapes of the absorbing regions. We can now use this equation to model the results of the graphite and diamond composites. The results are shown in Fig. 6. The best fit to the data was obtained for a graphite domain of 42μm, and SEM showed crystallites ranging from 40 to 60μm.

For the case of microcrystalline Si we can assume that α^{-1} of the crystalline regions is larger than the domain size, and it follows that the crystalline regions are fully illuminated. In contrast, the α of the amorphous regions is ~10 times greater than that of crystalline Si and cannot be neglected. The dependence of the observed Raman intensities for several different domain sizes are shown in Fig. 7. For the calculation the ratio of the Raman cross-sections was determined from the peak intensities of the materials, α for a-Si of 2×10^5 cm^{-1}, and the densities were assumed to be the same. From the figure it is clear that the observed Raman ratio cannot be directly related to the amorphous/microcrystalline ratio unless the domain size is known. Below ~500Å domain size, the model will break down, and in the limit of very small domains, the term, $1/\alpha_a l$ will go to 1. (In that limit, the straightforward analysis will apply.)

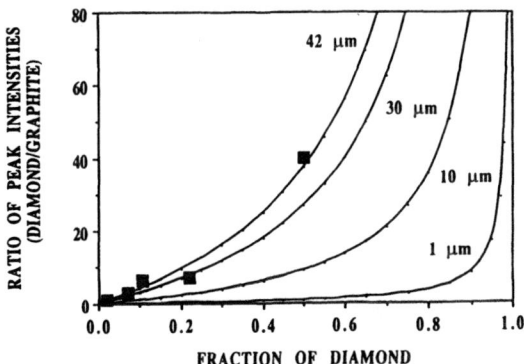

Fig. 6. The ratio of the peak intensities of the diamond-graphite composite samples. The solid lines are derived from Eq. 2 for different domain sizes.

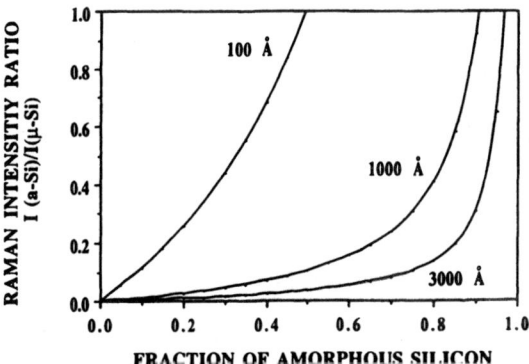

Fig. 7. The theoretical ratio of the Raman signal from the amorphous and crystalline regions of microcrystalline Si vs. the fraction of amorphous region. The calculations are for different domain sizes.

DISCUSSION

The above theory displays the importance of different optical absorption on the characterization of the films. A second aspect of the analysis that must be considered is that the spectral shape and scattering efficiency of the different regions can also vary as a function of the crystalline domain size.

This becomes apparent when the analysis is applied to determine the amount of diamond in the films shown in Fig. 1. TEM measurements indicate that there is less than 20% sp^2 regions in the films. In applying the analysis above even if we assume that the graphitic regions are much smaller than the α^{-1}, then the Raman data would indicate at most 20% diamond. Thus the Raman signal from the sp^2 bonded regions seems enhanced. Two posibilities to account for this are (1) the disorder allows scattering from more modes or (2) the Raman scattering efficiency is enhanced because of localization of the π electrons.

Spectral changes have also been observed for Si microcrystals [3,4]. It has been demonstrated that the linewidth and peak frequency changes as a function of the crystalline domain size for Si. These changes are most important in the range of 30 to 500Å domain sizes. It has been estimated that vibrational modes in amorphous materials are localized to regions ~50 Å, thus the Raman spectra of the amorphous regions should not change unless the domain size approaches this limit.

The most evident spectral change is the broadening of the 520 cm^{-1} Raman feature for microcrystalline domains. Inspection of the spectra shown in Fig. 3 indicates that the crystalline Si feature has a different linewidth for the different samples. The sample with the smallest amorphous component exhibits the narrowest linewidth. This aspect can account to large part for the deviation in the reconstructed spectra shown in Fig. 4 and 5.

CONCLUDING REMARKS

The results described here show that the Raman spectra of diamond and microcrystalline Si films can be described as a combination of the spectra of the components. Because of the large difference in the optical absorption of the crystalline and amorphous domains, the relative fraction of the two regions cannot be deduced unless the domain size has been determined. In addition, it has been established that the lineshape of the spectra of regions also changes, and this leads to deviations in the fitting procedure. In principle, these changes can also be accounted for by fitting the lineshape of the crystalline component. One aspect that is yet to be determined is the dependence of the absolute intensity vs. crystalline domain size.

ACKNOWLEDGEMENTS

We acknowledge Shannon Wells for her assistance on the data analysis. This work was partially supported by SERI under contract XM-9-18141-2, and SDIO/IST through the ONR.

REFERENCES

1. R.J. Nemanich, J.T. Glass, G. Lucovsky and R.E. Shroder, J. Vac. Sci. Tech. A6, 1783 (1988).
2. R.J. Nemanich, S.A. Solin, and R.M. Martin, Phys. Rev. B23 (1981) 6348.
3. Z. Iqbal and S. Veprek, J. Phys. C15 (1982) 377.
4. P.M. Fauchet and I.H. Campbell, CRC Critical Reviews in Solid State and Materials Sciences 14 (1988) S79.
5. R.E. Shroder, R.J. Nemanich, and J.T. Glass, Phys. Rev. B (in press).

PHONON STATES IN SiC SMALL PARTICLES

Y. SASAKI[*], C. HORIE[*], AND Y. NISHINA[+]
* Department of Basic Science, Ishinomaki Senshu University, Ishinomaki 986, JAPAN
+ Institute for Materials Research, Tohoku University, Sendai 980, JAPAN

ABSTRACT

Size dependence of optical phonon frequencies and that of phonon dampings of SiC small particles have been studied by analysing their Raman data. The particle size ranges from 30 nm to 1000 nm. Decrease in the TO-phonon frequency as well as the LT-splitting (the splitting between the LO- and TO-phonon) with decrease in the particle size are much larger than that expected from the spatial correlation model or that from the phonon confinement model. The phonon damping for the small particle consists of the usual temperature-dependent term and an excess damping term, which is independent of temperature. These results suggest that the scattering of the phonon at the particle surface plays a major role in determining the phonon states of the small particle. Our experimental data suggest that the surface phonon-polariton mode at the interface between the crystallites plays a minor role in the Raman spectrum of particles consisting of a number of crystallites.

INTRODUCTION

Geometrical boundary conditions for a thin film or for a small particle give rise to a set of quantized energy states for an elementary excitation. The eigenstate in the thin film is analogous to that for guided waves (guided wave mode) and that in the small particle can be described in terms of standing waves in a cavity (cavity mode). If the particle size is not large enough compared to the lattice spacing, the atoms at the surface of the particle play an important role for the energy states of the excitation. Atoms at the surface form an inter-atomic bonding different from the bonding scheme for atoms inside the particle. The surface vibrational modes are associated with the motion of these surface atoms. The concept of the (micro- or nano-) crystal is not valid if the particle size is so small that the lattice spacing, or the interatomic distance, is comparable to the particle size.

The purpose of this papaer is to discuss the size dependence of the phonon parameters in small SiC particles. The size ranges from 30 to 1000 nm. We discuss the rqle of the surface atoms and the origin of the size dependence of the phonon parameters in SiC.

EXPERIMENTAL ARRANGEMENT

The small particles of SiC have been made from polycarbosilane through its pyrolysis or from silica and carbon by their reaction [1,2]. Our samples belong to the 3C-polytype. Raman spectra are obtained with a double-grating monochromator and a photon-counting system. The light source is an Ar-ion laser beam (488 nm, 20 mW) focused to 0.1 x 2.5 mm^2. The sample is placed in the air except for the experiment above room temperature, where the sample is placed in the Ar gas. The particle size is estimated from the electron-microscope observation, and the size of crystallite from the width of the [111] line in the x-ray-diffraction spectrum. Details of the experimental procedures are given in ref. 2.

Mat. Res. Soc. Symp. Proc. Vol. 164. ©1990 Materials Research Society

RESULTS

Figure 1 shows the Raman spectra of SiC particles with the theoretical curve fit calculated in the framework of the Maxwell-Garnett theory [2-4]. Since our particle contains free carriers and there is a depletion layer at the surface [5], we use a model that the particle consists of a core with free carriers and a shell with no carrier. Different values of damping parameters for the core and for the shell are used in the curve fitting. The curve fit shows that the plasmon in the core is overdamped. The Raman line becomes broad for the LO-phonon - plasmon coupled mode if the damping of the plasmon is large. The Raman line due to the coupled mode is not resolved from the background in Figs. 1(a) and 1(b) because of the large plasmon-damping. The Raman line located at the LO-phonon frequency comes from the LO-phonon in the shell, where there is no free carriers (denoted as "Bare LO" in Fig. 1). The line(s) located between the TO-phonon and LO-phonon frequencies are due to the surface phonon-polariton (denoted as "SP"). Its frequency depends on the dielectric constant of the medium surrounding the particle, volume filling factor, f, of the particles in the medium, and the particle size [6]. The splitting of the SP line in Fig. 1(a) is attributed to spatial distribution of the filling factor, f, in the sample [2]. The line located at the TO-phonon frequency comes from the cavity-mode bulk-polariton. Its radial wavevector is quantized to approximately $(n+2)\pi/r$,

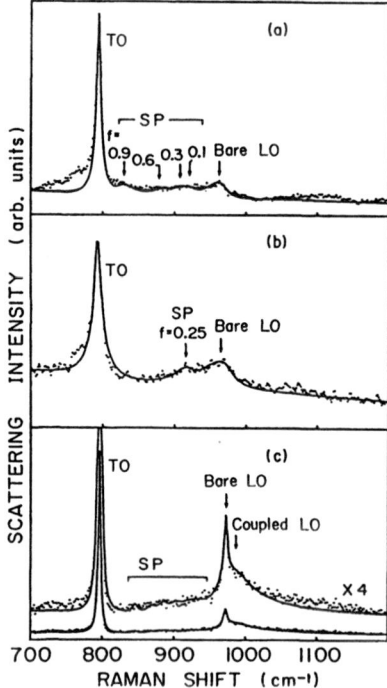

Fig. 1 Raman spectra of SiC small particles. Dashed lines are the calculated spectra. The particle/crystallite sizes are; (a) 85/22 nm, (b) 140/46 nm, and (c) ∿1000/160 nm.

where n is an integer and r the radius of the particle [7].

Figure 2 shows the TO-phonon frequency (ω_T), the LT-splitting (splitting between the LO- and TO-phonon; $\omega_L - \omega_T$), the damping in the core (Γ_c), and that in the shell (Γ_s) plotted as functions of r_s, which is the radius of the particle estimated by the curve fit. The data indicated by squares are obtained for samples prepared by different processes from the others; i.e. much lower pyrolysis temperatures or a different reaction process. The r_s turned out to be correlated well with the particle size rather than the crystallite size. The r_s is about a half of the average particle radius estimated by the electron-microscope observation. Note that there is an error in ref. 2 in the evaluation of the term express-ing relaxation of the momentum conservation, and consequently, the values of r_s have been underestimated by a factor of 2 to 4 in ref. 2.

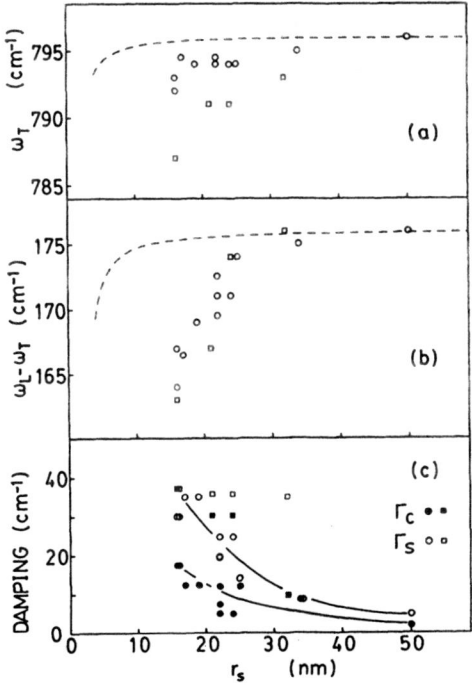

Fig. 2 (a) TO-phonon frequency, (b) $\omega_L - \omega_T$, and (c) phonon damping as obtained by the curve fit. Dashed lines represent the results calculated by the Gaussian attenuation function.

The temperature dependence of the TO-phonon frequency and that of the damping (Γ_c) are plotted in Fig. 3 and Fig. 4, respectively, for the same sample as in Fig. 1(a). The temperature coefficient of the frequency agrees with that reported for the single crystal [8]. The dashed line in Fig. 4 indicates the temperature dependence of the damping in the 3C-crystal [8]. The solid line in Fig. 4 is obtained by shifting the dashed line upward by 2.55 cm^{-1}. The solid line agrees with the experimental data quite well. This result indicates that the damping in the small SiC particle below 800 K is influenced by an excess damping mechanism independent of temperature besides usual temperature-dependent anharmonic damping.

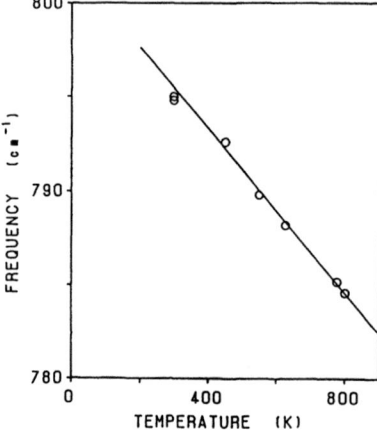

Fig. 3 Temperature dependence of the TO-phonon frequency of the small particle.

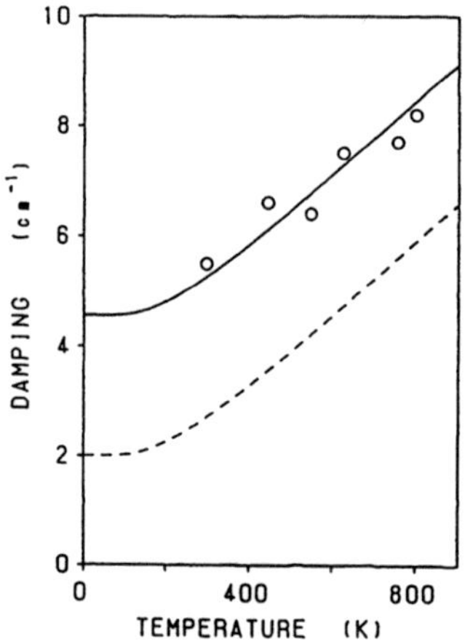

Fig. 4 Temperature dependence of the phonon damping of the small particle. The dashed line shows the result for the single crystal.

DISCUSSION

Figure 2 shows that the decrease in the particle size causes decrease in the phonon frequencies and increase in the damping. There are three possible causes for the origin of these size dependencies: (A) an increase in the anharmonicity in the vibrational states, (B) uncertainty of the k-selection rule in the Raman scattering process, and (C) the phonon scattering at the surface of the particle due to imperfections of atomic arrangement. If (A) were the case, the temperature-dependent term in the damping would become larger in smaller particles. The result shown in Fig. 4 indicates that (A) is not the case.

The small size in particles induces relaxation in the k-selection rule for the Raman scattering process. It is roughly π/r for the particle with radius r. Then phonon states with the wave vectors from zero to π/r can contribute to the Raman scattering. The dashed lines in Figs. 2(a) and 2(b) are obtained under the assumption of a Gaussian attenuation factor [9]

$$\exp(-2r^2/L^2),$$

with $L = r_s$. The quantization of k in small particles is not taken into account in this calculation. This effect should be smeared in the real sample because of the distribution in particle size. Here we use the phonon dispersions reported for the [111] direction [10]. The calculated values for the frequency shift and the damping are much smaller than those of the experimental results. The radius has to be reduced by a factor of 0.1 to 0.2 with respect to the experimental value in the calculation in order to get a good fit. Even if we perform the same calculation by using the perfect phonon confinement model (radius = r_s), the amount of the frequency shift and the damping increases by about 10 % at most compared with the Gaussian model. Thus the phonon confinement model is unable to give a quantitative explanation for our experimental result.

We obtain a good correlation between the damping and the phonon frequency by plotting the damping as a function of ω_T (Fig. 5(a)) and ω_L (Fig. 5(b)). Especially Γ_c is correlated with ω_T and Γ_s with ω_L. Since the LO-phonon - plasmon coupled line is quite broad, the width of the TO-line is reflected to the value of Γ_c. On the other hand, the width of the LO-line is related mainly to Γ_s. These results suggest that the decrease in the phonon frequency and the increase in the damping come from a common origin.

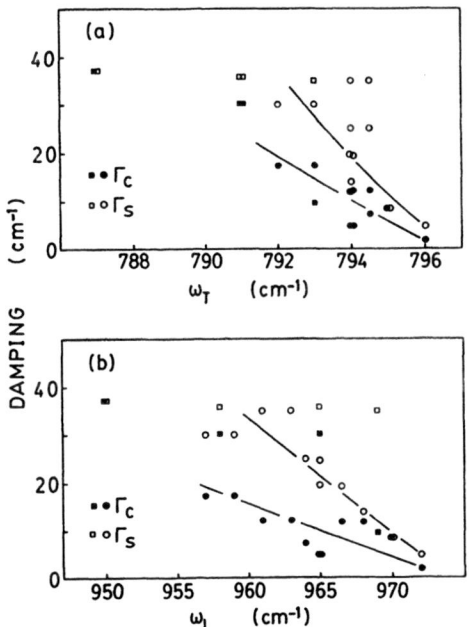

Fig. 5 Phonon damping plotted (a) as a function of ω_T and (b) as a function of ω_L.

The phonon confinement model explains the experimental result qualitatively but not quantitatively as already explained. If we consider a scattering of phonons due to imperfections in the atomic arrangement at the surface of the particle, it enhances the damping and the shift in the phonon frequency. The scattering probability is larger for smaller particles because the relative volume occupied by the surface atoms with respect to the inner atoms increases with decrease in the particle size. The damping of the LO-phonon (or Γ_s) is always larger than that of the TO-phonon. This partly comes from the fact that the dispersion of the LO-phonon is larger than that of the TO-phonon by a factor of 4.3.

The data points denoted by squares in Figs. 2(a) and 2(c) deviate from the others; i.e. they exhibit larger damping and lower ω_T. Since these data have been obtained for samples made with processes different from the others, the deviation in the phonon parameters are attributed to the difference in the impurity contents, shape of the particle, or the lattice imperfection in the particle. On the other hand, the data for the LT-splitting (Fig. 2(b)) are correlated well with the particle size. Thus this parameter depends strongly on the particle size and is not sensitive to other factors as is the case for ω_T or Γ. The decrease in the LT-splitting for small particles has been attributed to existence of a surface layer which does not contribute to the polarization [2]. It is not clear at the present stage which of the mechanism is dominant in lowering the LO-phonon frequency.

CONCLUSIONS

In small particles with a size ranging from 30 nm to 1000 nm, the most important process for affecting the phonon states is the scattering of the phonon by imperfections in atomic arrangements at near the surface. The present analysis shows that the Raman spectroscopy provides a powerful tool for estimating the particle size, carrier concentration, and their relaxation time in semiconductors and insulators.

ACKNOWLEDGMENT

The authors would like to thank Professor K. Okamura and Dr. M. Sato for cooporation in the experiment. This work was supported in part by the Grant-in-Aid for Scientific Research from the Ministry of Education, Science, and Culture.

REFERENCES

1. S. Yajima, Philos. Trans. R. Soc. London Ser. A 294, 419 (1980).
2. Y. Sasaki, Y. Nishina, M. Sato, and K. Okamura, Phys. Rev. B 40, 1762 (1989).
3. J.C. Maxwell-Garnett, Philos. Trans. R. Soc. London Ser. A 203, 385 (1904).
4. T.P. Martin and L. Genzel, Phys. Rev. B 8, 1630 (1973).
5. Y. Sasaki, Y. Nishina, M. Sato, and K. Okamura, Yogyo-Kyokai-Shi 94, 897 (1986).
6. R. Ruppin and R. Englman, Rep. Prog. Phys. 33, 149 (1970).
7. R. Ruppin, J. Phys. C 8, 1969 (1975).
8. D. Olego and M. Cardona, Phys. Rev. B 25, 3889 (1982).
9. J. Richter, Z.P. Wang, and L. Ley, Solid. State. Commun. 39, 625 (1981).
10. D.W. Feldman, J.H. Parker, Jr., W.J. Choyke, and L.Patrick, Phys. Rev. 173, 787 (1968).

NONLINEAR OPTICAL PROPERTIES OF STRUCTURED NANOPARTICLE COMPOSITES

MEYER H. BIRNBOIM* AND WEI PING MA*
*Rensselaer Polytechnic Institute, Troy, New York 12180

ABSTRACT

The effective nonlinear susceptibility $\chi^{(3)}$ for a composite that consists of a dilute suspension of structured nanoparticles that utilizes the surface mediated plasmon resonance can be enhanced by orders of magnitude compared to the intrinsic $\chi^{(3)}$ for a film of the same neat material. Here we report calculations for various multilayer spherical nanoparticle models in a host dielectric: i) polymeric core and metallic shell, ii) semiconductor core and metallic shell, iii) metallic core, polymeric shell and metallic second shell, and iv) metallic core, semiconductor shell and metallic second shell. The polymer is polydiacetylene, PDA, the semiconductor is Si, the metal is Ag, and the host is water, a GaAs$_2$ glass or Si. Enhancements as great as 10^4 can be obtained in both $\chi^{(3)}$ and the figure of merit with no degradation in the intrinsic speed of the nonlinearity.

The choice of geometrical parameters and of component materials permit tailoring of the wavelength dependence and bandwidth characteristics of the nonlinear response. However, fabrication of these structured nanoparticle suspensions remians the key challenge.

INTRODUCTION

Optical materials with a fast response time and a large nonlinear susceptibility, $\chi^{(3)}$, are central to optical domain computing devices. Some materials such as PDA [1] have a fast nonlinear mechanism that is electronic in origin with a response time in the picosecond range, but the small magnitude of $\chi^{(3)}$ would require a high laser power density to utilize these materials. Many classes of materials exhibit substantial nonlinear optical behavior. Polymers are competitive with semiconductors in this respect in magnitudes of $\chi^{(3)}$, speed of response, and figure of merit. In both classes of materials there is a tradeoff between magnitude and speed. This means that high laser power is required for for fast responding materials.

In the composites discussed here, the magnitude of $\chi_{eff}^{(3)}$ can be enhanced without sacrifice in speed by constructing a metal-polymer or metal-semiconductor composite. The principle is to locally concentrate the field in the neighborhood of the nanoparticle, and further increase the magnitude of the field through plasmon resonance. The effective $\chi_{eff}^{(3)}$ is proportional to the fourth power of local field.

Surface mediated plasmon resonance enhancement of the intrinsic $\chi^{(3)}$ of gold and of silver has been observed in suspensions of metallic nanospheres in water and also determined for ellipsoidal [2,3] nanoparticles in water. Composite materials of structured nanoparticles that consist of a metallic core and nonlinear shell or a nonlinear core and metallic shell have been shown [4,5] to provide substantial further enhancement in the speed/power performance of nonlinear optical devices. In each case the nonlinear behavior of the composite film is compared to a neat film of the nonlinear component material.

As an example, we emphasize the application of the nonlinear composite to the phase conjugate mirror (PCM) device. The performance of the PCM, and also of the nonlinear waveguide device, both depend on a figure of merit that is the ratio of $\chi^{(3)}$ to the linear absorption γ. Another exciting application of these composites to intrinsic optical bistability IOB devices has been reported [6,7].

THEORY

Single Particle Description

The model for the inhomogeneous spherical nanoparticle composite is described by the parameters radius r_q, dielectric permittivity ϵ_q, and third order susceptibility $\chi_q^{(3)}$, where $q=1,2,3,4$ refer to the core, first shell, second shell and host medium. The homogeneous equivalent of the composite is then described by $\bar{\epsilon}$ and $\chi_{eff}^{(3)}$.

The electric field distribution in the core, the first and the second shell and the region outside the particle is given respectively by

$$\vec{E}_1 = \frac{27\ \epsilon_2\ \epsilon_3\epsilon_4}{\epsilon_2\ \epsilon_a\ \epsilon_c + 2\epsilon_3\ \epsilon_b\ \epsilon_d}\ \{\ E_0 \cos\theta\ \hat{e}_r - E_0 \sin\theta\ \hat{e}_\theta\ ] \tag{1a}$$

$$\vec{E}_2 = \frac{9\ \epsilon_3\epsilon_4}{\epsilon_2\ \epsilon_a\ \epsilon_c + 2\epsilon_3\ \epsilon_b\ \epsilon_d}\ \{\ \epsilon_a(r)\ E_0 \cos\theta\ \hat{e}_r - \epsilon_b(r)\ E_0 \sin\theta\ \hat{e}_\theta\ \} \tag{1b}$$

$$\vec{E}_3 = \frac{3\ \epsilon_4}{\epsilon_2\ \epsilon_a\ \epsilon_c + 2\epsilon_3\ \epsilon_b\ \epsilon_d}\ \{\ [(\epsilon_2\epsilon_a + 2\epsilon_3\ \epsilon_b) + 2(\epsilon_2\ \epsilon_a - \epsilon_3\epsilon_b)(r_2/r)^3]\ E_0 \cos\theta\ \hat{e}_r$$

$$- [(\epsilon_2\epsilon_a + 2\epsilon_3\epsilon_b) - (\epsilon_2\epsilon_a - \epsilon_3\epsilon_b)(r_2/r)^3]\ E_0 \sin\theta\ \hat{e}_\theta\ \} \tag{1c}$$

$$\vec{E}_4 = \{\ 2\ \frac{\epsilon_2\epsilon_a[\epsilon_3(3-2Q) - \epsilon_4 Q] + \epsilon_3\epsilon_b[2\epsilon_3 Q - \epsilon_4(1+Q)]}{\epsilon_2\ \epsilon_a\ \epsilon_c + 2\epsilon_3\ \epsilon_b\ \epsilon_d}\ (\frac{r_3}{r})^3 + 1\}\ E_0 \cos\theta\ \hat{e}_r$$

$$+ \{\ \frac{\epsilon_2\epsilon_a[\epsilon_3(3-2Q) - \epsilon_4 Q] + \epsilon_3\epsilon_b[2\epsilon_3 Q - \epsilon_4(1+Q)]}{\epsilon_2\ \epsilon_a\ \epsilon_c + 2\epsilon_3\ \epsilon_b\ \epsilon_d}\ (\frac{r_3}{r})^3 - 1\}\ E_0 \sin\theta\ \hat{e}_\theta \tag{1d}$$

$$\epsilon_a = \epsilon_1(3-2P) + 2\ \epsilon_2 P \qquad\qquad \epsilon_b = \epsilon_1 P + \epsilon_2(3-P) \tag{2a,b}$$

$$\epsilon_c = \epsilon_3(3-2Q) + 2\ \epsilon_4 Q \qquad\qquad \epsilon_d = \epsilon_3 Q + \epsilon_4(3-Q) \tag{2c,d}$$

$$P = 1 - (r_1/r_2)^3 \qquad\qquad Q = 1 - (r_2/r_3)^3 \tag{2e,f}$$

The plasmon resonance condition is defined by

$$\text{Re}\ [\epsilon_2\ \epsilon_a\ \epsilon_c + 2\epsilon_3\ \epsilon_b\ \epsilon_d] = 0 \tag{3a}$$

with

$$\text{Im}\ [\epsilon_2\ \epsilon_a\ \epsilon_c + 2\epsilon_3\ \epsilon_b\ \epsilon_d] \tag{3b}$$

as a residual. The resonant denominator produces a high field in each region of the particle, hence produces a particle that behaves as a superdipole. For a particle with a single shell, the resonance condition equation 3a becomes quadratic in ϵ_2 and exhibits two resonant frequencies [8], whereas with two shells the equation is cubic in ϵ to permit three resonant frequencies. For the metal-dielectric interface, the key role of the metal is to provide the negative dielectric permittivity that permits the resonance condition to be satisfied. However the same metal that permits the resonance is generally the primary source of loss that prevents the residual denominator equation 3b from going to zero and limits the magnitude of the field enhancement. For the single shell particle there are two conditions [8] under which this loss can be minimized. One is to use a thin shell, therefore less metal; the other is a thick shell with $\epsilon_1 = -4\epsilon_3$. The corresponding conditions for the two shell model are not yet established.

These equations for the model with two shells reduce to the single shell model of references [4,5].

Composite Description

The complex dielectric permittivity of the composite is related to the single particle properties in the limit of small volume fraction ρ of particles by an equation of the Maxwell-Garnett form

$$\tilde{\epsilon} = \epsilon_4 + 3\rho\epsilon_4 \frac{\epsilon_2\epsilon_a[\epsilon_3(3-2Q)-\epsilon_4 Q] + \epsilon_3\epsilon_b[2\epsilon_3 Q-\epsilon_4(1+Q)]}{\epsilon_2\epsilon_a\epsilon_c+2\epsilon_3\epsilon_b\epsilon_d} \tag{4}$$

The optical absorption of the composite, if we neglect scattering, is related to $\tilde{\epsilon}''$, the imaginary part of $\tilde{\epsilon}$ by

$$\gamma = (2\pi/\lambda_0)\,\tilde{\epsilon}'' \tag{5}$$

The effective nonlinear susceptibility $\chi_{eff}^{(3)}$ for the homogeneous equivalent of the composite is related is related to the electric field E_q and the intrinsic $\chi_q^{(3)}$ in each region q of the nanoparticle by

$$\chi_{eff}^{(3)} = \overline{f_{1q}^2}\,\overline{f_{2q}}\,\overline{f_3}\,\chi_q^{(3)} \tag{6}$$

where $f_{1q} = E_q/E_0$, $f_{2q} = \partial\tilde{\epsilon}/\partial\epsilon_q$ and f_3 is the field outside the particle. The bar is the space average. This expression gives rise to a fourth power dependence on the local field. In the examples discussed herein, q=1 for the single shell model, and q=2 for the two shell model. That is, we consider only one material component to dominate the nonlinear response.

The reflectivity of a PCM fabricated from this composite material is given by

$$R = 1.86 \times 10^2\,\frac{\chi^{(3)2}}{\gamma^2}\,\frac{1}{n^4\lambda^2}\,I_p^2 \tag{7}$$

in which the film thickness is taken as $1.24/\gamma$, corresponding to the thickness for optimum reflectivity; so that the film thickness could range from a micron to meters. Indeed both $\chi_{eff}^{(3)}$ and γ are each proportional to the volume fraction ρ of particles, so the optimum reflectivity is independent of ρ, and to within the limits of linearity on concentration, any convenient thickness can be chosen by adjusting ρ. The units of $\chi^{(3)}$ are esu, $1/\gamma$ and λ are cm, and the pump power I_p is watts/cm^2.

DISCUSSION OF RESULTS

Comparisons have been calculated for two specific nonlinear materials, PDA and silicon, as examples. The dielectric data for silver and silicon is taken from Palik [9]. The method is generic and not limited to these materials. The table I is a comparison of the neat PDA film with various composites for which the effective $\chi_{eff}^{(3)}$, the γ and the reflectivity coefficient, R/I_p^2, are compared. In addition, a comparison is seen between

Table I. Nonlinearity in Organic Composites: Non-resonant PDA

	neat	composites			
λ nm	812	812	760	620	588
core	PDA	PDA	Ag	Ag	Ag
shell	PDA	Ag[1]	PDA	PDA	PDA
shell	PDA	—	Ag	Ag	Ag
host	PDA	H_2O	AsS_2	AsS_2	AsS_2
r_1/r_2	—	0.9	0.7	0.7	0.7
r_2/r_3	—	—	0.6	0.3	0.1
ρ	1	0.001	0.001	0.001	0.001
$\chi^{(3)}$ (esu)	1×10^{-9}	2×10^{-8}	3×10^{-6}	1×10^{-5}	1×10^{-4}
γ (cm^{-1})	8×10^{2}	3×10^{3}	1×10^{4}	1×10^{4}	1×10^{4}
R/I_p^2	7×10^{-15}	4×10^{-13}	6×10^{-11}	1×10^{-9}	6×10^{-8}

the one and two shell models. Also a few frequency ranges are examined. It is clear that reflectivity enhancements greater than 10^6 should be possible.

In table II, the comparisison is again for PDA; except in this case, the frequency was chosen to correspond to the intrinsic PDA resonance. The

Table II. Nonlinearity in Organic Composites: Resonant PDA

	neat	composites			
λ nm	673	673	654	663	673
core	PDA [r]	PDA [r]	Ag	Ag	Ag
shell	PDA	Ag[1]	PDA	PDA	PDA
shell	PDA	—	Ag	Ag	Ag
host	PDA	AsS_2	AsS_2	AsS_2	AsS_2
r_1/r_2	—	0.7	0.1	0.5	0.7
r_2/r_3	—	—	0.5	0.8	0.45
ρ	1.0	0.001	0.001	0.001	0.001
$\chi^{(3)}$ (esu)	1×10^{-7}	8×10^{-6}	1×10^{-5}	8×10^{-7}	1×10^{-4}
γ (cm^{-1})	1×10^{4}	2×10^{4}	2×10^{4}	2×10^{3}	8×10^{3}
R/I_p^2	6×10^{-13}	2×10^{-10}	7×10^{-10}	2×10^{-9}	1×10^{-7}

possible enhancement in reflectivity is greater than 10^5. Although the reflectivity coefficient of the neat polymer is 100 times higher in the resonant case than in the nonresonant case, attributable to the higher intrinsic resonant $\chi^{(3)}$; the composite reflectivity is comparable and large for both resonant and non-resonat PDA.

The next table III extends the comparison to the semicinductor Si [10].

Table III. Nonlinearity in Semiconductor Composites

λ nm	neat	composite			
	800	800	800	800	800
core	Si	Si	Ag	Ag	Ag
shell	Si	Ag	Si	Si	Si
shell	Si	—	Ag	Ag	Ag
host	Si	Si	Si	Si	Si
r_1/r_2	—	0.2	0.1	0.2	0.4
r_2/r_3	—	—	0.2	0.1	0.1
ρ	1	0.001	0.001	0.001	0.001
$\chi^{(3)}$ (esu)	1×10^{-7}	3×10^{-4}	5×10^{-4}	2×10^{-3}	4×10^{-4}
γ (cm^{-1})	1×10^3	2×10^4	2×10^4	2×10^4	2×10^4
R/I_g^2	2×10^{-12}	1×10^{-8}	1×10^{-7}	2×10^{-6}	6×10^{-8}

Once again large reflectivity enhancements of 10^6 and $\chi^{(3)}$ enhancements of nearly 10^4 are calculated. Notice that the comparison is for the region above the Silicon bap for the neat sample; in the composite with so little silicon, the loss is the resonantly enhanced metallic loss. This means that the composite is able to utilize materials with a high intrinsic $\chi^{(3)}$ but suffer from intrinsic loss.

The large enhancement of $\chi^{(3)}$ for the single shell model is further enhanced more than tenfold in the two shell model as seen for example in table II. This further enhancement is interpeted by the following heuristic argument. The curvature at the interface between shells establishes the denominator term that permits excitation of a plasmon resonance and enhancement of the external field. This local field is in turn further enhanced by a similar process at the core interface. This may be generalized, so that a multiple shell model may be considered as cascaded electric field amplifiers. Working from the outermost shell inwards, each is capable of successively enhancing the field under the suitable choice of parameters.

Another point of view is that the two shell model with cubic dependence on ϵ_q in the denominator can lead to three resonances, just as the one shell model with quadratic dependence was shown [8] to lead to two resonances. When two or more resonances are made to overlap, the cascading can occur.

CONCLUSIONS

Structured nanoparticle composites offer great flexibility in enhancement of nonlinear optical properties. Flexibility in tuning can be

282

achieved by choice of dielectric host medium, thickness of metallic shell, shape of particle (not discussed). The method is generic and applies to most nonlinear mechanisms. The concept of cascaded enhancement in multiple shell models is presented here. Fabrication remains a key challenge

Acknowlegement: This work was supported by NSF grant EET-8815141 and DARPA contract F19628-89-K-0045

REFERENCES

1. G. M. Carter, Y. J. Chen, M. F. Rubner, D. J. Sandman, M. K. Thakur and S. K. Tripathy, "Nonlinear Optical Properties of Organic Molecules and Crystals", Chemla and Zyss ed., Academic Press, 1987
2. P. Roussignol, D. Ricard, J. Lukasik and C. Flytzanis, J. Opt. Soc. Am. B4, 5 (1987)
3. J. W. Haus, N. Kalyaniwalla, R. Inguva,M. Bloemer and C. M. Bowden, J. Opt. Soc. Am B6, 797 (1989)
4. A.E. Neeves and M. H. Birnboim, Optics Letters 20, 1087 (1988).
5. A.E. Neeves and Meyer H. Birnboim, J. Opt. Soc. Am. B6, 787 (1989).
6. J. W. Haus, N. Kalyaniwalla, R. Inguva and C. M. Bowden, J. Appl. Phys. 65, 1420 (1989)
7. M. H. Birnboim, J. W. Haus, N. Kalyaniwalla, W. P. Ma and R. Inguva, OSA Technical Digest paper THJ4 (1989)
8. M. H. Birnboim and Wei Ping Ma, Am. Inst. Chem. Eng. Symposium Proc. (1989).
9. E. D. Palik, ed. "Handbook of Optical Constants of Solids" , Academic Press, 1985.
10. R. K. Jain and M. Klein in "Optical Phase Conjugation", R. A. Fisher ed., Academic Press 1983

ENHANCED NONLINEAR OPTICAL RESPONSE OF COATED NANOPARTICLES.

N. Kalyaniwalla[*], J.W. Haus[*], M.H. Birnboim[+], R. Inguva[**] and W.P. Ma[+]

* Rensselaer Polytechnic Institute, Physics Dept., Troy, NY 12180–3590

+ Rensselaer Polytechnic Institute, Dept. of M. E., A. E. and M., Troy, NY 12180–3590.

** University of Wyoming, Physics Dept., Laramie, WY

ABSTRACT

We study coated, nanometer–size, ellipsoidal particles that have a semiconductor or polymer core surrounded by a metal coating. We predict that composite materials containing these particles will have much larger enhancement of the nonlinear optical response than had previously been found by using semiconductor colloid suspensions or semiconductor − doped glasses. The enhancement is due to the surface plasmon resonance from the metal dielectric constant that increases the local field in the core material. The frequency of the resonance and the enhancement depend upon the particle shape and the coating thickness, as well as on the specific materials.

Also, we predict intrinsic optical bistability in these new materials and show that the threshold intensity for optical bistability can be greatly reduced by using the coated particles. We predict a switching intensity of silver coated GaAs particles below $100 W/cm^2$.

INTRODUCTION

The idea of engineered dielectric properties of heterogeneous materials has been a well developed field for many years [1]. Recent research has discovered the possibility that enormous enhancements can be expected for nonlinear optical response in composite materials [2–7]. These materials are made from small grains of semiconductor or metal particles that are dispersed in a glass matrix or a liquid. There are two reasons to expect enhanced nonlinearities; 1) the particles have nanometer sizes and therefore, the quantum confinement of the carriers can lead to increased binding and enhanced oscillator strengths; and 2) the local electric field inside the particles can be enhanced due to the difference between the dielectric properties of the particle and the dielectric material. In metal particles the so–called surface plasmon resonance is used to obtain a strong enhancement of the local field.

In previous papers, we reported calculations predicting intrinsic optical bistability (IOB) in a composite material consisting of silver particles embedded in a silica matrix near the surface plasmon resonance [5,8]. IOB occurs without a cavity because a resonance in the medium can be dynamically shifted by changing the intensity of the light in the medium. For silver particles, the threshold intensity can be reduced to about $20 MW/cm^2$ by optimizing the shape of the particles. There is not much more that can be achieved with metals alone because the nonlinear response of these materials is so weak.

The surface plasmon resonance has also been proposed as a mechanism to increase the intrinsic nonlinearity of coated spheres [7]. We have studied these particles and optimised the coatings to produce enhancements of the nonlinear Kerr coefficient of several orders of magnitude. The coating offers another important degree of freedom in engineering the properties of the medium to find the largest nonlinear effects. For instance, the core of the particle can be chosen as a material which has an intrinsically large nonlinear response, such as a semiconductor nonlinearity due to an exciton resonance. The metal coating material is then chosen to enhance the field at the resonance frequency by using a surface plasmon resonance effect.

Both the shape changes and the coating of the particles give us the freedom to optimise the threshold intensity for IOB in these new materials. This is reported in this paper for the case of spheroidal shaped GaAs particles coated with silver and embedded in a glass matrix.

THEORETICAL DEVELOPMENT

We consider prolate spheroidal particles. Each small particle is taken to be composed of a spheroidal core of aspect ratio, $r_c = (b_c/a_c)$, where b_c and a_c are the minor and major axes of the core. The particle has a metallic shell, which is confocal with the surface of the core. It has an aspect ratio, $r_s = (b_s/a_s)$, where b_s and a_s are the minor and major axes of the particle at the outer surface of the shell, respectively.

In the following development, the non–linearity is restricted to the core region and is taken to be of the Kerr–type. The dielectric function for the core material is thus

$$\epsilon_c = \epsilon_{cl} + \chi_c^{(3)} |\vec{E}_L|^2 . \qquad (1)$$

the dielectric function is ϵ_{cl} is the complex linear part of the core dielectric function and $\chi_c^{(3)}$ is the Kerr coefficient for the core material and $\vec{E}_L$ is the electric field in the core of the particle. The shell is metallic and has a complex dielectric coefficient, ϵ_s, which is assumed independent of the field. The particles are embedded in a host dielectric material, which has a linear, lossless dielectric response, ϵ_h.

Since the non–linearity is restricted to the core region, we need only calculate the local field in the core to find the enhancement of the optical nonlinear response. Since we assume that the particles are small compared to the wavelength (i.e $n \frac{2\pi a_s}{\lambda} << 1$; where n is the index of refraction of the metal), the quasi–static or Rayleigh approximations can be used. For the general case of the confocal spheroids we obtain the expression for the local field in the core material :

$$E_L^j = \gamma^j E_{0;}^j \qquad (2)$$

where $\vec{E}_0$ is the applied field in the host medium and γ^j is called the enhancement factor:

$$\gamma^j = \left\{ \frac{(1+A_s^j)\ (1+A_c^j)\ \epsilon_h\ \epsilon_s}{(\epsilon_c+A_c^j\ \epsilon_s)\ (\epsilon_s+A_s^j\ \epsilon_h)\ +\ \Gamma^j(\epsilon_c-\epsilon_s)(\epsilon_s-\epsilon_h)} \right\} .$$

$$(3)$$

We note that the denominator of Eq.(3) is a quadratic function of $|\vec{E}_L|^2$, through the field dependence of ϵ_c given in Eq.(1). The other functions in Eq.(3) are defined below and only depend on the geometry of the coated particles. Therefore, the relationship between the local field amplitude and the applied field amplitude is a cubic equation.

We label the nondegenerate axis with the superscript value j=0 and the degenerate axes are labeled with the superscript value j=1. The minor axes are degenerate in a prolate spheroid and the major axes are degenerate in an oblate spheroid.

There are three sets of coefficients that depend on the geometry of the particles: A_c^j, A_s^j and Γ^j. The constants A_c^j are shape factors for the inner spheroid. A_s^j are shape factors for the outer spheroid. Γ^j are the shape factors which determine the volume fraction of the shell.

$$A_\mu^j = -\frac{P_1^j(\xi_\mu)}{Q_1^j(\xi_\mu)} \frac{Q_1^{j'}(\xi_\mu)}{P_1^{j'}(\xi_\mu)} \; ; \tag{4}$$

where ξ_μ are spheroidal coordinates for the core ($\mu = c$) and shell ($\mu = s$) spheroids. When the core–shell interface is used the coordinate is ξ_c; and ξ_s denotes the interface between the shell and the dielectric host. The functions $P_n^m(x)$ and $Q_n^m(x)$ are the Legendre Polynomials of first and second kind. The primes denote derivatives of these functions with respect to the argument. The functions Γ^j defined in Eq.(3) are directly related to functions and variables that are previously defined:

$$\Gamma^j = -\frac{P_1^j(\xi_c)}{Q_1^j(\xi_c)} \frac{Q_1^{j'}(\xi_s)}{P_1^{j'}(\xi_s)} \; . \tag{5}$$

The coordinates ξ_μ are succinctly defined in terms of the eccentricity of each surface. The eccentricity of the core–shell interface $e_c = \sqrt{1 - r_c^2}$ and the eccentricity of the shell–host interface $e_s = \sqrt{1 - r_s^2}$. For a prolate spheroid the coordinates are:

$$\xi_\mu = 1/e_\mu \; . \tag{6}$$

For the oblate spheroids, the coordinates ξ_μ in Eqs. (4) and (5) are replaced by $i\xi_\mu$; we do not give those results here. In the specific case of coated spheres, the results are identical to those given earlier by Neeves and Birnboim [7].

We examine now the relationship between the applied field and the local field inside the particle. This field can have three real values for one value of the applied field. This is the IOB phenomenon because each particle exhibits the effect and it doesn't depend on feedback from external mirrors. For silver shell and GaAs core materials we demonstrate the existence of bistability and optimize the the geometric shape and the ratios of the core and shell volumes.

RESULTS AND DISCUSSION

Results of the optimization of IOB is shown in Fig.(1). For GaAs core, the operating wavelength was chosen to be 805nm; it is detuned from the exciton resonance at 820nm for which $\chi^{(3)} = -6 \cdot 10^{-2}$ esu [9]. The shell material is silver; the data for the linear dielectric constant of silver and GaAs was taken from Palik [10]. The threshold intensity for this case is reduced below $100 W/cm^2$. This represents a reduction of 10^5 over the threshold of uncoated spheroidal metal particles. Similar results, achieved for polydiacetylene core materials and other semiconductors, have been reported earlier [11]; for polydiacetylene we have found threshold intensities below $100 KW/cm^2$. More complete publications of these results are being prepared [12].

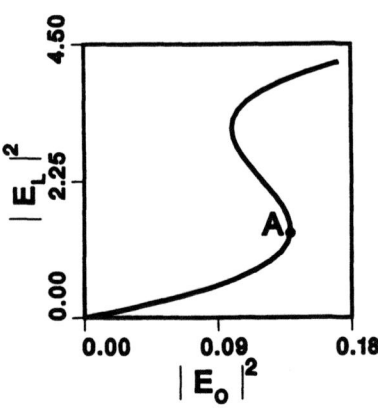

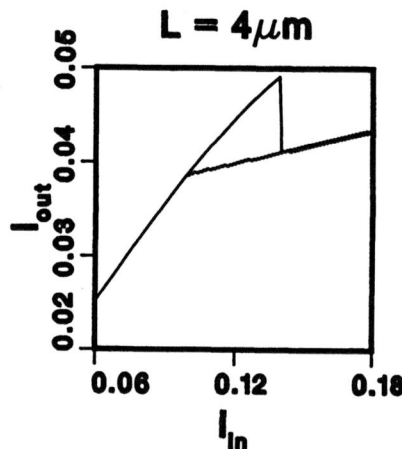

Fig. 1 The local field intensity versus applied field.

Fig. 2 The optical bistable output versus input intensity.

The relationship between the applied field $|E_0|$ and $|E_L|$ in Fig.(1) is obtained from Eqs.(2) and (3). The ratio of axes in the core is $r_c = 0.8$ and the core and shell volumes can be determined from the ratio $a_c/a_s = 0.83$. The field is applied along the minor axis of the spheroid. The shape and coating thickness was chosen to minimize the threshold intensity and the change in the index of refraction, $\Delta n \approx .3$.

The particles are embedded in a dielectric material at low volume fractions, f. In our case, the volume fraction of particles is $f = 10^{-3}$; therefore, a much simpler description based on the induced dipole moments of individual particles will suffice. Each of the coated inclusions is treated as a dipole; then their contributions are summed up and the net polarization is found for the medium. From this we determine the effective dielectric function of the inhomogeneous medium as

$$\overline{\epsilon} \simeq \epsilon_h + \frac{3f\epsilon_h}{(1-e_s^2)} \left[\frac{(\epsilon_s-\epsilon_h)(\epsilon_c+A_c^j\epsilon_s)+G^j(\epsilon_c-\epsilon_s)(\epsilon_h+A_s^j\epsilon_s)}{(\epsilon_c+A_c^j\epsilon_s)(\epsilon_s+A_s^j\epsilon_h)+\Gamma^j(\epsilon_c-\epsilon_s)(\epsilon_s-\epsilon_h)} \right]$$

$$(7)$$

where the coefficient G^j is

$$G^j = -\frac{P^j(\xi_c)}{Q^j(\xi_c)}\frac{Q^j(\xi_s)}{P^j(\xi_s)}$$

$$(8)$$

The effective dielectric function is used in Maxwell's equations to calculate the propagation of the electromagnetic waves in the medium [8]. In Fig.(2) we show the output intensity versus input intensity from oriented particles embedded in silica glass. The medium is $8\mu m$ in length. The output intensity jumps discontinuously to a lower value as the input intensity reaches the turning point labeled A in Fig.(1), which indicates an increase of the absorption in the medium. The increase of absorption is due to the large local field in the core, which adjusts the surface plasmon resonance of the particles.

The first experiment to investigate this phenomenon will undoubtedly be a degenerate four–wave mixing experiment. These experiments for metal particles embedded in a dielectric matrix have shown huge enhancements of the phase conjugate signal and a sensitive dependence on the particle shape [5,6]. This signal is a direct measure of the average nonlinear susceptibility of the medium. For coated particles, the nonlinear susceptibility is given by:

$$\overline{\chi}^{(3)} = \frac{3f\,\epsilon_h}{(1 - e_s^2)}\; \chi_c^{(3)}\,(\overline{\gamma}_1^j)^2\,|\gamma_1^j|^2 * \frac{N}{D}\,,\qquad (9)$$

where γ_1^j is the linear part of the enhancement factor in Eq. (3) and the two variables in Eq. (9) are

$$N = \Gamma^j\,(\epsilon_s - \epsilon_h)^2\,(1 + A_c^j)\,\epsilon_s - G^j\,(1 + A_c^j)\,\epsilon_s\,(\epsilon_h + A_s^j\,\epsilon_s)\,(\epsilon_s + A_s^j\,\epsilon_h);$$

and

$$D = [(\epsilon_s - \epsilon_h)\,(\epsilon_{cl} + A_c^j\,\epsilon_s) + G^j\,(\epsilon_{cl} - \epsilon_s)\,(\epsilon_h + A_s^j\,\epsilon_s)]^2.$$

The nonlinear susceptibility is plotted in Fig.(3) along with the expected linear absorption coefficient. $\overline{\chi}^{(3)}$ is sharper than the absorption. This occurs because the absorption is proportional to the square of the enhancement factor, whereas the nonlinear susceptibility is proportional to the <u>fourth</u> power of the enhancement factor.

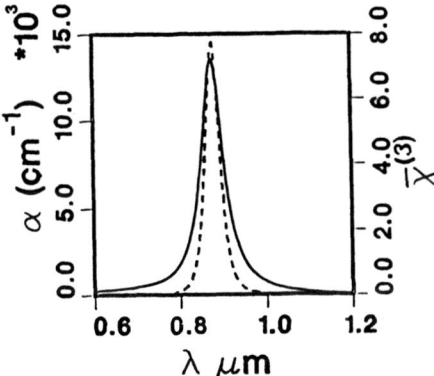

Fig. 3 The absorption (solid curve) and nonlinear susceptibility (dashed curve) versus wavelength.

We conclude by remarking that this new class of composite nonlinear materials can be developed for applications in optical communications and optical computing by using the IOB behavior. The shape of the particles and the ratio of core to shell volumes is critical and needs to be carefully controlled. For optimal results the particles need to be oriented relative to one another, but otherwise, they can be randomly dispersed in the medium. We emphasize that the challenge is in synthesis of these new materials; techniques for precise control of geometry and coating thickness need to be developed.

Acknowledgment: This research was supported by NSF grants. ECS–8813028 and EET–8815141 and DARPA contract: F19628–89–K–0045.

REFERENCES

1. L. K. H. Van Beek, Progress in Dielectrics, Volume 7, ed. J. B. Birks (CRC Press, Cleveland, 1967), p. 69.

2. R. K. Jain and R. C. Lind, J. Opt. Soc. Am 73, 647 (1983).

3. P. Rousignol, D. Ricard, J. Lukasik and C. Flytzanis, J. Opt. Soc. Am. B4, 5 (1987).

4. D. Ricard, P. Roussignol and C. Flytzanis, Opt. Lett 10, 511 (1985); F. Hache, D. Ricard and C. Flytzanis, J. Opt. Soc. Am B3, 1647 (1986); F. Hache, D. Ricard , C. Flytzanis and U. Kreibig, Appl. Phys. A47, 347 (1988).

5. J. W. Haus, N. Kalyaniwalla, R. Inguva, M. Bloemer and C. M. Bowden, J. Opt. Soc. Am 6, 797 (1989); J. W. Haus, R. Inguva and C. M. Bowden, Phys. Rev. A40, in press (1989);

6. M. J. Bloemer, J. W. Haus and P. R. Ashley, J. Opt. Soc Am B, in review (1989); M. J. Bloemer, P. R. Ashley, J. W. Haus and N. Kalyaniwalla,. IEEE J. Quant. Electron, accepted (1989).

7. A. E. Neeves and M. H. Birnboim, J. Opt. Soc. Am B6, 787 (1989); Opt. Lett 13, 1087 (1988).

8. J. W. Haus, N. Kalyaniwalla, R. Inguva and C. M. Bowden, J. Appl. Phys. 65, 1420 (1989); D. S. Chemla and D. A. B. Miller, Opt. Lett 11, 522 (1986).

9. D. S. Chemla, D. A. B. Miller, P. W. Smith, A. C. Gossard, W. Wiegmann, IEEE J. Quant. Electron QE–20, 2655 (1986).

10. E. D. Palik, ed. Handbook of Optical Constants of Solids, (Academic, New York, 1985).

11. M.H. Birnboim, J.W. Haus, N. Kalyaniwalla, W.P. Ma and R. Inguva, Optical Society of America, Technical Digest (1989), papaer THJ4.

12. M.H. Birnboim, W.P. Ma, J.W. Haus, N. Kalyaniwalla and R. Inguva, unpublished (1989); N. Kalyaniwalla, J.W. Haus, R. Inguva, M.H. Birnboim and W.P. Ma, unpublished (1989).

PART V

Silicon Alloys

OPTOELECTRONICS AND PHOTOVOLTAIC APPLICATIONS OF
MICROCRYSTALLINE SiC

Y. HAMAKAWA, Y. MATSUMOTO, G. HIRATA and H. OKAMOTO
Faculty of Engineering Science, Osaka University, Toyonaka, Osaka, 560 Japan

ABSTRACT

    A review is given of the electrical and optical properties of
hydrogenated microcrystalline silicon carbide (μc-SiC:H) films prepared by
ECR (Electron Cyclotron Resonance) plasma chemical vapor deposition. The
material produced with the ECR plasma technology has a very wide energy gap
from 2 to 2.8 eV with good valency electron controllability,e.g., a dark
conductivity as high as 10 Scm^{-1} which is more than seven orders of magnitude
larger than that of amorphous SiC:H.
    Employing this material as a wide gap heterojunction window, 15.4% and
12.0% conversion efficiencies have been achieved with the structures of ITO/p
type μc-SiC:H/n type poly-Si and p type μc-SiC:H/i type a-Si:H/n type μc-Si:H
heterojunction solar cells, respectively. The successful development of a
visible light thin film light emitting diode show the promise of
microcrystalline materials for optoelectronic applications.

INTRODUCTION

    Since the success of valency electron control in the glow discharge
produced amorphous silicon carbide (a-SiC:H) in 1981 (1), the era of the
amorphous silicon alloy has started. A group of new materials such as
amorphous silicon-germanium (a-SiGe:H), amorphous silicon-nitride (a-SiN:H)
and amorphous silicon-tin (a-SiSn:H) has been successively developed in the
following few years.
    The significance of this material innovation is that one can control
electrical, optical and also opto-electronic properties by controlling atomic
compositions in the mixed alloys. Therefore, a wide variety of application
fields are now pursued with this new electronic material.
    The microcrystallization of the material deposited from the reaction gas
sources was reported in 1968 by Veprek's group (2), but the great evolution
began in the early stage of this decade. For example, a series of
experimental studies have been made to establish the formation kinetics
through plasma diagnosis by Matsuda (3). Recently, an evolution in the
preparation technologies has occurred with use of ECR (Electron Cyclotron
Resonance) CVD. By this technique, we have succeeded in preparing
highly-conductive and wide-gap microcrystalline SiC:H (μc-SiC:H). The
extremely high conductivity of 10 S cm^{-1} has been obtained for boron doped
p-type μc-SiC with optical energy gap of 2.25 eV. The optical energy gap can
be increased up to 2.8 eV without any significant reduction of electrical
conductivity. We will present here the applicability of this highly
conductive p μc-SiC:H for the window electrode layer in actual μc-SiC/a-Si
and μc-SiC/poly-Si heterojunction solar cells, demonstrating the improved
photovoltaic performances. Employing p- and n-type materials as hole and
electron injectors in LED devices, the luminance has been increased by more
than one order of magnitude as compared with that of the LED prepared by
conventional RF plasma CVD. This paper presents a series of technical data
on the basic properties of microcrystalline SiC:H produced by ECR plasma CVD.
The improvement of device performances achieved by the use of this material
will be discussed.

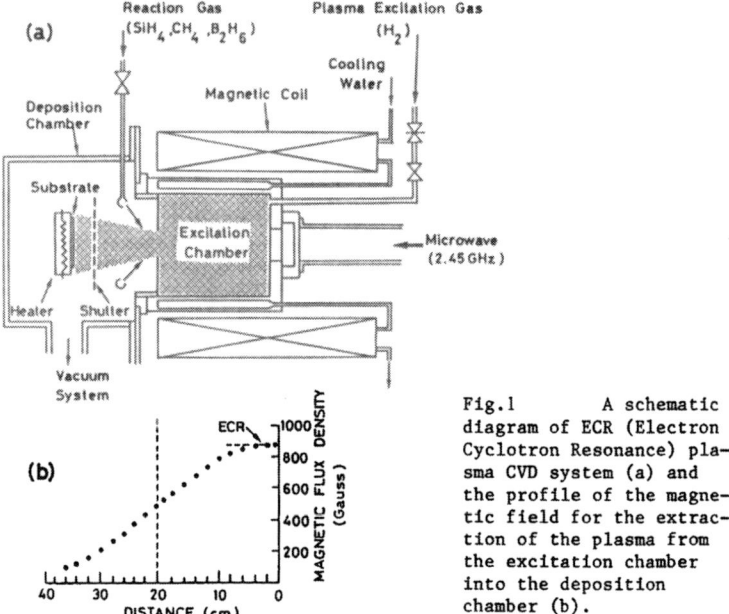

Fig.1 A schematic diagram of ECR (Electron Cyclotron Resonance) plasma CVD system (a) and the profile of the magnetic field for the extraction of the plasma from the excitation chamber into the deposition chamber (b).

ECR PLASMA CVD SYSTEM

The schematic diagram of the ECR CVD apparatus employed in this work is shown in Fig.1(a). Microwave power at 2.45 GHz is introduced into the ECR plasma excitation chamber through a rectangular wave guide and a window made of fused quartz plate. The ECR excitation chamber forms a cylindrical resonator of TE_{113} mode, surrounded by a magnetic coil. In the system, the magnetic flux required for satisfying the electron cyclotron resonance condition is about 875 Gauss. The ECR plasma is extracted from the ECR excitation chamber into the deposition chamber along with the gradient of dispersed magnetic field as shown in Fig.1(b). The extracted ECR plasma interacts with the reaction gas introduced into the deposition chamber to produce active species for film growth.

The unique advantage of the ECR CVD is that the growing surface is subject to bombardment by electrons and/or other heavy species having a uniform and moderate energy of several tens of eV (4). This effect not only prevents weak bonds from being introduced into the network but also promotes diffusion of long-lifetime radical species due to the raised surface temperature. It is expected that films with dense network and low defect density are formed.

In the present experiment, hydrogen was used as an ECR plasma excitation gas, and a mixture of SiH_4, CH_4 and B_2H_6 or PH_3 was used as a reaction gas for the growth of p- and n-type SiC:H. Details of the preparation conditions are summarized in Table I. Since the operation pressure is in the range of 10^{-3} and 10^{-4} Torr, the lifetime of chemically active hydrogen radicals is quite long, so that a large amount of hydrogen radicals will reach the growing surface and play a important role in determining the properties of growing films. Therefore, the dependence of the material properties on the hydrogen dilution ratio in the reaction gas has been investigated.

TABLE I.   Preparation conditions of p-type
amorphous and microcrystal SiC in ECR plasma CVD.

Substrate Temperature	: R.T. – 400°C
Microwave Power	: 150 – 400 W
Total Gas Pressure	: 10^{-3} – 10^{-4} Torr
Plasma excitation Gas (flow rate)	: H_2 (10 – 100 sccm)
Reaction Gas (Flow rate)	: SiH_4 (10 – 50 sccm)
	: CH_4 (10 – 50 sccm)
	: B_2H_6 (40 – 100 sccm)
Microwave Frequency	: 2.45 GHz
Magnetic Flux Density	: 875 gauss

VALENCY ELECTRON CONTROL AND OPTOELECTRONIC PROPERTIES OF THE μc–SiC:H

Both p- and n-type films were prepared by an ECR plasma CVD system
(ANELVA,310-D). The typical microwave power was 300 Watts, and the substrate
temperature 250 – 300°C. The total gas pressure during deposition was about
7.0×10^{-4} Torr. H_2 was used as the ECR plasma excitation gas and SiH_4 (10% in
H_2), CH_4 (10% in H_2), B_2H_6 or PH_3 (500 ppm in H_2) were used for the film
deposition.

The film properties are strongly dependent not only on the substrate
temperature and microwave power but also on the ratio of hydrogen to
reaction gases ($H_2/(SiH_4 + CH_4)$) during the deposition.

The formation of Si and SiC microcrystallites is confirmed by Raman
spectra as is shown in Fig.2. The Raman spectrum of the films prepared at
microwave powers higher than 250 Watts exhibits distinct structures at around
520 and 740 cm^{-1}, which correspond to TO phonon modes of crystalline Si and
crystalline SiC clusters.

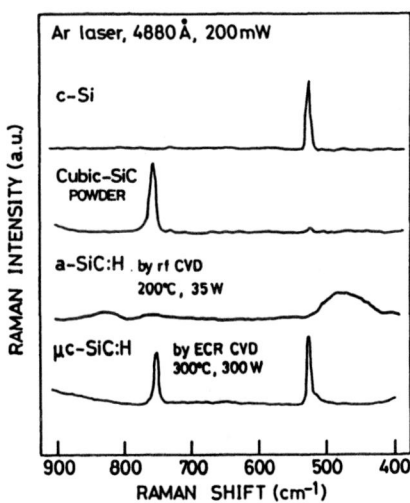

Fig.2   Raman spectrum of
microcrystal $Si_{1-x}C_x$:H
(for x 0.3, bottom line).
The Raman spectra of
μc–Si:H, cubic c–SiC powder
and a–SiC:H are also shown
for comparison.

As in the case of RF plasma CVD, hydrogen radicals make a large contribution to the formation of microcrystallites. In the case of ECR CVD using hydrogen as an excitation gas, the growing surface is subject to a large concentration of hydrogen radicals, which drive out weakly-bonded hydrogen at the surface and promote the formation of non-hydrogenated Si-related microcrystallites. Such a reaction is temperature activated, so that microcrystallization might be further enhanced by the elevated surface temperature due to the impact of electrons and/or other species. The key factors for the formation of μc-SiC:H are the density of hydrogen radicals reaching the growing surface and the surface temperature. Since the density of hydrogen radicals can also be controlled by the flow ratio of hydrogen gas to other reaction gases, it is instructive to present the flow ratio dependence of the film properties.

Figure 3 shows the dependence of the optical energy gap and dark conductivity of the samples on the H_2 dilution ratio. As the ratio increases the optical gap (E_o) and also the dark-conductivities (σ_d) of both p- and n-type films increase.

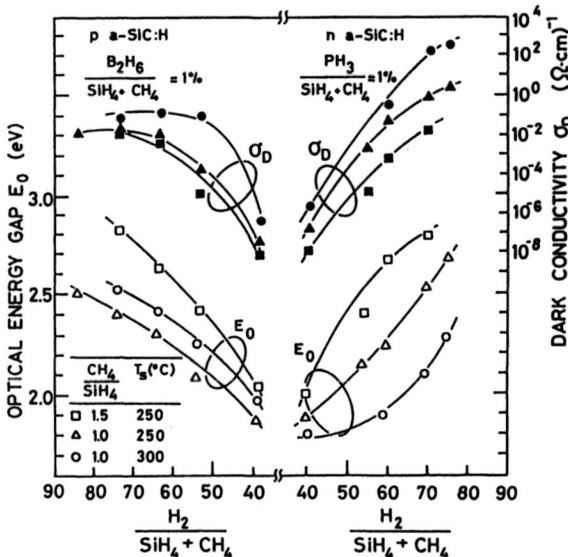

Fig. 3 Dependence of the optical energy gap and dark-conductivities of p- and n-aSiC:H on the hydrogen dilution ratio in the reaction gas in ECR CVD.

Figure 3 also demonstrates that the optical energy gap increases with increasing source gas ratio CH_4/SiH_4. There are two main factors which determine the optical energy gap; one is the composition ratio of Si:C:H, and the other is the details of the network structure. Although the optical energy gap increases with the flow rate of CH_4, the effect is not as remarkable as the dependence on hydrogen dilution. Hydrogen dilution has the effect of reducing the hydrogen content in the film, but of enhancing the degree of microcrystallinity.

TABLE II. Hydrogen dilution dependence on the μc-SiC composition.

Hydrogen dilution $H_2/(SiH_4+CH_4+B_2H_6)$	Carbon content (%)	Hydrogen content (%)	Optical Eg (eV)
39	19.3	27.5	1.98
54	11.7	10.7	2.26
74	16.1	10.6	2.53

Preparation condition : temperature = 300°C, microwave power = 300 W,

Gas pressure 7×10^{-4} Torr., $SiH_4/CH_4/B_2H_6/H_2$ = 1/1/0.025/78-148 (sccm).

In Table II, we show the hydrogen and carbon contents in films prepared with different hydrogen dilution. The Raman spectra analysis indicates that the microcrystals are formed for samples produced with 50 times or more dilution ratio in H_2.

Figure 4 demonstrates the temperature dependence of the dark conductivity for p μc-SiC:H samples. Sample (a) has an extremely high dark conductivity (10^2 Scm^{-1}), while that of sample (b) is on the order of 10^0 Scm^{-1}. In the inset of the figure, we indicate the values of the Hall mobility and carrier (hole) concentration measured using Van der Paw's method. The sample (b) exhibits a temperature-activated conductivity above about 250°K, while the Mott's variable-range hopping prevails in the lower temperature regime (5). By contrast the conductivity of the sample (a) shows no remarkable dependence over the temperature range investigated in this work.

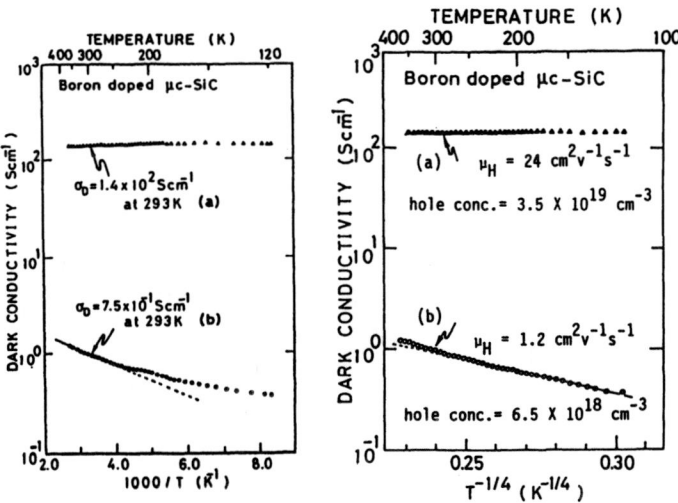

Fig. 4 Dark conductivity vs. inverse temperature for films prepared by ECR plasma CVD system. Each sample was prepared with :
(a) $SiH_4/CH_4/B_2H_6/H_2$ = 1/1/0.05/180 and (b) 1/1/0.025/150.

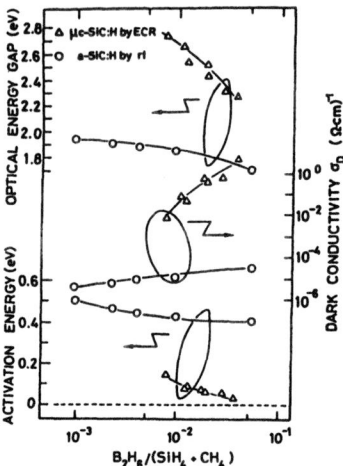

Fig.5 Dependence of dark conductivity, activation energy and optical gap on the dopant gas ratio in the reaction gases for p-type μc-SiC:H.

$SiH_4/CH_4/H_2 = 1/1/150$ (ECR)

$SiH_4/CH_4/H_2 = 1/2/150$ (RF)

The conductivity of μc-SiC:H can also be controlled by adjusting the flow ratio of dopant gas to host reaction gas. Figure 5 shows the dependence of the optical energy gap and dark conductivity of μc-SiC:H prepared by ECR CVD on the flow ratio of B_2H_6 p-type doping gas. Here, the hydrogen dilution ratio is kept constant at 74. The data for p a-SiC:H prepared by RF plasma CVD are also shown for comparison. The carbon content x in both cases is about 0.3. It is clear that the total doping efficiency in p μc-SiC:H is higher than that in p a-SiC:H by several orders of magnitude.

Figure 6 summarizes the relation between the dark conductivity and the optical energy gap of p- and n-type a-SiC:H prepared by conventional RF plasma CVD and p- and n-type μc-SiC:H prepared by ECR plasma CVD. As the optical energy gap increases, the dark conductivity of the films prepared by the RF plasma CVD rapidly decreases, while that of the films prepared by the ECR plasma CVD remains higher than 10^{-3} Scm^{-1} even when the optical energy gap exceeds 2.5 eV.

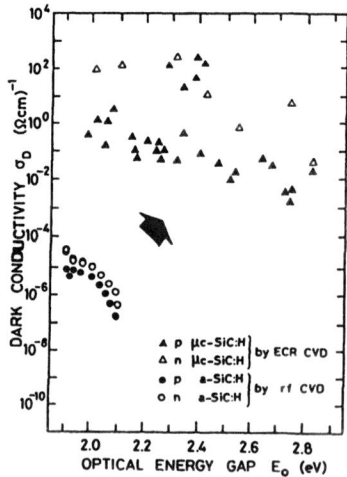

Fig.6 Relationship between dark conductivity and optical gap of amorphous and microcrystalline SiC:H prepared by RF and ECR plasma CVD.

APPLICATION OF μc-SiC:H TO OPTOELECTRONIC DEVICES

a-Si:H based Solar cell

As long as the basic properties are concerned, microcrystalline SiC:H prepared by ECR CVD undoubtedly satisfies the requirements for heterojunction window material. However, as is often experienced, a good film quality does not immediately result in good device performance. This is because the device performance is highly sensitive to any kind of perturbing factors which each layer of the device encounters during the fabrication processes (6).

The most severe technological problem we face when attempting to fabricate solar cells by using ECR CVD is that due to the presence of dense hydrogen radicals, the transparent conductive oxide (TCO) is reduced (by a reduction reaction) and its transparency as well as its conductivity is largely degraded. One possible way to prevent the damage is using a double layered TCO such as SnO_2/ZnO, which has a sufficient chemical stability for hydrogen plasma, whereas the optical transparency and resistivity are slightly inferior to that of conventional ITO or single SnO_2 layer. Another approach is to form a thin p-type a-SiC:H (as a buffer layer) prior to the deposition of microcrystalline p SiC:H.

We have adopted these two techniques to avoid the degradation of TCO property during deposition of microcrystalline p SiC:H layer. The cell structure was then glass/SnO_2/ZnO/p a-SiC:H(buffer)/p microcrystalline SiC:H/i a-Si:H/n microcrystalline Si:H/Ag. The SiC layers were formed by ECR CVD, while other layers were formed by conventional RF plasma CVD under the conditions described in our previous papers (7). The optical band gap and the dark conductivity of microcrystalline SiC:H layer were 2.25 eV and more than 1 Scm^{-1}, respectively. The thickness of each semiconductor layer was 2nm, 20nm, 500nm, and 30nm, respectively.

Up to the present, we have obtained a conversion efficiency η of 12.0% with a very large open circuit voltage, Voc = 967 mV, a short circuit current, Isc = 17.7 mA/cm^2 and a fill factor, FF = 70.3% for a cell of area 0.033 cm^2. In Figure 7, we demonstrate the current-voltage characteristics.

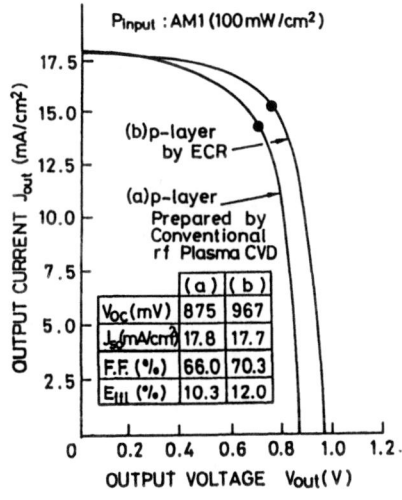

	(a)	(b)
V_{OC}(mV)	875	967
Isc(mA/cm²)	17.8	17.7
F.F. (%)	66.0	70.3
E_{ff} (%)	10.3	12.0

Fig.7 Output characteristics of solar cell measured under AM1 simulated sunlight. Curve (a) is for a cell using conventional p-type a-SiC:H prepared by RF CVD and curve (b) for p microcrystal-SiC:H prepared by ECR CVD.

For comparison, the output characteristics of a cell whose p a-SiC:H was prepared by RF plasma CVD are also shown in the figure (η = 10.3%, Voc = 875 mV, Isc = 17.8 mA/cm^2, FF = 66.0%). The increase in the Voc is primarily due to an increase in the built-in potential, which has been confirmed by direct measurement by means of the electroabsorption in the reflection geometry (8). Figure 8 shows the relation observed between the built-in potential and open circuit voltage.

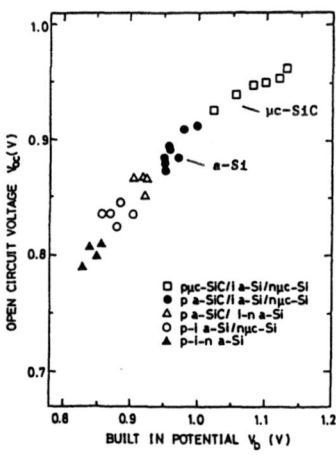

Fig.8 Open circuit voltage vs. built-in potential for several kinds of solar cells made with RF and ECR plasma CVD.

Polycrystalline based Solar Cell

The cell structure was ITO/p μc-SiC:H/n poly-Si/Al being the bulk material of 3 ohm-cm, 0.025 cm thick n-type cast polycrystalline silicon. For fabrication, first aluminum was deposited on acid etched poly-Si substrate for back ohmic contact while the p-n heterojunction was formed with deposition of p μc-SiC:H on this substrate. The process temperature and the microwave power were 300°C and 300 Watts respectively. Finally ITO was deposited by E.B. technique as antireflective coating and electrode. A systematic investigation has been carried out on the optimization of material properties and thickness of μc-SiC layer as well as the formation technology of the μc-SiC:H/poly-Si interface (9). Figure 9 shows the output characteristics of this heterojunction solar cell which has 15.4% in conversion efficiency with 70nm thickness of p microcrystalline SiC:H emitter layer. Recently we also developed a four-terminal tandem type heterojunction solar cells employing this poly-Si materials as a bottom cell and a-Si as a top cell reaching the highest conversion efficiency of 16.8% (10).

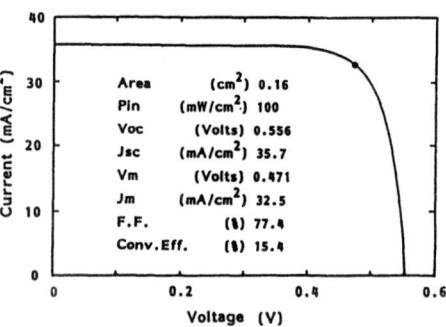

Fig.9   Performance of ITO/p μc-SiC/n poly-Si/Al heterojunction solar cell under AM 1,100mW/cm^2,25°C.

Thin Film Light Emitting Diode

We have developed a visible thin film light emitting diode, consisting of amorphous silicon carbide p-i-n junction (a-SiC TFLED) (11). The emitting colors can be varied from red to yellow by changing the optical gap of the luminescent i-layer. However, the luminance obtained so far was still insufficient for a practical application to a thin film flat display. A recent study on the carrier injection mechanism in the LED has revealed that the low luminance is due to the poor carrier injection efficiency from p- and n-type injectors into the luminescent i-layer. An approach to improve the injection efficiency is to utilize wide-band gap injection material possessing high conductivity.

The a-SiC TFLED has a typical structure of glass/ITO/SnO$_2$/p a-SiC:H/i a-SiC:H/n a-SiC:H/Al, with thicknesses of p-, i-, and n a-SiC:H layers of 15nm, 25-100nm and 30nm, respectively. Figure 10 shows the EL spectra of a-SiC TFLED's as a function of the optical energy gap of the p-layer.

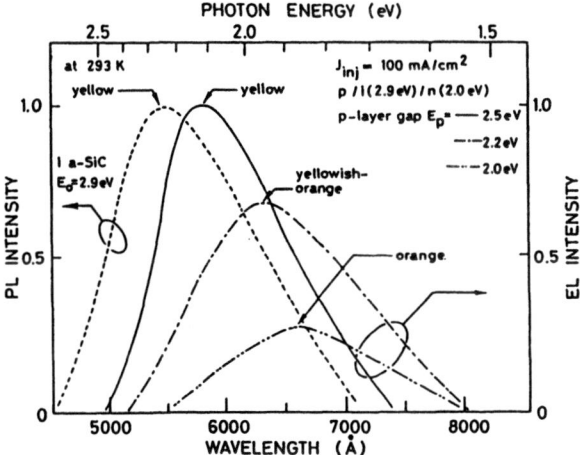

Fig.10 Dependence of emission spectra of a-SiC TFLED's on the optical energy gap of the p-layer.

In this case, the optical energy gaps of the i- and n-layers were kept constant at 2.92 and 2.00 eV, respectively. As the optical energy gap of the p-layer increases, the EL spectra shift towards shorter wavelength and the emitting color drastically changes from red to orange and then to yellow. The dash line shown in the figure is the PL spectrum (excited by 365 nm light) of the i a-SiC film. It should be noted that as the optical energy gap of the p-layer increases, the EL spectra shift closer to the position of the PL spectrum. The increment of the optical energy gap of the p-layer, decreases the barrier height for holes to tunnel, or in other words, lifts holes up to be injected into the localized tail states locating closer to the valence band edge of the i-layer and this results in the shift of the spectra as well as the improvement of the EL intensity. There is no significant change in the peak position of the spectra as the optical energy gap of the n-layer is increased.

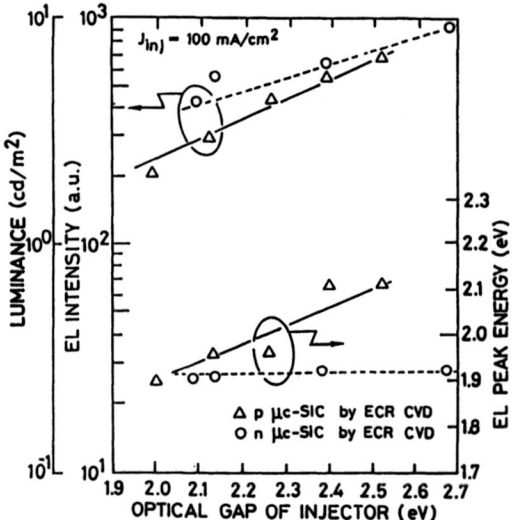

Fig.11 Summary of dependences of luminance and peak position of EL
spectra on the optical energy-gap of p and n type
microcrystalline SiC:H layers.

Figure 11 summarizes the EL intensity and peak energy dependence on the
optical gap of p-layer (triangle) and n-layer (circle). The optical energy
gap of the luminescent i-layer in this case is 2.9 eV. As the optical energy
gaps of p- and n-layers increase, the luminance gradually increases. When
the n-layer's gap increases from 2.0 to about 2.5 eV, the applied voltage
which is needed to achieve an injection current of 100 mA/cm^2 decreases from
15 to about 10 V. The increase in the luminance as a result of increment in
the n-layer optical gap is considered to be due to the decrement in the
electrical field in the i-layer, which weakens the field-quenching effect of
the luminescent efficiency (12). The highest luminance of a TFLED obtained
so far was about 13 cd/m^2 with an injection current density of 200mA/cm^2 and
bias voltage of 15 V.

CONCLUSION

ECR plasma CVD has been employed to prepare microcrystalline films in
low temperature process. Formation of Si and SiC microcrystalline phases
embedded in the a-SiC:H network structure is found to be largely enhanced by
hydrogen dilution of the reaction gas. Because of the inclusion of the
microcrystalline phase, the material exhibits excellent controllability of
the electrical properties by impurity doping. The optical band gap of
μc-SiC:H can be varied over a large range, from 2 to 2.8 eV, while retaining
good electrical conductivity. The outstanding features of p-type μc-SiC:H
satisfy the essential requirements for a wide-gap window material in
heterojunction solar cells as well as carrier injection layer in a-SiC p-i-n
junction light emitting diodes.

REFERENCES

1. Y. Hamakawa and Y. Tawada: Int.J.Solar Energy $\underline{1}$,(1982) 251
2. S. Veprek and V. Marecek, Solid State Electronics $\underline{11}$,(1968) 683.
3. A. Matsuda, J.of Non Crystalline Solids $\underline{59\&60}$,(1983) 767.
4. M. Matsuoka and K. Ono, J.Vac.Sci Technol.A6 $\underline{1}$,(1988) 25.
5. N.F. Mott and E.A. Davis, Electronic Processes in Non Crystalline Materials 2nd ed. (Oxford Press,1979).
6. Y. Hattori, D. Kruangam, K. Katoh, Y. Nitta, H. Okamoto and Y. Hamakawa, Proc. 19th IEEE Photovoltaic Specialists Conference,(1987) 689.
7. K. Hanaki, Z.Y. Xu, D. Kruangam, H. Okamoto and Y. Hamakawa, Proc. 18th IEEE Photovoltaic Specialists Conference,(1985) 394.
8. S. Nonomura, K. Fukumoto, H. Okamoto and Y. Hamakawa, J.of Non Crystalline Solids $\underline{59\&60}$,(1983) 1099.
9. M.K. Han, Y. Matsumoto, G. Hirata, H. Okamoto and Y. Hamakawa, First International Conference on Amorphous Semiconductor Technology, (1988) in press.
10. Y. Matsumoto, G. Hirata, K. Miyagi, H. Takakura, H. Okamoto and Y. Hamakawa, Int.Solar Energy Society, Solar World Congress,(1989) in press.
11. Y. Hamakawa, D. Kruangam, M. Deguchi, Y. Hattori, T. Toyama and H. Okamoto, Applied Surface Sc. $\underline{33/34}$,(1988) 1142.
12. D. Kruangam, T. Endo, W.G. Pu, S. Nonomura, H. Okamoto and Y. Hamakawa, J.of Non Crystalline Solids $\underline{97\&98}$,(1987) 293.

THE EFFECT OF HYDROGEN ON THE STRUCTURE OF AMORPHOUS AND
MICROCRYSTALLINE SIC PREPARED BY THE POLYMER ROUTE

C-J Chu, S-J. Ting, F. Bobonneau and J.D. Mackenzie,
Department of Material Science and Engineering, University of
California, Los Angeles, CA 90024

ABSTRACT

Amorphous SiC:H thin films have been deposited on different
substrates using metal-organic polymer solutions. The structure
of the amorphous phase has been proposed as the rings of Si and
C atoms with various sizes. Microcrystalline phase can be
produced when fired at temperatures higher than 1000°C. ^{13}C
MAS-NMR shows the evidence of the C=C double bonds. ESR results
show the major defects in this material are C-dangling bonds. As
predict, defects can be decreased by heat treatment in H2
atmosphere.

INTRODUCTION

Amorphous SiC has attracted much interest in recent years,
due to the potential applications in electronic devices. One of
the most significant applications is the replacement of p-type
a-Si:H with a-SiC:H for solar cell window materials, leading to
the increased efficiency from 8 to 12.5%[1]. Because of the
difference in chemical properties of constituent elements, it is
difficult to form defect-free amorphous network structure, thus
results in the low efficiency of the dopants. Understanding the
defect structures becomes very important in improving the
properties of the a-SiC:H. Electron spin resonance (ESR) is one
of the powerful tools that have been used to identify defects in
these materials.

An alternative method to prepare a-SiC:H with high doping
efficiency and wide optical band gap is to introduce
microcrystalline phase in the amorphous network. Theoretically,
the optical properties may be determined by the amorphous phase,
while the crystalline phase can contribute to the electrical
properties. In fact, many researches on the μc-SiC:H have been
carried out and used in different applications[2].

Recently, many advanced ceramic materials like carbides and
nitrides have been produced by the pyrolysis of some
metal-organic polymers[3]. By firing the polymer precursor in
the inert atmosphere up to the crystallization temperature, Tc,
a microcrystalline phase is formed and grow to become a
polycrystalline material. One of the advantages of this polymer
process is that the polymer can be easily fabricated into
different shapes, like fibers and films. The most well known
process is the fabrication of NICALON fiber from polycarbosilane
precursor which was invented by Yajima at 1975[4]. Although the
polymer process of the SiC fiber has been well established, the
structural evolution from the polymer to the polycrystalline
phase has not been well understood, especially in the amorphous
and the microcrystalline phase.

By the polymer route, films can also be easily fabricated
from polycarbosilane. After certain firing procedure, amorphous
and microcrystalline SiC films can be prepared. The structure
and some electrical and optical properties of this new SiC have

been studied by authors' group and compare to the SiC:H from vapor route[5].

FILM PREPARATION AND CHARACTERIZATION

The starting material, polycarbosilane (PC), is commerically avaliable from Dow Corning (X9-6348), which can then be dissolved in n-hexane or tolune to make a polymer solution.By dipping, spinning or spraying, polymer films can be coated on different substrates, like fused quartz, silicon and Al_2O_3. The samples were fired in Ar at different temperatures using a high temperature furance (Thermal Technology Inc. Astro 1000A). The structural evolution was studied using X-ray diffraction and TEM/SAED. The local environment changes of Si and C atoms were investigated using ^{29}Si and ^{13}C MAS-NMR spectra (MSL 300 Brucker Spectrometer). Electron spin resonance (ESR) has also been used to study the defect concentration. FTIR and UV visible spectroscopy have been used to characterized the optical properties.

Structure Evolution from Polymer to Amorphous Phase

The chemical analysis of the PC shows the composition to be $SiC_{2.0}H_{4.1}$. PC can be assumed to form chains of $Si-CH_2$ bonds with two kinds of Si units, $-(CH_2)-Si(CH_3)_2-(CH_2)-$ and $-(CH_2)-SiH(CH_3)-(CH_2)-$ in approximately one to one ratio. ^{29}Si MAS-NMR shows two peaks at 0 and -16 ppm representing these two units with approximately the same intensity. As the firing temperature increases, the disappearance of the peak at -16 ppm shows the cleavage of the Si-H bonds. Similarly, the FTIR result of the sample after 500°C firing shows a significant intensity decrease for the peak at 2100 cm^{-1}, which has been assigned to Si-H stretching mode vibration. In the 900°C sample the peak is still present, but shift to 2070 cm^{-1}. This may be due to the decreased number of hydrogen bonded to one silicon at higher temperatures.

Figure 1 shows the possible condensation reaction of polycarbosilane chains with evolution of H_2 and CH_4, and lead to the formation of rings of Si and C atoms. In the amorphous phase, the rings may have various sizes instead of the identical 6-member rings in the crystalline β-SiC. The appearance of Si-Si bonds or C-C bonds become possible. However the ^{29}Si MAS-NMR shows no evidence of Si-Si bonds, while ^{13}C MAS-NMR spectra shows peaks due to C-C bonds and C=C double bonds for the samples fired above 600°C. The structure of the amorphous phase of this new SiC from polymer route has been proposed as figure 2 [6].

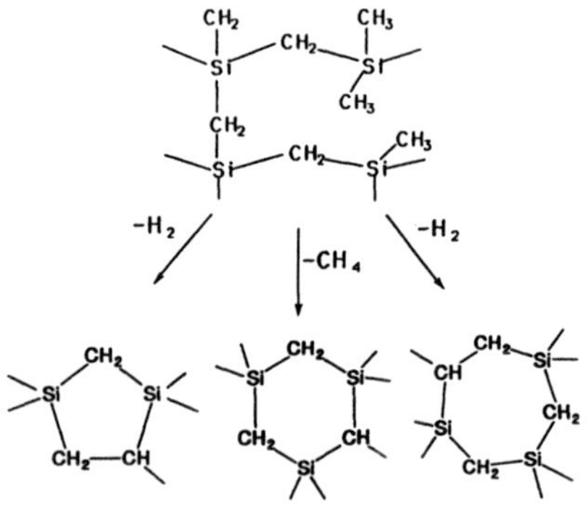

Fig.1 Possible condensation reactions of polycarbosilane

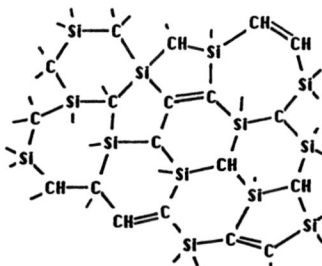

Fig.2 Proposed structure of the a-SiC from polymer route

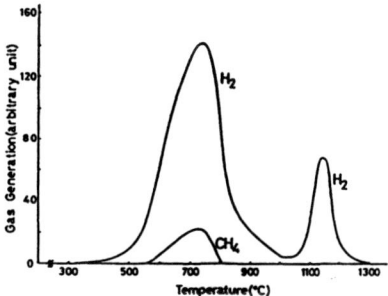

Fig.3 Gas chromatography for relased gas during heat- treatment of polycarbosilane

The pyrolytic gas during heat-treatment of PC under Ar gas flow has been studied by gas chromatography (figure3) [7]. From the composition formula SiC_2H_6 of PC, a simple presumption is SiC_2H_6 -- $SiC + CH_4 + H_2$. But from the GC results, the release of CH4 is very small compared with that of H_2. This appears to account for the excess carbon of the new a-SiC:H, and can also create defects as dangling bonds and double bonds. From the reactions in figure 1, samples fired in H_2 atmophere should suppress the evolution of H_2 and may even increase the release of CH_4. Defects in the SiC should also decrease.

Formation of Microcrystalline Phase

The new a-SiC:H can be converted into a μc phase when fired at higher temperatures. DTA study of the PC shows an exothermic peak at 1000°C, which may be due to the crystallization. In the XRD patterns, broad peaks due to crystalline β-SiC phases start to appear at 1000°C and sharpen with increasing temperature. The crystallite sizes can be evaluated by using the peak broadening procedure. Figure 4 shows the crystallite sizes at different firing temperatures. The DC conductivity of the samples fired at different temperatures has also been studied. The activation energy drops from 0.49ev to 0.12ev between 900 and 1000°C samples; which also correlated to the formation of the microcrystalline phase[8].

In the XRD pattern of the sample heated up to 1700°C, a small shoulder is seen at around $2\theta = 34°$. It has been assigned to α-SiC; suggesting that, at higher temperatures the crystalline phase consists of mixture of β-SiC with traces of the hexagonal form. In XRD patterns, a small peak around $2\theta = 26°$ is clearly visible. the intensity of this peak reaches a maximun value for 1200°C sample and decreases with increasing temperature. It could be assigned to the (002) line of

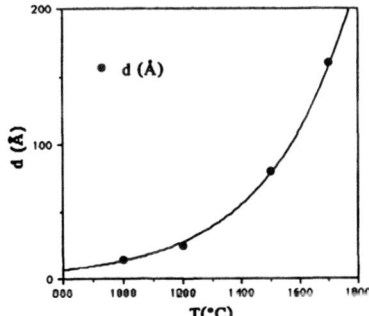

Fig.4 Crystallite sizes at different firing temperatures

carbon [9], which may be due to the presence of small clusters of graphite in the amorphous phase. As reported in literature[9], the graphite clusters should be present at the edge of SiC microcrystals and could prevent or slow the rate of crystal growth.

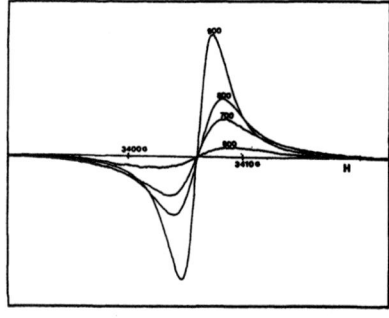

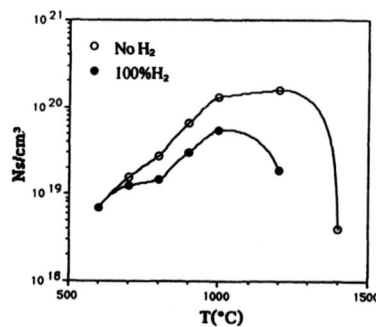

Fig.5 ESR results for the samples fired at 600-900°C

Fig.6 The spin density of the samples that fired at Ar and H$_2$

ESR characteristics of a-SiC:H

Figure 5 shows the ESR results of some samples fired at 600-900°C. All the samples show the same g-value as 2.003, which has been assigned to the C-dangling bonds. Liedtke et al. reported a similar result of a-SiC:H prepared by grow discharge with the same C/Si ratio (C/Si=1.5). Figure 6 compares the spin density of the samples fired at different temperatures. The spin density can go up to about 10^{19} cm^{-3}, which is in agreement with the number of defects founded in the a-SiC:H by glow-discharge[10]. The density of the C-dangling bonds reaches a maximum at 1200°C, which also correlates well with the XRD results of the maximum formation of graphite clusters at 1200°C.

Hydrogen effects

The samples have also been prepared at the same conditions by replacing Ar with 100% H$_2$ atmophere to confirm the theorical approach of H$_2$ effects on the defects. As shown in figure 6, the samples fired under 100%H$_2$ atmosphere show significant decrease in the number of C-dangling bonds. Some of the samples have also been studied by ^{13}C MAS-NMR. Figure 7 compares the NMR spectra of the samples fired at 700 and 800°C in Ar and H$_2$. The broad peak at around 140 ppm is assigned to C=C double bonds. However, the NMR peak almost disappears in the 700°C H2 treated sample. Similarly, in 800°C H$_2$ treated sample, the intensity of the NMR peak also decreases significantly. In the meantime, FTIR spectra of these samples also show a absorption peak at 1630 cm^{-1}, which has also been assigned to C=C stretching mode.

The 700°C sample shows a significant decrease of C=C double bonds with a small change of the density of C-dangling bonds; while the 800°C sample shows significant changes of both

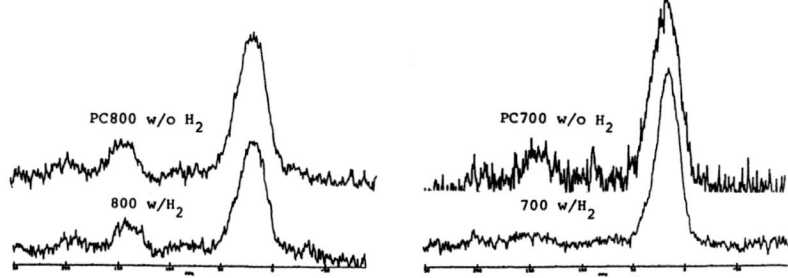

Fig.7 ^{13}C MAS-NMR spectra of 700°C and 800°C samples that fired with and without H_2

defects. Presumably the H_2 effect on the formation of C-dangling bonds is more temperature dependent than the formation of the C=C double bonds.

The C=C double bonds(sp^2 sites) have a tendency to form cluster. The decrease in C=C bonds should also introduce a drastic change in crystallization and the electrical and optical properties. Prelimary studies shows a significant change in optical gap. More detail studies are ongoing and will be published separately.

CONCLUSIONS

Microcrystalline phase of the new a-SiC by polymer route can be achieved by firing at temperatures higher than 1000°C. The defects of this new material at amorphous and microcrystalline phase have be identified as C-dangling bonds and C=C double bonds. As expected, the defect concentration decreases in the samples fired under H_2. However, the formation of C-dangling bonds shows more temperature dependence than the C=C double bonds.

ACKNOWLEDGEMENTS

The authors would like to thank Francis Taulelle and Mike Jackle for the MAS-NMR experiments, Jame Strouse for the ESR and FTIR experiments, Mary Colby and T-S Ho for the XRD and SEM investigations. This research was funded by NSF under grant No. DMR 87 06379.

REFERENCES

1. Y. Tawada, M. Kondo, H. Okamoto and Y. Hamakawa, Solar Energy Mater. _6_, 299 (1982).
2. Y. Hattori, D. Kruangam, T. Toyama, H. Okamoto and Y. Hamakawa, J. Non-Cryst. Solids _97&98_, 1079 (1987).
3. K.J. Wynne and R.W. Rice, Ann, Rev. Mater. Sci. _14_, 297 (1984).
4. S. Yajima, J. Hayashi and M. Omori, Chem. Lett. 931 (1975).
5. G.D. Soraru, F. Babonneau and J.D. Mackenzie,

Presented at the 7th International Symposium on Ceramics, Dec. 14-16, 1988, Bologna, Italy: in press in the proceeding.

6. G.D. Soraru, F. Babonneau and J.D. Mackenzie, J. Non-Cryst. Solids <u>106</u>, 256 (1988).

7. S. Yajima, Y. Hasegawa, J. Hayashi and M. Iimura, J. Mat. Sci. <u>13</u>, 2569 (1978).

8. C-J. Chu, G.D. Soraru, F. Babonneau and J.D. Mackenzie, Presented at the 2nd International Conference on Amorphous and Crystalline Silicon Carbide and Related Materials (ICACSC, 88), Dec. 15-16, 1988, Santa Clara, CA, U.S.A., in press in the proceeding.

9. J. Ayache, S. Bonnamy, X. Bourrat, A. Deurbergue, Y. Maniette, A. Oberlin, E. Bacque, M. Birot, J. Dunogues and J.P. Pillot, J. Mat. Sci. Lett. <u>7</u>, 885 (1988).

10. S. Liedtke, K. Jahn, F. Finger and W. Fuhs, J. Non-Cryst. Solids <u>77&78</u>, 849 (1985).

POWER DENSITY EFFECTS IN THE PHYSICAL AND CHEMICAL PROPERTIES OF SPUTTERED DIAMOND-LIKE CARBON THIN FILMS

N.-H. Cho*, K.M. Krishnan*, D.K. Veirs*, M.D. Rubin*, C.B. Hopper*, B. Bhushan*, and D.B. Bogy**
*Lawrence Berkeley Laboratory, One Cyclotron Rd., University of California, Berkeley, CA 94720
**Department of Mechanical Engineering, University of California, Berkeley, CA 94720

ABSTRACT

Thin films of diamond-like amorphous carbon were prepared by dc magnetron sputtering. A systematic variation in the physical properties of the films (mass density and electrical resistivity) was found as a function of sputtering power density. Chemical bonding and microstructure of the carbon thin films were investigated using electron energy loss spectroscopy (EELS) and Raman spectroscopy. Films grown at a lower power density were found to have more sp^3-bonded atomic sites and larger graphite microcrystals than films produced at higher sputtering power densities.

INTRODUCTION

Diamond-like amorphous carbon thin films have attracted much attention due to their useful and unique properties, e.g., high hardness, high electrical resistance, optical transparency in the infrared region, and chemical inertness. These films have been used as hard overcoats for magnetic storage media and as optical coatings, and in various other applications for wear and corrosion protection. Such carbon films are produced by a variety of techniques, such as hydrocarbon plasma-assisted chemical vapor deposition, ion-beam sputtering deposition, and magnetron sputtering techniques [1-4]. The macroscopic properties of the films have been reported to vary depending on the deposition method and conditions [5].

Carbon has two allotropic crystalline states, graphite and diamond, composed of sp^2 and sp^3 hybrid covalent bondings, respectively. The unique macroscopic properties of diamond-like amorphous carbon films may result from the microstructure and chemical bonding of the carbon atoms. The ratio of sp^2 to sp^3 chemical bondings has been regarded as a key issue in understanding the behavior of diamond-like carbon films. In order to identify chemical bonding distributions, electronic band structures and phonon density of states of solid film have been investigated using the interaction of lasers, incident accelerated electrons, and X-rays with solid carbon thin films [6-9].

In this study, the effect of sputtering power density has been investigated by growing diamond-like amorphous carbon films under well-controlled conditions. The main focus is on determining the microstructural differences between the carbon thin films prepared at different sputtering power densities and relating the differences to their properties.

EXPERIMENTAL

Sample preparation

Magnetron sputtering was used to produce diamond-like amorphous carbon thin films. A graphite target with a diameter of 3.0 inches was used as the carbon source. Pure Ar gas was introduced into the chamber and ionized by use of dc power. Total pressure was kept at 10 mTorr.

Samples were produced with sputtering power densities of 0.1, 1.1, 2.1, and 10 watts/cm^2. No substrate bias voltage was applied; no heat was supplied to the substrates from outside.

NaCl and Si were used as substrates. Carbon films with thicknesses of 300-400 Å deposited on NaCl were used for TEM and Raman spectroscopy studies. For the TEM studies, the films were separated from the NaCl substrates by dissolving the NaCl in distilled water. For mass density determinations, carbon films with thicknesses of about 1 μm were produced on Si substrates; the Si substrates were etched with HF acid in order to avoid delamination due to severe stress in these thick films.

Characterization techniques

Electron energy loss spectra were obtained using a JEOL 200 CX electron microscope operating at 200 kV. A parallel electron energy loss spectrometer (Gatan 666) with a resolution of 1.2 eV was used to analyze the energy distribution of electrons transmitted through the specimen.

Raman spectra were obtained between 1000-2000 cm^{-1} using the 488 nm Ar$^+$ laser line. The Raman spectra were fit using the sum of two Gaussian line shapes and a linear background.

Measurement of physical properties

The mass density of each film was determined by measuring its weight and volume. Film thickness was measured with a profilometer. The net weight of each film was determined from the difference between the weight of the substrate prior to deposition of the film, and the weight of the substrate and film after deposition. The error range of the balance was 10^{-5} g. A four-point probe apparatus was used to measure the electrical resistivity of the films.

RESULTS AND DISCUSSION

Electron energy loss spectroscopy

Electron energy loss spectra were collected from graphite and the thin carbon films. An example of the low-loss region of each spectrum is illustrated in Fig. 1. Spectra a, b, c, and d in Fig. 1 were obtained from graphite and samples prepared at 10, 2.1, and 0.1 watts/cm^2, respectively; these samples are referred to as samples A, B, C, and D in this paper. Two prominent peaks are seen at 6.4 eV and 27 eV in spectrum a. The losses associated with features in this energy range are normally related to either plasmon oscillations or interband transitions of valence electrons [10]. In particular for graphite, it is known that the peak at 6.4 eV corresponds to plasmon oscillations of the π electrons in the valence band, and the peak at 27 eV arises from the collective excitation of all the valence electrons (π + σ) [11,12].

The plasmon resonances of diamond-like carbon films appear at a lower energy value than those of graphite, as shown in Fig. 1. Strong π plasmon oscillations are observed at 6.0 eV, 6.0 eV, and 5.0 eV for samples B, C, and D, respectively, while the π + σ peaks are found at nearly the same energy, 24.5 eV.

Using a simple free electron gas model, one can establish a quadratic relationship between valence electron density and plasmon frequency [13]. Hence, samples B, C, and D appear to have nearly the same total density of π + σ electrons, but the density of π electrons for samples B and C is higher than that of sample D. These results indicate that more sp^2-bonded atomic sites are present in samples B and C than in sample D, while the sp^3/sp^2 ratio in sample B is similar to that in sample C.

Raman spectrometry

Raman spectra were obtained from samples A, B, C, and D. Fig. 2 shows spectra obtained from samples B and D for comparison. Each plot consists of the experimental data and the fitted curves. The "G" line positions for samples B and D are seen at 1558 cm^{-1} and 1541 cm^{-1} respectively, while the "D" line positions are observed at 1379 cm^{-1} and 1368 cm^{-1}, respectively. The integrated intensity ratio I_d/I_g for sample D is lower than that of sample B.

Raman spectra for the two crystalline forms of carbon are very well known; the first-order Raman spectra of highly ordered pyrolitic graphite and of diamond have peaks at 1580 cm^{-1} and 1332 cm^{-1}, respectively. A peak in the graphite spectrum near 1350 cm^{-1} appears in disordered

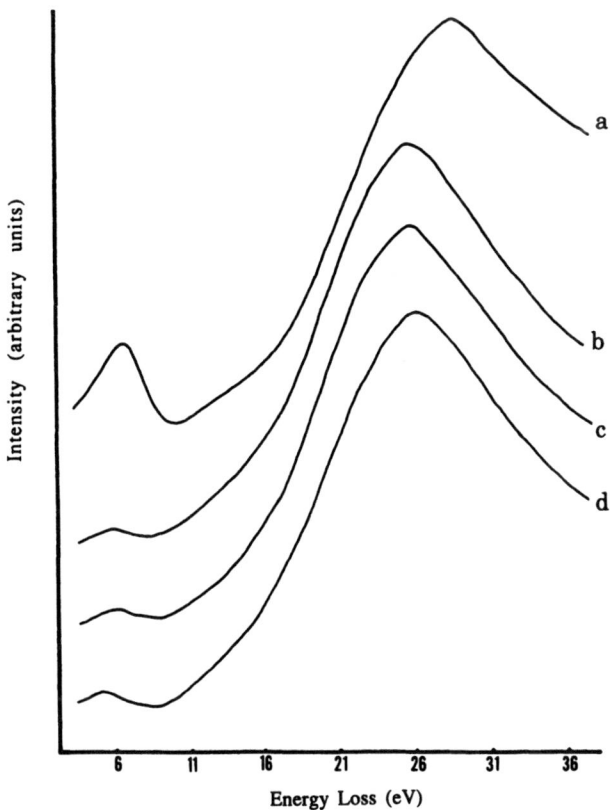

Fig. 1. Electron energy loss spectra in the plasmon region: (a) sample A (graphite); (b) sample B, prepared at 10.0 watts/cm^2; (c) sample C, prepared at 2.1 watts/cm^2; (d) sample D, prepared at 0.1 watts/cm^2.

graphite [6,14]. This peak increases in intensity relative to the intensity of the 1580 cm^{-1} peak as the microcrystallite size, L_a, decreases [15]. Raman spectra of amorphous solids feature broad bands related to the phonon density of states [16]. In particular, for amorphous carbon, the positions of the "G" and "D" peaks shift and the linewidth increase from those of graphite as disorder and the fraction of sp^3-bonded atomic sites increases [17,18]. Therefore, the band positions may yield information about bond-angle disorder and bonding in the amorphous carbon films.

The "G" band position of sample D in Fig. 2b is shifted to a lower frequency by about 17 cm^{-1} compared to that of sample B, which in turn is shifted by 22 cm^{-2} from the frequency in graphite. The shift may indicate that sample D has a higher concentration of sp^3-bonded atomic sites than sample B. A calculation by Richter et al. using a mixture of sp^2- and sp^3-bonded atomic sites and C-C force constants yields shifts in the "G" line frequency for sp^3-bonded fractions of 15% and 5%, which are similar to those observed in samples D and B, respectively.

If the linear relationship between I_d/I_g and $1/L_a$ continues beyond the experimentally observed region, then the ratios of I_d/I_g for samples B and D indicate that the size of graphite

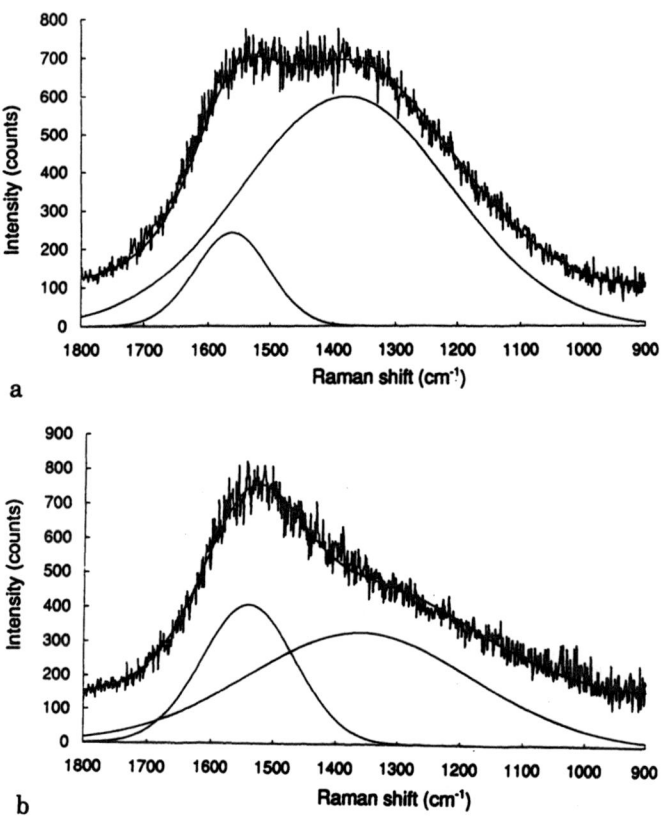

Fig. 2. Raman spectra of the diamond-like carbon films. Spectra for sample B, prepared at 10.0 watts/cm^2, and sample D, prepared at 0.1 watts/cm^2, are shown in (a) and (b), respectively. The experimental data and fit are illustrated.

microcrystallites decreases as the sputtering power density increases. These interpretations of the data suggest that the total volume of sp^3-bonded atomic sites in sample R is higher than in sample P.

Physical properties

Fig. 3 shows the electrical resistivity of carbon thin films grown at various sputtering power densities. A rapid decrease in electrical resistivity is observed as the power density increases below 2.1 watts/cm^2. As was discussed in the EELS and Raman spectroscopy sections, samples deposited at the relatively higher power densities have higher percentages of sp^2-bonded atomic sites. The increase in sp^2-bonded atomic sites and higher density of π electrons near the Fermi energy level is believed to result in lower electrical resistivity.

The mass density of carbon thin films is also shown in Fig. 3. Increasing the sputtering power density from 0.1 to 10 watts/cm^2 results in a decrease in film density, from 2.1 to 1.75 g/cm^3. The mass density of samples appears to vary from that of pyrolitic graphite (1.6-1.95

g/cm^3) to that of single crystal graphite (2.26 g/cm^3). The mass density of sample D is close to that of crystalline graphite; however, the results of EELS and Raman spectroscopy studies show

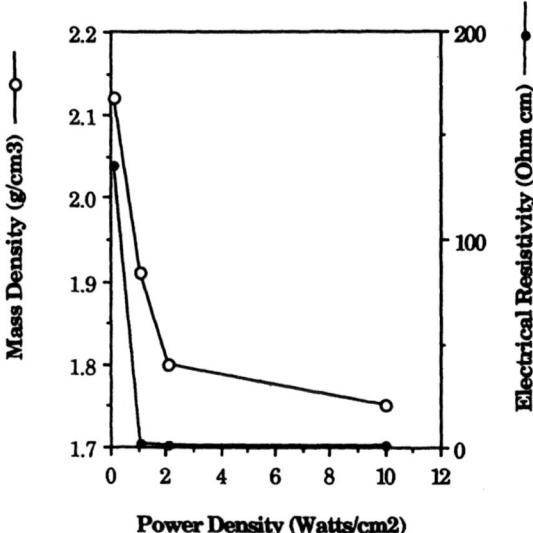

Fig. 3. Electrical resistivity and mass density of the diamond-like carbon films vs sputtering power density.

that sample D is amorphous. The relatively high mass density results from the particular arrangements and/or distributions of the sp^3- and sp^2-bonded atomic sites, and is not due to the enlargement or growth of graphite microcrystals.

CONCLUSION

EELS and Raman spectroscopy were used to determine the microstructural variation of carbon thin films depending on deposition conditions. It has been shown that the chemical bonding distributions and microstructure of diamond-like carbon thin films are affected by sputtering power density. Films produced at a lower sputtering power density have more sp^3-bonded atomic sites and larger graphite microcrystals than films grown at higher sputtering power densities. The electrical resistivity in diamond-like carbon films is consistent with a microstructure containing both sp^2- and sp^3-bonded carbon. As the fraction of sp^2-bonded atomic sites increases, the higher probability for orbital overlap and the higher density of π electrons results in the reduction of electrical resistivity. More efforts are being made to determine such physical properties as hardness, optical band gap, and residual stress in order to obtain a better understanding of the relationship between the microstructure and the physical properties of the films.

ACKNOWLEDGMENTS

The authors would like to thank Mr. C.J. Echer for assistance in using the microscope facilities at the National Center for Electron Microscopy. This work was supported by the National Science Foundation through Grant MSE-8800149 with the University of California at Berkeley, by the Computer Mechanics Lab. at UCB, and by the Director, Office of Energy

Research, Office of Basic Energy Sciences, Materials Sciences Division, of the U.S. Department of Energy under Contract No. DE-AC03-76SF00098

REFERENCES

1. D.A. Anderson, Phil. Mag., **35**, 17, 1977.
2. A. Bubenzer, B. Dischler, B. Brandt, and P. Koidl, J. Appl. Phys., **54**(8), 4590, 1983.
3. C. Weissmantel, K. Bewilogua, K. Breuer, D. Dietrich, U. Ebersbach, H.J. Erler, B. Rau and G. Riessen, Thin Solid Films, **96**, 31, 1982.
4. N. Savvides, J. Appl. Phys., **58**, 518, 1985.
5. H. Tsai and D.B. Bogy, J. Vac. Sci. Tech., A**5**(6), 3287, 1987.
6. R.O. Dillon, J.A. Woollam, and V. Katkanant, Phys. Rev.B, **29**(6), 3482, 1984.
7. N. Wada, P.J. Gaczi, and S.A. Solin, J. Non-Cryst. Solids, **35**, 543, 1980.
8. F.R. McFeely, S.P. Lowlaczyk, L. Ley, R.G. Cavell, R.A. Pollak, and D.A. Shirley, Phys. Rev., B**9**, 5268, 1974.
9. P.E. Batson and A.J. Craven, Phys. Rev. Lett., **42**, 893, 1979.
10. H. Raether, in "Excitation of Plasmons and Interband Transitions by Electrons," Springer-Verlag.
11. W.Y. Liang and S. Cundy, Phil. Mag., **19**, 1031, 1969.
12. E.A. Taft and H.R. Phillip, Phys. Rev., **138**, 197, 1965.
13. D.B. Williams and J.W. Edington, J. Microscopy, **108**(2), 113, 1976.
14. R.J. Nemanich and S.A. Solin, Phys. Rev.B, **20**(2), 392, 1979.
15. F. Tuninstra and J.L. Koenig, J. Compos. Mater., **4**, 492, 1970.
16. R. Zallen, in "The Physics of Amorphous Solids," Wiley-Interscience Pub., NY, 1983.
17. D. Beeman, J. Silverman, R. Lynds, and M.R. Anderson, Phys. Rev.B, **30**, 870, 1984.
18. A. Richter, H.-J. Scheibe, W. Pompe, K.-W. Brzezinka, and I. Muhling, J. Non-Cryst. Solids, **88**, 131, 1986.
* Permanent address of Dr. B. Bhushan: IBM research division, Almaden Research Center, San Jose, CA 95120

The Effect of Hydrogen on the Structure and Electrical
and Optical Properties of Silicon-Germanium Alloys

C.M. Fortmann and D.E. Albright
University of Delaware, Institute of Energy Conversion Newark,
Delaware 19716

I.H. Campbell and P.M. Fauchet
Princeton University, Department of Electrical Engineering
Princeton, NJ 08544

ABSTRACT

The transport properties of intrinsic micro-crystalline and
amorphous SiGe films are investigated. In the amorphous system
it is found that the electron lifetime is relatively independent
of band gap while the electron mobility of a-Ge:H is a factor of
ten less than that of a-Si:H. Micro-crystalline silicon films
have greater excess photo-conductivity as compared to a-Si:H
predominantly due to longer effective lifetimes. Films containing
micro-crystalline silicon in a matrix of a-SiGe:H have been prepared and are found to have comparatively poor transport.

INTRODUCTION

Amorphous silicon (and its alloys) are one of the few thin
film semiconductors to reach commercialization. Large area amorphous silicon based commercial products include solar cells and
thin film transistor arrays. It is natural to assume that the
ease of large area deposition will extend to the microcrystalline system.
To facilitate the discussion on the role of film hydrogen
content on the electronic properties of materials in the Si-Ge
amorphous and micro-crystalline systems it is necessary to summarize several key results of previously published work. Microcrystalline silicon occurs when the deposition rate of a-Si:H
film precursors is nearly the same as rate of film removal by
hydrogen etching (1). H radicals do not etch amorphous Germanium
(2) or micro-crystalline silicon at rates as large as those of
a-Si:H (1). In the amorphous phase, C_H (film hydrogen content)
increases with increasing H radical flux until the etch rate
becomes comparable to the deposition rate and a transition to
micro-crystalline structure occurs (1),(3).
In this work the majority carrier transport properties of
amorphous and micro-crystalline silicon-germanium alloys are
explored. The amorphous alloy system is comparatively well studied (4), however we will need to both reexamine and extend the
study of the amorphous alloys in order to provide a framework in
which the micro-crystalline system can be evaluated. The pro-
spect of using these materials in large area devices will depend
on the relative changes in lifetime, fermi level position and
mobility that accompany the alloying.

TRANSPORT IN AMORPHOUS ALLOY FILMS

Figure 1 shows the excess photo-conductivity, $AM1.5@100$ mW/cm^2
$(\triangle\sigma_p=\sigma_p-\sigma_d=eu G\overline{T})$ as a function of band gap for all films in our
data base. All films in this study had activation energies for

the dark conductivity within a few kT of mid-gap. In general the largest $\Delta\sigma_p$ for a given band gap decreases with decreasing band gap reaching a minimum at band gaps of ~1.3 eV. The dark conductivity (σ_d) was measured as a function of temperature to establish its activation energy (E_a). The $\Delta\sigma_p$ to σ_d ratios, R, for the same data base (Figure 2) shows far less scatter. This is interesting because R is independent of the electron mobility.

$$R=\Delta\sigma_p/\sigma_d=eG\tau u/(euN_c exp[-\{E_f+W\}/kT])=G\tau/(N_c exp[-\{E_f+W\}/kT]) \quad 1)$$

Where G is the generation rate, u is the effective electron mobility, τ is the effective electron lifetime, N_c is the density of states in the conduction band and W allows for the possibility of a thermally activated mobility ($u_\tau=u_o exp\{-W/kT\}$) as would be the case for conduction through localized states (5).

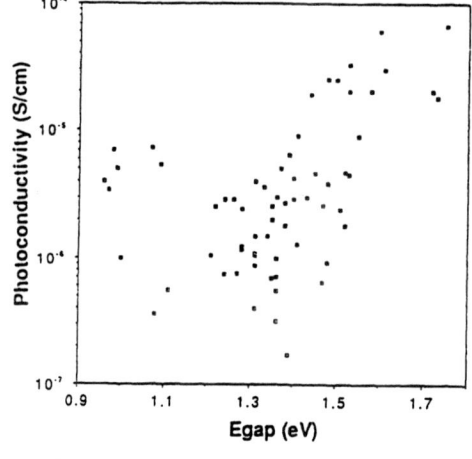

Figure 1
The $\Delta\sigma_p$ of a-SiGe:H films as a function of band gap.

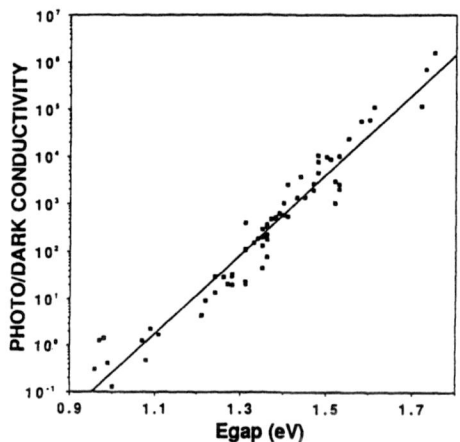

Figure 2.
The $\Delta\sigma_p$ to σ_d ratio for the films of figure 1.

If W=0 then E_f is equal to the E_a and multiplying R by exp[$-E_a/kT$] yields $G\tau/N_c$ (figure 3), thus the band gap dependance that enters through E_a is removed. It is interesting to note that both a-Si:H and a-Ge:H have the same value of $G\tau/N_c$. The relatively poor $\Delta\sigma_p$ of a-Ge:H (less than 10% of that of a-Si:H) together with a value of $G\tau/N_c$ that is equal to that of a-Si:H is consistent with u of a-Ge:H being only 10% of that of a-Si:H. Thus far, we have prepared alloys with band gaps as small as 1.44eV that have a $G\tau/N_c$ value as large as that of a-Si:H. Over the range of alloys investigated the minimum $G\tau/N_c$ occurs at the intermediate compositions ($1.1 < E_g < 1.4$ eV) but even here the change is a relatively small (~30% of the values for either a-Si:H or a-Ge:H). This decrease in $G\tau/N_c$ for the intermediate band gap alloys can be due to either a reduction in τ for these alloys or a non-zero value for W (that is the conductivity of the intermediate alloys has a larger thermal activation energy than either a-Si:H or a-Ge:H).

Figure 4 (left) shows the photo conductivity of amorphous films containing ~60% germanium as a function of C_H. It is important to note that for these set of films $\Delta\sigma_p$ deceases with increasing band gap (band gap increases with C_H) for a fixed Ge content as seen in figure 4 (right). In the amorphous alloy system C_H has a detrimental effect on electronic transport and along with the Ge content provides a range of transport properties for a given band gap. It is important to note that films grown under H radical etching (at levels not sufficient to result in detectable micro-crystalline structure), films grown from polymeric species as well as those grown at lower temperatures have relatively large C_H.

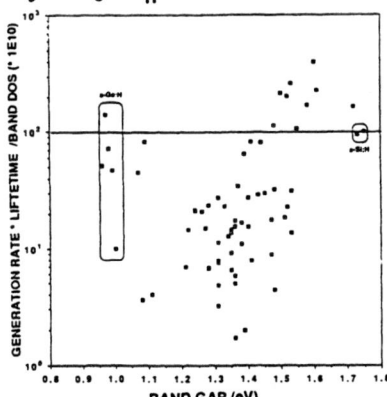

Figure 3
$G\tau/N_c$ versus E_g for the films of figure 1.

These results are consistent with an alloy transport model that proposes that conduction band disorder results in a thermally activated mobility (W>0). In terms of this model conduction band barriers result from the clustering of Si atoms, while wells result from the clustering of Ge atoms (6,7) as illustrated in figure 5. Both Ge and Si clusters are expected from a statistical distribution of Si-Ge, Si-Si and Ge-Ge bonds in a randomly mixed alloy. Since H bonds preferentially to the silicon atoms (8) and H increases the average band gap in a-Si:H it is assumed that the conduction band barriers will be heightened by H. The result is that alloys with intermediate band gaps are expected to have the smallest $\Delta\sigma_p$ and that $\Delta\sigma_p$ decreases further with increasing C_H as found experimentally. The small reduction in $g\tau/N_c$ seen in these alloys would be explained by a violation of the assumption that W~0, rather than a decrease in τ.

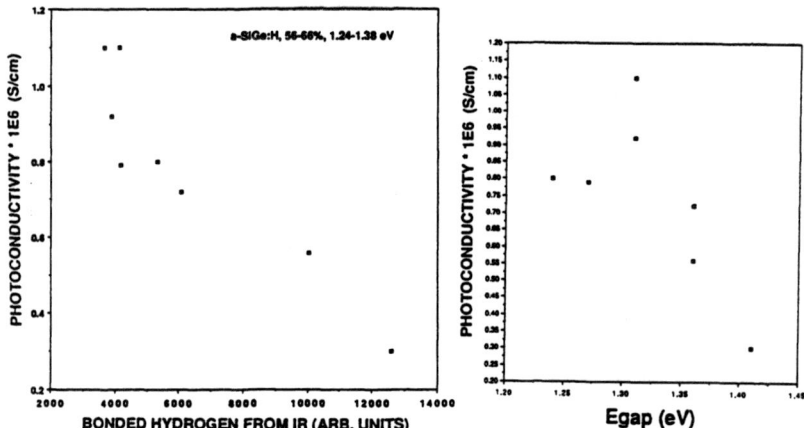

Figure 4. $\Delta\sigma_p$ versus C_H for films containing~60% Ge(left), and $\Delta\sigma_p$ versus E_g for the same films (right).

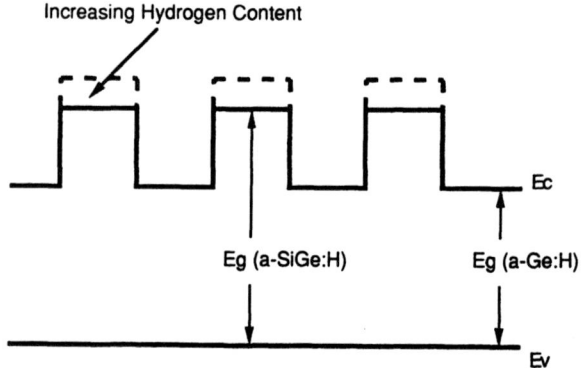

Increasing Hydrogen Content

Eg (a-SiGe:H)   Eg (a-Ge:H)

Figure 5 Idealized diagram of band gap fluctuation due to clustering and hydrogen on Si atoms in a-SiGe:H

TRANSPORT IN THE MICRO-CRYSTALLINE SI-GE SYSTEM

Raman analysis determined that the micro-crystalline content of the Si films was nearly 100%. The alloy films had a somewhat more complex structure as they were composed of several distinct phases including: micro-crystalline Si, micro-crystalline Ge and amorphous SiGe (figure 6). Table 1 shows the conductivity data of micro-crystalline films. It is clear that the large $\Delta\sigma_p$ observed in the micro-crystalline Si films deposited at 205°C is predominantly due to a large lifetime since $G\Upsilon/N_C$ is at least two order of magnitude greater than of a-Si:H. Consequentially, the change in the electron mobility must be relatively small (<10).

The $\triangle\sigma_D$ and $G\Upsilon/N_C$ of SiGe mixed structure alloys are both less than that of the analogous amorphous alloys indicating that both mobility and lifetime are less than those in the amorphous case. The reduction in electronic transport is predicted by the alloy transport model as band gap fluctuations arise due to the Si and Ge micro-crystalline clusters (clusters are sufficiently large so as to have Raman shifts characteristic of Si and Ge crystals).

Table 1
Micro-crystalline Film Properties

Alloy	DEP TEMP (°C)	$\triangle\sigma_D$ (S/cm)	σ_d (S/cm)	E_a (eV)	%Ge	R	$G\Upsilon/N_C$
Si	205	3.5e-4	3.0e-6	0.45	0	1.2e2	2.8e-6
SiGe	205	1.5e-7	1.3e-9	0.68	41	1.2e2	3.6e-10
SiGe	205	8.6e-6	1.6e-7	0.57	47	5.4e1	1.2e-8
SiGe	205	1.6e-6	1.7e-8	0.62	48	9.4e1	3.1e-9

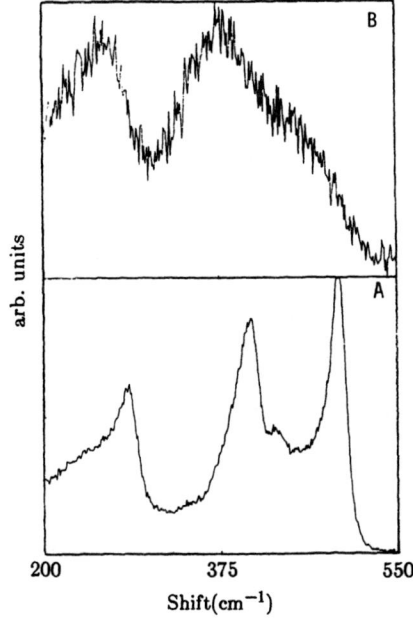

Figure 6
Raman spectrum
of mixed crystalline
phase SiGe (A) and
amorphous SiGe (B).

CONCLUSIONS

Previously we reported that the performance of graded amorphous SiGe solar cells could be explained in terms of mobility gradients arising due to the relatively poor electron mobility in a-SiGe:H (compared to that of a-Si:H)(9). This is consistent with the present results which show that the decrease in a-SiGe:H conductivity with increasing Ge content is predominantly due to a decrease in electron mobility. This result stands in contrast to those obtained by Aljishi et al. (10). For a given band gap increasing C_H also results in reduced conductivity. These results are consistent with an electron mobility model based on conduction band fluctuations caused by Ge and Si clustering.

In micro-crystalline silicon the increase in excess photoconductivity is predominantly due to an increase in the effective lifetime, indicating that this material will not drastically improve the performance of a majority carrier device such as thin film transistors. The transport properties of mixed phase (micro-crystalline and amorphous) SiGe alloys appear to both have poorer mobilities and lifetimes as compared to their amorphous counterparts.

REFERENCES

1. N. Saxena, C.M. Fortmann, T.W.F. Russell, Materials Research Society Proceedings Vol. 149, p.99, MRS, Pittsburgh, PA (1989)

2. D.E. Albright, C.M. Fortmann, T.W.F. Russell, Materials Research Society Proceedings Vol. 149, p. 521, MRS, Pittsburgh, PA (1989)

3. N. Saxena, D.E. Albright, C.M. Fortmann, T.W.F. Russell, Proc. of the 13th ICALS to be published in the J. Non. Cryst. Solids

4. K.D. Mackenzie, J.R. Eggert, D.J. Leopold, Y.M. Li, S. Lin, and W. Paul Phys. Rev. B31, p. 2198 (1985)

5. M.H. Brodsky, Topics in Applied Physics Vol. 36, Amorphous Semiconductors, p124 Springer-Verlag NY (1985)

6. C.M. Fortmann, Materials Research Society Proceedings Vol. 118, p.691, MRS, Pittsburgh, PA (1988)

7. W. Paul, Amorphous Silicon and Related Materials pp. 63-79 edited by H. Fritzsche, World Scientific, NJ (1988)

8. C.M. Fortmann and J. Tu Conference record of the 20th IEEE PVSC, p. 139, IEEE, NY (1989)

9. C.M. Fortmann, S.S. Hegedus, and W.A. Buchanan, Proc. of the 1st ICAST to be published in the J. Non. Cryst. Solids

10. S. Aljishi, Z E. Smith, V. Chu, J. Kolodzey, D. Slobodin, J.P. Conde, D.S. Shen, S. Wagner, AIP Conf. Proc. 157, p. 25, AIP, NY (1987)

RAMAN STUDIES OF MICROSTRUCTURAL CHANGES IN AMORPHOUS SILICON-BORON ALLOYS DUE TO ANNEALING

G. YANG*, P. BAI*, Y.-J WU*, B. Y. TONG**, S. K. WONG**, J. DU**, and I. HILL***
*Center for Integrated Electronics and Physics Department, Rensselaer Polytechnic Institute, Troy, NY 12180, USA
**Center of Chemical Physics; ***Surface Science Laboratory, University of Western Ontario, London, Ontario, Canada

ABSTRACT

Crystallization of amorphous $Si_{1-x}B_x$ alloy films by annealing is studied. Amorphous $Si_{1-x}B_x$ alloy films with composition of boron x ranging from 0.01 to 0.5 are deposited on Si substrates at a temperature of 480° in a low pressure chemical vapor deposition (LPCVD) system. Three films with the boron contents, 1%, 7% and 45%, are used in this study. The films are annealed in a nitrogen ambient for 30 minutes at temperatures between 600°C and 900°C. Raman spectra of the silicon vibrational mode serve as a indicator for the microstructure of the a-$Si_{1-x}B_x$ alloy films. Quantitative estimates of the volume fraction of the crystalline silicon component in respect to the amorphous silicon component in the films are calculated based on the silicon TO mode. The results show that while for the film with the boron content of 1% crystallization occurs at the annealing temperature of 500°C, the annealing temperature of 700°C is required to observe crystallization in the films with the boron contents of 7% and 45%. As the annealing temperature increases, the volume fraction of the crystalline component increases. For a given annealing temperature, the rate of crystallization depends inversely on the boron content in the films.

I. INTRODUCTION

Crystallization by heating to elevated temperatures has been studied for a number of silicon-base amorphous alloys, among which a-Si:B has also received a certain degree of attention.[1-9] The a-Si:B studies, however, have been mostly concentrated on materials with the boron content in the doping regime.[7-9] The crystallization process in an amorphous alloy is a complex process involving both the structural constraints of a amorphous network and the chemistry among the alloying components. Morimoto et al[2] found that the crystallization temperature is a function of the component composition in an alloy and segregation of the excess element was also observed. Gonzalez-Hernandez et al[1] pointed out that the nucleation process in the P doped a-Si:H solids has higher activation energy than the growth process. Therefore, a sample that has trace crystallinity behaves differently in the crystallization process compared to a purely amorphous sample.

The local structural order of the amorphous silicon-boron alloy films has been found to depend on the boron content in the films.[10] The variation in the local structural order is attributed to the mechanisms of the boron incorporation. In this paper we report our Raman study on the structural changes in the a-$Si_{1-x}B_x$ alloy films induced by the heat treatment at various temperatures.

II. EXPERIMENTAL

The a-$Si_{1-x}B_x$ films with x ranging from 0.01 to 0.5 are prepared using a low pressure chemical vapor deposition (LPCVD) system in which the silane $[SiH_4]$ and diborane $[B_2H_6]$ gasses are thermally co-pyroled.[11] The films are grown on crystalline silicon substrates at a temperature of 480°C. The thickness of the films is 3000 Å. The LPCVD system operates at a pressure of 3 Torr. The boron concentration in the films is controlled by the B_2H_6/SiH_4 gas ratio and measured using secondary ion mass spectrometry. The hydrogen content in all the films is found to be less than 0.1%. The relatively low hydrogen content is presumably due to the high substrate temperature in the film deposition. The oxygen content in the films is between 0.1-1%. The isochronical annealing of the films are performed in a nitrogen ambient for 30 minutes. The Raman spectra were measured in the back scattering geometry using a multichannel spectrometer (Dilor OMARS89) and an Ar ion laser at a wavelength of 514.5 nm with a typical output power of 100 mw.

Mat. Res. Soc. Symp. Proc. Vol. 164. ©1990 Materials Research Society

III. RESULTS AND DISCUSSION

Three boron contents, 1%, 7%, and 45%, are chosen, respectively, from three regions corresponding to, as is shown in a our previous paper, the different behaviors of the local structural disorder of the as-deposited a-$Si_{1-x}B_x$ films.[7] Figure 1-3 show, respectively, a set of Raman spectra for the a-Si:B film with the boron content of 1%, 7%, and 45%, the as-deposited and annealed at different temperatures. For the a-Si:B films with the boron content of 1% and 7% the spectral region shown is between 200cm^{-1} and 600cm^{-1} while for the a-Si:B film with the boron content of 45% the spectral region shown is from 50cm^{-1} to 550cm^{-1}. The structure below 100cm^{-1} is due to the air ambient of the Raman scattering experiment.

The unannealed films with the boron content of 1% and 7% show a spectrum, containing a broad peak at about 480cm^{-1}, very similar to that of amorphous silicon. No structure at 520cm^{-1}, which corresponds to crystalline silicon, is visible. For the unannealed film with the boron content of 45%, the spectrum deviates from a typical a-Si one, although a small broad peak at about 480cm^{-1} is still present suggesting the existence of silicon clusters in the films. Upon annealing at 500°C, the film with 1% boron start to crystallize as is indicated by a sharp peak at about 520cm^{-1} in figure 1. As the annealing temperature increases, the crystalline peak grows at the expense of the amorphous peak. For the films with 7% and 45% boron, crystallization is observed only at the annealing temperature of 700° (figures 2 and 3). An increase of the annealing temperature is also accompanied by an increase of the crystalline peak at about 520cm^{-1}.

It is generally believed that if the volume fraction of crystallinity is less than 3%, the TO mode crystalline peak height is below the tail of, and thus indistinguishable from, the amorphous peak.[1] The as-deposited amorphous films therefore may contain a small volume fraction of crystallinity that is below the Raman resolution. The ratio of the integrated intensity of the crystalline peak to the amorphous peak yields a measure of the relative volume fraction of the two components for silicon. The integrated intensity of the crystalline silicon TO peak divided by the total integrated intensity of the silicon TO band including the crystalline and amorphous peaks is calculated for the films and listed in table I. This quantity serves as the relative spectral contribution of the crystalline component in the film. To quantitatively relate the spectral contribution to the volume fraction, many factors, including the Raman scattering cross section and the absorption coefficient of the silicon crystallites and amorphous silicon, have to be taken into account. For small particles of microcrystalline silicon these two parameters have been shown to be not too different from amorphous silicon.[12] In this case, the spectral contribution can be viewed as being roughly equal to the volume fraction. The boron content dependence of the crystallization of the films is clearly shown by the crystallite volume fraction at the annealing temperature of 700°C where a value of 0.42 is obtained for the boron content of 1%, 0.36 for the boron content of 7%, and 0.13 for the boron content of 45%.

While there are obviously uncertainties in directly relating our results to other's, it is still useful to have a comparison. Bisaro et al[9] found that the crystallization rate increases with an increase of the boron content in the film. But the maximum boron content in their study was in the doping regime, that is, less than 1%. Ishiwara et al[7] in their lateral solid phase epitaxy study observed a similar increase in the crystallization rate with an increase of the boron content for the boron content below the solid solubility limit of boron in silicon. They suggested that the limit is about 7×10^{20} cm^{-3} which corresponds approximately 1.5-2% of boron in a-Si:B films. As the boron content is higher than the solid solubility limit, however, a decrease of crystallization rate was observed. The results seem to agree qualitatively with our results in table I. The high boron contents of 7% and 45% are definitely out of the solubility limit, therefore crystallization is hindered, while the film with 1% boron falls within the limit.

In figures 1 and 2, no structure is observed at around 620cm^{-1} or 643cm^{-1} which correspond to the vibrational modes of four-fold substitutional ^{11}B and ^{10}B in c-Si.[13] This suggests that in the annealing process the boron atoms are mostly not incorporated substitutionally in the c-Si crystallites. Figure 3 shows that for the a-Si:B film with the boron content of 45% the increase of the crystalline silicon Raman TO peak with the annealing temperature seems to be disproportionally higher than the decrease of the amorphous peak.

Table I

The integrated intensity of the crystalline silicon TO peak divided by the total integrated intensity of the TO band including both crystalline and amorphous peaks.

Annealing temperature	Boron content		
(°C)	1%	7%	45%
500	0.25		
600	0.27		
700	0.42	0.36	0.13
800		0.74	0.20
900			0.31

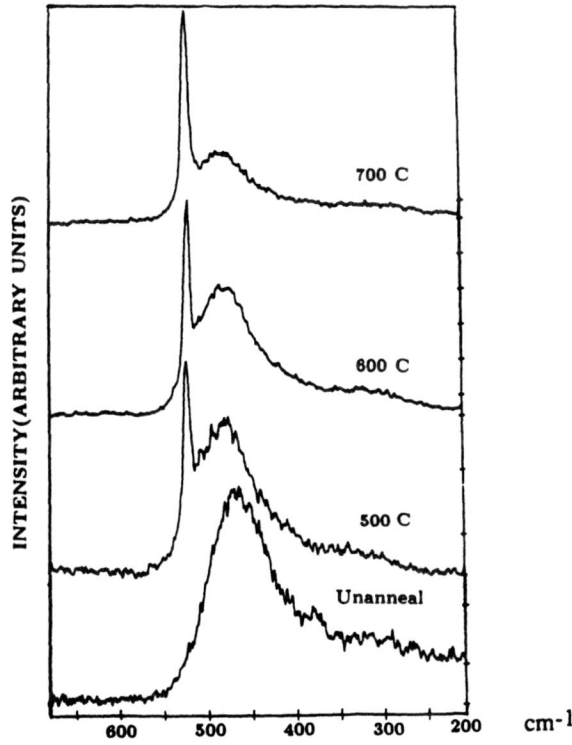

Figure 1: A set of Raman spectra for the a-Si:B film with the boron content of 1%, unannealed and annealed at different temperatures.

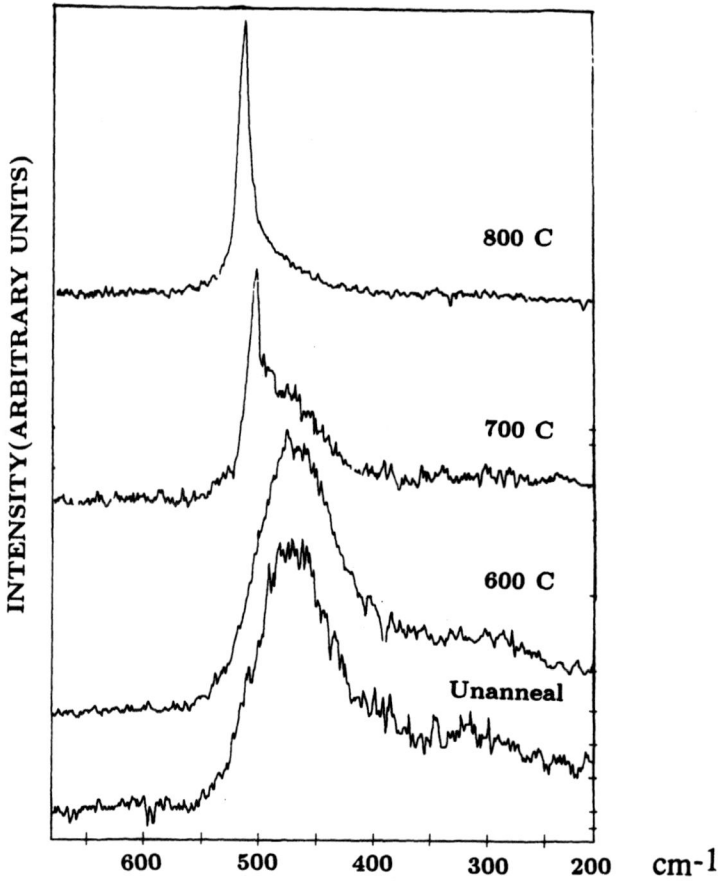

Figure 2: A set of Raman spectra for the a-Si:B film with the boron content of 7%, unannealed and annealed at different temperatures.

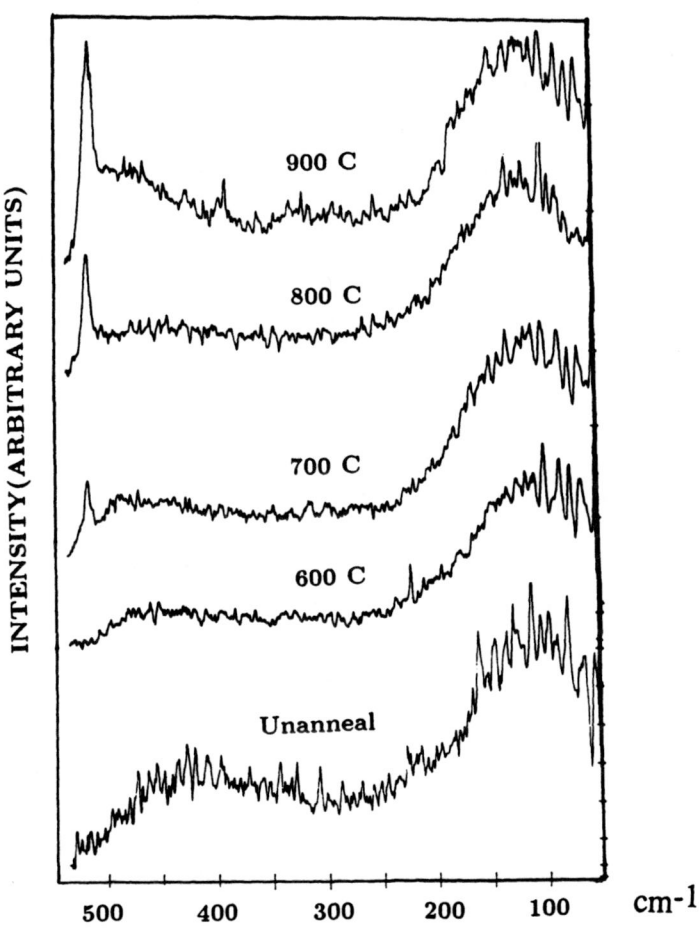

Figure 3: A set of Raman spectra for the a-Si:B film with the boron content of 45%, unannealed and annealed at different temperatures.

This might suggest a segregation of the silicon atoms from the boron atoms to form c-Si crystallites in the annealing process. This is quite plausible, given that the solid solubility of boron is small in silicon.[7,14] It is speculated that the excessive boron atoms form boron clusters with a predominantly three-fold bonding structure. In the phase diagram of the Si:B binary system there is a SiB_3 phase which is stable in the annealing temperature range of this study,[14] therefore, is expected to be present in the film too. It has been found, however, that it is difficult to identify these phases using the Raman spectra due to their complexity and a lack of standards to compare with. It would be of great interest to verify that these phases are indeed formed in the annealing process.

SUMMARY

Crystallization of a-Si:B alloy films with high and largely different boron contents are studied for the annealing temperature range between $500^\circ C$ and $900^\circ C$. The results show that the crystallization rate at a given annealing temperature depends inversely on the boron content for the boron content greater than the solid solubility limit of boron in silicon, in contrast to the case of the boron content in the doping regime where an increase of the crystallization rate is observed with the increase of the boron content.

ACKNOWLEGEMENT

The authors wish to thank J. Yao for his help in preparing the samples. Professor T.-M. Lu's support of this work is acknowledged.

REFERENCES

1. J. Gonzalez-Hernandez and R. Tsu, Appl. Phys. Lett., 42, 90(1983).
2. A. Morimoto, M. Kumeda, and T. Shimizu, J. Non-Cryst. Solids, 59&60, 537(1983).
3. F. Morin and M. Morrel, Appl. Phys. Lett., 35, 686(1979).
4. Z. Iqbal, A. P. Webb, and S. Veprek, Appl. Phys. Lett., 36, 163(19800.
5. T. Hamasaki, H. kurata, M. Hirose, and Y. Osaka, Appl. Phys. Lett., 37, 1084(1980).
6. R. Tsu, S. S. Chao, M. Izu, S. R. Ovshinsky, G. Jan, and F. H. Pollak, J. Phys. Paris Coll. C4, 269(1981).
7. H. Ishiwara, A. Tamba, H. Yamamoto, and S. Furukawa, Jpn. J. Appl. Phys., 24, L513(1985).
8. N. S. Alvi, S. M. Tang, R. Kwor, and M. R. Fulcher, J. Appl. Phys., 62, 4878(1987).
9 R. Bisaro, J. Magarino, K. Zellama, S. Squelard, P. Germain, and J. F. Morhange, Phys. Rev. B, 31, 3568(1985).
10. G. Yang, P. Bai, B. Y. Tong, S. K. Wong, and I Hill, Solid State Commun., 72, 159(1989).
11. B. Y. Tong, S. K. Wong, J. Yao, W. M. Lau, N. Du, and P. K. John, Proc. MRS Spring Meeting, 1987.
12. R. Tsu, J. Gonzalez-Hernandez, S. S. Chao, S. C. Lee, and K. Tanaka, Appl. Phys. Lett., 40, 534(1982).
13. M. Stutzmann, Phys. Rev. B, 35, 5921(1987).
14. T. B. Massalsky, editor-in-chief, Binary Alloy Phase Diagrams, vol. 1, American Society for Metals, October, 1986. p. 385-386.

PART VI

Devices and Applications

PREPARATION OF HIGH-QUALITY poly-Si AND μc-Si FILMS BY THE SPC METHOD

T. Matsuyama, M. Nishikuni, M. Kameda, S. Okamoto, M. Tanaka, S. Tsuda, M. Ohnishi, S. Nakano and Y. Kuwano
Sanyo Electric Co., Ltd., Functional Materials Research Center, 1-18-13, Hashiridani, Hirakata, Osaka, Japan

ABSTRACT

We have achieved the highest total area conversion efficiency for an integrated type 10cm X 10cm a-Si solar cell at 10.2%. This value is the world record for a 10cm X 10cm a-Si solar cell. For further improvement of conversion efficiency in a-Si solar cells, it is necessary to develop materials with high-photosensitivity in the long wavelength region and materials with high conductivity. We have developed a Solid Phase Crystallization (SPC) method of growing a Si crystal at temperatures as low as 600°C. Using this method, thin-film polycrystalline silicon (poly-Si) with high-photosensitivity in the long wavelength region and Hall mobility of 70cm^2/V · sec was obtained and quantum efficiency in the range of 800 ~ 1000nm was achieved up to 80% in the n-type poly-Si with grain size of about 2μm. We also succeeded in preparing a device-quality p-type microcrystalline silicon (μc-Si) using the SPC method at 620°C for 3 hours from the conventional plasma-CVD p-type amorphous silicon (a-Si) without using any post-doping process. Obtained properties of $\sigma_d = 2$ X 10^{-3} (·cm)$^{-1}$ and high optical transmittance in the 2.0 ~ 3.0 eV range are better as a window material than the conventional p-type μc-Si:H. Therefore, it was concluded that the SPC method is better as a new technique to prepare high-quality solar cell materials.

INTRODUCTION

In the near future, it will be necessary for us to establish a clean energy system that will not destroy the global environment. Solar cells are gathering much attention as a clean energy source. In order to make solar cells fit for practical use as a clean energy source, solar cells with a high conversion efficiency and low cost must be developed. For the development of solar cells with a much higher conversion efficiency, amorphous silicon (a-Si)/polycrystalline silicon (poly-Si) tandem solar cells are most feasible[1]. In order to fabricate thin-film P-doped poly-Si with large grains at a low temperature, a new Solid Phase Crystallization (SPC) method was applied to prepare solar cell materials for the first time. Furthermore, for the improvement of conversion efficiency of a-Si solar cells, a new p-type microcrystalline silicon (μc-Si) film was developed by using the SPC method from the conventional plasma-CVD p-type a-Si:H films, because the conductivity of the conventional p-type a-Si:H and p-type μc-Si:H is lower. This paper also discusses the potential of SPC method as a new fabrication technique for solar cell materials and solar cells.

HIGH CONVERSION EFFICIENCY A-SI SOLAR CELLS

In order to improve the conversion efficiency in a-Si solar cells, we have developed new technologies such as fabrication of high-quality a-Si:H and buffer layer using the super chamber and high-quality p-type a-SiC doped with B(CH$_3$)$_3$, which is a new p-type doping gas, and so on. This time, we have achieved the world record with a 10cm X 10cm a-Si solar cell. Fig.1 shows I-V characteristics of an integrated type 10cm X 10cm a-Si submodule. The p-layer is a-SiC doped with B(CH$_3$)$_3$. The p/i buffer layer and i-layer were fabricated using the super chamber method. By using highly textured and high-quality TCO and optimizing the p-layer and p/i buffer layer for the highly textured TCO, a total area efficiency of 10.2% has been achieved. Notably, short circuit current (Isc) was improved.

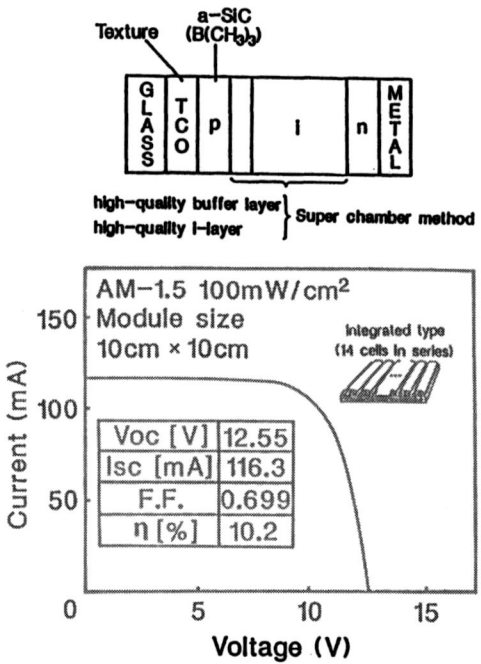

Fig.1 I-V characteristics of an integrated type a-Si submodule

SOLID PHASE CRYSTALLIZATION (SPC) METHOD

In order to improve the conversion efficiency of a-Si solar cells further, it is necessary to develop materials with high photocensitivity in the long wavelength region. We have given much attention to thin-film poly-Si as a material with high photosensitivity in the long wavelength region. There are fabrication methods such as Liquid Phase Growth, Chemical Vapor Deposition, Laser Recrystallization and Solid Phase Crystallization (SPC) for the preparation of thin-film poly-Si. Among these methods, the SPC method has many features fit for the fabrication process of solar cell materials; namely, large grain, large area, and simple process. The SPC method which we have developed is a way to prepare thin-film poly-Si or μc-Si from a-Si film made using the plasma-CVD method by thermal annealing. This SPC method has three features ; (1) The impurity atoms (phosphorus or boron) are in-situ doped into a-Si. (2) This process is very simple. (3) Thin-film poly-Si or μc-Si with a large area can be prepared. Fig.2 shows a diagram of the SPC method which consists of two processes. In the first process, phosphorus (P) or boron (B) doped a-Si films were deposited by the plasma-CVD method on a quartz substrate. In the second process, doped a-Si films were transferred to poly-Si or μc-Si by thermal annealing in a vacuum at various temperatures.

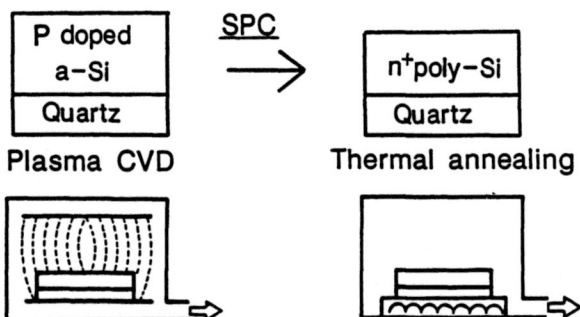

Fig.2 The diagram of the SPC method which consists of two processes

PREPARATION OF THIN-FILM POLY-SI AND ITS APPLICATION TO SOLAR CELLS

Properties of thin-film poly-Si prepared by the SPC method

P-doped a-Si films were prepared by the plasma-CVD method on quartz substrates. Phosphorus was in-situ doped by mixing PH_3 with SiH_4 as a source gas for the SPC process. The substrate temperature was in the range from 450°C to 650°C. Thermal annealing was performed for 3 ~ 100 hours at temperatures ranging from 570°C ~ 850°C in a vacuum. This shows the effect of phosphorus atoms on Solid Phase Crystallization (SPC). Fig.3 shows the cross sectional Scanning Electron Microscopy (SEM) photographs taken after SPC of non-doped a-Si film and P-doped a-Si film at the SPC temperature of 600°C. SPC of non-doped a-Si film at does not occur and the cross section after SPC has amorphous phase. On the other hand, the SPC of P-doped a-Si film does occur and the cross section after SPC has polycrystalline phase. So, it was found that SPC occured easily when phosphorus doping was used.

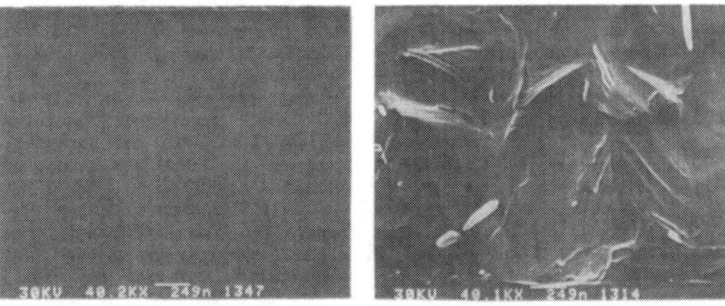

   (a) non-doped a-Si/quartz   (b) P-doped a-Si/quartz
Fig.3 Cross sectional SEM photographs after SPC of non-doped a-Si and P-doped a-Si

For preparation of high-quality n-type poly-Si as a photovoltaic active layer, the influence of phosphorus concentration and SPC temperature on the n-type poly-Si was investigated. Fig.4 shows cross-sectional Scanning Electron Microscopy (SEM) photographs of n-type poly-Si with phosphorus concentrations of $1.2 \times 10^{20} cm^{-3}$ ~ $1.7 \times 10^{21} cm^{-3}$, where the SPC condition is 700°C, for 10 hours. In Fig.4, the grain size of n-type poly-Si drastically increases with reduction of phosphorus concentration. In n-type poly-Si with phosphorus concentration of $1.2 \times 10^{20} cm^{-3}$, a maximum grain size of about 2.0μm was obtained.

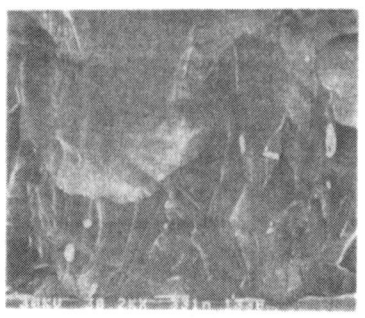 

① P:1.2×10²⁰cm⁻³(700 ℃,10H)   ② P:2.6×10²⁰cm⁻³(700 ℃,10H)

 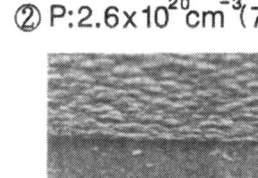 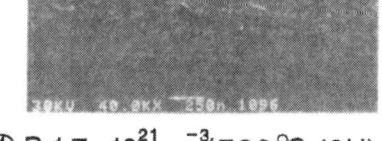

③ P:1.2×10²¹cm⁻³(700 ℃,10H)   ④ P:1.7×10²¹cm⁻³(700 ℃,10H)

Fig.4 Cross sectional SEM photographs of n-type poly-Si for various phosphorus concentration with SPC conditions of 700°C, 10H

Fig.5 shows cross-sectional SEM photographs of n-type poly-Si for two SPC temperatures of 600°C and 700°C, where phosphorus concentration is 1.2×10^{20} cm⁻³. The n-type poly-Si fabricated at 600°C has larger grains than that prepared at 700°C in the region near the substrate. And finally, the n-type poly-Si with larger grain is obtained at 600°C.

Fig.6 shows Hall mobility of n-type poly-Si prepared using the SPC method as a function of carrier concentration, compared with that of single-crystalline silicon (c-Si) [2] . Hall mobility of n-type poly-Si decreases with increases in carrier concentration, similar to c-Si. Hall mobility of 70cm²/V·sec was achieved at the carrier concentration of 1×10^{17} cm⁻³, which is 50% of the Hall mobility of c-Si.

Minority carrier characteristics are very important for solar cells. So, we analized the minority carrier trap density in terms of a trapping model [3]. According to the trapping model, the following equation (1) can be introduced;

$$\sigma T^{1/2} \propto \exp(-q^2 N t^2 / 8 \varepsilon N k T) \tag{1}$$

where Nt is minority carrier trap density, N phosphorus concentration, σ conductivity of poly-Si, q electronic charge, ε dielectric constant of silicon, k Boltzman's constant, and T absolute temperature. In P-doped poly-Si, the dependence of conductivitiy on the temperature was measured and the minority carrier trap density Nt was analyzed from the slope of the $\ln(\sigma T^{1/2})$ vs. $1/T$ relationship. Fig.7 shows the minority carrier trap density in P-doped poly-Si prepared by the SPC method as a function of the phosphorus concentration in poly-Si. As shown in Fig.7, minority carrier trap density decreased with decreases in phosphorus concentration. Especially, in the region of low phosphorus concentration, minority carrier trap density decreased significantly with a reduction in SPC temperature.

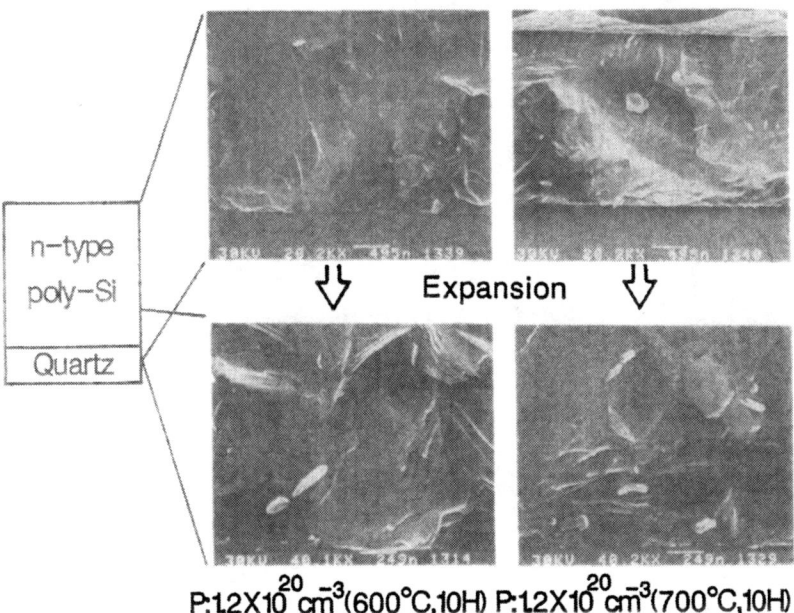

Fig.5 Cross sectional SEM photographs of n-type poly-Si for two SPC temperatures

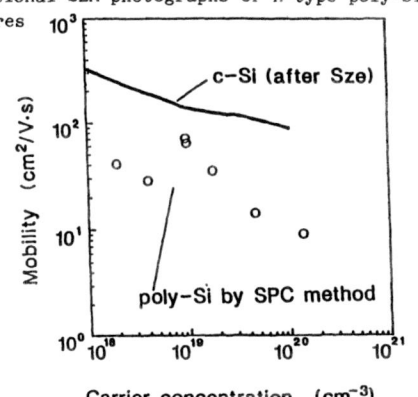

Carrier concentration (cm^{-3})

Fig.6 Hall mobility of n-type poly-Si prepared by the SPC method as a function of carrier concentration

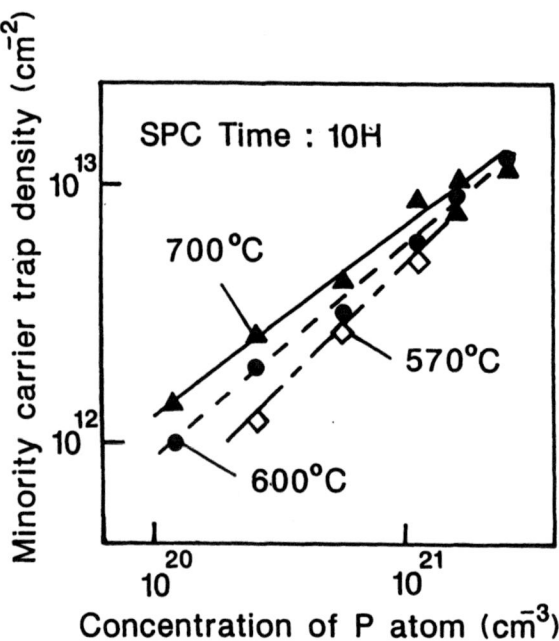

Fig.7 Minority carrier trap density of n-type poly-Si prepared by the SPC method as functions of phosphorus concentration and SPC temperature

There are two ways to prepare high-quality n-type poly-Si. One is the reduction of phosphorus concentration, as mentioned above. The other is control of polycrystalline nuclei. In order to control the polycrystalline nuclei, we used the partial doping method. The partial doping method is a way to prepare n-type poly-Si from quartz/P-doped a-Si/non-doped a-Si structure. P-doped a-Si has the role of growing polycrystalline nuclei. Fig.8 shows cross sectional SEM photographs of n-type poly-Si prepared by bulk doping and partial doping, where phosphorus concentration is $1.2 \times 10^{20} cm^{-3}$ and the SPC condition is 600°C, over 10 hours. In bulk doping and partial doping, n-type poly-Si after SPC have almost the same cross section and grain size. But phosphorus concentration in n-type poly-Si prepared by partial doping is lower than that prepared by the bulk doping and its concentration is about $2 \times 10^{18} cm^{-3}$. So, by using the partial doping method, n-type poly-Si with large grain was obtained in the region of low phosphorus concentration.

Next, we will show the minority carrier trap density of n-type poly-Si prepared by partial doping. Fig.9 shows the comparison of minority carrier trap density between partial doping and bulk doping. The minority carrier trap density of n-type poly-Si prepared by partial doping decreased further, compared to that prepared by bulk doping. A minimum value of $4.6 \times 10^{11} cm^{-2}$ was obtained.

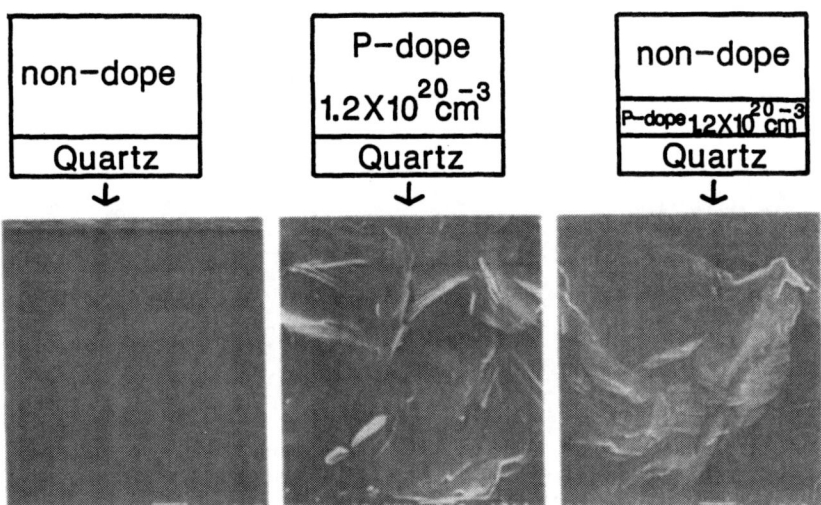

Fig.8 Cross sectional SEM photographs of non-doped and n-type poly-Si
prepared by bulk doping and partial doping.

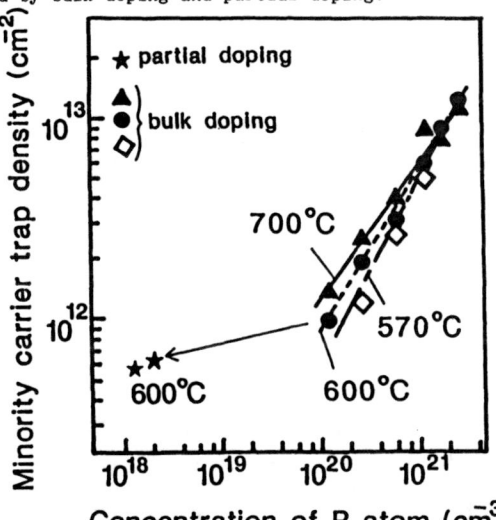

Fig.9 Comparison of minority carrier trap density in partial
doping and bulk doping as functions of phosphorus concentration
and SPC temperature

Fabrication of n⁺ poly-Si with high conductivity

In order to fabricate a poly-Si solar cell, we have tried to prepare
n⁺ poly-Si with high conductivity using the SPC method. Fig.10 shows the
conductivity after SPC as a function of the doping ratio. The conductivity
increases with increase in the doping ratio and has the maximum value of
$1.5 \times 10^3 (\Omega\text{-cm})^{-1}$. This conductivity is equal to a sheet resistance of about
$10\ \Omega\ /\square$ at $0.7\mu m$, which corresponds to that of TCO.

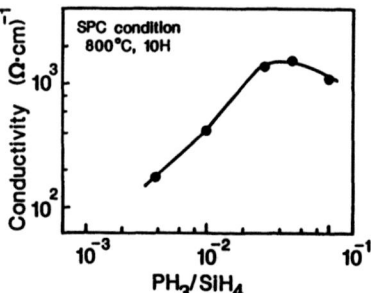

Fig.10 Conductivity of n$^+$poly-Si prepared by the SPC method as a
function of doping ratio

Application of n-type poly-Si to solar cells

By using the SPC method, it was found that n$^+$poly-Si with high
conductivity and n$^-$poly-Si with low minority carrier trap density could
be prepared. So, we have developed a poly-Si solar cell using the SPC
method. The structure of the poly-Si solar cell is
ITO/p·a-Si:H/n$^-$poly-Si/n$^+$poly-Si/quartz. The thickness of n$^-$poly-Si layers,
n$^+$poly-Si layers, and p·a-Si:H layers are 0.7 μm, 3.0 μm, and 900A,
respectively. Fig.11 shows the collection efficiency of this poly-Si solar
cell prepared by the SPC method. In the long wavelength region, collection
efficiency is about 20%. The inner collection efficiency calculated from
the collection efficiency and absorption coefficient is about 80% in the
long wavelength region. This value is high, considering that grain size
of n$^-$poly-Si is about 2 μm. Therefore, it was confirmed that n$^-$poly-Si
prepared by the SPC method is excellent as a material with
high-photosensitivity in the long wavelength region. Furthermore, it was
found that the SPC method has high potential as a new fabrication technique
for solar cell materials and solar cells.

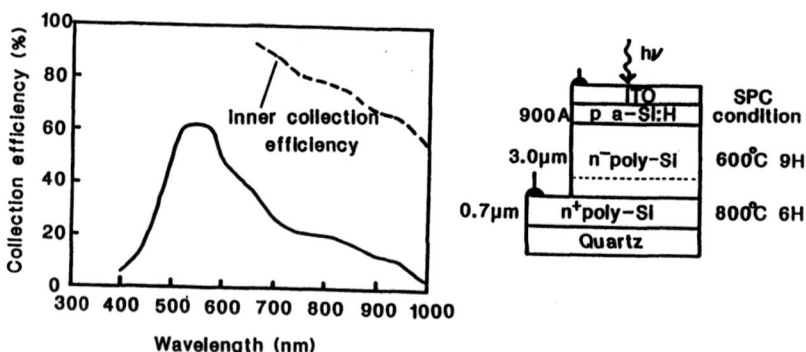

Fig.11 The collection efficiency of a poly-Si solar cell prepared
by the SPC method

PREPARATION OF HIGH-QUALITY P-TYPE μc-SI AND ITS APPLICATION TO SOLAR CELLS

For the developement of a high conversion efficiency a-Si solar cell, it is necessary to develop materials with very high conductivity. As mentioned above, n⁺poly-Si with high conductivity was obtained using the SPC method. So, we have been investigating the preparation of p-type μc-Si using the SPC method.

Properties of p-type μc-Si prepared by the SPC method

B-doped a-Si films were prepared by the plasma-CVD method on a quartz substrate. Boron was in-situ doped by mixing B_2H_6 with SiH_4 as a source gas for the SPC process. The substrate temperature was 200°C. Thermal annealing was performed for 1 ~ 7 hours at a temperature of 600 ~ 700°C in a vacuum. After the solid phase crystallization of p-type a-Si films, Raman spectroscopy was measured for the structural analysis. Fig.12 shows the Raman spectra of the films after SPC. As a comparison, the spectrum of n-type poly-Si is also shown. The peak was observed at 520 cm⁻¹ in the poly-Si. On the other hand, The peak was observed in about 512cm⁻¹ and a peak in the region of 470cm⁻¹ ~ 480cm⁻¹ was not observed in the films after SPC of p-type a-Si films. Therefore, it was confirmed that these films were microcrystalline.

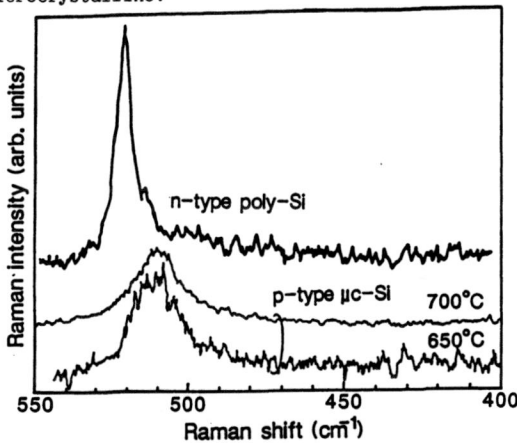

Fig.12 Raman spectra of the films after SPC

Fig.13 shows the conductivity before and after SPC as a function of doping ratio (B_2H_6/SiH_4). The SPC conditions are 600 ~ 700°C, 3 hours. The conductivity of p-type μc-Si increases with the SPC temperature for each doping ratio and has the maximum value at the doping ratio of 0.5%. Furthermore, a prominent feature is the increases in conductivity by 9 ~ 11 orders after SPC at the doping ratio of 0.5% and the maximum value of about 2×10^3 $(\Omega \cdot cm)^{-1}$ was achieved. High conductivity attributes to the high activation ratio of the impurity atom (boron) when using the SPC as shown in Fig.14. Therefore, by using the SPC method, it was confirmed that the activation ratio of boron (B) atom was also very high. Fig.15 shows the dependence of absorption coefficient of p type μc Si prepared by the SPC method on the doping ratio (B_2H_6/SiH_4), where SPC conditions are 650°C, 3H and 700°C,3H. The absorption coefficient of p-type μc-Si at the doping ratio of 0.5 % is the smallest and its value is nearest to that of c-Si. So, the p-type μc-Si at the doping ratio of 0.5% shows the highest-quality crystallization in these samples. Therefore, the conductivity of p-type µc-Si is the highest at the doping ratio of 0.5%.

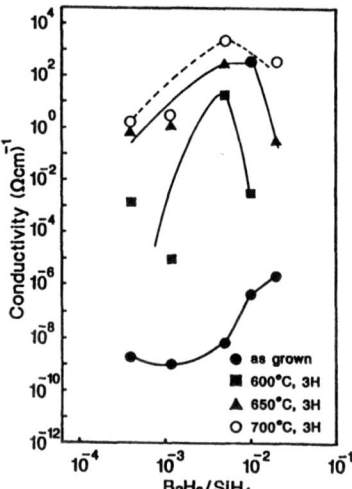

Fig.13 Conductivity before and after SPC as a function of the doping ratio.

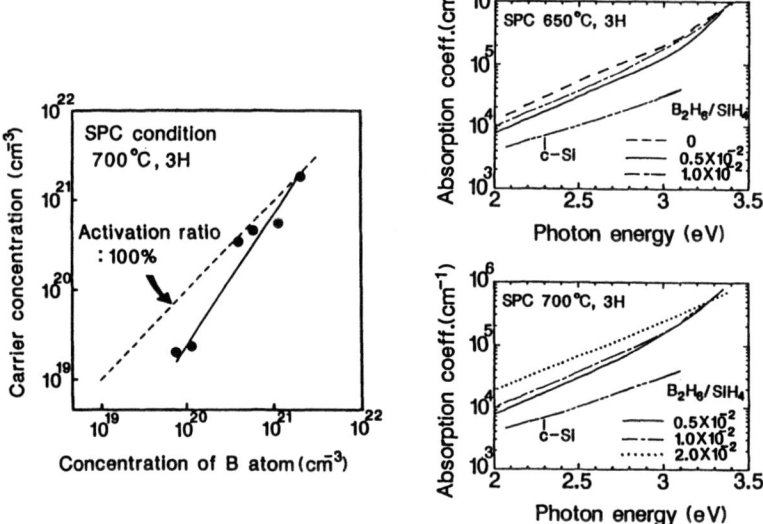

Fig.14 Carrier concentration in the p-type µc-Si films as a function of boron concentration

Fig.15 Absorption coefficient of p-type µc-Si films as a function of the doping ratio

Fig.16 shows comparisons of absorption coefficient and mobility between p-type µc-Si films prepared by the SPC method and the conventional plasma-CVD method. The absorption coefficient of p-type µc-Si prepared by the

SPC method is smaller than that prepared by the conventional plasma-CVD method in the region of 2.0eV ~ 3.0eV. So, by using the SPC method, p-type µc-Si with a smaller absorption coefficient can be prepared. As for mobility, in the region of carrier concentration as high as two orders, the mobility of p-type µc-Si prepared by the SPC method is almost equal to that of p-type µc-Si prepared by the plasma-CVD method and its value is about 7 ~ 13cm^2/V·sec. So, by using the SPC method, p-type µc-Si with a higher carrier concentration can be prepared.

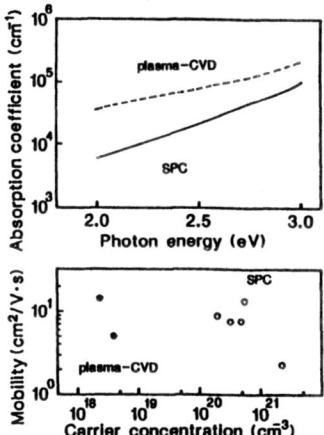

Fig.16 Comparisons of absorption coefficient and mobility between p-type µc-Si films prepared by the SPC method and the conventional plasma-CVD method

In order to analyze p-type µc-Si prepared by the SPC method as a window material for solar cells, the dependence of the conductivity on E_{04} was investigated, where E_{04} is the energy that the absorption coefficient is equal to 1 X 10^4 cm^{-1}. Fig.17 shows the dependence of the conductivity on EO_4. The E_{04} of p-type µc-Si prepared by the SPC method is almost 2.0eV ~ 2.2eV and the p-type µc-Si with E_{04} of 2.12eV and the conductivity of 2 X 10^3(Ω·cm)$^{-1}$ was obtained. So, p-type µc-Si film prepared by the SPC method is higher-quality material compared to the conventional p-type µc-Si as a window material for solar cells. Therefore, the SPC method is better for the preparation of a material with high conductivity and wide bandgap.

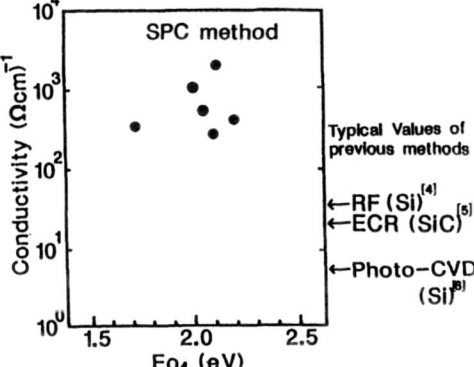

Fig.17 Conductivity of p-type µc-Si films as a function of E_{04}

CONCLUSION

We have achieved a total area efficiency world record of 10.2% in 10cm X 10cm a-Si solar cells. By using the SPC method, the n^+ poly-Si with high conductivity and n^- poly-Si with low minority carrier trap density as a photovoltaic active layer were prepared. Using these materials, we have developed the poly-Si solar cell using the SPC method for the first time. And a high inner collection efficiency of about 80% was achieved in the long wavelength region of 800nm ~ 1000nm, despite the grain size of about 2μm. Furthermore, we applied this SPC method to the preparation of p-type μc-Si from p-type a-Si and obtained p-type μc-Si with high conductivity of about $2 \times 10^3 (\Omega \cdot cm)^{-1}$. Therefore, the SPC method we have developed is better as a new technology for the preparation of solar cell materials with high photosensitivity for the long wavelength region or with high conductivity and wide bandgap. It is expected that this SPC method will gather much attentions as a new fabrication technology for high-quality solar cell materials and solar cells.

ACKNOWLEDGEMENT

This work is supported by NEDO (New Energy and Industrial Technology Development Organization) as a part of the Sunshine Project under the Ministry of International Trade and Industry.

REFERENCES

1. H. Takakura, K. Miyagi, T. Kanata, H. Okamoto, Y. Hamakawa, Proc. 4th Int'l PVSEC (1989) 403.
2. S. M. Sze, Physics of Semiconductor Devices, 2nd ed.
3. John Y. W. Seto, J. Appl. Phys. 46, 5247 (1975)
4. K. Mori et al.; Jap. J. Appl. Phys. 20, 2431 (1981)
5. Y. Hamakawa; Proceedings of 8th EC-PVSEC, 1211 (1989)
6. M. Konagai et al.; Private communication

CHARACTERISTICS OF μc-Si:H FOR Si
HETEROJUNCTION BIPOLAR TRANSISTORS

H. Fujioka, M. Ito, and K. Takasaki
Fujitsu Limited, 1015 Kamikodanaka, Nakahara, Kawasaki 211, Japan

ABSTRACT

To improve the thermal stability of the Si heterojunction bipolar transistors (HBTs), we studied the effect of
carbon and fluorine doping on μc-Si:H characteristics. We
found that carbon doping suppresses crystalline growth and
increases the hydrogen concentration in the film, and that
fluorine atoms are more thermally stable than hydrogen atoms.
We confirmed that carbon or fluorine doping is promising for
use with the μc-Si:H HBT.

INTRODUCTION

Recently, the Si heterojunction bipolar transistor (HBT)
has attracted much attention[1]. Si HBTs with wide gap
emitters are especially promising because of their
compatibility with the conventional bipolar transistors in
fabrication processes. To ensure high performance, the bandgap
of the wide gap emitters must be wide enough to suppress back
injection of holes from the base into the emitter. Electrical
properties of the heterojunction must also be good enough to
suppress the recombination base current, which can completely
negate the advantages of the HBT. Resistivity and contact
resistance with metal must also be low enough to reduce
emitter resistance. We chose hydrogenated microcrystalline
silicon (μc-Si:H) as the material best ensuring these
features.

HBTs with μc-Si:H emitters showed a much higher current
gain than homojunction transistors[2,3]. They also can be
operated normally even at liquid nitrogen temperature, which
is thought to be difficult for homojunction transistors[4].
Last year we fabricated a 400-gate ECL gate array that
operates normally at both room and liquid-nitrogen
temperatures[5]. However, severe degradation in current gain
was observed during double-layer metallization, even though
the maximum process temperature in metallization was 450°C.
This degradation is related to an increase in recombination
centers at the heterojunction due to the detachment of
hydrogen which terminates dangling bonds. We first tried
improving film's thermal stability by increasing the substrate
temperature to that of metallization during deposition. The
crystalline structure of the film deposited at 480°C was
columnar, and similar to that of poly Si, however. Because the
energy bandgap of such film is thought to be equal to that of
single-crystalline Si, such film is not suitable as an HBT
emitter. There are two possible solutions. One is to deposit
μc-Si:H at temperatures higher than those in metallization
while suppressing polycrystalline growth. We did this by
adding CH_4 to the growth atmosphere. We expected carbon atoms
in the film to inhibit polycrystalline growth even at higher
temperatures during film deposition. The other possible

solution is to find a terminator more stable than the hydrogen atom. We tried depositing fluorine-doped film by adding SiF_4 at lower substrate temperatures. We expected that fluorine atoms incorporated into the film as terminators would increase the thermal stability because the binding energy of the Si-F bond is greater than that of the Si-H bond.

EXPERIMENT

Table 1 summarizes growth conditions for μc-Si:H films. All films were deposited in a conventional parallel-plate plasma CVD apparatus in a gaseous mixture of SiH_4, PH_3, and H_2 on Si (111) substrates. Native oxides were removed from the wafer with diluted HF just before it was loaded into the reaction chamber. A small amount of CH_4 or SiF_4 was added for carbon or fluorine doping. During deposition, the substrate temperature was from 230°C to 480°C. The pressure was 0.9 Torr. μc-Si:H/crystalline Si heterodiodes were fabricated on p type 1 Ω cm Si substrates to investigate the heterojunction properties.

Table 1. Growth conditions for μc-Si:H.

Apparatus:	Parallel-plate plasma CVD
Gas:	SiH_4, PH_3, H_2
	SiH_4, PH_3, H_2, CH_4 (C doping)
	SiH_4, PH_3, H_2, SiF_4 (F doping)
Temperature:	230°C - 480°C
Pressure:	0.9 Torr

RESULTS AND DISCUSSION

Carbon doping

Figure 1 shows X-ray diffraction patterns for undoped and carbon doped μc-Si:H films. The substrate temperature was 480°C, which is higher than the maximum temperature in multilayer metallization. Sharp peaks for the undoped film indicate large grains like poly Si. Peaks for the carbon-doped film are weak and broad, indicating microcrystalline Si. Figure 2 shows cross-sectional TEM micrographs of undoped and carbon doped films deposited at 480°C on Si (111) wafers. Large columnar grains are clearly observable in the undoped but not carbon-doped film, indicating that carbon doping inhibits polycrystalline growth. IR absorption spectra in figure 3 show the effect of carbon doping on the incorporation of hydrogen atoms into the μc-Si:H film deposited at 480°C.

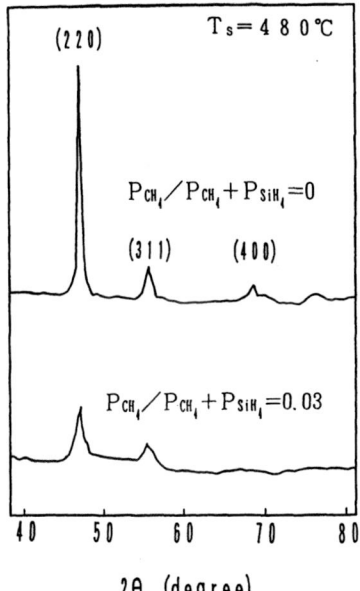

Fig. 1. X-ray diffraction patterns for (a) undoped and (b)
carbon doped μc-Si:H film.

(a)   (b)

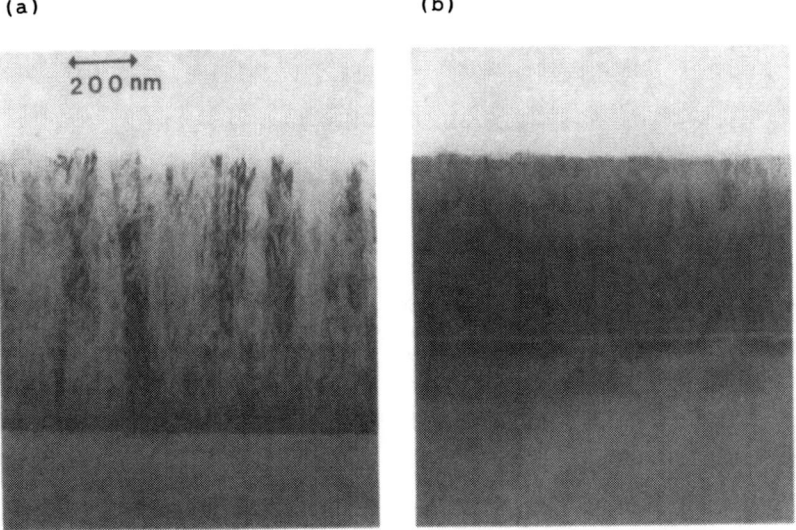

Fig. 2. Cross-sectional TEM micrographs of (a) undoped and
(b) carbon doped μc-Si:H film deposited at 480°C
on Si (111) substrates.

Note that the hydrogen concentration increases with the amount of CH_4. Hydrogen atoms incorporated into the μc-Si:H film increase the bandgap. The above results indicate that adding CH_4 to the growth atmosphere suppresses polysilicon growth and increases hydrogen concentration in the film.

We studied the effect of carbon doping on the film's electrical properties. Figure 4 shows the dependence of resistivity on carbon doping for a film prepared at 480°C. Carbon doping causes an increase in the resistivity of the μc-Si:H film, which increases emitter resistance. The optimum amount of carbon doping is that which maximizes the bandgap

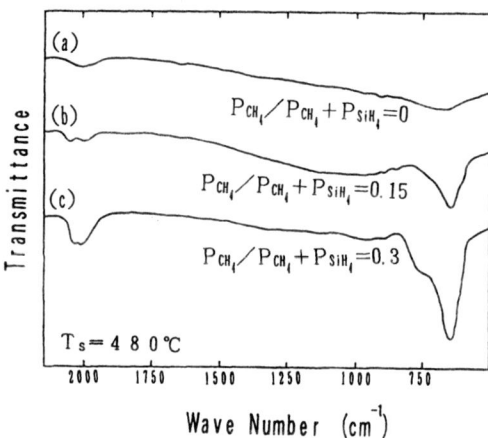

Fig. 3.   IR absorption spectra for μc-Si:H deposited with a CH_4-to-SiH_4 ratio of (a) 0%, (b) 15%, and (c) 30%.

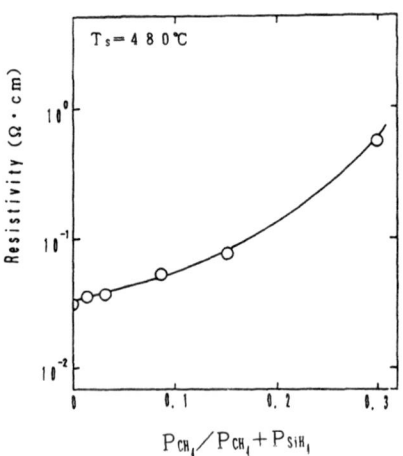

Fig. 4.   Resistivity of carbon doped μc-Si:H film as a function of the CH_4-to-SiH_4 ratio.

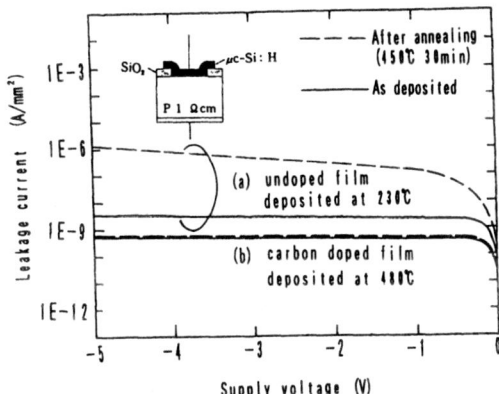

Fig. 5. I-V characteristics for heterodiodes with (a) undoped film deposited at 230°C and (b) carbon doped
film deposited at 480°C before and after 30minutes
annealing at 450°C.

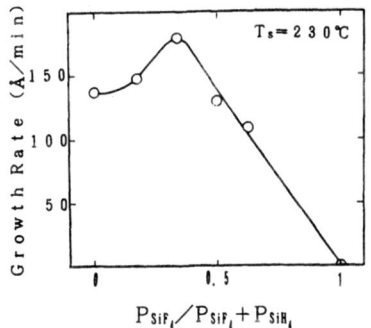

Fig. 6. Growth rate of fluorine doped μc-Si:H film as a
function of the SiF_4-to-SiH_4 ratio.

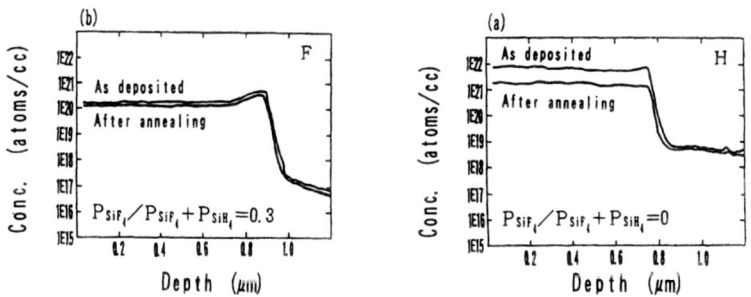

Fig. 7. SIMS profiles of (a) undoped and (b) fluorine doped
films before and after 30 minutes of annealing at
450°C in N_2 atmosphere.

and thermal stability while minimizing resistivity. Figure 5 shows the leakage current for the heterodiodes fabricated with undoped film deposited at 230°C and carbon doped film deposited at 480°C before and after 30 minutes of annealing at 450°C in N_2. Increase in the leakage current after annealing is clearly seen for the heterodiode with undoped μc-Si:H but not for the diode with carbon doped film. These results indicate that the carbon-doped μc-Si:H film is suitable as a Si HBT emitter because of both its wide bandgap and its stability during thermal processes up to about 450°C.

Fluorine doping

Figure 6 shows the dependence of the growth rate of μc-Si:H at 230°C on the amount of SiF_4 in the growth atmosphere. In the region where the flow rate of SiF_4 is high, addition of SiF_4 causes the growth rate of the film to decrease. This is due to the etching effect of fluorine atoms generated in the plasma. We confirmed the existence of Si-F bonds in the deposited film by IR spectrum. SIMS profiles in figure 7 show the changes in the fluorine concentration of a fluorine doped film and in the hydrogen concentration of an undoped film before and after 30 minutes of annealing at 450°C in N_2. The deposition temperature of these films was as low as 230°C. The decrease in fluorine concentration is much smaller than that in hydrogen concentration. This indicates that the fluorine atom is a better terminator for a dangling bond than the hydrogen atom in terms of thermal stability.

CONCLUSION

We studied carbon and fluorine doping to μc-Si:H film to improve thermal stability for use in Si HBTs. We found that carbon doping suppresses polycrystalline growth and enhances hydrogen incorporation into film even at relatively high deposition temperatures. The heterodiode fabricated with carbon doped μc-Si:H shows good thermal stability. The film shows promise as a Si HBT emitter because of its high thermal stability and a wide bandgap. Fluorine atoms incorporated into the μc-Si:H as terminators are more thermally stable than hydrogen and are thus better terminators for dangling bonds in μc-Si:H films for Si HBTs in terms of thermal stability.

REFERENCES

1. H. Kreomer, Proc. IEEE, 70, 1, p. 13 (1982).
2. K. Sasaki, S. Furukawa, and M. M. Rahman, IEDM Tech. DIg., 1987, p. 186.
3. H. Fujioka, S. Ri, K. Takasaki, K. Fujino, and Y. Ban, IEDM Tech. Dig., 1987, p. 190
4. H. Fujioka, T. Deguchi, K. Takasaki, and T. Takada, Extended Abstract, 20th SSDM Dig., 1988, p. 125
5. H. Fujioka, T. Deguchi, K. Takasaki, and T. Takada, IEDM Tech. Dig., 1988, p. 574

DOPANT SEGREGATION AT POLYCRYSTALLINE SILICON GRAIN BOUNDARIES IN DEVICE FABRICATION PROCESSES

M. ITOH, I. AIKAWA, N. HIRASHITA, AND T. AJIOKA,
Oki Electric Industry Co., Ltd., VLSI R&D Laboratory
550-1, Higashiasakawa, Hachioji, Tokyo 193, Japan.

ABSTRACT

The dopant segregation at the polycrystalline silicon grain boundaries in device fabrication processes has been studied with a new approach using spreading resistance(SR) measurement, SIMS and cross-sectional TEM(XTEM). Phosphorus implanted LPCVD poly-Si films were annealed at 900°C-1000°C in N_2 for 30min. Electrically active dopant concentrations obtained from SR measurements are constant in depth within the poly-Si films. On the other hand, the phosphorus concentration measured by SIMS is found to increase with increasing depth and to have linear relationships to reciprocal grain sizes observed by XTEM for all poly-Si films. The linear relationship indicates that the number of segregated phosphorus atoms per unit grain surface area at the grain boundaries is uniform throughout poly-Si films. Both phosphorus concentrations in the grains and at the grain boundaries are evaluated. The heat of segregation of 1.7eV is obtained from the annealing temperature dependence of the segregation ratio. Our results indicate that carrier concentration in the poly-Si film is more sensitive to annealing temperature in device fabrication processes. The carries concentration is determined by kinetics rather than by equilibrium segregation of dopants.

INTRODUCTION

It is well known that dopant segregation at the grain boundaries of polycrystalline silicon(poly-Si) films causes an unfavorable increase in resistivities of poly-Si films[1]. The dopant segregation is not a serious problem when poly-Si films are used for gate electrodes of MOS devices. However in resent several years, poly-Si films play new roles in VLSI technology, for examples buried poly-Si contacts in a DRAM[2] and poly-Si emitter contacts in high performance bipolar VLSI[3,4]. For these uses, resistivities of poly-Si films are the main factor which determines device parameters such as access time of DRAM's and switching speed of transistor. The dopant segregation of poly-Si films has been widely studied by a number of techniques including Hall effect[5], energy dispersion X-ray[6] and RBS[7] measurements. These previous works only dealt with the dopant segregation at equilibrium condition. Since thermal annealing processes of device fabrication is relatively short time and low temperature, the dopant segregation does not reach equilibrium condition. It is important to determine a ratio of dopant segregation in device fabrication processes.

In this paper, we describe a new method of determination of the dopant segregation at the grain boundaries in the similar condition of device fabrication process.

EXPERIMENT

Poly-Si films were deposited to a thickness of approximately 1μm by low pressure chemical vapor deposition(CVD) at 625°C onto thermally oxidized silicon wafers. Phosphorus atoms with dose of $1.5 \times 10^{16}/cm^2$ were implanted into the films at an energy of 40keV. After the ion implantation, 100nm of low temperature CVD oxide were deposited on the wafers to prevent out-diffusion of phosphorus atoms during subsequent annealing. Then the wafers were annealed in dry nitrogen at various temperatures between 900 and 1000°C for 30 min.

Spreading resistance(SR) measurements were carried out to obtain electrically active dopant distribution in poly-Si films. The total dopant concentration profiles were also measured by secondary-ion mass spectrometry(SIMS). The grain sizes of poly-Si films were determined from cross-sectional transmission electron microscopy(XTEM).

EXPERIMENTAL RESULTS

Resistivity depth profiles of the poly-Si films obtained by SR measurement are shown in Fig.1 for the samples annealed at 900°C and 1000°C. It is found that the resistivities are constant within the films for both samples. At sufficiently high dopant concentration in the poly-Si films as high as $10^{19} cm^{-3}$, potential barriers at the grain boundaries have very little effect on movement of free carriers[5]. Therefore, the electrically active dopant concentrations, related to resistivities, are considered to be uniform throughout the poly-Si films for each sample.

Figure 2 shows SIMS depth profiles of phosphorus concentration for 900°C and 1000°C. The phosphorus concentrations gradually increase with increasing depth. As projection range Rp equals to 48.6nm at 40keV, these phosphorus profiles can not be simply explained by diffusion theory. SIMS results indicate that electrically inactive dopant concentrations increase with increasing depth, as active dopant concentrations are uniform within the poly-Si films.

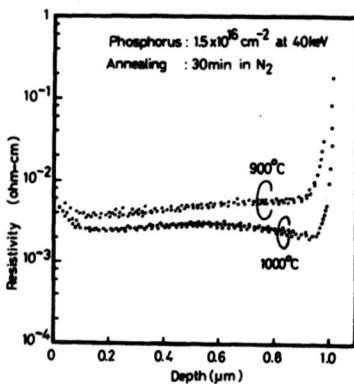

Figure 1. Resistivity profiles in the poly-Si films annealed at 900° and 1000°C, which are obtained by spraeding resistant mesurement.

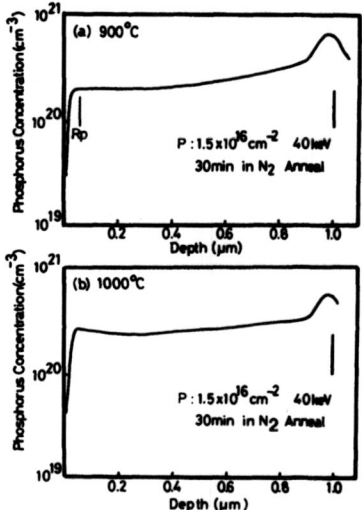

Figure 2. Phosphorus profiles in the poly-Si films measured by SIMS. The sample were annealed at (a) 900°C and at (b) 1000°C.

Figure 3 shows XTEM micrographs of as-deposited poly-Si film and phosphorus ion implanted poly-Si films that were annealed for 30min at 900°C and 1000°C. Although the grain structures are columnar type in the poly-Si films, fine grain structures are observed near the poly-Si/SiO2 interface in the as-deposition sample. The growth of the fine grains is seen for the annealing samples. Furthermore, the columnar type grains also grow in the sample of 1000°C annealing. Any precipitates like silicon-phosphorus complexes were not observed at the grain boundaries, because the dopant concentration is much less than solid solibility of phosphorus in silicon.

Figure 4 shows the grain sizes in the poly-Si films obtained from XTEM observation as shown in Fig.3 as a function of depth for 900°C and 1000°C. In this figure, a vertical axis is shown by reciprocal grain size for later discussion. The reciprocal grain sizes corresponding to the grain boundary surface area per unit volume increase with increasing depth. Therefore, the electrically inactive dopant concentration seems to have a relationship to the grain boundary surface area from comparing phosphorus profiles to grain size profiles.

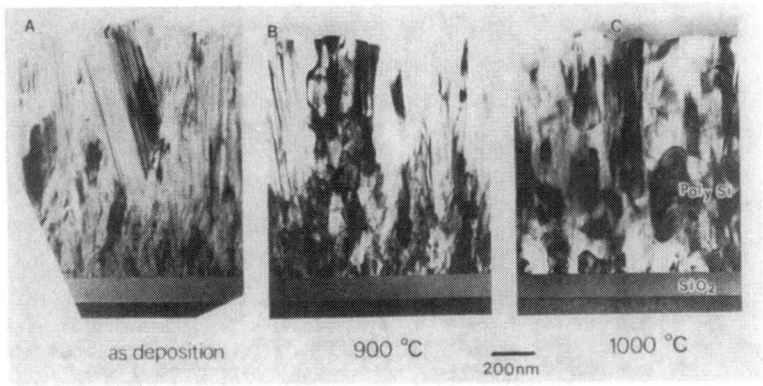

Figure 3. Cross-sectional TEM micrographs of poly-Si films:
(a) as-deposition, (b) 900°C annealing, and (c) 1000°C annealing samples.

Figure 4. Reciprocal grain sizes as a function of depth for 900°C and 1000°C annealing samples.

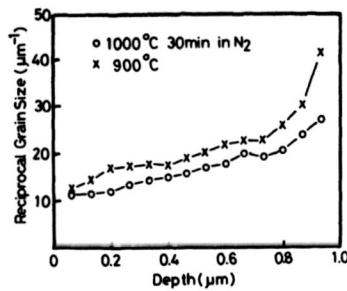

DISCUSSION

Phosphorus atoms can be divided into two categories. One is within the grain bulk, and another is at the grain boundaries.

The phosphorus concentration per unit volume at the depth of x from the surface is given by

$$N(x)=N_G+N_{GB}(x),\qquad (1)$$

where N_G and N_{GB} are the phosphorus concentrations presented in the grain bulks and segregated at the grain boundaries per unit volume, respectively. According to results of SR measurements, the phosphorus concentration in the grain bulks is constant in the poly-Si films. On the other hand, the phosphorus concentration segregated to the grain boundaries N_{GB} depend on the grain size. Therefore, N_{GB} can be expressed as

$$N_{GB}(x)=\int Q_S dS(x)/\int dV(x),\qquad (2)$$

where Q_S is surface dopant segregation density per unit area. dS and dV are the grain surface area and the grain volume for each grain respectively. Q_S is onsidered to be independent of the grain size. Eq.(2) can be rewritten as

$$N_{GB}=Q_S\cdot\int dS(x)/\int dV(x).\qquad (3)$$

The term of $\int dS(x)/\int dV(x)$ is related to the grain size and the grain structure, then the integration term can be replaced by the structural factor, A, and the grain size, r(x). Equation (3) is modified to

$$N_{GB}=Q_S\cdot A\cdot 1/r(x).\qquad (4)$$

Combining Eqs.(1) and (4), we can obtained

$$N(x)=N_G+Q_S\cdot A\cdot 1/r(x).\qquad (5)$$

The phosphorus concentrations $N(x)$ measured by SIMS are plotted as a function of $1/r(x)$ given by XTEM in Fig.5 for various annealing temperatures of 900, 950 and 1000°C. Linear relationship between $N(x)$ and $1/r(x)$ is obtained for each curve. This linear relationships ensure that A and Q_S are constant within the poly-Si films.

Figure 5. Phosphorus concentration as a function of reciprocal grain size for various annealing temperatures.

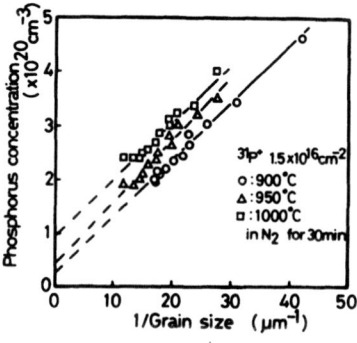

Table I. Obtained values from figure 5 for various annealing temperatures.

	900°C	950°C	1000°C
N_G ($\times 10^{20}$ cm^{-3})	0.244	0.497	0.979
$A \cdot Q_S$	0.105	0.112	0.104
r=100nm			
N ($\times 10^{20}$ cm^{-3})	1.301	1.617	2.020
N_{GB} ($\times 10^{20}$ cm^{-3})	1.057	1.120	1.040
N_{GB}/N	0.810	0.692	0.515
N_{GB}/N_G	4.330	2.250	1.062

Using Eq.(5), N_G and $Q_S \cdot A$ can be evaluated from Fig.5 for each temperature. The obtained values are summarized in Table I. Also presented in Table I are N, N_{GB}, N_{GB}/N, and N_{GB}/N_G at the grain size of 100nm.

The segregation ratios of N_{GB}/N_G range from 4.33 to 1.06 between annealing temperatures of 900°C and 1000°C. Figure 6 is a logarithmic plot of N_{GB}/N_G versus 1/T. A linear dependence on 1/T is obtained. Mandurah et.al.[5] calculated the heat of segregation, Q_0, from the slop of the line using following equation;

$$N_{GB}/N_G = Q_{S0}/N_{Si} \cdot \exp(Q_0/kT), \qquad (6)$$

where Q_{S0} is the density of dopant segregation sites at the grain boundaries per unit volume, N_{Si} is the silicon atom concentration per unit volume and T is annealing temperature. The heat of segregation is found to be equal to 1.7eV in this work, and four times as large as the previously reported values[5,6]. These works performed annealing of 1000°C for 1hr to stabilize and dopant redistribution annealing of 800 to 900°C for >12hr in order to eliminate grain growth effects from the dopant segregation. As a result of these annealing, dopant distribution in the poly-Si films was in a equilibrium condition. In device fabrication processes, however, only shorter time and lower temperature anneals are accomplished than these annealing conditions. Because in this work the dopant segregation is not in the equilibrium conditions, the heat of segregation includes grain growth effects and becomes larger value. Consequently, the dopant segregation ratio, hence carrier concentration, is more sensitive to annealing temperature than that in the equilibrium condition.

Figure 6. Segregation ratio as a function of reciprocal annealing temperature.

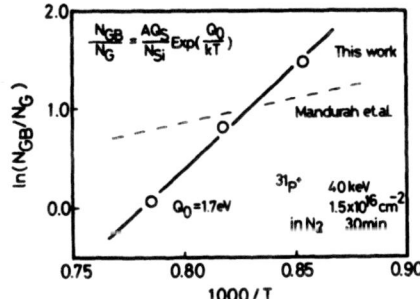

CONCLUSION

The dopant segregation at the poly-Si grain boundaries in device fabrication processes has been studied with a new approach using SR measurement, SIMS and TEM. Although SR measurements exhibit that electrically active dopant concentrations are uniform within the poly-Si films, phosphorus concentrations, measured by SIMS, increase gradually with increasing depth. XTEM observations reveal that the grain sizes decrease with increasing depth. Phosphorus depth profiles are found to be related linearly with reciprocal grain sizes corresponding to the same depth. Since the reciprocal grain size is proportional to surface area of the grain boundaries, the linear relationship indicates that the number of segregated phosphorus atoms per unit area is unique throughout poly-Si films. Segregation ratios of the phosphorus concentrations in the grain bulks to these at the grain boundaries are found to range from 4.33 to 1.06 between annealing temperatures of 900°C and 1000°C, and the heat of segregation of 1.7eV is obtained. Our results indicate that carrier concentration in the phosphorus doped poly-Si film is more sensitive to annealing temperature in device fabrication processes than that in equilibrium conditions.

ACKNOWLEDGMENTS

The authors wish to thank Masako Kinoshita for her technical assistance. They would also like to thank Shintaro Ushio and Taturo Miyoshi for many stimulating discussions.

REFERENCES

[1] M.E.Cowher, and T.O.Sedgwick,
    J.Electrochem.Soc., 119, 1565(1972)
[2] T.Hamajima, and Y.Sugano,
    6th VLSI Multilevel Interconnection Conf. Proc., 144(1989)
[3] J.M.C.Stork, E.Ganin, J.D.Cressler, G.L.Patton, and G.A.Sai-Halasz
    IBM J.RES.DEVELOP., 31, 617(1987)
[4] V.Probst, H.J.Bohm, H.Schaber, H.Oppolzer and I.Weitzel,
    J.Electrochem.Soc., 135, 671(1988)
[5] M.M.Mandurah, K.C.Saraswant, C.R.Helms, and T.I.Kamins,
    J.Appl.Phys., 51, 5755(1980)
[6] C.Y.Wong, C.R.M.Grovenor, P.E.Batson and D.A.Smith
    J.Appl.Phys., 57, 438(1985)
[7] B.Swaminathan, E.Demoulin, T.W.Sigmon, R.W.Dutton, and R.Reif
    J.Electrochem.Soc., 127, 2227(1980)

CORRELATIONS BETWEEN OPTICAL, ELECTRICAL, AND STRUCTURAL PROPERTIES
OF IN-SITU PHOSPHORUS-DOPED HYDROGENATED MICROCRYSTALLINE SILICON —
EFFECTS OF RAPID THERMAL ANNEALING ON MATERIAL PROPERTIES

David E. Kotecki, Shwu J. Jeng, Jerzy Kanicki*, Christopher C. Parks, Werner Rausch, Krishna Seshan,
and Jean Tien
IBM, Semiconductor Development Laboratory, Hopewell Junction, NY 12533
(*) IBM, T.J. Watson Research Center, Yorktown Heights, NY 10598

ABSTRACT

    Films of in-situ phosphorus-doped hydrogenated microcrystalline silicon (n—μc-Si:H) were deposited by
plasma enhanced chemical vapor deposition (PECVD) on Si(100) and fused quartz substrates over a range
of substrate temperatures (100 - 500°C) and reactant gas dilutions (1 - 100% of 1% PH_3/SiH_4 in H_2)
while maintaining a constant RF power density (0.1 W-cm^{-2}) and total gas pressure (1 Torr). Some of
the films were subjected to a rapid thermal anneal (RTA) at temperatures between 600-1000°C for a duration of 10 seconds. The n—μc-Si:H films were characterized, before and after RTA, in terms of their
microstructure, optical band-gap, electrical conductivity, and hydrogen and phosphorus content. The
deposition rate was determined to be insensitive to substrate temperature and to decrease with increasing
H_2 gas dilution indicating that deposition kinetics are dominated by plasma chemistry and are not
thermally activated. For pre-annealed films, cross-sectional TEM confirmed the presence of a mixed
phase material at all deposition temperatures with gas dilutions ≤10%. The surfaces of thick films
(>0.15 μm) were rough, giving them a hazy appearance, while thin n—μc-Si:H films (<0.15 μm) were
smooth and mirror-like. The rough surfaces were correlated with voids and microcracks in the n—μc-Si:H
films observed by TEM. The optical band-gap of all pre-annealed films was $\simeq$1.8eV and the electrical
conductivity varied between 1 and 20 (Ω-cm)$^{-1}$. The H content was found to be independent of gas
dilution but decreased with increasing substrate temperature; the P content depended on both the gas
dilution and substrate temperature, decreasing at high deposition temperatures. RTA was observed to
significantly alter film morphology and microstructure, increase electrical conductivity, and decrease the
optical band-gap.

INTRODUCTION

    The mixed-phase (α-Si, c-Si) microstructure of microcrystalline silicon makes this an interesting material
from a basic physics point of view, as well as giving it technological importance. The microstructure of
n—μc-Si:H offers significantly greater numbers of dopant atom four-fold coordination sites resulting in
higher conductivities than obtainable from n—α-Si:H. Applications of n—μc-Si:H include thin film transistors [1], bipolar transistors [2], and photovoltaics. Previous investigations of the properties of
n—μc-Si:H have been reported for material prepared by both DC discharge [3,4] and RF discharge [5,6].
In this paper, we report on the relationship between the structural, electrical, and optical properties of
n—μc-Si:H prepared by RF-PECVD and discuss the effects of rapid thermal anneal (RTA) on material
properties.

EXPERIMENTAL

    n—μc-Si:H films were deposited by plasma enhanced chemical vapor deposition (PECVD) from a reactant
gas mixture of 1% PH_3 in SiH_4 and 100% H_2. During deposition, the substrates were situated face down
in a stainless steel, parallel plate RF (13.56 MHz) powered reaction chamber. A constant electrode separation of 1.0 cm was maintained and the reactant gases were injected through a showerhead at the lower
electrode. The vacuum assembly was equipped with a turbomolecular pump which provided a chamber
base pressure of 1×10^{-6} Torr. The flow rates of 1% PH_3 in SiH_4 and 100% H_2 were independently controlled. The total chamber pressure during deposition was also controlled by a throttle valve located prior
to a roots blower.
    For the purpose of this study, all n—μc-Si:H films were deposited at a constant RF power density of
0.1 W-cm^{-2} and a constant total reactant gas pressure of 1.0 Torr. The flow rate of 1% PH_3 in SiH_4 was

Table 1. Parameter space investigated for depositing n–μc-Si:H films. Process conditions labeled with a "M" produced films with a microcrystalline structure; those labeled with a "A" produced films with an amorphous structure. The structure was determined TEM.

Gas Dilution (%)	Substrate Temperature (°C)				
	100	200	300	400	500
1	M	M	M	M	M
2		M			
5		M			
10	M	M	M	M	M
∞		A			
RF power density = 0.1 W-cm^{-2}					
Total pressure = 1.0 Torr					
1% PH$_3$ in SiH$_4$ flow = 10 sccm					

held constant at 10 sccm. The substrate temperature and the reactant gas dilution (defined as the ratio of the flow rates of 1%PH$_3$ in SiH$_4$ to H$_2$) were varied as shown in Table 1. As indicated in the table, films were deposited at substrate temperatures from 100°C to 500°C in 100°C increments with reactant gas dilutions of 1% and 10%. Furthermore, films were deposited at a constant substrate temperature of 200°C with reactant gas dilutions of 1%, 2%, 5%, and pure 1%PH$_3$ in SiH$_4$. For all deposition conditions, n–μc-Si:H films were simultaneously deposited on both Si(100) and fused quartz substrates. Films with a thickness ranging from 0.03μm to 1.0μm were investigated with the film thickness being controlled by the duration of the deposition.

The n–μc-Si:H films deposited at 200°C and 1% dilution were subjected to rapid thermal anneal (RTA) at temperatures between 600°C and 1000°C for a duration of 10 seconds. Films deposited on both Si(100) and fused quartz substrates were simultaneously annealed for all RTA conditions in order to investigate any substrate dependence on recrystalization, while minimizing run-to-run temperature variations within the RTA. The temperature of the wafer during the anneal was monitored via a pyrometer. All anneals were performed in a N$_2$ ambient.

RESULTS AND DISCUSSION

Fig. 1a shows the measured deposition rate as a function of substrate temperature for gas dilutions of 1% and 10%. As can be seen, for a constant gas dilution, the deposition rate is insensitive to substrate temperature. This indicates that for these deposition conditions, the rate limiting step in the deposition process is not thermally activated (i.e., limited by surface desorption or migration). Rather, deposition is controlled by plasma chemistry and limited by the generation of Si containing radicals by collisional dissociation of the SiH$_4$. The deposition rate as a function of the reactant gas dilution, for a constant substrate temperature of 200°C, is shown in Fig. 1b. Here, the deposition rate is found to monotonically decrease from a maximum value of 245Å/minute using pure 1% PH$_3$/SiH$_4$ to a rate of 45Å/minute when using a 1% dilution in H$_2$. This monotonic decrease in deposition rate with increasing H$_2$ flow is most likely caused by the increase competition between the deposition reaction and an etching reaction mediated by the presence of increased atomic H. It has previously been shown [1] that under similar deposition conditions, the thickness of the n–μc-Si:H films is linearly dependent on time, and incubation times associated with nucleation effects are negligible.

The detailed microstructure of the n–μc-Si:H films was investigated by high resolution cross-sectional transmission electron microscopy (HRTEM). Process conditions listed in Table 1 labeled with a "M" were found to produced films which were microcrystalline; the condition labeled with an "A" produced an amorphous film. Thus, films with a microcrystalline structure were obtained over a wide process window and at deposition temperatures as low as 100°C. Previous work on intrinsic films deposited with the same RF power and total pressure were found to be amorphous for substrate temperatures ≤200°C and 10% dilution. Thus, the addition of small quantities of PH$_3$ helps to initiate microcrystalline growth. The average c-Si grain size in the n–μc-Si:H films was found to monotonically increase with both increasing substrate temperature and increasing film thickness.

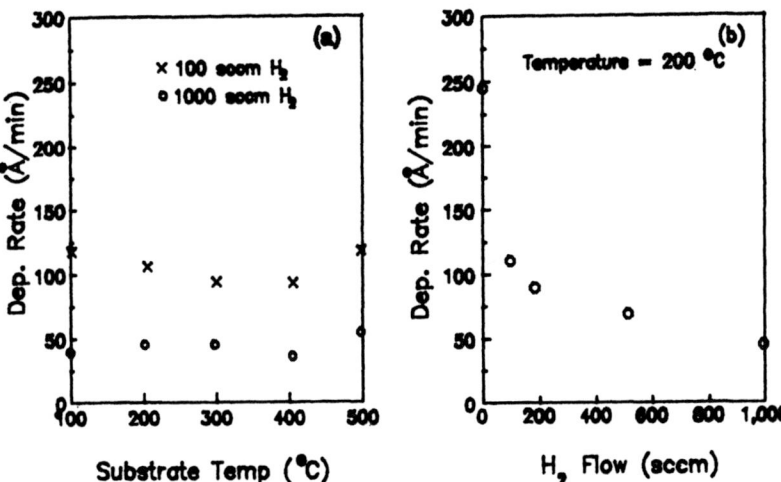

Figure 1. Measured deposition rate of n—μc-Si:H. (a) As a function of substrate temperature at constant reactant gas dilution of 1% and 10% and (b) as a function of gas dilution at a constant substrate temperature of 200°C.

Figs. 2a,b show the microstructure of a n—μc-Si:H film deposited at a substrate temperature of 200°C and 1% dilution for two different film thicknesses. For the ≃ 1μm thick film shown in Fig. 2a, voids are observed at the n—μc-Si:H/c-Si interface. The presence of large numbers of voids and microcracks were observed, either within the bulk of the film or at the n—μc-Si:H/c-Si interface, for all thick (> 0.15μm) films examined. These defects are thought to be stress-related brought about by the presence of a mixed phase of α-Si and c-Si in a common matrix. In addition, the surfaces of these thick films were rough, giving them a "hazy" appearance; the surface roughness was directly correlated to the presence of voids and microcracks. For thin (< 0.15μm) n—μc-Si:H films, such as the 0.03μm thick film shown in Fig. 2b,

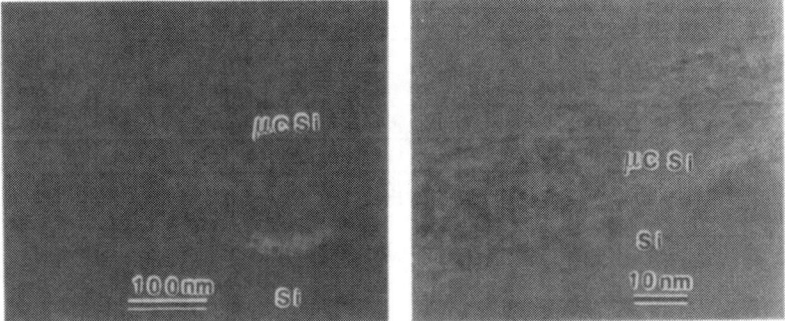

Figure 2. High resolution cross-sectional TEM of n—μc-Si:H films. (a) Thick (≃1μm) film showing voids at the n—μc-Si:H/c-Si interface. (b) Thin (0.03μm) film showing a homogeneous mixed phase material with good structural integrity, good adhesion, and a smooth surface. The c-Si grain size is ≃ 40Å.

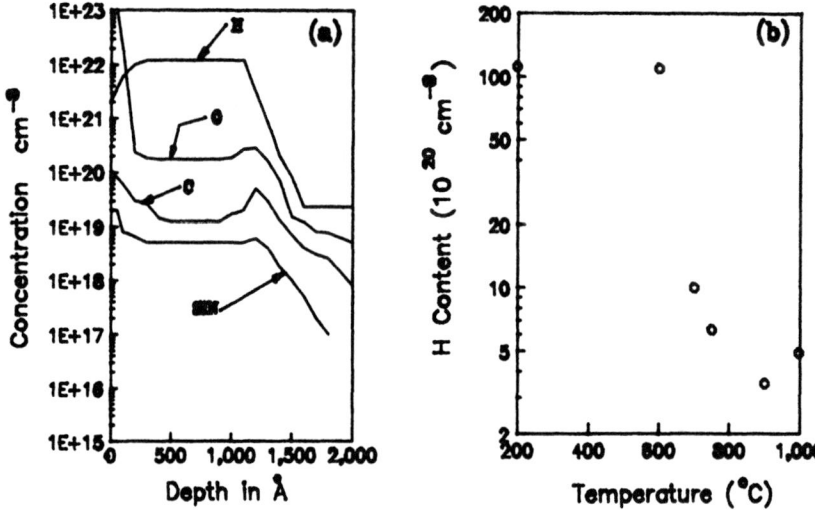

Figure 3. SIMS characterization of n−μc-Si:H films. (a) Typical impurity levels of H, O, N, and C found in n−μc-Si:H films and (b) measured H concentration as a function of RTA temperature. The films subjected to RTA were deposited at 200°C with 1% dilution. The duration for all anneals was 10 seconds.

no voids or microcracks are detected. All thin films had good structural integrity and were extremely uniform and homogeneous. The film surface appeared mirror-like.

The microstructure of n−μc-Si:H can be strongly effected by RTA. Thin n−μc-Si:H films deposited on c-Si substrates which were annealed at temperatures ≥900°C were found to recrystallize into smooth films with the crystal orientation aligned to the substrate. Thick n−μc-Si:H films annealed under the same conditions resulted in recrystallized regions with large voids. Additional voids were produced during the RTA and are believed due to trapped bubbles of H_2 within the film.

The chemical constituents of the deposited films were determined by SIMS. A typical SIMS result is shown in Fig. 3a. The concentrations of O, C, and N were measured to be 1.3×10^{20}, 2×10^{19}, and 5×10^{18} cm⁻³ respectively, and were found to be insensitive to the deposition conditions. The source of the high level of O has not been identified. But, the in-situ deposition of a SiN_x cap over the n−μc-Si:H film was used to rule out the possibility of post-deposition O incorporation. The measured H and P contents are

Table 2. Measured H and P content in the n−μc-Si:H films as measured by SIMS. (cm⁻³)

Gas Dilution (%)	Substrate Temperature (°C)				
	100	200	300	400	500
1	H: 1.3×10^{22}	H: 1.3×10^{22}	H: 8.1×10^{21}	H: 1.4×10^{21}	H: 8.0×10^{19}
	P: 1.5×10^{20}	P: 2.0×10^{20}	P: 3.3×10^{20}	P: 2.4×10^{20}	P: 1.0×10^{20}
2		H: 9.9×10^{21}			
		P: 4×10^{19}			
5		H: 9.9×10^{21}			
		P: 5×10^{19}			
10	H: 1.3×10^{22}	H: 1.2×10^{22}	H: 5.6×10^{21}	H: 9.9×10^{20}	
	P: 4.7×10^{20}	P: 4.7×10^{20}	P: 4.3×10^{20}	P: 3.1×10^{20}	
∞		H: 1.1×10^{22}			
		P: 5×10^{19}			

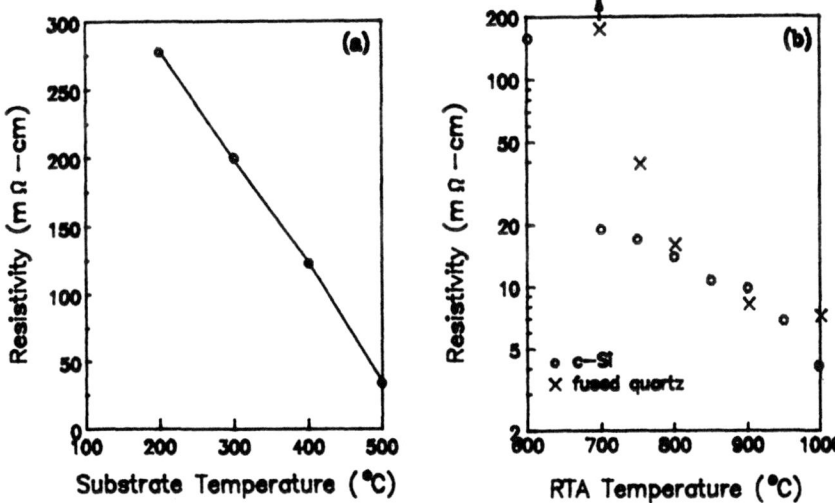

Figure 4. Measured film resistivity (a) as a function of deposition temperature and (b) as a function of RTA temperature. The films subjected to RTA were deposited at 200°C with 1% dilution. All anneals were for 10 seconds.

summarized in Table 2. The H content was $\simeq 1 \times 10^{20}$ cm^{-3} for substrate temperatures $\leq 300°C$, and was found to decrease by an order of magnitude for each 100°C increase in substrate temperature. The decrease in H content could not be modeled by a simple Arrhenius function with a single activation energy. At a deposition temperature of 200°C, the H content was found to be independent of the gas dilution. Even though the addition of H$_2$ gas was found to etch the surface, decreasing the deposition rate, no additional H is incorporated in the film. The P content was found to depend on both substrate temperature and gas dilution as indicated in Table 2. A maximum P incorporation of 4.5×10^{20} cm^{-3} was obtained and the P content decreased with increasing deposition temperatures $\geq 300°C$ for both 1% and 10% dilutions. SIMS results on samples subjected to RTA indicated that the C, O, N, and P contents remained unchanged. The H content decreased with increasing anneal temperatures as shown in Fig. 3b.

The resistivity of thick n$-\mu$c-Si:H films deposited at 1% dilution as a function of deposition temperature is shown in Fig. 4a. The resistivity was found to monotonically decrease with increasing deposition temperature, from 280 mΩ-cm at 200°C to 40 mΩ-cm at 500°C, even though SIMS results indicated that the amount of P incorporation decreased at the high temperatures. These resistivities are over an order of magnitude lower than those reported for in-situ doped α-Si [7]. The decrease in resistivity at higher deposition temperatures is due to greater electrical activation and larger mobilities associated with the observed increase in c-Si grain size with increasing deposition temperature. In addition, for a given deposition condition, the resistivity is observed to decrease with increasing film thickness. This thickness dependence of the resistivity has been correlated with the observed increase in grain size with increasing film thickness observed by TEM.

Fig. 4b shows the measured resistivity for films deposited at 200°C on both fused quartz and c-Si substrates as a function of the RTA temperature. The resistivity was found to decrease with RTA temperature and a minimum resistivity of 42 mΩ-cm was obtained for films on c-Si substrates after a 1000°C, 10 second anneal. Films deposited on the fused quartz substrates were found to have higher resistivities than those deposited on c-Si substrates after RTA due to increased numbers of grain boundaries in the films annealed on the amorphous substrates.

The optical band-gap of the n$-\mu$c-Si:H films was estimated using the Tauc model. The optical reflectivity and transmissivity were measured on n$-\mu$c-Si:H films deposited on fused quartz substrates. Fig. 5a shows the Tauc plot of a n$-\mu$c-Si:H film deposited at 200°C and 1% dilution both prior to any anneal, and after a 600°C, 10 second RTA. The optical band-gap is $\simeq 1.8$eV in both cases. Even though the absorption data for these conditions could be well fitted by the Tauc model, this was not the case at higher RTA temperatures. Nevertheless, an estimate of the optical band-gap as a function of the RTA temperature was obtained and the result is shown in Fig. 5b. For 10 second anneals, the band-gap was

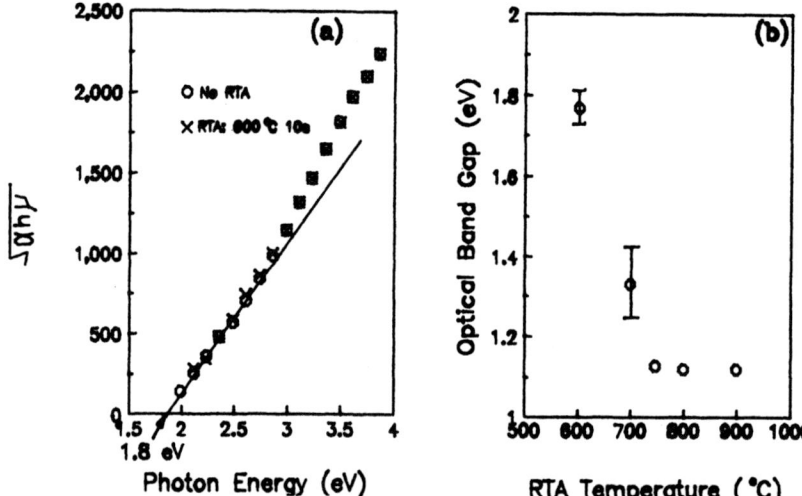

Figure 5. Optical band-gap. (a) Tauc plot for films deposited on fused quartz substrates at a deposition temperature of 200°C and 1% dilution. Films annealed at 600°C for 10 seconds show the same absorption characteristics. (b) Optical band-gap, determined from the Tauc model, as a function of the RTA temperature.

found to decrease with increasing RTA temperatures and approach that of c-Si for anneals at temperatures ≥750°C.

CONCLUSIONS

Films of n—μc-Si:H were deposited by PECVD under a range of substrate temperatures and reactant gas dilutions shown in Table 1. The conclusions of this paper are summarized below:

1. The deposition rate was found to be independent of substrate temperature indicating that the rate limiting step is not thermally activated.

2. Films of n—μc-Si:H with an average c-Si grain size $\simeq 40$Å can be obtained at deposition temperatures as low as 100°C.

3. Thin films (< 0.15μm) had good morphology and smooth surfaces, while thick films (> 0.15μm) had voids and microcracks both in the bulk of the n—μc-Si:H film and at the n—μc-Si:H/c-Si interface.

4. The presence of voids and microcracks produced films with rough surfaces giving them a "hazy" appearance.

5. No change in film properties were observed after temperature cycling to 600°C for 10 seconds.

6. n—μc-Si:H films annealed at temperatures ≥700°C, even for durations as short as 10 seconds, are unstable resulting in the loss of hydrogen, loss of α-Si regions, and a decrease in optical band-gap.

REFERENCES

1. J. Kanicki, E. Hasan, J. Griffith, T. Takamori and J.C. Tsang, Mat. Res. Soc. Symp. Proc. 149, p239 (1989).
2. H. Fukioka and K. Takasaki, Fujitsu J. Sci.tech. 24, 391 (1988).
3. R.V. Kruzelecky, S. Zukotynski, and J.M. Perz, J. Non-Cryst. Solids, 103, 221 (1988).
4. S. Usui and M. Kikuchi, J. Non-Cryst. Solids, 34, 1 (1979).
5. A. Matsuda, S. Yamasaki, K. Nakagawa, H. Okushi, K. Tanaka, S. Iizima, M. Matsumura and H. Yamamoto, Jap. J. Appl. Phys., 19, L305, (1980).
6. S. Hasegawa, S. Narikawa and Y. Kurata, Phil. Mag. B., 48, 431 (1983).
7. M.J.M. Pruppers, K.H.M. Maessen, J. Bezemer, F.H.P.M. Habraken and W.F. Van der Weg, Mat. Res. Soc. Symp. Proc. 95, p131 (1987).

SELECTIVE DEPOSITION OF N+ DOPED MC-SI:H:F BY RF PLASMA CVD ON SI AND SIO2 SUBSTRATES.

K. Baert, P. Deschepper, H. Pattyn, J. Nijs and R. Mertens
I.M.E.C. Kapeldreef 75, 3030 Heverlee, Belgium.

ABSTRACT

μc-Si:H:F can be deposited by rf plasma-CVD of fluorinated gas sources like SiF_4 and SiH_2F_2, mixed with H_2 and/or SiH_4. However, this growth process is usually not selective: the layers are deposited on SiO_2 as well as on Si substrates. In this paper, selective growth from $SiF_4 + SiH_4$ is reported. N^+ Si layers were deposited on <100> Si and poly-Si with a conductivity up to 300 resp. 100 S/cm. The selective growth process was applied for Source and Drain regions of poly-Si thin film transistors on insulating substrate.

INTRODUCTION

μc-Si:H:F can be grown by rf-PECVD from $SiF_4 + H_2$, $SiF_4 + SiH_4$, $SiH_2F_2 + H_2 + SiH_4$ etc., [1,2, 5-7], and growth of μc-Si:H:F was also demonstrated using other techniques like photo-CVD [3] and Hydrogen-Radical Enhanced CVD [4]. Its in-situ **phosphorus** doping process is characterized by a very high dopant incorporation ratio (up to 50) and a high doping efficiency (up to 100 %) [2, 8]. The (limited) long range order introduced by the crystallites results in a reduction of the tail states [9], and by consequence, much higher active carrier concentrations can be reached in μc-Si:H:F than in a-Si:H. The carrier mobility on the other hand is less influenced by the microcrystallites [9]. The net result is that conductivities of up to 50 S/cm can be obtained for heavily phosphorus doped μc-Si:H:F [3].

N^+ μc-Si:H:F material has been applied as an n^+ layer in a-Si:H solar cells. Another application is the deposition of n^+ layers as Source and Drain regions to poly-Si thin-film transistors (TFTs) on glass substrates [10]. At present, several technological problems limit the scalability of poly-Si TFTs to larger areas: in particular, ion implantation is one of the technologies which is not well suited to large area processing. There is therefore a need for alternative techniques which allow large area processing at temperatures compatible with low cost glass substrates. Two problems have to be solved however in order to allow the use of n^+ μc-Si:H:F in this application:
(1) The conductivity should be higher than 100 S/cm. This can be realized by improving the crystallite size, which will increase the mobility.
(2) The process should be selective (e.g. deposit only on poly-Si and not on the SiO_2 sidewall of the gate and the glass substrate) in order to make a self-aligned TFT structure.
In previous work, we have demonstrated that the first problem can be solved by applying *epitaxial* rf-PECVD growth technology to *poly-Si* films, which results in poly-Si n^+ layer with conductivities up to 500 S/cm [12]. Fig. 1 shows the transfer characteristic of a poly-Si TFT with a 200 Å

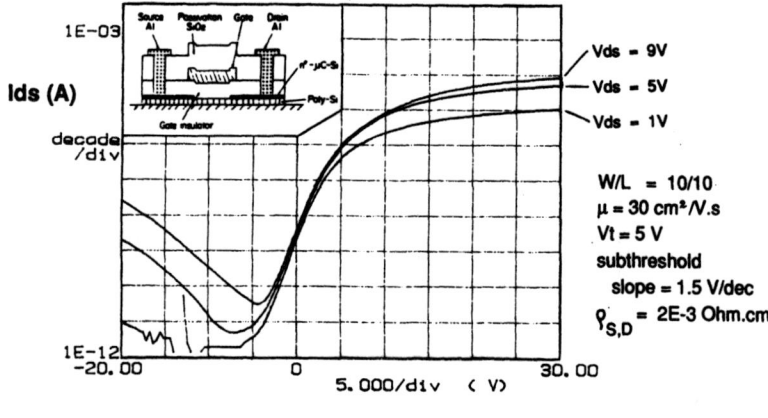

Vg (V)

Fig. 1: Low temp. poly-Si TFT with highly conductive Source and Drain deposited by rf-PECVD epitaxial growth technique (non-self aligned).

thick Source and Drain realized by this technique. However, for making a self-aligned TFT structure, the selective growth of n^+ μc-Si:H:F by rf-PECVD has to be realized.

In order to realize selective growth in conventional high temperature CVD epitaxial growth, a balance is established between deposition and etching. In this study, SiH_4 has been used for deposition and SiF_4 for etching. SiF_4 was preferred to more common etching gases like CF_4, because it was expected not to leave residues (e.g. C) which might inhibit crystalline film growth, and because rf-PECVD epitaxial growth has been reported using SiF_4 [4,8].

SIF4 ETCHING

Etching of Si and SiO_2 by SiF_4 has been used previously as a in-situ cleaning technique for epitaxial growth [13]. In this study, the etch rate of <100> p-type Si (1-100 Ohm.cm), thermal oxide and densified APCVD oxide was investigated as a function of temp., rf power and flow rate. The etch rate of oxides was measured by spectroscopic optical reflection. For the measurement of the etch rate on <100> Si, a mask of thermal or APCVD oxide was used. Since we observed that <100> Si etch rates were dependent on the type of oxide mask used, it can be concluded that loading effects influenced the etch rate, and the Si etch rates should therefore be regarded as relative values. From the dependence of etch rate on power, flow rate and temp. (Fig. 2-4), it can be seen that SiO_2 etching by SiF_4 is a dissociation-limited process with a small activation energy. On the other hand, the etch rate of Si increases with both power and flow rate, which cannot be interpreted as either a dissociation-limited or supply-limited regime. Also, the temperature dependence of the Si etch rate is more pronounced than that of SiO_2, but no activation energy can be determined.

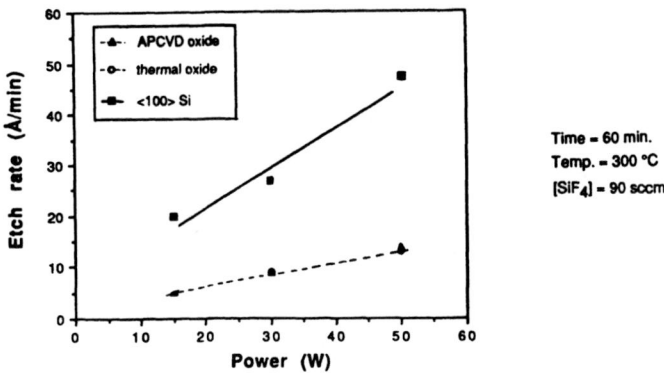

Fig. 2: SiF₄ etch rate vs. power (p=50 mTorr, d_{el}=20mm)

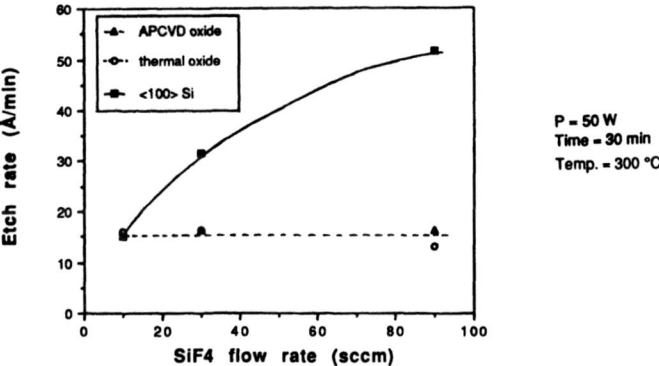

Fig. 3: SiF₄ etch rate vs. flow rate (p=50 mTorr, d_{el}=20mm)

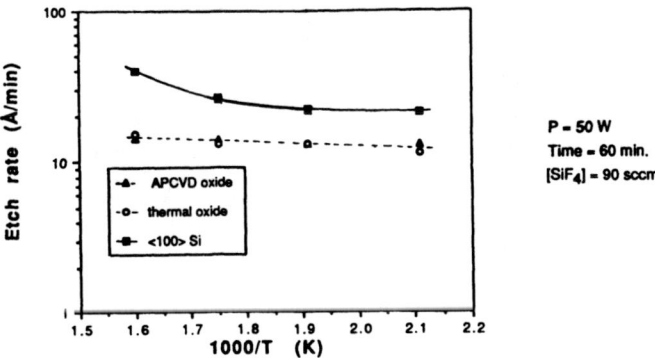

Fig. 4: SiF₄ etch rate vs. temp. (p=50 mTorr, d_{el}=20mm)

SELECTIVE DEPOSITION OF N+ μC-SI:H:F

To realize selective deposition, we have adapted conditions of high Si etch rate, namely high SiF4 flow rate (90 sccm), high rf power (50-70W), and high temp. (300 - 350 °C). The etch rate and deposition rate was measured on thermal oxide, <100> Si and poly-Si (recrystallized LPCVD a-Si). The selective growth was realized by adding a small amount of SiH4 (+ 0.1% PH3) to the SiF4 plasma (fig. 5). The growth condition could be controlled from etching over selective deposition to non-selective deposition by controlling the SiH4 flow rate. It can be explained that, under the condition of high power and low SiH4 flow rate, the deposition rate from SiH4 is a supply-limited process. Therefore, the SiH4 flow rate controls the deposition rate of SiH4 and hence the balance between deposition and etching. Since the Si etch rate from SiF4 is much more temperature dependent than the deposition of e.g. a-Si:H from SiH4, it can be expected that a similar control of the selective growth condition can be realized by the temperature. This is illustrated in fig. 6. At 300 °C, non-selective deposition occurs. At 315 - 330 °C, the deposition is selective.

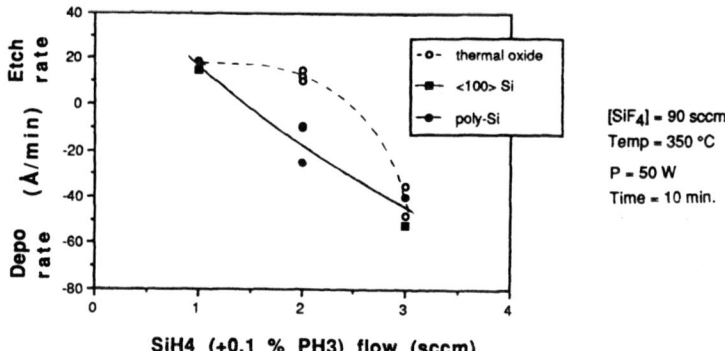

Fig. 5: Selective growth vs. SiH4 flow (p=50 mTorr, d_{el}=20mm)

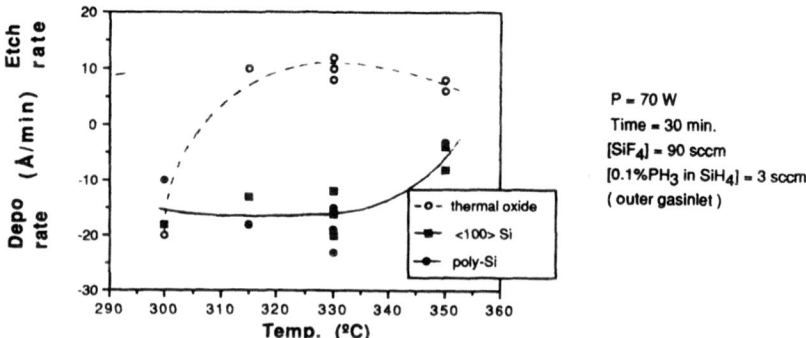

Fig. 6: Selective growth vs. temp. (p=50 mTorr, d_{el}=20mm)

The film resistivity of the selectively deposited n+ films on <100> Si ranged from 50 to 300 S/cm. Since the conductivity did not increase upon further increase of the PH3/SiH4 ratio up to 1%, it is not likely that the films are epitaxially grown, since for <100> Si, epitaxially grown by rf-PECVD, conductivities up to 3000 S/cm can be obtained [12]. On the other hand, the conductivity is significantly higher than previously reported values for n+ μc-Si:H:F. Also, the conductivity of films deposited on poly-Si was lower (3 - 100 S/cm), which could mean that the crystallography of the film is dependent on that of the substrate. We can conclude that the film grows to some extent epitaxially on the <100> Si and the poly-Si.

Although a detailed analysis of the growth process has to take into account gas phase reactions between SiH4 and SiF4 fragments, as well as the interaction observed in the etching experiments between SiO2 and Si, our data indicate that selective growth in rf-PECVD can be realized by a balance between etching and deposition. It can be explained that the role of the SiF4 etching is to etch away initial nuclei (formed from SiH4 dissociation), and that nuclei on SiO2 are less strongly bonded to the substrate than those on Si, and therefore more easily etched. This model agrees with the epitaxial growth observed in our experiments. Also, it could indicate that one of the roles of F-species in μc-Si:H:F growth is to etch the less stable (amorphous) phase, and promote the growth of the more stable (crystalline) phase. Such a model has previously been proposed for explaining the role of H-radicals in μc-Si:H grown from SiH4 and H2 [14].

SELECTIVE SOURCE AND DRAIN DEPOSITION FOR POLY-SI TFT

This selective growth process was applied to Source and Drain formation of a poly-Si TFT. In our experiments, a thermally oxidized Si wafer was used as a substrate, but the maximum process temp. was 630 ºC in order to be compatible with glass substrates. Preliminary data, for a non-optimized TFT process, are given in fig. 7. From the transfer characteristic, it is observed that the selective deposition does not result

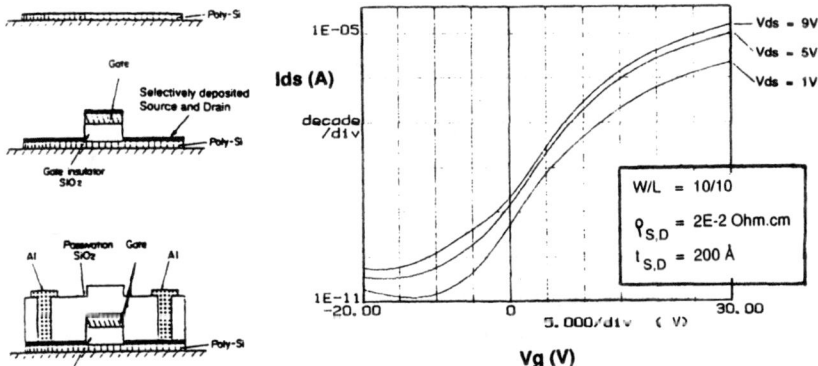

Fig. 7 Low temp. poly-Si TFT with selectively deposited Source and Drain by rf-PECVD growth (self-aligned)

in a short-circuit between Gate and Source/Drain. The conductivity of the Source/Drain was measured independently and was about 50 S/cm for a 200 Å n^+ layer.

CONCLUSIONS

We have demonstrated selective growth of n^+ μc-Si:H on crystalline Si and poly-crystalline Si versus thermal oxide. The selective growth condition was controlled by balancing etching (from SiF_4) versus deposition (from SiH_4). It was put forward that the selective deposition occurs because initial nuclei on SiO_2 are less stable, and therefore etched preferentially to those on Si. Phosphorus doped films with a conductivity up to 300 resp. 100 S/cm were deposited on <100> Si resp. poly-Si. The selective growth process was applied to Source and Drain formation in a low temperature (< 630 ºC) poly-Si TFT process.

ACKNOWLEDGEMENTS

The authors like to thank Mr. W. Laureys for the lithographic processing, and mention the contribution of Ms. A. Van Ammel and Mr. P. Laermans in TFT fabrication.

REFERENCES

1. S. Ovshinsky and A. Madan, Nature 276 , 483 (1978)
2. A. Matsuda, S. Yamasaki, K. Nakagawa, H. Okushi, K. Tanaka, S. Izima, M. Matsumura and H. Yamamoto, Jap. J. Appl. Phys. 19 (6) , L305 (1980)
3. S. Nishida, H. Tasaki, M. Konagai and H. Takahashi, Mater. Res. Soc. Proc. 49, 47 (1985)
4. S. Oda, S. Ishihara, N. Shibata, H. Shirai, A. Miyauchi, K. Fukuda, A. Tanabe, H. Ohtoshi, J. Hanna and I. Shimizu, Jap. J. Appl. Phys. 25 (3) , L188 (1987)
5. K. Nakazawa, K. Tanaka and N. Yamauchi, Jap. J. Appl. Phys 28 (4), 569 (1989)
6. S. Ray, S. De, G. Ganguky, A. Barua, A. Mascarenhas, M. Al-Jassim, S. Deb, J. Appl. Phys. 65 (10), 4024 (1989)
7. Y. Okada, J. Chen, I. Campbell, P. Fauchet and S. Wagner, presented at the 1985 MRS Spring Meeting (1985) (unpublished)
8. N. Shibata, K. Fukuda, H. Ohtoshi, J. Hanna, S. Oda and I. Shimizu, Jap. J.Appl. Phys. 26 (1) , L10 (1987)
9. W. Spear, G. Willeke and P. Lecomber, Physica 117B & 118B, 1983 ,908
10. K. Kobayashi, J. Nijs and R. Mertens, J. Appl. Phys. 65 (6), 2541 (1989)
11. J. Nijs, K. Baert, J. Symons, K. Kobayashi and P. Deschepper, Appl. Surf. Sc. 36 , 22 (1989)
12. K. Baert, P. Deschepper, H. Pattyn, A. Rodriguez, J. Nijs and R. Mertens, Thin Solid Films, 182 (in print)
13. A. Yamada, A. Satoh, M. Konagai and K. Takahashi, Proc. of the 34th Nat. Symp. of the Am. Vac. Soc., Anaheim, CA, Nov. 2-6 (1987)
14. R. Van Oort, M. Geerts, J.van den Heuvel and J. Metselaar, Electronics Letters 23 (18), 967 (1987)

SELECTIVE GROWTH OF Si CRYSTALS OVER AMORPHOUS SUBSTRATES SEEDED BY SOLID-STATE AGGLOMERATION OF PATTERNED Si

K. YAMAGATA and T. YONEHARA
R/D Headquarters, Canon Inc., 6770 Tamura Hiratsuka-city, Kanagawa, 254, Japan

ABSTRACT

Selective growth of Si crystals over amorphous substrates, seeded by agglomerated single domained Si crystals is demonstrated. In this method, Si crystal seeds are periodically placed and selectively overgrown until impingement upon adjacent crystals, resulting in a matrix of large Si islands with controlled grain boundary locations. Si seeds are formed over amorphous SiO_2 by the solid-state agglomeration phenomenon, and grown selectively up to 100 μm by CVD selective epitaxial growth technique. The grown crystals are classified in three crystalline forms of single crystals, primary twins, and multiple twins. However, most are single crystals with a specific orientation of (110) normal to the substrate surface.

INTRODUCTION

In general, overlayers as-deposited on amorphous substrates are amorphous or at best, polycrystalline. However, it would be possible to overgrow a single crystal layer on amorphous surface by selective growth technique if single crystal seeds were formed on its surface. Sentaxy[1-3] (Selective Nucleation based Epitaxy) originated by us, and demonstrated in silicon and diamond[4,5], is a crystal growth technique in which a Si nucleus is formed exclusively at an artificial nucleation site of an amorphous material, and a matrix of large Si crystals is overgrown selectively at the predicted locations on the amorphous substrate. Eventually the grain boundaries are formed at the center between adjacent sites. In this paper, we will demonstrate improved method of sentaxy which is seeded by "solid-state agglomeration phenomenon" in small portions of fine polycrystalline Si films on the amorphous substrates in order to produce device-quality Si layer over amorphous substrate.

The phenomenon of agglomeration is known such that continuous films are altered to discrete islands or beads in solid state, well below melting point. The phenomenon is driven by surface-energy minimization followed by large mass-transport, and has been studied extensively in semiconductors and metals. It was found that ultra thin films (less than 100 nm) of Ge readily agglomerated by annealing in Ar or vacuum[6]. In investigation of zone melting recrystallization of Si[7], a capping system on the Si films has been studied in order to prevent the beads-up of melted Si. Reports on metal-agglomeration have concerned with thermal stability of IC's electrode materials in Al, Cu, and Pt-Si[8]. Most of the investigations have been done for the purpose of impeding the agglomeration phenomenon. We propose to make use of the agglomeration phenomenon in crystal seed formation on amorphous surface.

It is revealed that annealing ambients strongly affect the temperatures at which the agglomeration starts, and that impurity in the initial films influences the mass-transport. Patterning size of initial films determines the number of agglomerated beads. Most significantly, the agglomerated Si semi-spherical beads are single domained crystals (i.e. no grain boundary in the beads).

EXPERIMENTAL DETAILS

SiO_2 was grown at 1000°C to thickness of 0.1 μm by pyrogenic oxidation of Si (511) wafers, 4 inches in diameter. Polycrystalline Si was formed over SiO_2 surface at 620 °C to thickness of

0.1 and 0.44 μm by LPCVD (Low Pressure Chemical Vapor Deposition). Phosphorus impurity was doped into the polycrystalline Si films by thermal diffusion or ion implantation. P_2O_5 glass formed by thermochemical reaction in $POCl_3$ and O_2 was deposited on the polycrystalline Si films at 950°C for 20 minutes, and simultaneously phosphorus were diffused into the Si films. $^{31}P^+$ ions were implanted onto the 0.1 μm and 0.44 μm thick-Si films at 30 KeV with dose of 7.5×10^{15} cm^{-2} and at 70 KeV with dose of 3.3×10^{16} cm^{-2} respectively, resulting in concentration of 7.5×10^{20} cm^{-3}. The phosphorus doped Si films were patterned into 1, 2, and 4 μm squares, 50 and 100 μm in period by conventional lithography and reactive ion etching.

The doped and pre-patterned Si films were annealed in H_2 and N_2 ambients at 1000–1200°C, for 5–20 minutes in a conventional epitaxial reactor or a furnace tube. Some non-doped continuous films were also annealed in H_2 ambient in order to observe the details of agglomeration phenomenon. Selective growth was carried out by seeding the agglomerated Si which was formed from the phosphorus doped and pre-patterned films by H_2 annealing in the same apparatus. The gas composition was Si source gas of SiH_2Cl_2, HCl additional etching gas, and H_2 dilution gas. Gas flow rates were 0.53, 2.0, 100 (l/min.) respectively. Temperature and pressure were 1030°C and 100 torr. It took 90 minutes to obtain a matrix of Si crystals up to 50 μm in diameter.

RESULTS AND DISCUSSION

Solid-state agglomeration

There are many factors to influence degrees of agglomeration, e.g. film thicknesses, annealing ambients, annealing temperatures, annealing times, pressures of the ambient, and impurities. The most important factor to give rise to agglomeration is annealing ambients.

50 nm-thick polycrystalline Si films deposited by LPCVD on SiO_2 surface and then doped phosphorus impurity were annealed in H_2 and N_2 ambients at 1000°C, for 5 and 30 minutes. The films were completely agglomerated by annealing in H_2 and altered discontinuous beads, less than 1 μm in diameter, as shown in Fig.1(a), while the films annealed in N_2 did not change at all in spite of longer annealing time as shown in Fig.1(b). In addition, the films annealed in N_2 were still continuous even at 1200°C for 20 minutes. The non-doped films were agglomerated by annealing in H_2 at higher temperatures of 50°C than phosphorus doped films, and non-doped ones annealed in N_2 did not change at all again. H_2 is known to be reactive with Si producing volatile SiH_x, so that its reactivity with Si presumably would enhance initiation of agglomeration.

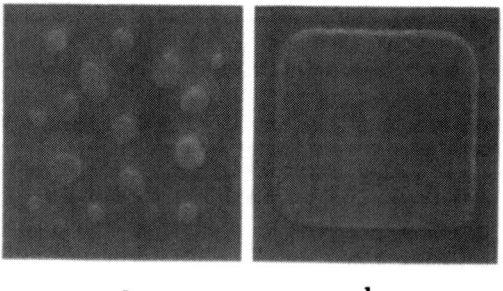

Fig.1. (a) SEM micrographs of agglomerated Si by annealing in H_2 ambient, and (b) non-agglomerated Si by annealing in N_2 ambient.

a   b   1 μm

Phosphorus impurity in initial Si films is the secondary crucial factor. The heavily doped phosphorus impurity in the Si films accelerates agglomeration and lower the temperature at which agglomeration starts in H_2 ambient. When the heavily doped films are annealed in N_2 ambient, the impurity initiates abnormal grain growth[9] instead of agglomeration. Abnormal grain growth is the phenomenon in which the grain size in the polycrystalline films is enlarged to several tens or about one hundred times larger than the film thickness. Therefore, it is considered that phosphorus impurity in polycrystalline Si films strongly affect Si migration rate.

The changing surface morphology was observed in Si films from as-deposited continuous surface to the agglomerated discontinuous states as a function of time. Fig.2 is the SEM (Scanning Electron Microscope) micrographs presenting time dependence on surface morphology in non-doped 50 nm-thick polycrystalline Si films annealed in H_2 ambient at 1045°C. In one minute, there is no change of surface flatness compared with as-deposited films, but some holes are observed in some places ((a) in Fig.2). In five minutes, Si surface start changing rough, and holes expand leaving small beads ((b) in Fig.2). In ten minutes, the surface becomes more rough, and the parts of film start dividing from the edge of holes ((c) in Fig.2). In twenty minutes, the films agglomerate completely all over the substrate ((d) in Fig.2). These SEM images also show the appearance of large mass-transport in agglomeration phenomenon.

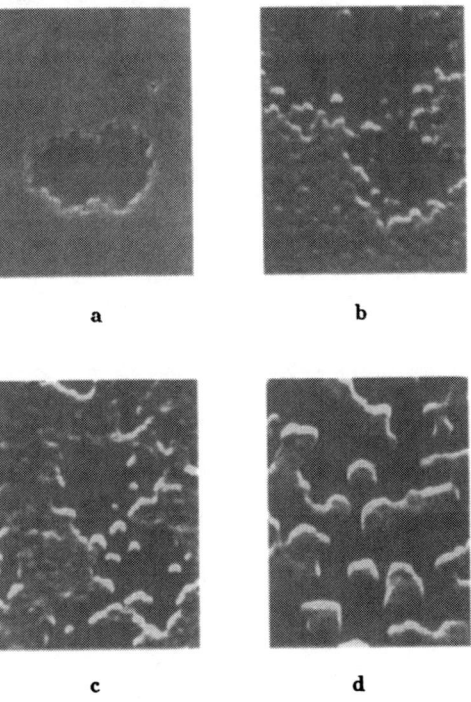

a   b

c   d   1 μm

Fig.2. SEM micrographs of the changing surface morphology of Si films by agglomeration as a function of time. The films annealed at 1045°C in H_2 for 1 min. (a), for 5 min. (b), for 10 min. (c), and for 20 min. (d). The film is not doped.

Single seed formation

In this Si selective growth by seeding the Si crystals over amorphous surface, the most important thing is that the seeds are single domained mono crystals. Whether the agglomerated Si become single domain or not is strongly dependent on patterning size of initial polycrystalline Si films. In order to determine the patterning size of polycrystalline films, pre-patterned 0.1 μm-thick films in three sizes of 4, 2, and 1 μm squares, were annealed in H_2 ambient at 1000°C. 4×4 μm^2 and 2×2 μm^2 sized films agglomerated in the multiple semi-spherical beads as shown in Fig.3(a) and (b), however, a single bead was selected in 1×1 μm^2 by agglomeration (Fig.3(c)).

The structure of an agglomerated single bead was observed by TEM (Transmission Electron Microscope), and it was revealed that some crystals included twin boundaries or other defects, however the most were the single crystals, in other words, no distinct grain boundary existed in the beads as shown in Fig.4.

Initial fine polycrystalline films had a few tens nm-grained structure, and small portions of phosphorus doped fine polycrystallin films showed abnormal grain growth after annealing in N_2 ambient, however grain boundaries were left even in the 1×1 μm^2 area.

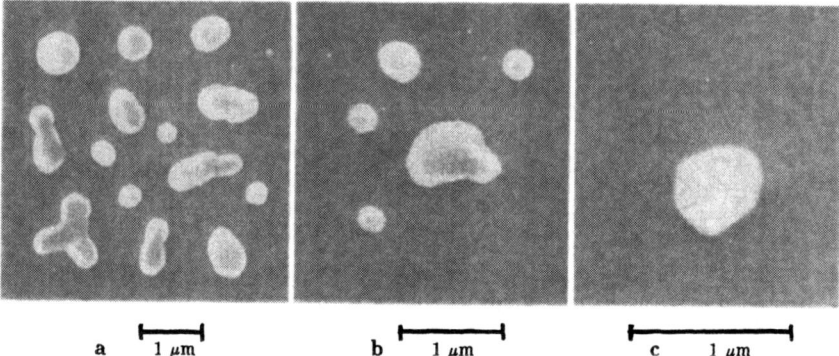

a   1 μm   b   1 μm   c   1 μm

Fig.3. SEM micrographs of agglomerated beads formation from the phosphorus doped poly-Si films patterned in 4 μm square(a), 2 μm square(b), and 1 μm square(c), which were annealed in H_2 ambient at 1000°C for 5min.

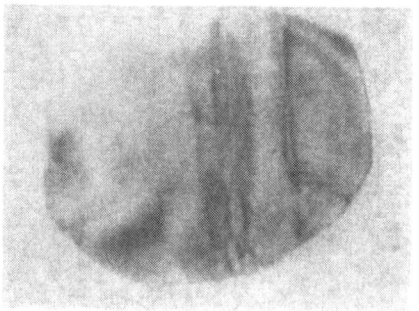

Fig.4. TEM micrograph of an agglomerated bead which was produced by annealing 1 μm square P–doped poly-Si at 1000°C for 5min.

0.1 μm

Selective growth from the seeds

The Si crystals were selectively grown seeded by the single domained beads which were placed 50 μm and 100 μm in period and impinged upon adjacent crystals at the center of seeding points, resulting in a matrix of large Si cystals. The grain boundaries were aligned as shown in Fig.5. The most crystals grown from the agglomerated seeds in the area of 1×1 μm^2 were surrounded by the facets peculiar to Si crystals. However, in contrast, all of crystals grown from patterned but non-agglomerated polycrystalline Si or the agglomerated seeds from 4×4 μm^2 patterned films were polycrystalline, and no clear facets were observed. In growth from the seed which agglomerated in 2×2 μm^2 patterned films, a part of grown crystals (several percent) was single crystalline. These results indicate that the large single crystals can not be grown from the multiple single crystal seeds as shown in Fig.3(a) and (b) or polycrystalline Si.

The outer shapes of 850 crystals were observed by SEM. It was found that the grown crystals were classified into three types, "single crystal" with 24 {311} and 8 {111} surfaces, "primarily twinned crystal" with a {111} twin plate, and "multiply twinned crystal" icosahedron[10] bounded by {111} planes only. The emerging ratio in forms was, single crystals : primary twins : multiple twins = 68 : 25 : 7. This ratio was expected to change to some extent by various annealing conditions, but the single crystals ratio was much higher than that in previous Sentaxy[1-3] (The Si nucleus was formed on small portions of amorphous materials by CVD, and then selective growth was carried out ; the emerging ratio was, single crystals : primary twins : multiple twins = 32 : 44 : 24.) The increased emerging ratio of single crystals may be due to agglomerated seed structures.

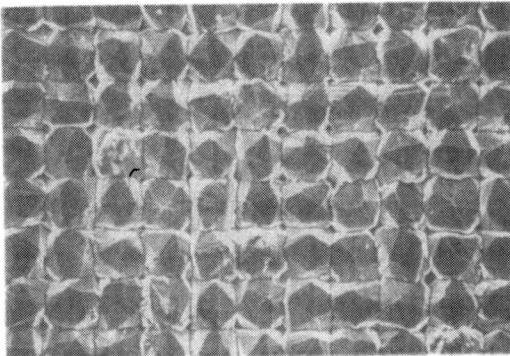

Fig.5. SEM micrograph of a matrix of large Si crystals grown from the seeds selectively by CVD. Initial films were patterned in 1×1 μm^2 and placed 50μm in period.

|———————————|
100 μm

Orientations normal to the substrate (texture) were observed by X-ray diffraction in the matrix of large Si crystals grown from various seeds which had been formed by changing substrate materials, Si film thickness, and phosphorus impurity doping methods. The <220> diffraction intensity was the strongest among the other peaks in all specimens. The <220> peak intensity occupied 80 % diffraction intensity in the entire peaks as shown in Fig.6 when Si crystal were grown from the agglomerated seeds which had been formed by annealing of 0.44 μm thick polycrystalline Si film doped with phosphorus by ion implantation over SiO$_2$.

Investigation on the cause of (110) texture is going on. It should be considered that the orientation was determined by complicated factors as followed : the interface energy minimization between Si and amorphous insulator, change in the internal energy, and surface energy minimization.

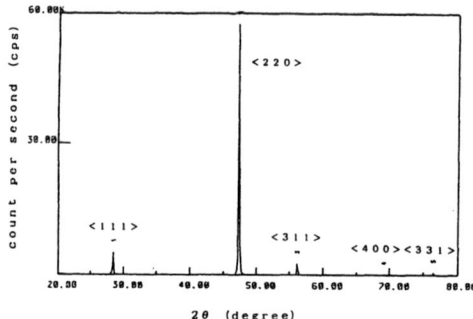

Fig.6. X-ray diffraction chart of a matrix of large Si crystals.

CONCLUSIONS

Small portions of polycrystalline Si films are altered to single beads by agglomeration phenomenon. A single bead is selected by decreasing the patterning size of initial films, and a matrix of large Si crystals is grown over SiO_2 by CVD selective growth technique from the agglomerated beads as seeds. The Si crystals are grown up to large islands of 100 μm square with controlled grain boundaries' location. These highly facetted islands are single crystals, primary twins, and multiple twins, however, most of islands are single crystalline. The dominant (110) texture is observed in the matrix of islands. This method is uniformly reproducible in a whole wafer and would be applied to silicon on insulator (SOI) structures and devices. Device fabrication is now under way.

ACKNOWLEDGEMENT

The authors would like to thank T. Noma and K. Sakaguchi of Waseda University for TEM and X-ray diffraction observations. We are also grateful to Y. Nishigaki, H. Kumomi, T. Ichikawa, and N. Sato for their helpful discussions.

REFERENCES

1) T.Yonehara, Y.Nishigaki, H.Mizutani, S.Kondoh, K.Yamagata and T.Ichikawa, Ext. Abstr. the 19th Conf. SSDM, Tokyo, p.191, (1987)

2) T.Yonehara, Y.Nishigaki, H.Mizutani, S.Kondoh, K.Yamagata and T.Ichikawa, MRS Symp. Proc., Vol.106, p.21,(1988)

3) T.Yonehara, Y.Nishigaki, H.Mizutani, S.Kondoh, K.Yamagata, T.Noma and T.Ichikawa, Appl. Phys. Lett., Vol.52, p.1231,(1988)

4) J.S.Ma, H.Kawarada, T.Yonehara, J.Suzuki, J.Wei, Y.Yokota and A.Hiraki, Appl. Phys. Lett., Vol.55, p.1071, (1989)

5) K.Hirabayashi, Y.Taniguchi, O.Takamatsu, T.Ikeda, K.Ikoma, and N.Iwasaki-Kurihara, Appl. Phys. Lett., Vol.53, p.1815, (1988)

6) T.Yonehara, C.V.Thompson and Henry I.Smith, MRS Symp. Proc. Vol.25, p.517, (1984)

7) John C.C.Fan, B.-Y.Tsaur and M.W.Geis, J. of Crystal Growth, Vol.63, p.453,(1983)

8) A.K.Shinha, S.E.Haszko and T.T.Sheng, J. of Electrochemical Soc., Vol.122, No.12, p.1714, (1975)

9) Y.Wada and S.Nishimatu, J. of Electrochemical Soc., Vol.125, No.9, p.1499, (1978)

10) T.Yonehara, Y.Nishigaki and H.Mizutani, OYO BUTURI (A monthly publication of the Japan Soc. of Appl. Phys.), Vol.57, No.9, p.105 (1988)

11) T.I.Kamins and T.R.Cass, Thin Solid Films, Vol.16, p.147, (1973)

CRYSTAL-AXIS-ROTATION OF LASER-RECRYSTALLIZED SILICON ON INSULATOR

K.Sugahara, T.Ippōshi, Y.Inoue, T.Nishimura, and Y.Akasaka
LSI R&D Laboratory Mitsubishi Electric Corporation
4-1 Mizuhara Itami 664 JAPAN

ABSTRACT

The relation between the seed pitch and defect density of the laser-recrystallized SOI film was investigated. It was found that the defect density of the SOI increases as the seed pitch increases. The dependences of the laser scan speed and laser power on rotation angle of the SOI film were experimentally and numerically investigated. The crystal-axis-rotation of the SOI film was considered to be due to the difference of the temperature between the top and bottom surface of the SOI film near the liquid-solid interface. A polysilicon heat sink structure with high thermal conductivity was newly proposed and was found to reduce the rotation in a small angle.

INTRODUCTION

Laser-recrystallized silicon on insulator (SOI) films are indispensable materials for the practical application of 3-D IC's [1]. We reported that the lateral seeded growth of SOI in the <100> direction from (001) Si seed provides better quality of single crystalline SOI film than any other crystalline directions, and also reported that a large area single crystal SOI film could be obtained by using the sample structure with the (001) seed and patterned stripes of anti-reflecting film in the <100> direction, and raster scanning of laser beam in the <110> direction [2]. The SOI film was applicable for some test devices of 3-D IC with multilayer structures [3]. However, the quality of crystal was still insufficient for the application of actual LSI chips.

In this paper, we demonstrated the detailed evaluation of crystalline defects of laser recrystallized SOI and the relation between the defects and crystal-axis-rotation. And we also provide the newly developed polysilicon heat sink cap structure to reduce the axis rotation, which was designed by using the 2 dimensional temperature simulation for various sample structures and laser irradiation conditions.

EVALUATION OF CRYSTALLINE DEFECTS

The basic sample structure used in this study was a combination of the (001) Si seed and selective anti-reflecting stripes [4,5]. The underlying insulator was an 1 μm thick thermally oxidized silicon dioxide (SiO$_2$) layer by LOCOS process. The thickness of polycrystalline silicon (polysilicon) film deposited by LPCVD technique was 0.6 μm. The anti-reflecting stripes (55 nm-thick silicon nitride film) was formed in a <100> direction with line and space of 6 and 9 μm respectively. The seed

Mat. Res. Soc. Symp. Proc. Vol. 164. ©1990 Materials Research Society

openings with size of 2 μm square were located between anti-reflecting stripes. A cw argon laser was scanned in the <110> direction for the recrystallization of polysilicon film. The substrate was heated at 450°C. Figure 1 shows the cross sectional TEM photographs of the laser-recrystallized SOI structure near a seed region. Any crystalline defects was not observed in the SOI film expect a sub-grain boundary at 94 μm apart from the seed, which was formed underneath an anti-reflecting stripe patterns. Interface between the Si and SiO$_2$ layers was very sharp and flat.

Figure 2 shows the optical photographs (OM) and TEM of several regions of laser recrystallized samples in which the sample surfaces for OM are Secco etched to delineate crystalline defects. Figure 2(a) is the OM of the same region as shown in Fig.1, where is adjacent to a seed area. Few crystalline defect was observed except the straight and parallel lines of sub-grain boundaries as a mark of anti-reflecting patterns. However as shown in Fig.2(b) of OM and TEM which are the area at about 800 μm part from a seed, sub-grain boundaries accompanied by some stacking faults were observed. The density of stacking faults increased as the distance from a seed increased, and finally the grain boundary was formed as shown in Fig.2(c).

Defect density in the SOI film as a function of the seed pitch and the ratio of the single crystal area to the whole recrystallized area were evaluated as shown in Fig.3. Laser beam scanning for these measurements were both <100> and <110> directions. The density of the stacking faults near the seed was 2×10^4 /cm^2 and gradually increased as the seed pitch increased. The ratio of single crystal area was 95-98 % at short duration of seed pitch, and also decreased down to 60-70 % as the seed pitch became longer than 500 μm.

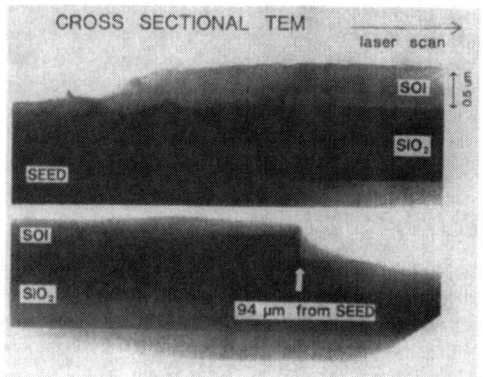

Fig.1 Cross sectional TEM photogragh
of laser-recrystallized SOI

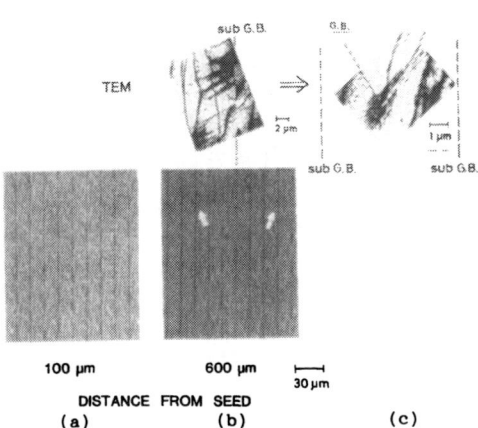

100 μm   600 μm   ├──┤
                               30 μm
DISTANCE FROM SEED
  (a)   (b)   (c)

Fig.2

Crystalline defects appeared
in the SOI films.

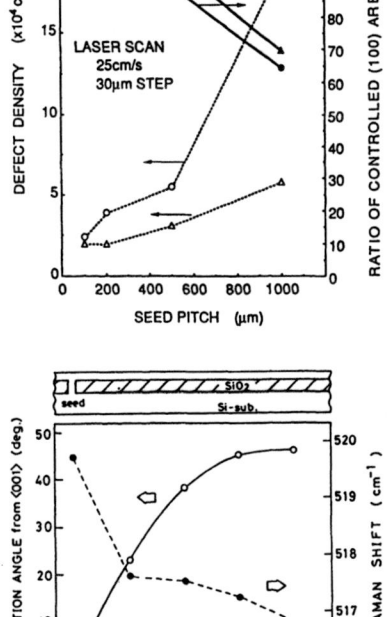

Fig.3 Defect density and
ratio of controlled (100)
area as a function of seed
pitch

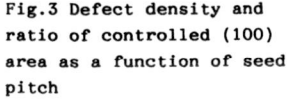

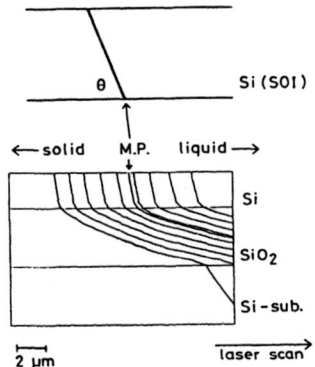

Fig.4 Rotation angle from a (001)
and Raman peak shift as a function
of a distance from the seed.

Fig.5 Cross-sectional thermal
profile of a laser-recrystallized
SOI film calculated by 2-dimensional
heat flow analysis

Table 1 Relation between the rotation angle and process condition

	CONDITION	ROTATION
SCAN SPEED	SLOW	SMALL
LASER POWER	HIGH	SMALL
SUBSTRATE TEMPERATURE	HIGH	SMALL

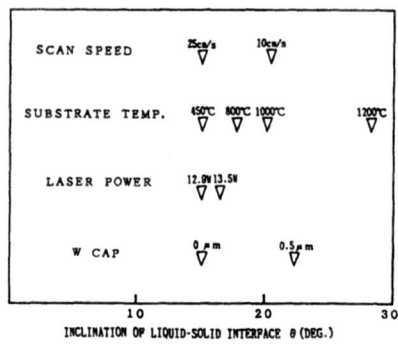

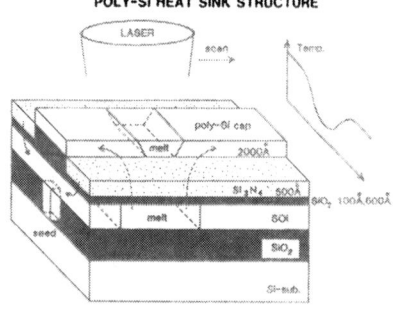

Fig.6 Calculated inclination angles
of liquid-solid interfac in SOI
films under various conditions

Fig.7 Schematic illustration of
polysilicon heat sink structure
for suppression of the
crystal-axis-rotation.

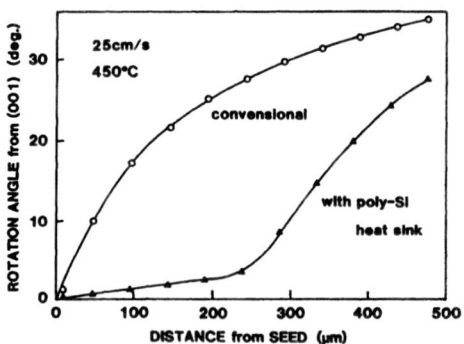

Fig.8 Rotation angle of the SOI film
with polysilicon heat sink structure.

CRYSTAL-AXIS-ROTATION

As one of the origins causing the crystalline defects, it was
expected the affect of the crystal-axis-rotation reported by us [2], in
which the crystallographic axis of <001>, controlled by the (001) seed,
was gradually rotated to <011> as a distance about 600   μm apart from a
seed. Figure 4 shows the measured rotation angle of the crystal axis, and
the peak shift of the Raman scattering measured for stress in the film as
a function of the distance from a seed. As the distance from a seed
increased, the rotation angle and the tensile stress increased. This
indicated that the crystalline defects became easy to formed as the
distance increased, and the similar appearances could be observed between
Fig.3 and Fig.4.

Table 1 summarizes the relations between the rotation angle and
process condition. Higher laser power, slower scan speed and higher
substrate temperature [8] gave the smaller rotation angle. To assess the
subject in common among those conditions, the temperature distribution of
the SOI structure during laser recrystallization was numerically
simulated for those process conditions. A typical result of simulation is
shown in Fig.5. This indicated that the liquid-solid interface was tilted
and the temperature of the top surface was higher than that of bottom
surface of the silicon film. The tilt angles became smaller (apart from
perpendicular position) for process conditions giving the larger rotation
angle, for example higher scan speed, lower laser power and higher
substrate temperature. Therefore it was speculated that the bottom region
solidified with the bending position of silicon atoms due to the
different conditions of thermal expansion in the SOI film, and was fixed
on the SiO_2 surface. The rotation was considered to be caused by the
accumulation of this small change in bond angles. Figure 6 shows the
summary of the relation between simulated inclination angles and process
conditions, in which the relative validity of the process condition for
crystal-axis-rotation was demonstrated.

STRUCTURE FOR SUPPRESSION OF THE ROTATION

For the suppression of the axis-rotation, the reduction of the
temperature distribution between the top and bottom surface of SOI is
considered to be important. The higher substrate temperature than 1000°C,
which gives the large inclination angle as shown in Fig. 8 can not be
applied to the 3-D IC's because such high temperature treatment changes
the previously defined impurity profiles of the underlying devices.

Among various parameters, the cap structure with high thermal
conductivity on the SOI film was considered to be most effective for
reduction of the axis-rotation. The inclination angle in the case of
using the cap film assumed to be transparent to the laser beam and to
have the same thermal conductivity as a tungsten film was simulated as
shown in Fig. 8, In the case of this virtual tungsten film, the
inclination angle of the liquid-solid interface became larger than the
cases of scan speed and laser power. However, from the practical point of
view, the tungsten film can not be used because of its high reflectivity

of laser beam and moreover its inconvenience as the starting material of IC's.

A new polysilicon heat sink cap structure was developed for the reduction of the temperature distribution as shown in Fig. 7. 500 Å-thick nitride stripes were formed as anti-reflecting films in the same manner as the conventional structure in order to make thermal profile for recrystallization of SOI film. A 2000 Å-thick polysilicon film was deposited and patterned on the SOI film as a heat sink in order to let escape the heat from the SOI film. Figure 8 shows the rotation angle of the SOI film with this polysilicon heat sink structure. The sample was recrystallized under the same conditions as the conventional structure. It was found that the rotation angle with this new structure decreased to 4 degrees at a distance of 200 μm from the seed.

CONCLUSION

The defect density in the laser-recrystallized SOI film increased as the seed pitch increased. The relation between the crystal-axis-rotation and inclination angle of liquid-solid interface in the SOI film was investigated. It was found that the rotation was caused by the difference of temperature between the top and bottom surface of SOI films. For the suppression of the rotation, a polysilicon heat sink structure on the SOI film was developed in order to reduce the difference of temperature. By this method, the rotation angle decreased to 4 degrees at a distance of 200 μm from the seed.

ACKNOWLEDGMENTS

The authors are grateful to Dr.H.Komiya and Dr.T.Nakano for their interest and support of this research program. This work was performed under the management of the R&D Association for Future Electron Devices as a part of the R&D of Basic Technology for Future Industries supported by New Energy and Industrial Technology Development Organization.

REFERENCES

1. Y.Akasaka, Proceeding of the IEEE 74 1703 (1986)
2. K.Sugahara, S.Kusunoki, Y.Inoue, T.Nishimura, and Y.Akasaka, J. Appl. Phys. 62 4178 (1987)
3. T.Nishimura, Y.Inoue, K.Sugahara, S.Kusunoki, T.Kumamoto, S.Nakagawa, M.Nakaya, Y.Horiba, and Y.Akasaka, Proceedings of the IEDM 87 111 (1987)
4. J.P.Colinge, E.Demoulin, D.Bensahel, and G.Auvert, Appl. Phys. Lett. 41 346 (1982)
5. S.Kusunoki, K.Sugahara, T.Nishimura, T.Kumamoto, M.Nakaya, N.Yazawa, and Y.Horiba, Proceedings 1987 Symp. VLSI Tech., Karuizawa, p107 (1987)
6. K.Shirakawa, M.Maekawa, O.Yamazaki, H.Tsuji and M.Koba, Extended Abstracts of the 19th Conference on Solid State Devices and Materials, Tokyo, Japan, p175 (1987)

A NEW MODEL FOR THE POLY-SILICON THIN FILM TRANSISTOR FOR USE WITH SPICE.

Izzard, M.J., Migliorato, P., Milne, W.I.
Cambridge University, Department of Engineering, Trumpington Street, Cambridge, CB2 1PZ, England.

ABSTRACT

A set of analytical equations that accurately model the large-signal current-voltage and charge-storage characteristics of the poly-silicon thin film transistor (TFT) have been developed. The model is based on theories of conduction in materials with a bulk-distribution of gap states[1]. The equations are suitable for incorporation into a circuit simulator capable of transient-analysis of circuits, such as SPICE.

INTRODUCTION

Advances in the field of large area microelectronics, especially in the area of active matrix displays, has brought about the need for circuit design tools that are able to model the building block of these circuits, the TFT. Such tools are required to aid the design of large area integrated logic and analogue circuits, such as shift registers and amplifiers or level-shifters, as well as new active matrix architectures[2]. The typical TFT displays four distinct features that must be modelled: post-threshold conduction, subthreshold conduction, anomalous off-current, and charge storage (see figure 1). Post-threshold conduction is similar to crystalline MOSFET conduction; the subthreshold conduction region is extended and thus has a large impact on the transfer characteristics of circuits; the

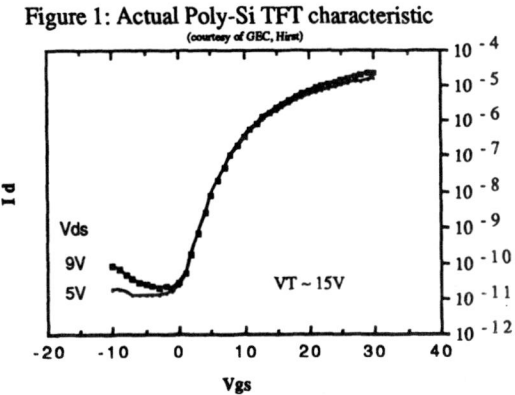

Figure 1: Actual Poly-Si TFT characteristic
(courtesy of GEC, Hirst)

anomalous off-current, or reciprocal characteristic region, is important in dynamic and memory-type circuits, where leakage is critical.

I-V CHARACTERISTICS

The conductance model of the poly-silicon is based on work by Migliorato et al [3]. They treat the poly-silicon as a semiconductor with a uniform bulk-distribution of gap states[1] and derive a single-carrier model of conductance. The resultant integral equations are simplified to derive analytical conductances by use of the Boltzman approximation of the Fermi integral, the use of a zero-temperature approximation when necessary, and an assumed exponential density of states in the band gap characterized by N_G and T_G in,

$$N(E) = N_G \exp\left[\frac{E - E_c}{kT_G}\right] \tag{1}$$

This results in two equations for the conductance of poly-silicon as a function of gate voltage V_G. For the subthreshold region,

$$G_1 = G_{subth} - G_0 = K_0(V_G - V_{FB})^{2\beta - 1} \tag{2}$$

in which, $G_0 = q\mu n_0 d$ represents the intrinsic conductance of the film (μ is the effective, constant, mobility, n_0 is the equilibrium number of carriers in the conduction band, and d is the thickness of the poly-silicon film) , β is T_G/T, and

$$K_0 = \mu \, N_c \frac{T}{2T_G \cdot T} \left(\frac{2\varepsilon_s}{N_G}\right)^{\frac{1}{2}} \left(\frac{\varepsilon_{ox}^2}{2\varepsilon_s t_{ox}^2 \, N_G \, k^2 T_G^2}\right)^{\frac{T}{T_G} - \frac{1}{2}} \tag{3}$$

For the post-threshold region

$$G_2 = \mu \frac{\varepsilon_{ox}}{t_{ox}} (V_{gs} - V_{FB} - V_T) \tag{4}$$

where

$$V_T = \frac{\varepsilon_{ox}}{t_{ox}} \left(\frac{N_c q}{kT_G N_G}\right)^{\frac{T}{2(T - T_G)}} \left[\frac{2\varepsilon_s N_G}{q} (kT_G)^2 \left(1 + \frac{T}{T_G}\right)\right]^{\frac{1}{2}} \tag{5}$$

Now, five regions of operation of the TFT may be identified in order to integrate the conductance along the channel, and hence, to obtain drain current expressions:

region 1: subthreshold	$V_{gs} - V_{FB} < V_T, V_{ds} < V_{gs} - V_{FB}$
region 2: subthreshold saturated	$V_{gs} - V_{FB} < V_T, V_{ds} > V_{gs} - V_{FB}$
region 3: all post-threshold	$V_{gs} - V_{FB} > V_T, V_{ds} < V_{gs} - V_{FB} - V_T$
region 4: semi-sat.	$V_{gs} - V_{FB} > V_T, V_{gs} - V_{FB} - V_T < V_{ds} < V_{gs} - V_{FB}$
region 5: saturated	$V_{gs} - V_{FB} > V_T, V_{ds} > V_{gs} - V_{FB}$

Performing the integral,

$$I_d = \frac{W}{L} \int_{V_1}^{V_2} G_i(V_{gs}, V_y) \, dV_y \tag{6}$$

where V_1 and V_2 are the appropriate channel boundary voltages for the appropriate G_i, results in five expressions for drain current I_d ,

$$I_d = I_{di} + \frac{W}{L} G_0 V_{ds} \tag{7}$$

$$I_{d1} = \frac{W}{L} \frac{K_0}{2\beta} \left[\left(V_{gs}\text{-}V_{FB} \right)^{2\beta} - \left(V_{gs}\text{-}V_{FB}\text{-}V_{ds} \right)^{2\beta} \right] \tag{8}$$

$$I_{d2} = \frac{W}{L} \frac{K_0}{2\beta} \left(V_{gs}\text{-}V_{FB} \right)^{2\beta} \tag{9}$$

$$I_{d3} = \frac{W}{L} \frac{\mu}{2} \frac{\varepsilon_{ox}}{t_{ox}} \left[2 \left(V_{gs}\text{-}V_{FB}\text{-}V_T \right) V_{ds}\text{-}V_{ds}^2 \right] \tag{10}$$

$$I_{d4} = \frac{W\mu}{L2} \frac{\varepsilon_{ox}}{t_{ox}} \left(V_{gs}\text{-}V_{FB}\text{-}V_T \right)^2 + \frac{W}{L} \frac{K_0}{2\beta} \left[\left(V_T \right)^{2\beta} - \left(V_{gs}\text{-}V_{FB}\text{-}V_{ds} \right)^{2\beta} \right] \tag{11}$$

$$I_{d5} = \frac{W\mu}{L2} \frac{\varepsilon_{ox}}{t_{ox}} \left(V_{gs}\text{-}V_{FB}\text{-}V_T \right)^2 + \frac{W}{L} \frac{K_0}{2\beta} \left(V_T \right)^{2\beta} \tag{12}$$

Figure 2a: Id versus Vgs (Vd=1V)

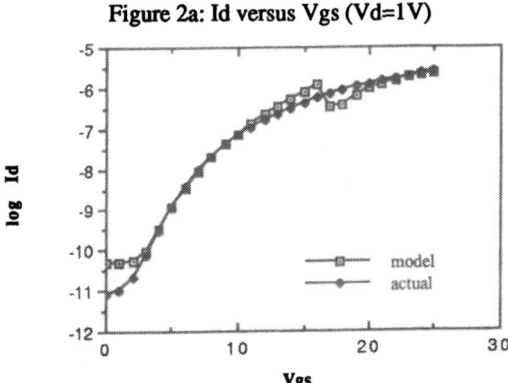

Figure 2b: Id versus Vgs (Vds=10V)

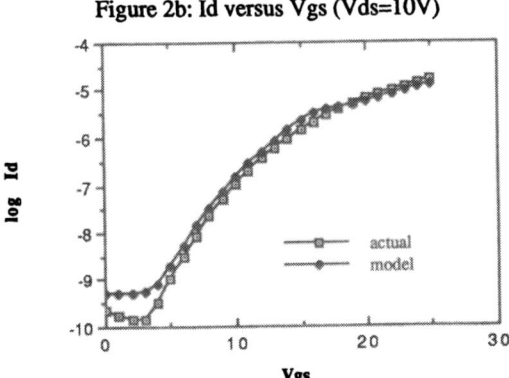

The different terms in the expressions represent the contributions of the separate conduction modes in various parts of the channel.

Figure 2 shows a set of transfer characteristics which demonstrate agreement with an actual device. The discontinuity is a result of the transition from region 1 to 4.

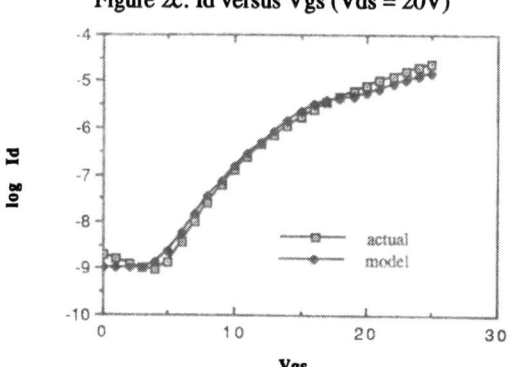

Figure 2c: Id versus Vgs (Vds = 20V)

The TFT used has a W/L of unity and the following parameters were assumed:
N_G= 1.5e+20cm^{-3}eV^{-1}, T_G= 1000K, V_T= 16V, μ= 8 cm^2/Vs.

The reciprocal characteristic region of operation of the TFT can easily be dealt with by applying the same theory about the mirror point of the flat band (V_{FB}) and using different parameters, especially N_G, T_G and μ.

Q-V CHARACTERISTICS

The TFT is similar in construction to a self-aligned silicon-on-insulator MOSFET. Unlike the MOSFET, however, it does not operate in inversion mode, but in enhancement mode, and it does not possess blocking junctions at the drain and source. In the light of this it is possible to say that the only real charge storage mechanism is the induced channel charge (Q). Assuming that all the field passes through the insulator-semiconductor interface, we can write,

$$Q = \int_s \varepsilon_{ox} F_{ox} \tag{13}$$

neglecting the effect of interface states we have, at each point in the channel[3]:

$$(V_{gs}-V_{FB}-V_y)\frac{1}{t_{ox}} = F_{ox} \tag{14}$$

Since V_y is constant across the width of the channel, we may write, for L the length of the channel:

$$Q = W\frac{\varepsilon_{ox}}{t_{ox}} \int_0^L (V_{gs}-V_{FB}-V_y)\, dy \tag{15}$$

and, as in the case of a conventional MOSFET[4],

$$dV_y \frac{W}{dy} G = I_d \qquad (16)$$

resulting in:

$$Q = W \frac{\varepsilon_{ox}}{t_{ox}} \int_{v_1}^{v_2} (V_{gs} - V_{FB} - V_y) \ W \frac{G}{I_d} \ dV_y \qquad (17)$$

where V_1 and V_2 again are the boundary channel voltages for the appropriate G, given the region of operation of the device. Integration yields five polynomial ratios in V_{gs} and V_{ds} or V_{gd}. Letting, $C_0 = WL \frac{\varepsilon_{ox}}{t_{ox}}$, if small contributions of charge are ignored, the charge in the different regions of operation simplify to:

$$Q_1 = C_0 \frac{2\beta}{2\beta+1} \left[\frac{(V_{gs}-V_{FB})^{2\beta+1} - (V_{gd}-V_{FB})^{2\beta+1}}{(V_{gs}-V_{FB})^{2\beta} - (V_{gd}-V_{FB})^{2\beta}} \right] \qquad (18)$$

$$Q_2 = C_0 \frac{2\beta}{2\beta+1} (V_{gs}-V_{FB}) \qquad (19)$$

$$Q_3 = C_0 \frac{2}{3} \left[\frac{(V_{gs}-V_{FB}-V_T)^3 - (V_{gd}-V_{FB}-V_T)^3}{(V_{gs}-V_{FB}-V_T)^2 - (V_{gd}-V_{FB}-V_T)^2} \right] \qquad (20)$$

$$Q_4 = C_0 \frac{2\beta}{2\beta+1} \left[\frac{(V_T)^{2\beta+1} - (V_{gd}-V_{FB})^{2\beta+1}}{(V_T)^{2\beta} - (V_{gd}-V_{FB})^{2\beta}} \right] + C_0 \frac{2}{3} (V_{gs}-V_{FB}-V_T) \qquad (21)$$

$$Q_5 = C_0 \frac{2\beta}{2\beta+1} (V_T) + C_0 \frac{2}{3} (V_{gs}-V_{FB}-V_T) \qquad (22)$$

Figure 3: q versus Vgs (Vds = 1V)

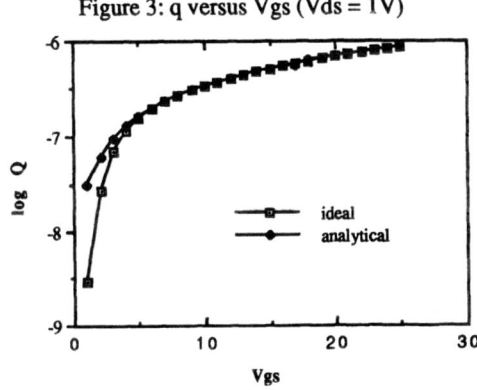

The charge equations or their simplifications may be used directly to implement a Ward-type charge model[5] in SPICE. Alternatively, the charge equations may be differentiated with respect to V_{gs} and V_{ds} to obtain the Meyer capacitances[6] which may be used to implement a modified Meyer model that is capable of conserving charge[7].

Figure 3 shows the ideal charge, Q, derived by numerical integration of the full theoretical expressions, and charge given by the analytical model, versus V_{gs}.

It is interesting to note that the Meyer capacitance C_{gs} for the above threshold, saturated TFT is $2/3C_0$, like a MOSFET. Also, the Meyer C_{gs} below threshold is $(2\beta/2\beta+1)C_0$ which is very high and corresponds to the steep increase of charge in the channel below threshold; this assumes quasi-static operation, the time constant for which will be governed by the time constant of the gap states if the drain voltage is low.

CONCLUSION

A model for the poly-silicon TFT has been presented in the form of a set of analytic equations suitable for use in a circuit simulator with transient analysis capabilities. The equations are in the form of polynomials in the terminal voltages of the device, which may be simplified to suit the required accuracy of the model. It is, however, necessary to pay further attention to the continuity of the expressions when moving between regions of operation of the device. The authors are in the process of implementing and testing the model in SPICE.

REFERENCES

1. P. Migliorato and D.B. Meakin, Applied Surface Science 30, 353-371 (1987).
2. M.J. Izzard, W.I. Milne and P. Migliorato, presented at the International Topical Conference on hydrogenated amorphous silicon devices and technology, New York, 1988.
3. G. Fortunato, and P. Migliorato, to be published.
4. S.M. Sze, Physics of Semiconductor Devices, 2nd ed. (John Wiley & Sons Publishers, U.S.A., 1981), p. 440.
5. D.E. Ward and R.W. Dutton, IEEE Journal of Solid-state Circuits SC-13 (5), 703-707 (1978).
6. J.E. Meyer, RCA Review, Vol. 32, 42-63 (1971).
7. M.A. Cirit, IEEE Transactions on Compuet-aided Design Vol.8 (10), 1033-1037 (1989).

ELECTRICAL PROPERTIES OF SIPOS FILMS
DEPOSITED ON CRYSTALLINE SILICON

TIEN-MIN CHUANG, KENNETH ROSE AND RONALD J. GUTMANN
Center for Integrated Electronics, Rensselaer Polytechnic Institute, Troy, NY 12180

ABSTRACT

Transport and trapping effects on undoped and phosphorus-doped SIPOS films on n-type and p-type Si have been characterized by C-V-ω, I-V, DLTS and photo-induced microwave reflection techniques. Our C-V and DLTS data on undoped SIPOS indicate that interface traps with a density of $2 \times 10^{12} cm^{-2} eV^{-1}$ and a response time between 1 and 10 μsec exist near the midgap of the SIPOS. Photo-induced microwave reflection transient-waveforms show an order of magnitude increase in photoconductivity decay time with doped SIPOS films. Our data shows that trapping effects exist in doped SIPOS and that these traps could reduce the cutoff frequency of SIPOS emitter transistors.

INTRODUCTION

SIPOS is composed of silicon microcrystals in an oxide matrix [1] [2]. Undoped SIPOS is important as a passivant in high-voltage planar devices [3], and doped SIPOS shows promise as an electron emitter for BiCMOS devices [4]. Carrier transport and trapping are more complex and less understood in these two-phase materials than in simple oxides with both bulk and interface properties important. In this paper, we report on the characterization of SIPOS deposited on crystalline Si by LPCVD, using SiH_4 and N_2O as the reactant gasses. In this study, the gas ratio of SIPOS films γ, defined as $[N_2O]/[SiH_4]$ is either 0.1 or 0.8. Physical mechanisms and device implications of the data are also discussed.

SAMPLE PREPARATION

Two test structures, namely an MIS diode and a p-n junction diode as shown in Fig. 1 are fabricated for C-V, I-V and DLTS measurements. The sample preparation procedure is described as follows:

(i) Oxidation: RCA cleaned p/p+ and n/n+ epi-wafers were thermally oxidized (dry) to a thickness of 140 nm. A dot matrix of contact windows with 10mil diameter was delineated. (For the MIS samples, the oxidation step is omitted.)

(ii) SIPOS film growth: The thickness is 50nm for SIPOS films with γ of 0.8, and 150nm for SIPOS films with γ of 0.1. The growth temperature is 610°C. SIPOS films were then annealed at 950°C in nitrogen for 20 min.

(iii) $POCl_3$ diffusion: The diffusion time is 2 to 4 minutes at 950°C for SIPOS films with γ of 0.8, and 5 minutes for SIPOS films with γ of 0.1. PSG was then stripped in 2% HF solution. (For the MIS samples, this step is omitted.)

(iv) Aluminum electrode delineation: The pattern is a dot matrix with 20mil diameter. The Al film was sintered at 400-450°C for 20 min.

The n-type samples for photo-induced microwave reflection measurements are prepared in a similar process. The various samples are designated as: N3 - RCA cleaned Si wafer, N18 - oxidized sample, N21 - undoped SIPOS sample and N24 - n+SIPOS sample.

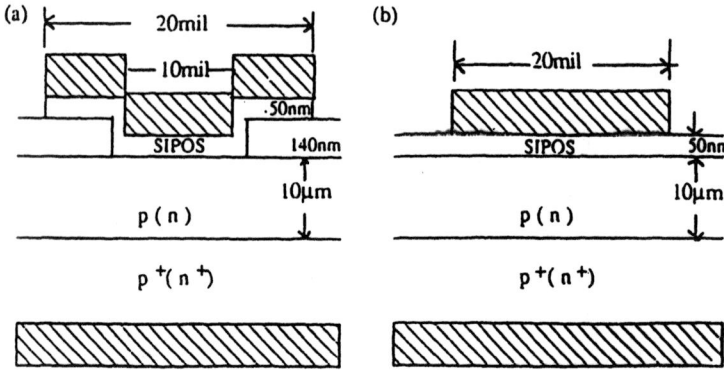

Fig. 1 Cross-Section Views of (a) the p-n junction and (b) the MIS diode incorporating SIPOS films

RESULTS AND DISCUSSION

C-V Measurements

The C-V curve of a SIPOS/p-Si sample at 1MHz testing frequency is shown in Fig. 2(a). Two differences between our data and that of an ideal MIS C-V curve are observed: (i) The capacitance in inversion is further reduced after the applied voltage is larger than 2V. This type of C-V curve was also observed by Black, who attributed the capacitance reduction at high voltage to the field dependent conductivity of the SIPOS film [5]. We further explain this phenomena quantitatively. For example, the leakage current of our sample at 2V is approximately 10nA, which implies a charge flow of 6×10^{10} carriers per second through the SIPOS/Si interface. The minority carrier generation rate within the depletion region is approximately equal to $AW_d n_i/2\tau = 10^9$ electrons/sec (with $\tau = 1\mu s$). Under steady state conditions, these generated carriers will be extracted from the interface. Lack of an inversion layer at high field will cause the depletion width to increase further and the capacitance to decrease correspondingly. (ii) The C-V curve is extended along the gate bias axis. This phenomena is attributed to the existence of interface traps. Namely, the capacitance changes much more slowly with gate bias as the abrupt increase in interface trap level density is swept past the Fermi level at the Si surface.

The C-V data on the SIPOS/n-Si sample at 1MHz testing frequency is shown in Fig. 2(b) and can be explained in a similar way. The only difference is the bias polarity and the trap center is donor-like, rather than acceptor-like. The C-V data for SIPOS on p-Si and n-Si under 10 kHz or 100 kHz test frequencies are also shown in Fig. 2(a) and (b), respectively. As we vary the testing frequency, the capacitance of the SIPOS/p-Si diode under large negative bias is enhanced appreciably at low testing frequencies. This indicates that the response time of trap states in the SIPOS is larger than 10 μsec. The enhancement of capacitance of the SIPOS/n-Si diode at low frequency also occurs under large positive bias, and the response time of the trap states is shorter, between 1 μsec and 10 μsec.

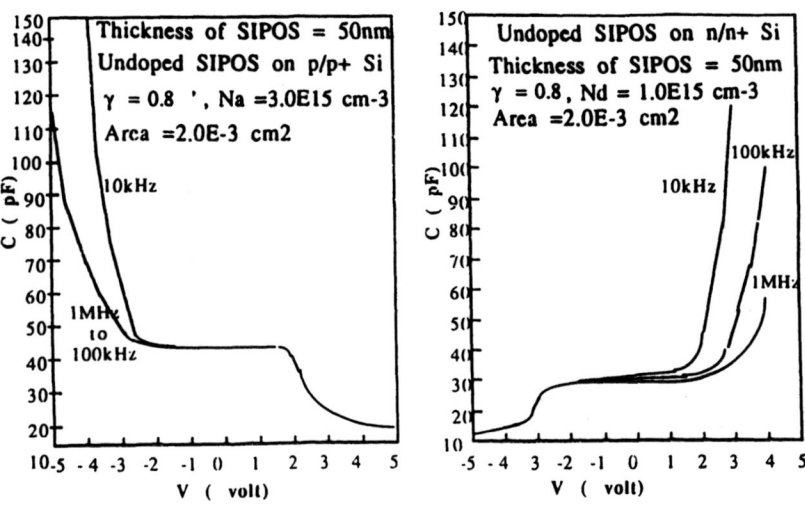

Fig. 2 C-V curves of (a) Al/SIPOS/p-Si MIS sample and
(b) Al/SIPOS/n-Si sample under different test frequencies

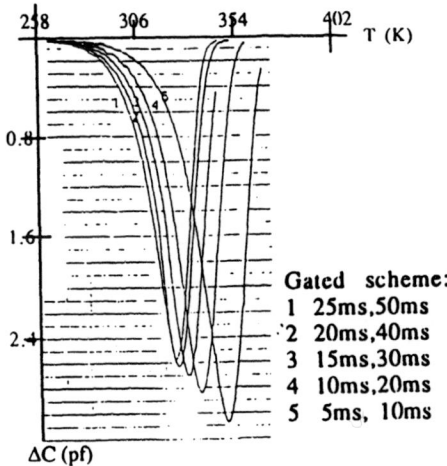

Fig. 3 DLTS curves of SIPOS/p-Si sample P11 with 5V applied voltage
steps.

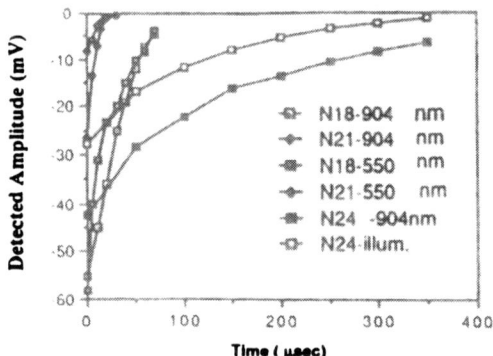

Fig. 4 The measured recovery waveforms of samples after 904nm laser or 550nm flash lamp illumination. "N24-illum." is the waveform of sample N24 with background illumination.

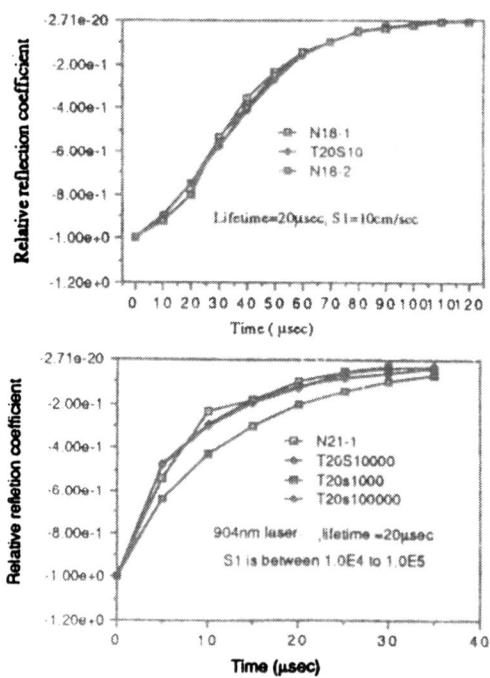

Fig. 5 Curve fitting with one-region model for (a) oxidized sample N18 and (b) undoped-SIPOS sample N21.

DLTS Measurements

Both DLTS of undoped-SIPOS/p-Si Sample P11 and n⁺-SIPOS/p-Si sample P3 are taken. However, we find that there is no significant peak with the P3 sample while a large negative voltage peak appears for P11 sample. The measured DLTS curves of sample P11, ΔC vs. T under different gating conditions, are shown in Fig. 3. The conclusions from DLTS data on samples P3 and P11 are as follows:

(i) There is a discrete hole trap interface state in the bandgap with the undoped SIPOS. Because the interface state disappears after doping the SIPOS, we conclude that the interface state is located in the SIPOS side of the SIPOS/Si interface.

(ii) The density of traps and its energy level can be estimated as follows: By the following relationship of ΔC and interface state density N_{ss}[6]

$$\Delta C = -\frac{C_1(t_1)^3}{\varepsilon_s N_d(W)} \frac{kT}{C_{ox}} N_{ss} ln\left(\frac{t_2}{t_1}\right) \tag{1}$$

where C_{ox} is the oxide capacitance, ε_s the dielectric constant of Si, t_1 and t_2 the gate times corresponding to ΔC, and $N_d(W)$ the doping concentration at the edge of the space charge region. N_{ss} is calculated to be 2×10^{12} $cm^{-2}eV^{-1}$. From an Arrhenius plot, the interface state is 0.53V above the valence band, i.e. almost at midgap, indicating that these deep states should be effective recombination centers. This result is comparable with previously reported data [7], but our data shows a sharper distribution.

The DLTS data is compatible with the high turn-on voltage in the I-V curve of an undoped-SIPOS/Si diode and the stretch-out in the C-V curve. The large unsaturated density of states at the SIPOS/Si interface results in a DC voltage drop and pins the capacitance of the MIS capacitor during the gate bias sweep.

Photo-Induced Microwave Reflection

Samples, N2, N18, N21 and N24, are illuminated with either a pulsed 904nm laser or a Xenon flash lamp with a bandpass optical filter (passband from 545nm to 554nm). Fig. 4 shows the recovery waveforms of the above samples after illumination by the 904nm laser or flashlamp with filter. The recovery times, defined as the time for the waveform to change from -90% to -10%, calculated from these curves are summarized in Table 1. The measured recovery time on the doped SIPOS sample, #N-24, is longer than the data on the rest of samples by one order of magnitude. There are two possibilities to explain this photoconductivity-decay enhancement phenomena: (i) the existence of deep trap centers due to the POCl doping process, and (ii) the bulk lifetime enhancement due to PSG gettering. In order to differentiate these two mechanisms, we prepared a n⁺-SIPOS sample with the SIPOS layer etched off in a CF_4 plasma, and optically illuminated with a background light to saturate any traps in the doped SIPOS sample. The waveform comparison on samples with and without background illumination is also shown in Fig. 4, which indicates that traps exist in the doped SIPOS sample. That is, the waveform of the SIPOS stripped-off sample is similar to that of a bare Si wafer, indicating that PSG gettering is not the dominant mechanism.

Table 1. Recovery time from the transient waveforms of Si samples

Excitation source	N-3	N-18	N-21	N- 24
904nm laser	$15\mu sec$	$60\mu sec$	$25\mu sec$	$350\mu sec$
Flashlamp	$25\mu sec$	$120\mu sec$	$45\mu sec$	$600\mu sec$
Flashlamp + filter	$20\mu sec$	$90\mu sec$	$35\mu sec$	$500\mu sec$

The extraction of the effective recombination velocity (ERV) for a SIPOS/Si sample is another application of this measurement and is described as follows: (i) using a one region model [8] with an ERV or S_1 of 10 cm/sec, we obtain a bulk lifetime of 20 μsec. This comparison of the simulated waveform (T20S10) and the measured waveform on the thermally oxidized sample is shown in Fig. 5(a). (ii) By comparing the measured waveform on the undoped SIPOS sample with the simulated family of curves with 20 μsec bulk lifetime and various ERV's, we extract the ERV in the SIPOS/Si interface to be between 1×10^4 and 1×10^5 cm/sec. Fig. 5(b) shows the result of this data fitting.

SUMMARY

In an MIS structure with an undoped SIPOS film, deep depletion occurs due to the nonlinear conduction of the SIPOS under high electric field. Photo-induced microwave reflection transient-waveforms show an order of magnitude increase in photo- with the doped SIPOS films. These traps cause a lower cutoff frequency for a SIPOS emitter transistor. The ERV data indicates that the gain enhancement of the SIPOS emitter over a conventional homojunction bipolar transistor is due to the relatively low interface recombination velocity at the SIPOS/Si interface compared to a metallic contact.

ACKNOWLEDGEMENTS

We would like to thank Dr. W. I. Lee for DLTS measurement, Ms. M. Tait for photo-induced microwave measurement and CSIST, Taiwan, ROC for financial support.

REFERENCES

1. M. Hamasaki, T. Adachi, S. Wakayma and M. Kikuchi, J. Appl. Phys., 49, 3987 (1978).

2. K. T. Chang, Ph.D. Thesis, RPI (1987).

3. J. N. Sandoe, J. R. Hughes and J. A. G. Slatter, IEE Proc. Pt. 1, 132, 281 (1985).

4. T. M. Chuang, Ph.D. Thesis, RPI (1989).

5. R. D. Black, J. Appl. Phys., 63, 2458 (1988).

6. K. Yamasaki, M. Yoshida and T. Sugano, Japan. J. Appl. Phys., 18, 113 (1979).

7. E. P. Burte, J. Electrochem., Soc., 135, 1017 (1988).

8. C. S. Lo, Ph.D. Thesis, RPI (1988).

DEVELOPMENT OF THE VERY THIN MICROCRYSTALLINE N-LAYER AND ITS APPLICATION TO THE STACKED SOLAR CELL.

F. Nakabeppu, T. Ishimura, K. Kumagai and K. Fukui,
CRDL TONEN Corp., 1-3-1, Nishi-turugaoka, Ohi-machi, Iruma-gun, Saitama-ken, 354, Japan

ABSTRACT

In order to improve the efficiency of the amorphous silicon stacked solar cells, we have developed the preparation method of highly conductive very thin microcrystalline silicon n-layers. We have found that the addition of a small amount of Ar gas to deposition gases is effective to make microcrystallite size small. The obtained thin films were characterized by conductivity measurement, R-HEED observation and TEM observation.

This newly developed thin microcrystalline n-layers have been applied to the stacked solar cells. Increase of Jsc by 5-6% has been achieved because of the reduction of light absorption loss in n-layer without decrease of Voc and FF.

INTRODUCTION

The amorphous silicon (a-Si:H) stacked solar cell has been very interesting because it shows low light-induced degradation[1]. Formation of the n/p ohmic connection is one of the key technologies to fabricate the high efficiency a-Si:H stacked solar cells. Several kinds of method to form the ohmic connection have been investigated[2]. Recently n-type amorphous-microcrystalline mixed phase hydrogenated silicon (μc-Si:H) films are widely used as the ohmic connection layers of stacked solar cells. It is known that in order to form the n/p ohmic connection, n-layer should be highly conductive, namely sufficiently microcrystallized.

Improvement of the efficiency of the a-Si:H stacked solar cell is realized by reducing the light absorption loss in p- and n-layers because they don't work as the active layers. But in general, as the microcrystalline film becomes thinner, its conductivity rapidly decreases. Consequently, if n-layer becomes thinner, Jsc increase but Voc and FF decrease due to the lack of the good ohmic connection. Therefore, establishing the preparation method for a highly conductive very thin μc-Si:H film is essential for the improvement of the efficiency of the a-Si:H stacked solar cells.

Our conventional conditions to deposit the μc-Si:H n-layers for single solar cells were optimized around several thousand angstrom thickness. It becomes clear that less than 300A thick films are not microcrystallized in this condition. Then we considered that in order to obtain a very thin μc-Si:H film, it was important to make the crystallite size smaller and the crystalline fraction higher.

Crystallite size control in μc-Si:H was reported by A. Matsuda[3], which showed that the crystallite size becomes smaller as the amount of ions impinging to the substrate surface increases. According to Matsuda's guiding principle, we tried to increase the amount of ions in plasma by control of PCVD condition and dilution gases. We found that the addition of a small amount of Ar gas to the deposition gases is effective to make crystallite size smaller, and the thin μc-Si:H films were obtained with this method.

In order to characterize the properties of these thin films, we carried out the dark conductivity measurement, R-HEED (Reflection High Energy Electron Deflection) observation and TEM (Transmission Electron Microscope) observation. And this newly developed thin n-layer was applied to the amorphous silicon stacked solar cells.

Table I Deposition conditions.

	SiH$_4$/H$_2$	total flow rate (sccm)	rf power density(W/cm^2)	total gas pressure(torr)
condition 1	1/10-50	5-50	0.01-0.7	0.1-0.3
condition 2	1/100-400	50-200	0.2-0.5	1-2
condition 3	Ar/H$_2$ = 2-4 %			

(For all conditions PH$_3$/SiH$_4$ =1-5% and Ts=160-250°C.)

EXPERIMENT

Deposition of n-type films was carried out using a conventional capacitively-coupled glow discharge system. Table I shows deposition conditions; Condition 1 is our conventional condition to deposit the μ c-Si:H films for single solar cells. Condition 2 is the optimized one in this work in order to obtain thin μ c-Si:H films. And in case of condition 3, a small amount of Ar gas is added to deposition gases and other parameters are the same as condition 2.

RESULTS AND DISCUSSIONS

Conductivity measurement

The dark conductivity versus the thickness of films for each deposition condition is shown in figure 1. Each sample was deposited on glass substrate and a-Si:H film and both types of samples revealed no difference in conductivity, which was measured with the coplanar method.

In case of condition 1 the conductivity of about 300A thick film is about 0.1 S/cm and as the film becomes thinner than 300A, its conductivity rapidly decreases. But in case of condition 2 the film thickness which keeps 0.1 s/cm in conductivity is about 200A. Furthermore in case of condition 3, where a small amount of Ar gas is added to the feed gases, even a 150A thick film has a high conductivity of about 0.1 S/cm. These results show that the addition of Ar gas to deposition gases is effective to make the highly conductive thin films.

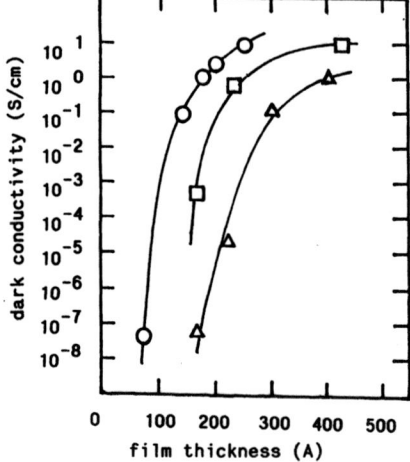

Fig. 1 Dark conductivity vs film thickness for condition 1($\triangle$), 2($\square$) and 3($\bigcirc$).

R-HEED observation

R-HEED patterns of the films deposited in each condition are shown in figure 2 . Each sample is about 200A thick and deposited on an a-Si:H layer.

In case of condition 1, halo pattern appears (see figure 2(A)) and this means the film structure is amorphous. On the other hand, both figure 2(B) and 2(C) show ring pattern and this result means that these films are microcrystallized. These results are consistent with the results of conductivity measurement, which show that in 200A thickness the films of condition 2 and 3 are highly conductive but the film of condition 1 is not.

TEM observation

We investigated the structure of the μ c-Si:H films deposited in condition 2 and 3 through TEM and cross sectional TEM observation. Figure 3 shows the TEM images and each sample was about 250A thick and deposited on glass substrate. In case of condition 2, as shown in figure 3(A) the typical size of crystallite is about 200A in diameter. And in case of condition 3, where a small amount of Ar gas was added to deposition gases, the typical size of crystallite is about 100A in diameter.

(A)

(B)

(C)

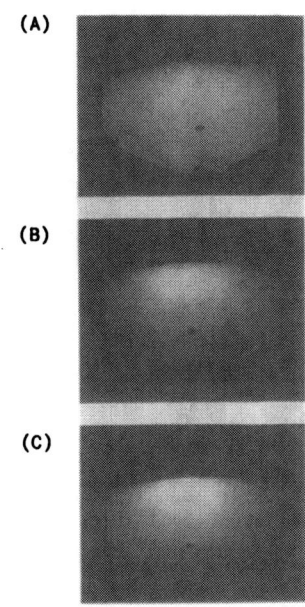

Fig. 2 R-HEED pattern images of films deposited in condition 1 (A), 2 (B), and 3 (C).

(A) condition 2   (B) condition 3

Fig. 3 TEM images of μ c-Si:H n-type films deposited in condition 2 (A) and 3 (B)

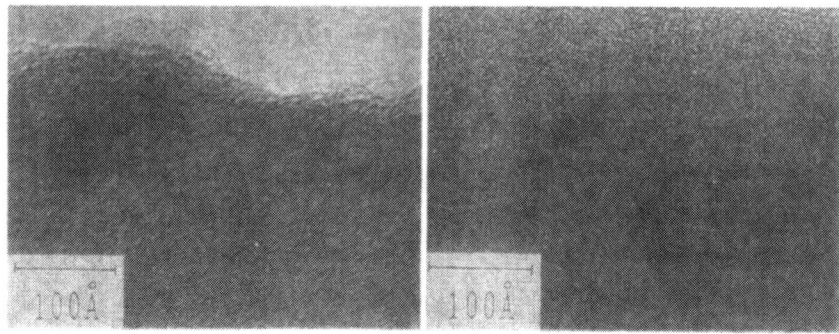

(A) condition2   (B) condition 3

Fig. 4   Cross sectional TEM images of μ c-Si:H films deposited
         in condition 2 (A) and 3 (B).

Figure 4 shows the cross sectional TEM images where each n-layer was deposited about 200A thick on a-Si:H layer. The crystal lattice image is observed in both cases and the typical size of crystallite is about 100A in case
of condition 2 and about 70A in case of condition 3.
     These results show that the thin μ c-Si:H films deposited in condition 3
is highly conductive in spite of the small crystallite size and this supposes
that the crystalline fraction of these films is high.

Stacked solar cell

     We tried to apply the newly developed μ c-Si:H film to a-Si:H stacked so-
lar cells. In order to investigated the effect of this film on the characteristics of the stacked solar cells we fabricated three kinds of double
stacked solar cells.
     In order to avoid the fluctuation of Jsc, namely the thickness of
i-layer, we fabricated the stacked solar cells in following process; (1)
p_1i_1-layers were deposited on the glass/SnO$_2$ substrate, (2) this substrate was
taken out of the chamber and divided into three pieces, (3) n_1-layers were deposited on each piece respectively, and (4) $p_2i_2n_2$-layers were deposited on
these three pieces simultaneously. Then the configuration of the stacked solar
cell was glass/SnO$_2$/$p_1i_1n_1$/$p_2i_2n_2$/metal. The thickness balance of i_1/i_2 was
1000A/1500A to restrict Jsc to the current of the bottom cell, which reflected
the amount of incident light to i_2-layer of the bottom cell. The n-type films
used in the devices were as follows;

     1. condition 2   300A
     2. condition 2   150A
     3. condition 3   150A

     The I-V characteristics of the stacked solar cells are shown in figure 5.
As compared with cell 1, cell 2 shows higher Jsc by 5-6% due to decrease of
light absorption loss in n_1-layer, but there appears a kink near the Voc point
and then Voc and FF decrease because of the lack of ohmic connection.
     On the other hand, in case of cell 3 Jsc becomes as large as that of cell
2 and Voc and FF keep the same value as cell 1. This means that our newly developed very thin n-layer satisfies both requirements for decrease of light
absorption loss in n-layer and formation of ohmic n/p connection.

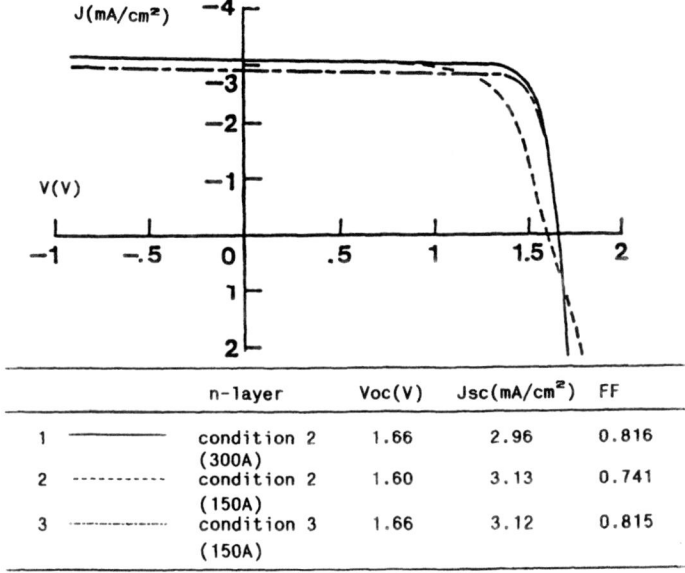

		n-layer	Voc(V)	Jsc(mA/cm²)	FF
1	———	condition 2 (300A)	1.66	2.96	0.816
2	-----------	condition 2 (150A)	1.60	3.13	0.741
3	·—··—··—··	condition 3 (150A)	1.66	3.12	0.815

Fig.5 I-V characteristics of double stacked cells.
Thickness balance i_1/i_2 is 1000A/1500A.

We have applied these technologies to fabricate a triple stacked solar cell in which the optical band gap is 1.75eV/1.75eV/1.50eV. And we have obtained the energy conversion efficiency of 10.27% (1cm²).

SUMMARY

We have developed the preparation method of the very thin μ c-Si:H n-layers, where the addition of a small amount of Ar gas to the deposition gases made the crystallite size smaller and the crystallite fraction higher. And then the newly developed 150A highly conductive μ c-Si:H n-layer has been applied to the stacked solar cells and Jsc increased 5-6% keeping Voc and FF constant because of reduction of light absorption loss and formation of n/p ohmic connection.

ACKNOWLEDGEMENT

This work is supported by New Energy and Technology Development Organization under the Sunshine project of Agency of Industrial Science and Technology.

REFERENCE

1. M. S. Benett and K. Rajan, Proc. of 20th IEEE PVSC, pp. 87-72 (1988)
2. Y. Tawada, J. Takada, N. Fukada, M. Yamaguchi, H. Yamagishi, K. Nishimura, M. Kondo, Y. Hosokawa, K.Tsuge, T. Nakayama and I. Hatano, Appl. Phys. Lett. 48, pp 584-586 (1986)
3. A. MATSUDA, J. Non-Cryst. Solids. 59&60, pp. 767-774 (1983)

Author Index

Abeles, B., 217
Aikawa, I., 347
Ajioka, T., 347
Akasaka, Y., 371
Albright, D.E., 315
Albu-Yaron, A., 81
Aljishi, Samer, 51
Anderson, G.B., 183
Arai, Toshihiro, 99
Ashok, S., 239

Bachrach, R.Z., 183
Baert, K., 359
Bai, P., 321
Bandyopadhyay, A.K., 69
Banerjee, Ratnabali, 69
Barua, A.K., 69
Batabyal, A.K., 69
Bauman, J., 27
Bhushan, B., 309
Birnboim, Meyer H., 277, 283
Bobonneau, F., 303
Bogy, D.B., 309
Bouldin, D.H., 33
Boyce, J.B., 183, 265
Breitschwerdt, A., 189
Buehler, E.C., 21, 265
Bustarret, Etienne, 211, 235

Campbell, I.H., 15, 259, 315
Chen, Lingrong, 223
Chen, Shuguang, 75
Chenglu, Lin, 147
Chianelli, R.R., 153
Cho, N.-H., 309
Chu, C-J, 303
Chuang, Tien-Min, 383
Chung, Kwan Soo, 171
Cumberbatch, Toby J., 129
Curtins, H., 27

Deschepper, P., 359
Dias, A.G., 57
Dong, Qin, 147
Driscoll, T., 105
Du, J., 321

Engelhard, T., 81

Fauchet, P.M., 15, 259, 315
Figueiredo, J., 57
Finger, F., 27
Fortmann, C.M., 315
Fujioka, H., 341
Fukui, K., 389

Furukawa, Shoji, 247

Geerts, M.J., 177
Gibson, J.M., 161
Gier, T.E., 123
Goto, T., 3
Gutmann, Ronald J., 383

Hachicha, M.A., 211, 235
Hamakawa, Y., 291
Hanna, Jun-ichi, 63, 195
Harrison, W.T.A., 123
Hatalis, Miltiadis K., 87
Haus, J.W., 283
Heiman, D., 141
Herrero, C.P., 189
Hill, I., 321
Hiraki, Akio, 205
Hirashita, N., 347
Hirata, G., 291
Hodes, G., 81
Hopper, C.B., 309
Horie, C., 271
Huber, C.A., 117, 141
Huber, T.E., 117
Hui, Tan, 147
Hwang, Jung Tae, 171

Ingels, Martin, 189, 229
Inguva, R., 283
Inokuma, Takao, 99
Inoue, Y., 371
Ipposhi, T., 371
Ishikawa, Mitsuru, 99
Ishimura, T., 389
Ito, M., 341
Ito, Toshimichi, 205
Itoh, M., 347
Iwami, Motohiro, 205
Izzard, M.J., 377

Jang, Jin, 171
Jeng, Shwu J., 353
Jin, Shu, 51
Johnson, R.I., 183
Jung, Chang Young, 171

Kalyaniwalla, N., 283
Kameda, M., 329
Kamo, Akira, 63
Kanicki, Jerzy, 353
Kim, Sung Chul, 171
Koh, Sung Ok, 171
Komiya, Tohru, 63
Kotecki, David E., 353

Krishnan, K.M., 309
Kujirai, Hiroshi, 63
Kumagai, K., 389
Kumar, Binod, 135
Kuwano, Y., 329

Lam, C.H., 33
Lee, Seung Kyu, 171
LeGrice, Y.M., 265
Lewis, M., 105
Ley, Lothar, 51
Lin, Fuyu, 87
Lin, Sam Shuhan, 75
Lu, Enlian, 93, 153
Lucovsky, G., 21, 265

Ma, Wei Ping, 277, 283
MacDougall, J.E., 123
Mackenzie, J.D., 303
Matsuda, A., 3
Matsumoto, Y., 291
Matsuyama, T., 329
Mertens, R., 359
Migliorato, P., 377
Milne, W.I., 377
Mingde, Tao, 147
Miyasato, Tatsuro, 247
Mo, Dang, 75
Moran, K.L., 123

Nakabeppu, F., 389
Nakano, S., 329
Nemanich, R.J., 21, 265
Nijs, J., 359
Nishikuni, M., 329
Nishimura, T., 371
Nishina, Y., 271

Ohnishi, M., 329
Okada, Y., 15
Okamoto, H., 291
Okamoto, S., 329

Parks, Christopher C., 353
Parsons, G.N., 21, 265
Pattyn, H., 359
Perez, J.A., 117
Persans, Peter D., 93, 105,
    153, 217
Pettford-Long, A., 81
Prabhakar, E.N., 141
Prasad, Kshem, 27
Putnis, Andrew, 129

Rai, D.K., 135
Rausch, Werner, 353
Ready, S.E., 183
Redwing, R., 105
Rose, Kenneth, 33, 383
Rubin, M.D., 309

Ruppert, A.F., 153

Salzberg, A.P., 117
Sasaki, Y., 271
Schroeder, John, 93
Seshan, Krishna, 353
Shah, A., 27
Sharma, S.N., 69
Shichang, Zou, 147
Shimizu, Isamu, 63, 195
Shirai, Hajime, 195
Shroder, R.E., 265
Soe, Sung Moo, 171
Srikanth, K., 239
Stucky, G.D., 123
Stutzmann, Martin, 51, 189,
    229
Sugahara, K., 371

Takasaki, K., 341
Tanaka, Akinori, 99
Tanaka, M., 329
Tauc, J., 223
Tien, Jean, 353
Ting, S-J., 303
Tong, B.Y., 321
Tsuda, S., 329
Tu, An, 105

van den Heuvel, J.C., 177
van Oort, R.C., 177
Vardeny, Z., 223
Veirs, D.K., 309
Veprek, Stan, 39

Wagner, S., 15, 161
Wang, Cheng, 21, 265
Wang, S.-L., 217
Watabe, Hirokuni, 205
Westcott, Michael R., 87
Winer, K., 183
Wolff, S.H., 161
Wong, S.K., 321
Wraback, M., 223
Wu, Y.-J., 217, 321

Yamagata, K., 365
Yang, G., 321
Yasumatsu, Tatsuro, 205
Yonehara, T., 365

Zhao, Xue-Shu, 93
Zollner, Stefan, 229

Subject Index

a-Ge:H, 315
a-Si:H, 315
alloy, 57
ambient effect, 69
amorphous
  hydrogenated carbon
    films, 75
  Si, 265
  SiC:H, 303
  SiGe, 315
  silicon-boron alloys, 321
  substrates, 365
annealing, 183, 321
aqueous, 129
atomic hydrogen, 195
Auger electron spectroscopy
  (AES), 33

B-doped microcrystalline
  silicon, 235
binary Si:H, 247
bulk, 51

C=C double bonds, 303
C-dangling bonds, 303
carbon doping, 341
CdS crystallites, 81, 135
CdS_xSe_{1-x}, 105
CdSe, 81, 99, 141
chalcogen-containing, 81
characterization, 87
chemical
  equilibrium, 39
  vapor deposition, 15
chemically vapour
  deposited, 57
clusters, 99
collection efficiency, 329
colloidal
  CdS, 93
  dispersion, 153
colloids, 129
composites, 265
conversion efficiency, 329
crystalline Si, 63
crystallization, 39, 161,
  183, 321

dc dark conductivity, 183
deactivation, 239
device fabrication
  processes, 347
diamond, 265
diamond-like carbon
  thin films, 309

dopant segregation, 347
doped hydrogenated micro-
  crystalline silicon, 27

edge emission, 99
effective medium
  approximation, 33
electrical conductivity, 75
electrodeposited, 81
electron
  diffraction, 21
  energy loss spectroscopy
    (EELS), 309
  lifetime, 315
electronic
  conductivity, 39, 183
  mobility, 183, 315
  transport, 183
ellipsoidal particles, 283
ellipsometric measurement, 33
ellipsometry, 75
enhancement of the local
  field, 283
epitaxial, 63
  (epi)-Si, 195
ESR, 303
excimer laser, 183
excitonic absorption, 135

fast pulse, 183
film composition, 33
fluorine, 63
  doped silicon oxide, 57
  doping, 341
  enhanced silicon oxidation,
    57
fractal-like structures, 177

glass matrix, 99
grain
  boundaries, 347
  boundary model, 15
  sizes, 171
graphitization, 75
growth dynamics, 51

H2, 303
Hall effect, 235
heat of segregation, 347
heterogeneous material, 283
high P_H, 21
high T_S, 21
highly conductive, 27
HRTEM, 33

hydrogen
  amorphous silicon, 177
  atoms, 161, 205, 239
  dilution, 51
  effusion, 189
  evolution, 183
  plasma, 189
    sputtering, 247
  rf plasma, 177
hydrogenated
  amorphous silicon, 183
  defect structure, 51
  microcrystalline
    films, 177
    silicon, 353

II-VI semiconductors, 123
III-V semiconductors, 123,
  129
impurity concentration, 183
infrared
  absorption, 75, 235
  spectroscopy, 189
ion implantation, 239

laser-induced, 183
low temperature photo-
  luminescence, 141
LPCVD, 383

magnetron sputtering, 69
mesh-like, 117
metal-organic polymer, 303
microcrystalline, 69, 265,
  315, 389
  phase, 303
  Si, 15, 21, 39, 189,
    205, 217, 229, 315,
    329
microcrystallite CdS, 93
microcrystals, 141, 171
microstructures, 117
model, 377
molecular beam epitaxy, 161
molybdenum disulfide
  microcrystals, 153

nanocrystalline
  silicon, 39, 211
  structure, 81
nanometer-size, 283
nanoparticle composites,
  277
nanoparticles, 105
  coated, 283
non-aqueous electrolytes,
  81
nonlinear optical, 277
  response, 283

optical
  absorption, 39, 51, 105
    spectra, 211
  band-gap, 353
  bistability, 283
  gap, 247
  properties, 229
opto-electronic properties, 3
optoelectronics, 291

p-doped microcrystalline
  silicon, 235
particle-size distribution,
  141
passivation, 189, 205
PECVD, 353
permeation, 239
phase transition, 93
phonon
  spectra, 93
  states, 271
phosphorus, 347
  -doped SIPOS films, 383
  doped μc-Si:H films, 27
photo-conductivity, 315
photo-oxidation processes, 57
photocrystalline silicon, 329
photoelectrochemical activity,
  81
photoinduced effects, 223
photoluminescence, 93, 105,
  247
photovoltaic, 291
picosecond photomodulation,
  223
plasma
  -enhanced, 15, 57
  frequency, 189
polycrystalline
  cadmium selenide films, 87
  silicon, 347
poly-Si thin film transistors,
  359
poly-silicon, 377
porous vycor glass, 117
post-deposition processing,
  183
power
  density, 309
  -modulated multi-layered
    structure, 247
pressure, 117
  dependence, 93

Raman
  absorption, 51
  scattering, 21, 39, 105,
    153, 235
    resonant, 93
  spectroscopy, 75, 189,
    259, 265, 309
  spectrum, 271
remote plasma CVD, 171
RF
  glow discharge decom-
    position, 75
  plasma CVD, 359
  reactive magnetron
    sputtering, 21
RTA, 353
Rutherford back scattering
  (RBS), 33

selective growth, 359, 365
semiconductor
  -impregnated, 117
  microcrystals, 259
  quantum dots, 141
Si crystals, 265
Si microcrystals, 247
SiC small particles, 271
SiF_4, 15
silane, 63
silicon
  clusters, 57
  rich oxides, 33
SIPOS films, 383
size dependence, 271
size of 50Å, 117
solid phase crystalliza-
  tion, 271
solid-state
  agglomeration, 365
  chemistry, 123
Source & Drain, 359
stacked solar cell, 389
subgap absorption, 189
superlattice, 247
suppression, 239
surface, 51
  plasmon resonance, 283
  reactions, 3
switching intensity, 283

TEM, 353
  bright-field imaging, 21
  dark-field imaging, 21
thermal
  annealing, 75, 147
  stability, 341
thin film transistor (TFT),
  377
3-D quantum size effects, 81
time-resolved spectra, 99

total yield photoelectron
  spectroscopies, 51
transport properties, 235
trapping effects, 383
triode PECVD, 235
two phases, 223

μc, 195
μc-Si:H, 3, 27, 63, 183, 341
μc-Si:H:F, 359
ultrafast
  carrier recombination, 223
  carrier trapping, 223
ultrathin layers, 217
undoped SIPOS films, 383

x-ray diffraction, 39, 105

zeolites, 123

MATERIALS RESEARCH SOCIETY SYMPOSIUM PROCEEDINGS

ISSN 0272 - 9172

Volume 1—Laser and Electron-Beam Solid Interactions and Materials Processing, J. F. Gibbons, L. D. Hess, T. W. Sigmon, 1981, ISBN 0-444-00595-1

Volume 2—Defects in Semiconductors, J. Narayan, T. Y. Tan, 1981, ISBN 0-444-00596-X

Volume 3—Nuclear and Electron Resonance Spectroscopies Applied to Materials Science, E. N. Kaufmann, G. K. Shenoy, 1981, ISBN 0-444-00597-8

Volume 4—Laser and Electron-Beam Interactions with Solids, B. R. Appleton, G. K. Celler, 1982, ISBN 0-444-00693-1

Volume 5—Grain Boundaries in Semiconductors, H. J. Leamy, G. E. Pike, C. H. Seager, 1982, ISBN 0-444-00697-4

Volume 6—Scientific Basis for Nuclear Waste Management IV, S. V. Topp, 1982, ISBN 0-444-00699-0

Volume 7—Metastable Materials Formation by Ion Implantation, S. T. Picraux, W. J. Choyke, 1982, ISBN 0-444-00692-3

Volume 8—Rapidly Solidified Amorphous and Crystalline Alloys, B. H. Kear, B. C. Giessen, M. Cohen, 1982, ISBN 0-444-00698-2

Volume 9—Materials Processing in the Reduced Gravity Environment of Space, G. E. Rindone, 1982, ISBN 0-444-00691-5

Volume 10—Thin Films and Interfaces, P. S. Ho, K.-N. Tu, 1982, ISBN 0-444-00774-1

Volume 11—Scientific Basis for Nuclear Waste Management V, W. Lutze, 1982, ISBN 0-444-00725-3

Volume 12—In Situ Composites IV, F. D. Lemkey, H. E. Cline, M. McLean, 1982, ISBN 0-444-00726-1

Volume 13—Laser-Solid Interactions and Transient Thermal Processing of Materials, J. Narayan, W. L. Brown, R. A. Lemons, 1983, ISBN 0-444-00788-1

Volume 14—Defects in Semiconductors II, S. Mahajan, J. W. Corbett, 1983, ISBN 0-444-00812-8

Volume 15—Scientific Basis for Nuclear Waste Management VI, D. G. Brookins, 1983, ISBN 0-444-00780-6

Volume 16—Nuclear Radiation Detector Materials, E. E. Haller, H. W. Kraner, W. A. Higinbotham, 1983, ISBN 0-444-00787-3

Volume 17—Laser Diagnostics and Photochemical Processing for Semiconductor Devices, R. M. Osgood, S. R. J. Brueck, H. R. Schlossberg, 1983, ISBN 0-444-00782-2

Volume 18—Interfaces and Contacts, R. Ludeke, K. Rose, 1983, ISBN 0-444-00820-9

Volume 19—Alloy Phase Diagrams, L. H. Bennett, T. B. Massalski, B. C. Giessen, 1983, ISBN 0-444-00809-8

Volume 20—Intercalated Graphite, M. S. Dresselhaus, G. Dresselhaus, J. E. Fischer, M. J. Moran, 1983, ISBN 0-444-00781-4

Volume 21—Phase Transformations in Solids, T. Tsakalakos, 1984, ISBN 0-444-00901-9

Volume 22—High Pressure in Science and Technology, C. Homan, R. K. MacCrone, E. Whalley, 1984, ISBN 0-444-00932-9 (3 part set)

Volume 23—Energy Beam-Solid Interactions and Transient Thermal Processing, J. C. C. Fan, N. M. Johnson, 1984, ISBN 0-444-00903-5

Volume 24—Defect Properties and Processing of High-Technology Nonmetallic Materials, J. H. Crawford, Jr., Y. Chen, W. A. Sibley, 1984, ISBN 0-444-00904-3

Volume 25—Thin Films and Interfaces II, J. E. E. Baglin, D. R. Campbell, W. K. Chu, 1984, ISBN 0-444-00905-1

Volume 26—Scientific Basis for Nuclear Waste Management VII, G. L. McVay, 1984, ISBN 0-444-00906-X

Volume 27—Ion Implantation and Ion Beam Processing of Materials, G. K. Hubler, O. W. Holland, C. R. Clayton, C. W. White, 1984, ISBN 0-444-00869-1

Volume 28—Rapidly Solidified Metastable Materials, B. H. Kear, B. C. Giessen, 1984, ISBN 0-444-00935-3

Volume 29—Laser-Controlled Chemical Processing of Surfaces, A. W. Johnson, D. J. Ehrlich, H. R. Schlossberg, 1984, ISBN 0-444-00894-2

Volume 30—Plasma Processing and Synthesis of Materials, J. Szekely, D. Apelian, 1984, ISBN 0-444-00895-0

Volume 31—Electron Microscopy of Materials, W. Krakow, D. A. Smith, L. W. Hobbs, 1984, ISBN 0-444-00898-7

Volume 32—Better Ceramics Through Chemistry, C. J. Brinker, D. E. Clark, D. R. Ulrich, 1984, ISBN 0-444-00898-5

Volume 33—Comparison of Thin Film Transistor and SOI Technologies, H. W. Lam, M. J. Thompson, 1984, ISBN 0-444-00899-3

Volume 34—Physical Metallurgy of Cast Iron, H. Fredriksson, M. Hillerts, 1985, ISBN 0-444-00938-8

Volume 35—Energy Beam-Solid Interactions and Transient Thermal Processing/1984, D. K. Biegelsen, G. A. Rozgonyi, C. V. Shank, 1985, ISBN 0-931837-00-6

Volume 36—Impurity Diffusion and Gettering in Silicon, R. B. Fair, C. W. Pearce, J. Washburn, 1985, ISBN 0-931837-01-4

Volume 37—Layered Structures, Epitaxy, and Interfaces, J. M. Gibson, L. R. Dawson, 1985, ISBN 0-931837-02-2

Volume 38—Plasma Synthesis and Etching of Electronic Materials, R. P. H. Chang, B. Abeles, 1985, ISBN 0-931837-03-0

Volume 39—High-Temperature Ordered Intermetallic Alloys, C. C. Koch, C. T. Liu, N. S. Stoloff, 1985, ISBN 0-931837-04-9

Volume 40—Electronic Packaging Materials Science, E. A. Giess, K.-N. Tu, D. R. Uhlmann, 1985, ISBN 0-931837-05-7

Volume 41—Advanced Photon and Particle Techniques for the Characterization of Defects in Solids, J. B. Roberto, R. W. Carpenter, M. C. Wittels, 1985, ISBN 0-931837-06-5

Volume 42—Very High Strength Cement-Based Materials, J. F. Young, 1985, ISBN 0-931837-07-3

Volume 43—Fly Ash and Coal Conversion By-Products: Characterization, Utilization, and Disposal I, G. J. McCarthy, R. J. Lauf, 1985, ISBN 0-931837-08-1

Volume 44—Scientific Basis for Nuclear Waste Management VIII, C. M. Jantzen, J. A. Stone, R. C. Ewing, 1985, ISBN 0-931837-09-X

Volume 45—Ion Beam Processes in Advanced Electronic Materials and Device Technology, B. R. Appleton, F. H. Eisen, T. W. Sigmon, 1985, ISBN 0-931837-10-3

Volume 46—Microscopic Identification of Electronic Defects in Semiconductors, N. M. Johnson, S. G. Bishop, G. D. Watkins, 1985, ISBN 0-931837-11-1

Volume 47—Thin Films: The Relationship of Structure to Properties, C. R. Aita, K. S. SreeHarsha, 1985, ISBN 0-931837-12-X

Volume 48—Applied Materials Characterization, W. Katz, P. Williams, 1985, ISBN 0-931837-13-8

Volume 49—Materials Issues in Applications of Amorphous Silicon Technology, D. Adler, A. Madan, M. J. Thompson, 1985, ISBN 0-931837-14-6

Volume 50—Scientific Basis for Nuclear Waste Management IX, L. O. Werme, 1986, ISBN 0-931837-15-4

Volume 51—Beam-Solid Interactions and Phase Transformations, H. Kurz, G. L. Olson, J. M. Poate, 1986, ISBN 0-931837-16-2

Volume 52—Rapid Thermal Processing, T. O. Sedgwick, T. E. Seidel, B.-Y. Tsaur, 1986, ISBN 0-931837-17-0

Volume 53—Semiconductor-on-Insulator and Thin Film Transistor Technology, A. Chiang. M. W. Geis, L. Pfeiffer, 1986, ISBN 0-931837-18-9

Volume 54—Thin Films—Interfaces and Phenomena, R. J. Nemanich, P. S. Ho, S. S. Lau, 1986, ISBN 0-931837-19-7

Volume 55—Biomedical Materials, J. M. Williams, M. F. Nichols, W. Zingg, 1986, ISBN 0-931837-20-0

Volume 56—Layered Structures and Epitaxy, J. M. Gibson, G. C. Osbourn, R. M. Tromp, 1986, ISBN 0-931837-21-9

Volume 57—Phase Transitions in Condensed Systems—Experiments and Theory, G. S. Cargill III, F. Spaepen, K.-N. Tu, 1987, ISBN 0-931837-22-7

Volume 58—Rapidly Solidified Alloys and Their Mechanical and Magnetic Properties, B. C. Giessen, D. E. Polk, A. I. Taub, 1986, ISBN 0-931837-23-5

Volume 59—Oxygen, Carbon, Hydrogen, and Nitrogen in Crystalline Silicon, J. C. Mikkelsen, Jr., S. J. Pearton, J. W. Corbett, S. J. Pennycook, 1986, ISBN 0-931837-24-3

Volume 60—Defect Properties and Processing of High-Technology Nonmetallic Materials, Y. Chen, W. D. Kingery, R. J. Stokes, 1986, ISBN 0-931837-25-1

Volume 61—Defects in Glasses, F. L. Galeener, D. L. Griscom, M. J. Weber, 1986, ISBN 0-931837-26-X

Volume 62—Materials Problem Solving with the Transmission Electron Microscope, L. W. Hobbs, K. H. Westmacott, D. B. Williams, 1986, ISBN 0-931837-27-8

Volume 63—Computer-Based Microscopic Description of the Structure and Properties of Materials, J. Broughton, W. Krakow, S. T. Pantelides, 1986, ISBN 0-931837-28-6

Volume 64—Cement-Based Composites: Strain Rate Effects on Fracture, S. Mindess, S. P. Shah, 1986, ISBN 0-931837-29-4

Volume 65—Fly Ash and Coal Conversion By-Products: Characterization, Utilization and Disposal II, G. J. McCarthy, F. P. Glasser, D. M. Roy, 1986, ISBN 0-931837-30-8

Volume 66—Frontiers in Materials Education, L. W. Hobbs, G. L. Liedl, 1986, ISBN 0-931837-31-6

Volume 67—Heteroepitaxy on Silicon, J. C. C. Fan, J. M. Poate, 1986, ISBN 0-931837-33-2

Volume 68—Plasma Processing, J. W. Coburn, R. A. Gottscho, D. W. Hess, 1986, ISBN 0-931837-34-0

Volume 69—Materials Characterization, N. W. Cheung, M.-A. Nicolet, 1986, ISBN 0-931837-35-9

Volume 70—Materials Issues in Amorphous-Semiconductor Technology, D. Adler, Y. Hamakawa, A. Madan, 1986, ISBN 0-931837-36-7

Volume 71—Materials Issues in Silicon Integrated Circuit Processing, M. Wittmer, J. Stimmell, M. Strathman, 1986, ISBN 0-931837-37-5

Volume 72—Electronic Packaging Materials Science II, K. A. Jackson, R. C. Pohanka, D. R. Uhlmann, D. R. Ulrich, 1986, ISBN 0-931837-38-3

Volume 73—Better Ceramics Through Chemistry II, C. J. Brinker, D. E. Clark, D. R. Ulrich, 1986, ISBN 0-931837-39-1

Volume 74—Beam-Solid Interactions and Transient Processes, M. O. Thompson, S. T. Picraux, J. S. Williams, 1987, ISBN 0-931837-40-5

Volume 75—Photon, Beam and Plasma Stimulated Chemical Processes at Surfaces, V. M. Donnelly, I. P. Herman, M. Hirose, 1987, ISBN 0-931837-41-3

Volume 76—Science and Technology of Microfabrication, R. E. Howard, E. L. Hu, S. Namba, S. Pang, 1987, ISBN 0-931837-42-1

Volume 77—Interfaces, Superlattices, and Thin Films, J. D. Dow, I. K. Schuller, 1987, ISBN 0-931837-56-1

Volume 78—Advances in Structural Ceramics, P. F. Becher, M. V. Swain, S. Sōmiya, 1987, ISBN 0-931837-43-X

Volume 79—Scattering, Deformation and Fracture in Polymers, G. D. Wignall, B. Crist, T. P. Russell, E. L. Thomas, 1987, ISBN 0-931837-44-8

Volume 80—Science and Technology of Rapidly Quenched Alloys, M. Tenhover, W. L. Johnson, L. E. Tanner, 1987, ISBN 0-931837-45-6

Volume 81—High-Temperature Ordered Intermetallic Alloys, II, N. S. Stoloff, C. C. Koch, C. T. Liu, O. Izumi, 1987, ISBN 0-931837-46-4

Volume 82—Characterization of Defects in Materials, R. W. Siegel, J. R. Weertman, R. Sinclair, 1987, ISBN 0-931837-47-2

Volume 83—Physical and Chemical Properties of Thin Metal Overlayers and Alloy Surfaces, D. M. Zehner, D. W. Goodman, 1987, ISBN 0-931837-48-0

Volume 84—Scientific Basis for Nuclear Waste Management X, J. K. Bates, W. B. Seefeldt, 1987, ISBN 0-931837-49-9

Volume 85—Microstructural Development During the Hydration of Cement, L. Struble, P. Brown, 1987, ISBN 0-931837-50-2

Volume 86—Fly Ash and Coal Conversion By-Products Characterization, Utilization and Disposal III, G. J. McCarthy, F. P. Glasser, D. M. Roy, S. Diamond, 1987, ISBN 0-931837-51-0

Volume 87—Materials Processing in the Reduced Gravity Environment of Space, R. H. Doremus, P. C. Nordine, 1987, ISBN 0-931837-52-9

Volume 88—Optical Fiber Materials and Properties, S. R. Nagel, J. W. Fleming, G. Sigel, D. A. Thompson, 1987, ISBN 0-931837-53-7

Volume 89—Diluted Magnetic (Semimagnetic) Semiconductors, R. L. Aggarwal, J. K. Furdyna, S. von Molnar, 1987, ISBN 0-931837-54-5

Volume 90—Materials for Infrared Detectors and Sources, R. F. C. Farrow, J. F. Schetzina, J. T. Cheung, 1987, ISBN 0-931837-55-3

Volume 91—Heteroepitaxy on Silicon II, J. C. C. Fan, J. M. Phillips, B.-Y. Tsaur, 1987, ISBN 0-931837-58-8

Volume 92—Rapid Thermal Processing of Electronic Materials, S. R. Wilson, R. A. Powell, D. E. Davies, 1987, ISBN 0-931837-59-6

Volume 93—Materials Modification and Growth Using Ion Beams, U. Gibson, A. E. White, P. P. Pronko, 1987, ISBN 0-931837-60-X

Volume 94—Initial Stages of Epitaxial Growth, R. Hull, J. M. Gibson, David A. Smith, 1987, ISBN 0-931837-61-8

Volume 95—Amorphous Silicon Semiconductors—Pure and Hydrogenated, A. Madan, M. Thompson, D. Adler, Y. Hamakawa, 1987, ISBN 0-931837-62-6

Volume 96—Permanent Magnet Materials, S. G. Sankar, J. F. Herbst, N. C. Koon, 1987, ISBN 0-931837-63-4

Volume 97—Novel Refractory Semiconductors, D. Emin, T. Aselage, C. Wood, 1987, ISBN 0-931837-64-2

Volume 98—Plasma Processing and Synthesis of Materials, D. Apelian, J. Szekely, 1987, ISBN 0-931837-65-0

Volume 99—High-Temperature Superconductors, M. B. Brodsky, R. C. Dynes, K. Kitazawa, H. L. Tuller, 1988, ISBN 0-931837-67-7

Volume 100—Fundamentals of Beam-Solid Interactions and Transient Thermal Processing, M. J. Aziz, L. E. Rehn, B. Stritzker, 1988, ISBN 0-931837-68-5

Volume 101—Laser and Particle-Beam Chemical Processing for Microelectronics, D.J. Ehrlich, G.S. Higashi, M.M. Oprysko, 1988, ISBN 0-931837-69-3

Volume 102—Epitaxy of Semiconductor Layered Structures, R. T. Tung, L. R. Dawson, R. L. Gunshor, 1988, ISBN 0-931837-70-7

Volume 103—Multilayers: Synthesis, Properties, and Nonelectronic Applications, T. W. Barbee Jr., F. Spaepen, L. Greer, 1988, ISBN 0-931837-71-5

Volume 104—Defects in Electronic Materials, M. Stavola, S. J. Pearton, G. Davies, 1988, ISBN 0-931837-72-3

Volume 105—SiO$_2$ and Its Interfaces, G. Lucovsky, S. T. Pantelides, 1988, ISBN 0-931837-73-1

Volume 106—Polysilicon Films and Interfaces, C.Y. Wong, C.V. Thompson, K-N. Tu, 1988, ISBN 0-931837-74-X

Volume 107—Silicon-on-Insulator and Buried Metals in Semiconductors, J. C. Sturm, C. K. Chen, L. Pfeiffer, P. L. F. Hemment, 1988, ISBN 0-931837-75-8

Volume 108—Electronic Packaging Materials Science II, R. C. Sundahl, R. Jaccodine, K. A. Jackson, 1988, ISBN 0-931837-76-6

Volume 109—Nonlinear Optical Properties of Polymers, A. J. Heeger, J. Orenstein, D. R. Ulrich, 1988, ISBN 0-931837-77-4

Volume 110—Biomedical Materials and Devices, J. S. Hanker, B. L. Giammara, 1988, ISBN 0-931837-78-2

Volume 111—Microstructure and Properties of Catalysts, M. M. J. Treacy, J. M. Thomas, J. M. White, 1988, ISBN 0-931837-79-0

Volume 112—Scientific Basis for Nuclear Waste Management XI, M. J. Apted, R. E. Westerman, 1988, ISBN 0-931837-80-4

Volume 113—Fly Ash and Coal Conversion By-Products: Characterization, Utilization, and Disposal IV, G. J. McCarthy, D. M. Roy, F. P. Glasser, R. T. Hemmings, 1988, ISBN 0-931837-81-2

Volume 114—Bonding in Cementitious Composites, S. Mindess, S. P. Shah, 1988, ISBN 0-931837-82-0

Volume 115—Specimen Preparation for Transmission Electron Microscopy of Materials, J. C. Bravman, R. Anderson, M. L. McDonald, 1988, ISBN 0-931837-83-9

Volume 116—Heteroepitaxy on Silicon: Fundamentals, Structures,and Devices, H.K. Choi, H. Ishiwara, R. Hull, R.J. Nemanich, 1988, ISBN: 0-931837-86-3

Volume 117—Process Diagnostics: Materials, Combustion, Fusion, K. Hays, A.C. Eckbreth, G.A. Campbell, 1988, ISBN: 0-931837-87-1

Volume 118—Amorphous Silicon Technology, A. Madan, M.J. Thompson, P.C. Taylor, P.G. LeComber, Y. Hamakawa, 1988, ISBN: 0-931837-88-X

Volume 119—Adhesion in Solids, D.M. Mattox, C. Batich, J.E.E. Baglin, R.J. Gottschall, 1988, ISBN: 0-931837-89-8

Volume 120—High-Temperature/High-Performance Composites, F.D. Lemkey, A.G. Evans, S.G. Fishman, J.R. Strife, 1988, ISBN: 0-931837-90-1

Volume 121—Better Ceramics Through Chemistry III, C.J. Brinker, D.E. Clark, D.R. Ulrich, 1988, ISBN: 0-931837-91-X

Volume 122—Interfacial Structure, Properties, and Design, M.H. Yoo, W.A.T. Clark, C.L. Briant, 1988, ISBN: 0-931837-92-8

Volume 123—Materials Issues in Art and Archaeology, E.V. Sayre, P. Vandiver, J. Druzik, C. Stevenson, 1988, ISBN: 0-931837-93-6

Volume 124—Microwave-Processing of Materials, M.H. Brooks, I.J. Chabinsky, W.H. Sutton, 1988, ISBN: 0-931837-94-4

Volume 125—Materials Stability and Environmental Degradation, A. Barkatt, L.R. Smith, E. Verink, 1988, ISBN: 0-931837-95-2

Volume 126—Advanced Surface Processes for Optoelectronics, S. Bernasek, T. Venkatesan, H. Temkin, 1988, ISBN: 0-931837-96-0

Volume 127—Scientific Basis for Nuclear Waste Management XII, W. Lutze, R.C. Ewing, 1989, ISBN: 0-931837-97-9

Volume 128—Processing and Characterization of Materials Using Ion Beams, L.E. Rehn, J. Greene, F.A. Smidt, 1989, ISBN: 1-55899-001-1

Volume 129—Laser and Particle-Beam Modification of Chemical Processes on Surfaces, A.W. Johnson, G.L. Loper, T.W. Sigmon, 1989, ISBN: 1-55899-002-X

Volume 130—Thin Films: Stresses and Mechanical Properties, J.C. Bravman, W.D. Nix, D.M. Barnett, D.A. Smith, 1989, ISBN: 1-55899-003-8

Volume 131—Chemical Perspectives of Microelectronic Materials, M.E. Gross, J. Jasinski, J.T. Yates, Jr., 1989, ISBN: 1-55899-004-6

Volume 132—Multicomponent Ultrafine Microstructures, L.E. McCandlish, B.H. Kear, D.E. Polk, and R.W. Siegel, 1989, ISBN: 1-55899-005-4

Volume 133—High Temperature Ordered Intermetallic Alloys III, C.T. Liu, A.I. Taub, N.S. Stoloff, C.C. Koch, 1989, ISBN: 1-55899-006-2

Volume 134—The Materials Science and Engineering of Rigid-Rod Polymers, W.W. Adams, R.K. Eby, D.E. McLemore, 1989, ISBN: 1-55899-007-0

Volume 135—Solid State Ionics, G. Nazri, R.A. Huggins, D.F. Shriver, 1989, ISBN: 1-55899-008-9

Volume 136—Fly Ash and Coal Conversion By-Products: Characterization, Utilization and Disposal V, R.T. Hemmings, E.E. Berry, G.J. McCarthy, F.P. Glasser, 1989, ISBN: 1-55899-009-7

Volume 137—Pore Structure and Permeability of Cementitious Materials, L.R. Roberts, J.P. Skalny, 1989, ISBN: 1-55899-010-0

Volume 138—Characterization of the Structure and Chemistry of Defects in Materials, B.C. Larson, M. Ruhle, D.N. Seidman, 1989, ISBN: 1-55899-011-9

Volume 139—High Resolution Microscopy of Materials, W. Krakow, F.A. Ponce, D.J. Smith, 1989, ISBN: 1-55899-012-7

Volume 140—New Materials Approaches to Tribology: Theory and Applications, L.E. Pope, L. Fehrenbacher, W.O. Winer, 1989, ISBN: 1-55899-013-5

Volume 141—Atomic Scale Calculations in Materials Science, J. Tersoff, D. Vanderbilt, V. Vitek, 1989, ISBN: 1-55899-014-3

Volume 142—Nondestructive Monitoring of Materials Properties, J. Holbrook, J. Bussiere, 1989, ISBN: 1-55899-015-1

Volume 143—Synchrotron Radiation in Materials Research, R. Clarke, J. Gland, J.H. Weaver, 1989, ISBN: 1-55899-016-X

Volume 144—Advances in Materials, Processing and Devices in III-V Compound Semiconductors, D.K. Sadana, L. Eastman, R. Dupuis, 1989, ISBN: 1-55899-017-8

Recent Materials Research Society Proceedings listed in the front.

Tungsten and Other Refractory Metals for VLSI Applications, Robert S. Blewer, 1986; ISSN 0886-7860; ISBN 0-931837-32-4

Tungsten and Other Refractory Metals for VLSI Applications II, Eliot K. Broadbent, 1987; ISSN 0886-7860; ISBN 0-931837-66-9

Ternary and Multinary Compounds, Satyen K. Deb, Alex Zunger, 1987; ISBN 0-931837-57-X

Tungsten and Other Refractory Metals for VLSI Applications III, Victor A. Wells, 1988; ISSN 0886-7860; ISBN 0-931837-84-7

Atomic and Molecular Processing of Electronic and Ceramic Materials: Preparation, Characterization and Properties, Ilhan A. Aksay, Gary L. McVay, Thomas G. Stoebe, J.F. Wager, 1988; ISBN 0-931837-85-5

Materials Futures: Strategies and Opportunities, R. Byron Pipes, U.S. Organizing Committee, Rune Lagneborg, Swedish Organizing Committee, 1988: ISBN 1-55899-000-3

Tungsten and Other Refractory Metals for VLSI Applications IV, Robert S. Blewer, Carol M. McConica, 1989; ISSN 0886-7860; ISBN 0-931837-98-7

Tungsten and Other Advanced Metals for VLSI/ULSI Applications V, S. Simon Wong, Seijiro Furukawa, 1990; ISSN 1048-0854; ISBN 1-55899-086-2

High Energy and Heavy Ion Beams in Materials Analysis, Joseph R. Tesmer, Carl J. Maggiore, Michael Nastasi, J. Charles Barbour, James W. Mayer, 1990; ISBN 1-55899-091-7

Physical Metallurgy of Cast Iron IV, Goro Ohira, Takaji Kusakawa, Eisuke Niyama, 1990; ISBN 1-55899-090-9

CPSIA information can be obtained at www.ICGtesting.com
Printed in the USA
LVOW12s0839230514

386805LV00012BA/486/P